Brent Wilson.

Confined Turbidite Systems

Geological Society Special Publications

Society Book Editors

R. J. PANKHURST (CHIEF EDITOR)
P. DOYLE
F. J. GREGORY
J. S. GRIFFITHS
A. J. HARTLEY
R. E. HOLDSWORTH
P. T. LEAT
A. C. MORTON
N. S. ROBINS
M. S. STOKER
J. P. TURNER

Special Publication reviewing procedures

The Society makes every effort to ensure that the scientific and production quality of its books matches that of its journals. Since 1997, all book proposals have been refereed by specialist reviewers as well as by the Society's Books Editorial Committee. If the referees identify weaknesses in the proposal, these must be addressed before the proposal is accepted.

Once the book is accepted, the Society has a team of Book Editors (listed above) who ensure that the volume editors follow strict guidelines on refereeing and quality control. We insist that individual papers can only be accepted after satisfactory review by two independent referees. The questions on the review forms are similar to those for *Journal of the Geological Society*. The referees' forms and comments must be available to the Society's Book Editors on request.

Although many of the books result from meetings, the editors are expected to commission papers that were not presented at the meeting to ensure that the book provides a balanced coverage of the subject. Being accepted for presentation at the meeting does not guarantee inclusion in the book.

Geological Society Special Publications are included in the ISI Index of Scientific Book Contents, but they do not have an impact factor, the latter being applicable only to journals.

More information about submitting a proposal and producing a Special Publication can be found on the Society's web site: www.geolsoc.org.uk.

It is recommended that reference to all or part of this book should be made in one of the following ways:

LOMAS, S. A. & JOSEPH, P. (eds) 2004. *Confined Turbidite Systems*. Geological Society, London, Special Publications, **222**.

ZINK, C. & NORRIS, R. J. 2004. Submarine fans within small basins: examples from the Tertiary of New Zealand. *In*: LOMAS, S. A. & JOSEPH, P. (eds) 2004. *Confined Turbidite Systems*. Geological Society, London, Special Publications, **222**, 229–240.

GEOLOGICAL SOCIETY SPECIAL PUBLICATION NO. 222

Confined Turbidite Systems

EDITED BY

S. A. LOMAS

Baker Atlas Geoscience, UK

and

P. JOSEPH

Institut Français du Petrole, France

2004

Published by

The Geological Society

London

THE GEOLOGICAL SOCIETY

The Geological Society of London (GSL) was founded in 1807. It is the oldest national geological society in the world and the largest in Europe. It was incorporated under Royal Charter in 1825 and is Registered Charity 210161.

The Society is the UK national learned and professional society for geology with a worldwide Fellowship (FGS) of 9000. The Society has the power to confer Chartered status on suitably qualified Fellows, and about 2000 of the Fellowship carry the title (CGeol). Chartered Geologists may also obtain the equivalent European title, European Geologist (EurGeol). One fifth of the Society's fellowship resides outside the UK. To find out more about the Society, log on to www.geolsoc.org.uk.

The Geological Society Publishing House (Bath, UK) produces the Society's international journals and books, and acts as European distributor for selected publications of the American Association of Petroleum Geologists (AAPG), the American Geological Institute (AGI), the Indonesian Petroleum Association (IPA), the Geological Society of America (GSA), the Society for Sedimentary Geology (SEPM) and the Geologists' Association (GA). Joint marketing agreements ensure that GSL Fellows may purchase these societies' publications at a discount. The Society's online bookshop (accessible from www.geolsoc.org.uk) offers secure book purchasing with your credit or debit card.

To find out about joining the Society and benefiting from substantial discounts on publications of GSL and other societies worldwide, consult www.geolsoc.org.uk, or contact the Fellowship Department at: The Geological Society, Burlington House, Piccadilly, London W1J 0BG: Tel. +44 (0)20 7434 9944; Fax +44 (0)20 7439 8975; E-mail: enquiries@geolsoc.org.uk.

For information about the Society's meetings, consult *Events* on www.geolsoc.org.uk. To find out more about the Society's Corporate Affiliates Scheme, write to enquiries@geolsoc.org.uk.

Published by The Geological Society from:
The Geological Society Publishing House
Unit 7, Brassmill Enterprise Centre
Brassmill Lane
Bath BA1 3JN, UK

(*Orders*): Tel. +44 (0)1225 445046
Fax +44 (0)1225 442836
Online bookshop: http://bookshop.geolsoc.org.uk

The publishers make no representation, express or implied, with regard to the accuracy of the information contained in this book and cannot accept any legal responsibility for any errors or omissions that may be made.

© The Geological Society of London 2004. All rights reserved. No reproduction, copy or transmission of this publication may be made without written permission. No paragraph of this publication may be reproduced, copied or transmitted save with the provisions of the Copyright Licensing Agency, 90 Tottenham Court Road, London W1P 9HE. Users registered with the Copyright Clearance Center, 27 Congress Street, Salem, MA 01970, USA: the item-fee code for this publication is 0305-8719/04/$15.00.

British Library Cataloguing in Publication Data
A catalogue record for this book is available from the British Library.

ISBN 1-86239-149-1

Typeset by J W Arrowsmith Ltd, Bristol, UK
Printed by Cromwell Press, Trowbridge, UK

Distributors

USA
AAPG Bookstore
PO Box 979
Tulsa
OK 74101-0979
USA
Orders: Tel. +1 918 584-2555
Fax +1 918 560-2652
E-mail bookstore@aapg.org

India
Affiliated East-West Press PVT Ltd
G-1/16 Ansari Road, Daryaganj,
New Delhi 110 002
India
Orders: Tel. +91 112 327-9113
Fax +91 112 326-0538
E-mail affiliat@nda.vsnl.net.in

Japan
Kanda Book Trading Co.
Cityhouse Tama 204
Tsurumaki 1-3-10
Tama-shi
Tokyo 206-0034
Japan
Orders: Tel. +81 (0)423 57-7650
Fax +81 (0)423 57-7651
E-mail geokanda@ma.kcom.ne.jp

Contents

Preface   vii

LOMAS, S. A. & JOSEPH, P. Confined turbidite systems   1

BOUMA, A. H. Key controls on the characteristics of turbidite systems   9

SMITH, R. Silled sub-basins to connected tortuous corridors: sediment distribution systems on topographically complex sub-aqueous slopes   23

AL JA'AIDI, O. S., MCCAFFREY, W. D. & KNELLER, B. C. Factors influencing the deposit geometry of experimental turbidity currents: implications for sand-body architecture in confined basins   45

GERVAIS, A., SAVOYE, B., PIPER, D. J. W., MULDER, T., CREMER, M. & PICHEVIN, L. Present morphology and depositional architecture of a sandy confined submarine system: the Golo turbidite System, eastern margin of Corsica   59

BABONNEAU, N., SAVOYE, B., CREMER, M. & BEZ, M. Multiple terraces within the deep incised Zaire Valley (ZaïAngo Project): are they confined levees?   91

CIBIN, U., DI GIULIO, A., MARTELLI, L., CATANZARITI, R., POCCIANTI, S., ROSSELLI, C. & SANI F. Factors controlling foredeep turbidite deposition: the case of Northern Apennines (Oligocene–Miocene, Italy)   115

HODGSON, D. M. & HAUGHTON, P. D. W. Impact of syndepositional faulting on gravity current behaviour and deep-water stratigraphy: Tabernas-Sorbas Basin, SE Spain   135

ESCHARD, R., ALBOUY, E., GAUMET, F. & AYUB, A. Comparing the depositional architecture of basin floor fans and slope fans in the Pab Sandstone, Maastrichtian, Pakistan   159

CRABAUGH, J. P. & STEEL, R. J. Basin-floor fans of the Central Tertiary Basin, Spitsbergen: relationship of basin-floor sand-bodies to prograding clinoforms in a structurally active basin   187

SMITH, R. Turbidite systems influenced by structurally induced topography in the multi-sourced Welsh Basin   209

ZINK, C. & NORRIS, R. J. Submarine fans within small basins: examples from the Tertiary of New Zealand   229

SATUR, N., CRONIN, B., HURST, A., KELLING, G. & GÜRBÜZ, K. Down-channel variations in stratal patterns within a conglomeratic, deepwater fan feeder system (Miocene, Adana Basin, Southern Turkey)   241

CORNAMUSINI, G. Sand-rich turbidite systems of the Late Oligocene Northern Apennines foredeep: physical stratigraphy and architecture of the 'Macigno costiero' (coastal Tuscany, Italy)   261

FELLETTI, F. Spatial variability of Hurst statistics in the Castagnola Formation, Tertiary Piedmont Basin, NW Italy: discrimination of sub-environments in a confined turbidite system   285

CONYBEARE, D. M., CANNON, S. KARAOĞUZ, O. & UYGUR, E. Reservoir modelling of the Hamitabat Field, Thrace Basin, Turkey: an example of a sand-rich turbidite system   307

Index   321

Preface

The papers in this volume consider the general themes of flow confinement and topographic control on processes and sedimentary architecture in deepwater clastic systems.

This publication grew out of an international workshop on Confined Turbidite Systems, held in Nice (France) in September 2001. Many of the papers presented at that meeting related to case studies of the Grès d'Annot turbidites, which are so spectacularly exposed in the region north of Nice. Hence we have also produced a companion volume in the same series (Geological Society Special Publication 221) specifically focusing on the Grès d'Annot.

The editors would like to thank the following colleagues for reviewing papers in this volume: Greg Browne, Mary Carr, David Conybeare, Gianluca Cornamusini, Jeff Crabaugh, Bryan Cronin, Brian Dade, Andrea Di Giulio, Trevor Elliott, Bill Galloway, Andy Gardiner, Peter Haughton, Richard Hiscott, Dave Hodgson, Richard Holmes, Jon Ineson, Juraj Janocko, Gilbert Kelling, Ben Kneller, Doug Masson, Stephen Morris, Kevin Pickering, Nick Satur, Erik Scott, Pete Sixsmith, Ru Smith, Ron Steel, Pete Talling, Nigel Woodcock, Russell Wynn and Christoph Zink, plus two anonymous sedimentologists.

Simon Lomas
Philippe Joseph

Confined turbidite systems

SIMON A. LOMAS[1] & PHILIPPE JOSEPH[2]

[1] *Baker Atlas Geoscience, Stoneywood Park North, Aberdeen AB21 7EA, UK*
(e-mail: simon.lomas@bakeratlas.com)
[2] *IFP School/Geology-Geochemistry Division, Institut Français du Pétrole,*
228–232, avenue Napoleon Bonaparte, 92852 Rueil Malmaison Cedex, France
(e-mail: philippe.joseph@ifp.fr)

Abstract: Deep-water sedimentation is currently a major focus of both academic research and industrial interest. Recent studies have emphasized the fundamental influence of sea-floor topography on the growth and morphology of submarine 'fans': in many turbidite systems and turbidite hydrocarbon reservoirs, depositional system development has been moderately to strongly confined by pre-existing bounding slopes. This publication examines aspects of sediment dispersal and accumulation in deepwater systems where basin-floor topography has profoundly affected deposition, and the associated controls on hydrocarbon reservoir architecture and heterogeneity. The papers herein offer a global perspective which is wide-ranging in terms of both approach and location, including contrasting case studies of outcrop, subsurface, modern and experimental systems.

Confined turbidite systems are those deepwater clastic depositional systems whose development has been fundamentally constrained by pronounced basin-floor topography. Many early conceptual and experimental models of turbidity current deposition and submarine fan development envisaged essentially unconfined radial development of lobe and fan-shaped depositional bodies. However, a wealth of case studies now available from many contrasting turbidite systems worldwide makes it clear that, in many basins, both sediment dispersal patterns and the geometries of depositional bodies have been profoundly affected by pre-existing or developing basin-floor relief. It seems that perhaps most large natural turbidity currents, particularly in intracontinental basins, are not free to spread radially across a uniform basin floor.

Van Andel & Komar (1969) used the term *ponding* to refer to a situation in which turbidity currents of sufficiently large volume are fully contained by an area of enclosed bathymetry. Similarly, Pickering & Hiscott (1985) used the term *contained turbidites* to describe beds deposited from turbidity currents that were confined within a basin too small to permit sustained unidirectional flow. The terms *ponding* and, to a lesser extent, *containment*, are now generally used to indicate a situation in which sediment gravity flows fill the floor of an enclosed depression and are unable to surmount the bounding slopes (or 'sills') which bound that depositional area ('mini-basin', 'silled sub-basin' or 'ponded depocentre'). Thus, essentially all of the sediment carried into the depositional area by that flow will be trapped and deposited there. The same terms are applied equally to process (ponded turbidity current), depositional product (ponded turbidite) and basin/stratigraphic behaviour (ponded depocentre). We favour the more general term *confinement* to describe situations where sediment gravity flows and their deposits are appreciably affected by the presence of significant basin-floor topography, but without the connotation of complete containment.

Closely associated with the concept of confined deposition are the notions of flow reflection and deflection. In their study of ponded Quaternary sediments in valleys on the flanks of the Mid-Atlantic Ridge, Van Andel & Komar (1969) speculated that the rebound of turbidity currents off the bounding slopes was responsible for repeated grading profiles seen in their piston cores. Pickering & Hiscott (1985) recognized containment of turbidity currents in the Cloridorme Formation (Ordovician of Québec) from reversals of palaeocurrent indicators, which they interpreted as the result of deflection and reflection of part of large-volume turbidity currents from confining slopes. Similar evidence of divergent palaeoflow indicators has subsequently been recognized at outcrop in various other successions representing depositional systems where turbidity currents have interacted with basin-bounding or intrabasinal slopes (e.g. Kneller *et al* 1991; Smith, this volume), and the likely effects of such interaction have been widely investigated through physical experiments (e.g. Simpson 1987; Muck & Underwood 1990; Alexander & Morris 1994; Kneller 1995; Al Ja'Aidi *et al.*, this volume).

Implicit in the concept of confinement is the relationship of flow magnitude to size of

From: LOMAS, S. A. & JOSEPH, P. (eds) 2004. *Confined Turbidite Systems*. Geological Society, London, Special Publications, **222**, 1–7. 0305-8719/04/$15.00 © The Geological Society of London 2004.

depocentre. The general concept of *flow efficiency* (e.g. Mutti 1979) is helpful here, describing a flow's ability to deliver sediment, especially sand, in a basinward direction. Thus, small volume, low-efficiency sediment gravity flows may behave in an effectively unconfined manner even in small basins. Conversely, very large magnitude flows may interact with marginal confining slopes even in large basins. Of course, if a confined depocentre is filled more quickly than the topographic relief grows, the surface area of the basin floor will increase with time. Hence, if sediment supply characteristics remain similar, successive depositional systems would be expected to reflect progressively decreasing confinement. The reverse scenario could develop in certain compressional settings.

In most of the studies detailed in this volume, the origin of the confining topography is tectonic: large-scale structural features, local fault scarps, folds or more subtle perturbations of basin-floor gradients with associated tilting and faulting. In other instances, basin-floor relief is the result of slope collapse masses, halokinesis, mud diapirism and, to a limited extent, differential compaction.

The collection of papers hosted in this volume is not intended to provide an exhaustive treatment of the subject of confined deepwater sedimentation, but rather to bring together a diverse assortment of approaches and observations which have a bearing on this topic. We have ordered the contributions to reflect the varying approaches of the researchers. Here we present an overview of those contributions and the new perspectives they afford, and draw attention to promising channels (and lobes?) for future research. We are pleased that several of the systems described in the following papers are not widely known. It is hoped that the juxtaposition of conceptual reviews, experimental work and case-studies of outcrop, subsurface and modern systems will stimulate new insights and connections and highlight key general questions which surmount the usual bounding slopes of these scientific sub-communities.

Overviews of confined turbidite systems

Two review papers introduce the broad-scale context of controls on, and patterns of, confined turbidite system development from contrasting viewpoints.

Bouma sets the scene with an overview of the principal external controls on turbidite systems (tectonics, climate, sedimentary characteristics and processes, and sea-level fluctuations), illustrating the interplay between these key controls in varied modern and ancient examples. The author emphasizes the contrasting behaviour and architectural characteristics of coarse-grained/sand-rich and fine-grained/mud-rich end-member systems but reminds us of the inadequacy of general models in capturing the variability of these real cases. The practical limitations of idealised models arise either from emphasis on only one or two aspects or from over-simplification (being too general to display architectural details). Bouma advocates the construction of case-specific models based on a reconstruction of the specific natural controls and their interactions in the generation and delivery of sedimentary particles, the evolving basin configuration and processes of deposition and possible redeposition in that basin.

Smith takes an alternative look at patterns of sediment dispersal on topographically complex slopes, distinguishing two conceptual end-member systems, which he refers to as 'cascades of silled sub-basins' and 'connected tortuous corridors'. The first scenario is essentially a refinement of the 'fill-and-spill' model that emerged from studies of the northern Gulf of Mexico slope (e.g. Winker 1996; Prather *et al* 1998). In the 'connected tortuous corridors' scenario, flows partly circumvent bathymetric obstacles, but follow a (laterally confined) continuous tortuous path down the slope. From two-dimensional sections, these two scenarios may be difficult to distinguish (i.e. apparently separate sub-basins may be directly connected out of the plane of section), although recognition of stratal onlap against a confining frontal slope should be diagnostic of the fill of a true silled sub-basin.

Experimental studies

Laboratory experiments have offered crucial insights into the potential effects of basin-floor topography on sedimentation (e.g. Simpson 1987; Kneller *et al.*, 1991; Alexander & Morris 1994). **Al Ja'Aidi *et al.*** outline a series of scaled experiments (using ballotini and silica flour to model sand and mud) to investigate how flow volume, density and grain-size distribution affect the transport efficiency of turbidity currents (i.e. their ability to carry 'sand' basinward) and the geometric characteristics of the resulting deposits. In general, these experiments reaffirm intuitive expectations (cf. Mutti 1979; Mutti *et al.* 1999): flow efficiency correlates directly with flow volume (which increases the initial potential energy and lessens the rate of velocity decrease due to gravitational spreading), sediment concentration (which increases the potential energy

of the flow, and hence the run-out distance), and proportion of fine-grained material (which serves to maintain overall negative buoyancy and also decreases the settling velocity of the coarser fraction). When an arcuate obstacle was placed in the path of the flows, the proportion of sediment reaching the topographic obstruction, and the proportion of it able to surmount the topography, increased as flow efficiency increased. The most provocative outcome of these experiments concerns the influence of solids concentration on deposit geometry. With increasing initial flow densities, the resulting deposits generally show an unsurprising trend of increasing length to width ratios (i.e. denser flows produce longer and thinner deposits). However, above a certain density threshold ($c.13\%$ initial sediment concentration by mass), the deposits of successively denser flows tend to show *decreasing* length-to-width ratios. The authors interpret this intriguing result in terms of inhibition of turbulence at high concentrations.

Modern turbidite systems

Shallow penetration geological and geophysical data from modern turbidite systems offer the clearest perspective on the morphology and planform arrangement of the components of turbidite depositional systems. Two detailed studies of contrasting modern systems are presented here that document remarkable datasets.

Gervais et al. describe the modern sandy Golo turbidite system, developed in a partly confined setting on the eastern margin of Corsica. The upper Quaternary deepwater deposits have been imaged using a closely spaced grid of sparker seismic-reflection profiles, covering a total survey length of some 1000 km. These high-resolution data give a 2 m vertical resolution of a turbidite system composed of four non-coalescent fans characterized by stacked depositional bodies. Canyon, gully, sandy channel, muddy levee, and sandy lobe architectural elements are identified on this high-resolution dataset at a scale which should allow direct comparison with outcrop examples. Each of the four fans appears to have been fed by a separate canyon, ultimately linked up-dip to a single fluvio-deltaic point-source; the canyons were apparently active simultaneously. Structural confinement of the system as a whole has led to a predominance of aggradation, and smaller-scale slope variations have constrained the specific morphology and architecture of depositional lobes. Interestingly, it seems that, notwithstanding the setting adjacent to a tectonically active, uplifting margin, the dominant influences on depositional patterns appear to have been the pre-existing basin-floor morphology, sediment source characteristics and eustatic sea-level variations.

Babonneau et al. describe a very different confined system from the deeply incised Zaire submarine valley. They focus on a striking array of multiple terraces imaged along the inner flanks of the upper-fan valley, adjacent to the deeply incised sinuous thalweg. A range of possible origins for these confined terraces is evaluated (e.g. shallow-rooted faulting, channel-wall slumping, repeated sedimentary infilling and incision, or levee aggradation). Morphological analysis and seismic character favour an interpretation of these remarkable features as levees, confined within the incised valley. Many terraces correspond to the infill of abandoned meander loops, and are believed to record aggradation of relatively dilute turbidite current deposits associated with spillover and flow-stripping processes. Terrace initiation appears to correspond to phases of vertical incision of the thalweg, occurring concurrently with meander migration.

Outcrop case studies

The remainder of this volume is devoted to a contrasting collection of outcrop-based case studies, many focusing on the disproportionately numerous Eocene–Oligocene–Miocene turbidite basin-fills exposed around the world.

In a large-scale investigation of Oligocene–Miocene turbidite wedges exposed in the Northern Apennines, **Cibin et al.** find a repeated tripartite stacking motif involving an upward transition from thick-bedded, sandstone-rich units ('depocentre stage'), into thin-bedded mudstone-rich units, and ultimately grading up to mudstone-dominated units ('abandonment stage'). This rhythm is interpreted in terms of a sediment input signal: the abrupt 'switching on and off' of coarse-grained input. Biostratigraphic data allow discrimination of regional effects (overall decrease in sediment input) from intrabasinal changes in sediment dispersal (i.e. depocentre migration resulting in the 'switching-off' of one turbidite system as a new system is 'switched-on' basinward). Thus, Cibin et al. are able to recognize distinct pulses driven by hinterland tectonics (Chattian; Middle Burdigalian), and to distinguish these from episodes of depocentre-shift related to basin reorganization (Chattian, Langhian), and from phases of diminished regional sediment supply (Late Aquitanian,

Middle Burdigalian, Early Serravallian), perhaps driven by elevated sea levels.

Hodgson & Haughton detail the interplay between deepwater sedimentation and syn-depositional faulting in the Neogene fill of the Tabernas-Sorbas Basin, SE Spain, documenting the effects of a fault which appears to have propagated through to the sea-bed during turbidite deposition. The El Cautivo Fault is an oblique strike-slip fault that accommodated an area of deeper ponded bathymetry (a 'mini-basin') on its southern flank. The sea-bed expression of this fault varied between a blind fold, associated with local wedging and onlap of stratigraphy, and an unstable fault scarp with local collapse features. Its influence on sedimentation is apparent in terms of both sediment dispersal (sea-floor warping associated with the fault trace largely determined the flow paths of incident turbidity currents) and stratigraphic growth (evolving patterns of thickness variations associated with changing rates of differential subsidence across the fault). Characteristic sedimentation in the fault-controlled ponded depocentre involved distinctive thick sheet-like sandstone–mudstone couplets, interpreted as the deposits of large-volume turbidity currents that were entirely confined within the structurally defined mini-basin.

Eschard et al. analyse architectural relationships in the remarkable large-scale exposures of the Maastrichtian Pab Sandstone in Pakistan. The successions they describe record initial hinterland uplift with shelf erosion, canyon incision, and the development of an extensive high-efficiency basin-floor fan dominated by channel complexes. During subsequent transgression, a more restricted turbidite system developed on the slope, consisting of proximal conglomeratic channel fills, aggradational 'mid-fan' channel fills and tabular lobe deposits. In contrast to the older basin-floor fan, this slope fan testifies to reduced flow transport efficiencies, and hence less confinement (fewer flows reaching the bounding slopes) and less systematic internal organization. The evolution in depositional processes and architectural styles seems to have been driven by a general change in slope morphology, from an initially well-defined shelf-slope break to a smoother, lower-angle ramp profile developed through slope progradation.

Crabaugh & Steel describe architectural and facies relationships between shelf-slope clinoforms and basin-floor fan sandstone bodies from extensive and beautiful exposures of Eocene successions in the Central Tertiary Basin of western Spitsbergen. Where clinoforms are sand-prone they include a deepwater sand complex: submarine fans represent an early, basin-floor aggradational phase of clinoform growth, whereas later development of the same clinoform involves a phase of shelf-margin accretion. The component sandbodies each display an upward increase in bed thickness, involving a transition from predominantly thin-bedded, sheet-like ripple- to parallel-laminated turbidites (interpreted as channel-mouth lobes, sandy levees and crevasse splays) upwards into thickly bedded, erosive-based, parallel-laminated and structureless sandstones (interpreted as constructional channel-fills). Flows reaching the basin floor were apparently deflected laterally by anticlinal sea-floor topography, resulting in restricted basinward extent and longitudinal palaeocurrent indicators in the youngest fan sandbodies.

The Ordovician–Silurian fill of the Welsh Basin contains a diverse range of deepwater depositional systems that record the influence of basin-floor topography on sediment distribution patterns. **Smith** synthesizes extensive datasets to describe the evolution of these successive turbidite systems in the context of evolving topographic controls. An initial base-of-slope system became overstepped by a channel-fed lobe system once the inboard accommodation space became filled. All systems exhibit evidence that flows were deflected into near-parallelism with the tectonically generated regional slopes. Parts of these systems are also characterised by large azimuthal disparities between the trends of sole structures and ripple cross-laminae, which is regarded as direct sedimentological evidence of partial flow reflection from containing slopes. A forward geometric model is used to evaluate plausible stratal architectures in relation to the main bounding slope. Smith also champions the analogue value of lessons learned from this less fashionable outcrop system to subsurface prediction.

Zink & Norris describe stacking patterns in the Tertiary fill of two small fault-bounded basins, the Te Anau and Waiau basins, of southwest New Zealand. These basins, initially extensional but subsequently dominated by strike-slip tectonics, were only a few tens of kilometres wide yet each accumulated sedimentary successions 6–8 km thick before undergoing transpressive inversion in late/post-Miocene times. Marked lateral and vertical changes in facies are characterised, but, despite their limited lateral extent, these basins developed a full range of sedimentary facies and architectural elements, with pronounced proximal to distal transitions.

Satur et al. describe the proximal, channelized section of a topographically constrained deepwater fan in the Miocene Adana Basin of southern Turkey. Their study documents basinward facies changes along a 10 km transect within one of four fan-feeder channels. A well defined down-channel transition is observed from proximal, ungraded disorganized conglomerates to more organized, inverse-to-normal and normally graded conglomerates. Concomitantly, bedding style changes from scour-based, sheet-like units up-dip into a complex of small channels and bar-forms mid-way along the transect, into thick-bedded, structureless facies further downcurrent. Channel gradients can be tentatively reconstructed and changes in channel gradient are considered to constitute a key control on the hydrodynamics within the gravity flows, and thus on the resulting stratal patterns.

Cornamusini describes the sedimentary organization of the Oligocene 'Macigno Costiero' siliciclastic turbidites exposed in coastal Tuscany. These are the remnants of a sand-rich, low-efficiency turbidite system developed within a partially confined setting in the Northern Apennines foreland basin system. The system comprises unchannelized and channelized sheet sandstone bodies, a main channel-fill unit, together with distributary channel-fills, overbank and channel margin deposits. Stacking patterns indicate five stages of depositional system development. The development of sheet (lobe) stages is characterized by negative vertical trends (upward coarsening/thickening), which can be interpreted as a retrogressive transition from distal lobe to proximal lobe up to a channel-lobe transition zone.

Quantitative characterization of turbidite systems

Two final studies address different issues of quantitative characterization of turbidite systems. This is an area of increasing significance given the growing economic importance of deepwater hydrocarbon reserves, and the corresponding predictive and modelling needs of the industry.

Felletti uses quantitative outcrop data from the Castagnola Formation in the Tertiary Piedmont Basin of NW Italy to investigate the potential of geostatistical tools for discrimination of depositional sub-environments in a confined turbidite system. Three variables were studied: thickness of sandstone and siltstone beds, a grain-size index, and sand and silt thickness percentage (i.e. the ratio of coarse division to overlying mudstones). Felletti applies statistical tests (the Hurst tests) to assess vertical and lateral distributions of these variables. Significant clustering is found to occur in almost all measured sections, with systematic spatial variations in the degree of clustering, such that different parts of the basin are characterized by different Hurst statistics. The statistical clustering is associated with lateral and vertical facies variations in turn related to depositional sub-environments within the Castagnola Basin. Although the origins of this clustering are not fully constrained in terms of primary sedimentary processes, this type of analysis highlights the potential for semi-deterministic predictions of facies away from sample points in systems which are locally well characterized.

Conybeare et al. detail an approach to geocellular modelling of turbidite systems. Their case study is the Eocene Hamitabat Formation, which forms an important gas reservoir in the large fault-bounded Thrace Basin of northwestern Turkey, adjacent to the North Anatolian Fault System. The main reservoir comprises gravity flow sandstones sourced from fluvio-deltaic systems to the north and northeast. Correlation, based mainly on upward-fining cyclical successions, indicates lobate depositional geometries and subtle topographic control on sediment dispersal (compensation). Construction of the stratigraphic framework for the 3D geocellular reservoir model of the field relied heavily on the correlation of depositional cycles. A mathematical approach to correlation was adopted, employing frequency and difference spectra to characterize cyclicity in wireline log curves. The model itself was populated with facies objects whose dimensions were based on analogue data together with direct knowledge of the depositional systems. Upscaling of the static geological model and fluid-flow simulation allowed comparison with the actual production history and hence refinement of the model. Although the resulting model is very much a pragmatic rather than a scientifically-driven realization, the modelling process does afford new insights into the range of possible geometries of the sedimentary systems.

Closing remarks

The range of studies presented here reaffirms the diversity of deepwater clastic systems, but suggests a broad consensus that overriding controls on stratal patterns and architecture are sediment supply character and slope gradients in the receiving basin. In particular, flow volume and efficiency relative to the scale of the depocentre may largely determine the character and distribution of the deposit. A related point emphasized here is the appreciation of the three-dimensionality of sediment dispersal pathways in confined systems. Future studies on sediment supply and gravity flow initiation processes will be strongly influenced by the emerging importance of hyperpycnal flow as a likely major supply mechanism (e.g. Mulder & Syvitiski 1995), which will require development of new depositional models.

More generally, the case studies compiled here highlight the continuing importance of outcrop-derived datasets in providing ground-truth detail to constrain conceptual models. A critical impetus behind acquisition of detailed outcrop datasets has been the requirement of the hydrocarbon industry for generic and numerical constraints on correlation and modelling of deep-water reservoirs. In turn, geocellular and layered 3D geological models are now offering insight and constraints on sedimentary architectures and evolution. In reaching for outcrop analogues to help understand the architecture and heterogeneity of turbidite reservoirs, industry geoscientists have focused heavily on a surprisingly small number of outcrop favourites (the 'top ten' is perhaps: Ainsa, Annot, Brushy Canyon, Cerro Torro, Jackfork, Karoo, Marnoso-Arenacea, Mt Messenger, Ross, Tabernas). Although all of these systems certainly still have a great deal to teach us, it is to be hoped that compilations such as this volume will help to raise the profile of the many other informative systems available.

Previous authors have noted the difficulties of reconciling interpretations of modern and ancient turbidite systems (e.g. Normark *et al.* 1993), which arise partly from the limitations of the Holocene sea-level regime but more fundamentally from the differences in scale and resolution of outcrop, subsurface and marine data. Several contributions here indicate encouraging convergence in recent years between outcrop, subsurface, modern and also experimental results. Integration of modern datasets (which image morphology and spatial distribution) with appropriate outcrop data (giving the internal character and vertical organization) is perhaps one underexploited avenue for research.

The inherent variability of these sedimentary systems ensures that a unified deterministic understanding of their development remains elusive, and hence our ability to determine geometries and facies away from sample points remains limited. Central to furthering this predictive capacity will be an improvement in understanding of flow and depositional processes at real scale: large natural sediment gravity flows remain difficult to observe or model, but major progress with the design of physical experiments with particulate materials offers an obvious (if difficult) route to fresh insights into the behaviour of variably confined flows. A new emphasis here may be attempts to simulate erosion beneath large sediment gravity flows: bed contacts in many sand-rich confined systems show evidence of substantial erosion, but the mechanics of this erosion, and the resulting geometries and their implications (e.g. for flow transformations and for reservoir connectivity), have been little studied to date. Integration of experimental work with outcrop studies targeted to test developing hypotheses of sediment gravity flow processes may enhance the extent to which flow properties and efficiency be deduced from deposit characteristics.

We are grateful to Vincent Hilton and Jonathan Turner for constructive comments on the first draft of this paper.

References

ALEXANDER, J. & MORRIS, S. 1994. Observations on experimental, nonchannelized, high-concentration turbidity currents and variations in deposits around obstacles. *Journal of Sedimentary Research, Section A: Sedimentary Petrology and Processes,* **64**, 899–909.

KNELLER, B. C. 1995. Beyond the turbidite paradigm: physical models for deposition of turbidites and their implications for reservoir prediction. *In:* HARTLEY, A. & PROSSER, D. J. (eds) *Characterisation of Deep Marine Clastic Systems,* Geological Society, London, Special Publications, **94**, 29–46.

KNELLER, B. C., EDWARDS, D., MCCAFFREY, W. & MOORE, R. 1991. Oblique reflection of turbidity currents. *Geology,* **19**, 250–252.

MUCK, M. T. & UNDERWOOD, M. B. 1990. Upslope flow of turbidity currents: a comparison among field observations, theory and laboratory experiments. *Geology,* **18**, 54–57.

MULDER, T. & SYVITSKI, J. P. M. 1995. Turbidity currents generated at river mouths during exceptional discharges to the world oceans. *Journal of Geology,* **103**, 285–299.

MUTTI, E. 1979. Turbidites et cones sous-marins profonds. *In:* HOMEWOOD, P. (ed.) *Sedimentation detritique (fluvial, littoral et marine).* Institut de Géologie, Université de Fribourg, Suisse, 353–419.

MUTTI, E., TINTERRI, R., REMACHA, E., MAVILLA, N., ANGELLA, S. & FAVA, L. 1999. An Introduction to the Analysis of Ancient Turbidite Basins from an Outcrop Perspective. *AAPG Continuing Education Course Note Series,* **39**.

NORMARK, W. R., POSAMENTIER, H. & MUTTI, E. 1993. Turbidite systems: state of the art and future directions. *Reviews of Geophysics,* **31**, 91–116.

PICKERING, K. T. & HISCOTT, R. N. 1985. Contained (reflected) turbidity currents from the Middle Ordovician Cloridorme Formation, Quebec, Canada: an alternative to the antidune hypothesis. *Sedimentology,* **32**, 373–394.

PRATHER, B. E., BOOTH, J. R., STEFFENS, G. S. & CRAIG, P. A. 1998. Classification, lithologic calibration, and stratigraphic succession of seismic facies of intraslope basins, deep-water Gulf of Mexico. *AAPG Bulletin*, **82**, 701–728.

SIMPSON, J. E. 1987. *Gravity Currents in the Environment and the Laboratory*. Cambridge University Press, 24 pp.

VAN ANDEL, TJ. H. & KOMAR, P. D. 1969. Ponded sediments of the Mid-Atlantic Ridge between 22° and 23° North latitude. *Geological Society of America Bulletin*, **80**, 1163–1190.

WINKER, C. D. 1996. High-resolution seismic stratigraphy of a late Pleistocene submarine fan ponded by salt-withdrawal mini-basins on the Gulf of Mexico continental slope. *Offshore Technology Conference Proceedings*, 619–628.

Key controls on the characteristics of turbidite systems

ARNOLD H. BOUMA

Department of Geology and Geophysics, Louisiana State University, Baton Rouge, LA 70803-4101, USA (e-mail: bouma@geol.lsu.edu)

Abstract: Four main controls (tectonics, climate, sedimentary characteristics and processes, and sea-level fluctuations) commonly interact with each other and do so at varying intensities. This results in a wide variety of basin types and shapes, timing of transport within the sequence stratigraphy framework, transport and depositional processes, grain size ranges, and distribution of sediment within a basin. Two major end members of turbidite systems can be recognized: coarse-grained/sand-rich and fine-grained/mud-rich. Coarse-grained fans typically belong to active margin settings. They prograde gradually into a basin and show a decrease in thickness and grain size in the downflow direction. The sediment source is near the coastline, and the turbidite basins are commonly small to medium in size. The fine-grained fans occur on passive and active margins, prograde rapidly into a basin, and deposit most of the input sand in the distal fan as oblong sheet sands. Tectonically confined basins normally have their sediment source nearby, and therefore will be filled with coarse-grained fans. Most of the open (unconfined) basins are medium to large in size, have their sediment source far from the coast, and therefore lose the coarser fractions during continental transport. Diapirically controlled basins are small- to medium-sized confined basins that have a fine-grained turbidite fill, but may not reveal the bypassing of the majority of the sand to the outer fan because of the abundance of sediment transport to the basin.

Placing all known submarine fans/turbidite systems (single turbidite fans) in a few detailed models is not possible at this time, and probably never will be. In natural systems there are too many factors influencing the final depositional characteristics. Many geoscientists look for simple systems with well-defined boundaries. Some authors base a detailed model on one or a few locations, whereas others use many examples to provide a rather simple 'unifying' model that normally lacks necessary detail. Mahaffie (1994) made it clear in his study on Mars Field, Gulf of Mexico, that the thickness of an individual fan can be covered by a single seismic wavelet. Therefore, acceptable information required to understand its reservoir characteristics and to construct some general models is possible only when that information is properly integrated with well logs, microresistivity or dip-meter data, and some core. It would be ideal to have only a handful of simplistic models such as Reading & Richards (1994) attempted. Once selected, the applicable model can be expanded significantly to include the special characteristics of the example under study. Normark (1970), Mutti & Ricci Lucchi (1972), Walker (1978) and some others provided excellent models that can be used as a guide when dealing with sand-rich and gravel-rich fans.

Gradually it has become obvious that the models indicated above do not cover all turbidite systems, and forcing the geology into an existing model is not desirable. Later publications, such as those by Mutti & Normark (1987, 1991), Pickering *et al.* (1986, 1989), and Reading & Richards (1994), either were too general to display architectural details, or emphasized only one or two aspects. The objective of this paper is to take a step back by discussing the various natural controls on turbidite systems and their interactions that have an influence on the generation of sedimentary particles, their transport to the coast, the basin setting, deposition in that basin, and possible redeposition. Once those controls are better understood for a specific turbidite system/complex, a site-specific model can be constructed.

Defining some terminology

All processes involved in moving sediments from the mountains to a basin follow a logical progression; however, confusion caused by different uses of terminology is common.

The word *slope* can refer to two different seafloor gradients in the same area. One use refers to a continental slope with or without small basins, and another use of the word refers to the local slope leading to the basin floor of an individual sub-basin. Because the term *sub-basin* is correct but less frequently used, one may consider the overall area to fall under the term *continental slope*, and the individual basin slope as *sub-basin slope*, or just *slope*. Each individual small basin has its own shaping and filling histories (e.g. Prather et al. 2000). Because its own slopes and basin floor have to be referred to quite often, the simplest terms are preferred.

Debris flow deposit (debrite) has become a term with several different meanings. A debris flow is defined here as a high-density, plastico-viscous flow in which the larger particles are supported by the matrix strength of the fluid component and its incorporated fine-grained sediment (Hampton 1972). The final deposit is characterized by having most of the coarser particles 'floating' in the matrix.

A *turbidity current deposit* (turbidite) was transported by a flow in which the sediment was supported by the upward-moving component of fluid turbulence. The original definition results from outcrop observations. It indicates that a turbidite is characterized by an upward-fining grain size (graded bedding), and by a sequence of five divisions, each with its own characteristic sedimentary structure (Bouma 1962). The complete sequence is known as the *turbidite sequence*, or *Bouma Sequence*. Most turbidite deposits show an incomplete sequence with missing bottom and/or top divisions. Further studies show that many fine-grained turbidites do not have a sufficient grain size range to form visible graded bedding, and some authors (e.g. Shanmugam 2000) use the confusing term *sandy debrites*.

Amalgamated surfaces are contacts within what seems to be a homogeneous sandstone layer resulting from erosion and deposition by successive turbidity currents. These contacts are often characterized by a somewhat coarser sediment overlying a locally preserved finer-grained layer. In fine-grained turbidite systems a grain size difference may not be present or is not visible. However, a contact may be identified by a film of mica flakes, clay minerals, and/or organic matter. Laterally, such a film can be absent because of erosion. One should be very careful when comparing layer thickness measurements in fine-grained turbidites with those observed in coarse-grained turbidites.

The oil industry commonly uses the name *shale* for unconsolidated, consolidated, and nearly fissile clay-bearing fine-grained deposits. The proper name for unconsolidated shale should be *mud* (clayey mud, silty mud or silt-rich mud, sandy mud or sand-rich mud, or with the terms 'very' or 'slightly' in front). The consolidated version should be called *mudstone*. Because the misuse of the word 'shale' is so common, many authors mix the terms. Rather than using sand/mud ratio, or sandstone/mudstone ratio, it might be better to stick with sand/shale ratio, or sandstone/shale ratio, or ss/sh ratio, because these terms are common to everyone.

Less difficulty exists with the terms *turbidite system* and *turbidite complex*. These terms were introduced by Mutti & Normark (1987, 1991), and they remain very useful. *Turbidite system* is a body of genetically related turbidite facies and facies associations that were deposited in virtual stratigraphic continuity. A turbidite system, either coarse-grained or fine-grained, represents an individual fan that consists of alternating sand and mud layers, and is characterized by a relatively high sandstone/shale (sand/mud) ratio. A *turbidite complex* refers to a basin-fill succession, made up by several turbidite systems stacked only partially on top of one another. Coarse-grained complexes have relatively thin sand-rich mudstones between their systems. Fine-grained turbidite complexes often have thick mudstones between individual systems.

Interactive natural forces

Stow et al. (1985) indicated three primary controls on fan development:

(1) sediment type and supply;
(2) tectonic setting and activity; and
(3) sea-level variations.

In addition to those, climate, basin characteristics and sedimentary processes should also be taken into account. Each of these primary factors is composed of a number of secondary components that are interactive within the entire set of controls (see below).

Tectonics

Tectonics is the major primary natural force that influences almost all aspects of submarine fan geology affecting source, slope, and sink. With regard to turbidite complexes, tectonics can have a major influence on:

(1) the type of continental margins, their locations, orientation of basins with regard to the coast and their sizes;
(2) the latitude and height of mountains and their specific climate that dictates type and supply of the sediment;
(3) the rate of sediment supply;
(4) the distance from those mountains to the sea shore and the various gradients encountered (Fig. 1);
(5) the width and breadth of the coastal plain;
(6) the width and morphology of the shelf and continental slope; and
(7) the type, size and morphology of the receiving basin.

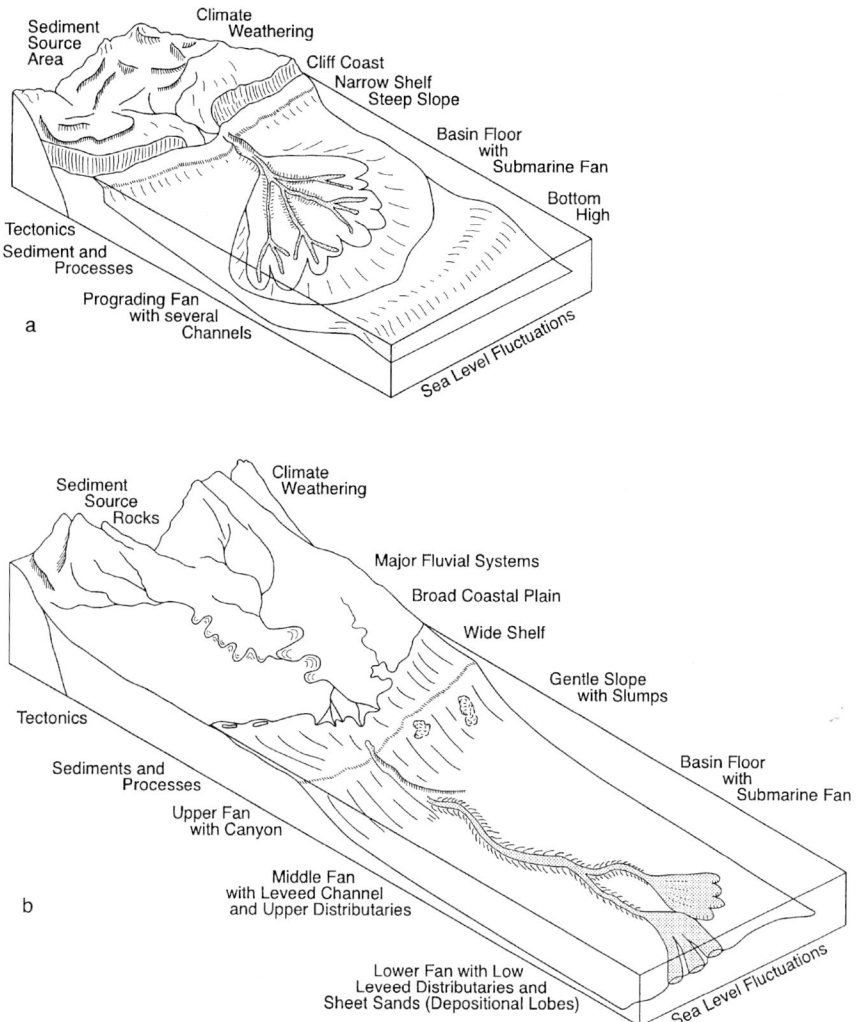

Fig. 1. Schematic block diagrams showing two end models with different relative distances from the sediment-producing mountains, to the coast, the relative width of the shelf, and the shape and location of the submarine fan. Major forces that produce transport and deposit sediment are mentioned. Morphological differences are shown. (**a**) Typical situation for coarse-grained/sand-rich fans fans; (**b**) for fine-grained/mud-rich fans. Based on Stow *et al.* (1985), Reading & Richards (1994) and Bouma (2000a,b).

Major fault and fold patterns, as well as diapirism, are the result of tectonic activity. Slopes continuing to trenches or other deep areas, as well as gentle continental slopes, can be segmented by faults in such a manner that each segment shows a back-tilting, resulting in a set of tectonically shaped confined basins (Fig. 2a). The more a segment of a continental slope moves downward along a fault plane, the more phases of fill can be discerned because complete filling seldom takes place in one depositional period. The most elevated slope basin has to be filled to the level of its lowest spillpoint before sediment can be transported to the next basin. The continental slope may contain a number of rotational basins of different size, shape, and location (Fig. 2b). The entrance to a slope basin can be anywhere and does not have to be at an end of a sub-basin. Consequently, palaeocurrents can run in one or two directions, rendering the fill characteristics of one slope basin different from the others. Correlation between such sub-basins, therefore, is not possible, except when volcanic ash layers or identifiable

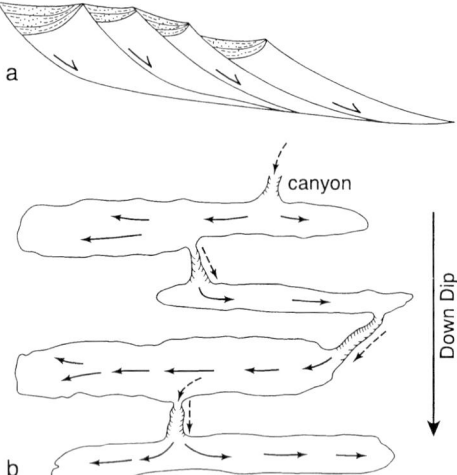

Fig. 2. Schematic dip section (**a**) and plan view (**b**) of faulted blocks, resulting in individual tilting of adjacent turbidite sub-basins. The overlying formations, commonly grading upward into shallow marine and continental deposits, are not shown. Spilling over occurs only when the updip basin is filled. The location of the lowest spot in the outside wall of an individual sub-basin, and the gradients in each sub-basin, can result in one or two transport and filling directions in the next sub-basin. Based on Dickinson (1980), Apps *et al.* (1994), van der Pluijm & Marshak (1997), and Wood & Roberts (2000).

condensed sections are present. Any single basin or set of tectonically generated basins can become a receiving basin. Therefore, shape and size vary tremendously.

Climate

Climate is determined by the amount of solar radiation received by the continents and water masses. In addition, its influence is dependent on:

(1) latitude and height of the mountains producing the bulk of the sediment;
(2) major wind directions that move moisture from the water masses to those mountains; and
(3) ratio of evaporation of ocean and lake water versus melting of ice on the mountains (Fig. 1).

Climate has a direct effect on:

(1) the freezing/thawing balance in the sediment-supplying mountains;
(2) the seasonal variations, types and quantities of fluvial runoff;
(3) the amount and types of vegetation that will grow at certain times;
(4) the amount of water available at a certain time to transport material to the base of the mountains and finally to the coast;
(5) the mechanical abrasion of sediment during transport;
(6) the chemical/diagenetic changes that take place when sediment is temporarily stored in updip settings; and
(7) the nature of hydraulic processes operating in the different parts of the lower fluvial and deltaic systems (hypo-, homo-, hyperpycnal).

Climate is the major control on sediment transport and non-tectonic sea-level fluctuations. It also controls mechanical and diagenetic changes of the sediment along its continental transport route, with the result that many feldspars will break down, and the resulting sediment becomes mineralogically mature. The shorter the transport distance, the more mineralogically immature the sediment reaching the sea will be.

There is also a relationship between the amount of runoff and the flow conditions in the more distal part of the fluvial system before it reaches the coastline. Hypopycnal flow, where the density of the fluvial water with sediment in transport is less dense than that of the ocean water, is a normal transport mechanism at the present time. When the density of the river water with suspended sediment is higher than the density of the ocean water (Wright *et al.* 1986; Chikita 1990), a turbulent bottom current (hyperpycnal) will prevail. Such a flow may be able to bypass a narrow continental shelf and slope, and thus can transport sediment directly to the receiving basin. In other words, maximum transport takes place during maximum fluvial runoff. This should happen as soon as the melting of continental ice is stronger than freezing in areas with continental ice, which is typical for the beginning of a climate-controlled transgressive systems tract (Bouma 2001).

Sea-level fluctuations

Fluctuations in sea level can be caused by tectonics, climate, major epochs of volcanism, bolide impacts, or a combination of the first two. Relative sea-level fluctuations can be caused by local forces (e.g. subsidence, salt or shale diapirism, and compaction; Revelle 1990).

A global decrease in temperature causes the accumulation of considerable amounts of ice on mountains and surrounding areas, and results in a global sea-level lowering. Fluvial runoff will exist but will be relatively small. The lowering sea-level forces the coastline to move seaward, and the active deltaic distributary to cut into its underlying deposits (incised valley), resulting in the transport of sediment to the shelfbreak area. However, those amounts of transported sediment are small compared with what will happen when the Earth's temperature rises, and the freezing/thawing balance changes in favour of thawing. When dealing with major mountain ridges and large river systems, large amounts of water become available, creating gigantic fluvial systems, and transporting enormous amounts of sedimentary material to the coastal area. Large volumes of material will be rapidly deposited at the end of the incised valley. Instability will result, followed by slumping. Those slumps commonly transfer into debris flows and/or turbulent flows. These initial clastic sediments on the basin floor are likely to be high in sand and shaped with sheet-sand geometries. Continuation of global warming, with its sea-level rise, may result in hyperpycnal flows that leave the mouth of the incised valley and continue to the basin floor without stopping. This implies that deposition of a submarine fan may occur in a very short time during the initial transgressive systems tract (Bouma 2001).

Sediment character and sedimentary processes

The size distribution and petrological composition of the sediment formed in the source area (e.g. alluvial fans) are very important. When the on-land transport distance is rather short, the mechanical breakdown will be incomplete and, therefore, partially influence the grain size distribution, depending on hardness, size, and composition of the grains. The chance for significant chemical breakdown of feldspars is very small, which means that clay minerals may not be an important part of the sediment when it reaches the coast.

The terms *efficient transport* and *non-efficient transport* were introduced by Mutti (1985) to indicate the character of transport into a basin. Efficient transport commonly relates to long transport, often fine-grained sediment, and is typical for a passive margin. Non-efficient relates to short transport, coarser sediment, and active margin. There are too many exceptions to that rule, especially when dealing with active margins. Therefore it has been suggested (Scott *et al.* 2000) that the terms 'active' and 'passive' margins should not be used when dealing with submarine fans. The descriptive terms *coarse-grained/sand-rich* and *fine-grained/mud-rich*, or just *coarse-grained* and *fine-grained*, are generally more useful.

Processes of transport and deposition can range from rockfall to a slide, slump, debris flow, or turbidity current when dealing with a mountainous coast. A passive margin coast typically consists of a large coastal plain with a wide shelf. A relative sea-level lowering is often required in such a setting to move the active delta distributary to the shelfbreak area. Common transport mechanisms are slump, slide, debris flow, and turbidity current.

Most active or structural margins are characterized by a sediment source on or near the coast. Longshore transport of the particles, crossing the shoreline, commonly causes trapping in depressions on the shelf. When the shelf is narrow and shallow, wave activity can move sediment to the shelfbreak. This has an erosional effect that results in tributary canyon heads that coalesce on the upper slope, forming a main canyon. Loose sediments, often containing significant amounts of organic matter, pile up until instability is reached and slumping results, moving that sediment to the basin.

Coarse-grained and fine-grained turbidite systems

There are a number of general models that can serve as starting points for the development of any site-specific working model. Reading & Richards (1994) published an extensive summary of what was known at that time. It is their opinion that grain size (mud-rich, mud-rich/sand-rich, sand-rich, gravel-rich) and type of feeder system (point source, multiple-source submarine ramp, linear-source slope ramp) produces 12 models needed to categorize all possibilities.

When studying deltas it is suggested that only one distributary channel is active, except during the rapid switch to a new one. In practice, multiple source and linear source often represent a series of successive point sources. The change from one point source to another seems to go rather fast, comparable to the change from one deltaic distributary channel to its successor. Gravel-rich fans are rather small and occur only on some active margins. Steep, short fluvial systems, very narrow shelves, and rather steep gradients are required to move this grain size.

In the mud-rich, mud/sand-rich and sand-rich systems it seems more common that only one point source is active at a given time. The transport of sediment is dictated primarily by the

Table 1. *Important differences between coarse-grained, sand-rich and fine-grained, mud-rich turbidite complexes*

Key controls and main characteristics	Coarse-grained, sand-rich complex	Fine-grained, mud-rich complex
Length of transport overland	Relatively short	Relatively long
Type of margin	Active	Passive and active
Location of receiving basin	Typically on continental crust	Typically on oceanic crust
Main grain size at the coast	Medium sand and coarser	Fine sand and mud
Feeding system to the fan	Canyon fed	Delta fed
Shelf width	Relatively narrow	Wide
Size of submarine fan	Often small to medium	Often medium to very large
Sandstone/shale ratio of the turbidite system	High	High
Sandstone/shale ratio of the turbidite complex	High: relatively thin shales between the complexes	Low: relative thick shales between complexes
Type of fan building	Prograding	Partial bypassing
Sandstone/shale ratio in dip direction	Decreasing	Increasing
Influence of sea-level variations	Not a major effect	Critical
Canyon fill	Gravel, sand, mud	Mainly mud, some olistholites and shelf sand bodies
Petrography	Often immature	Often mature

Gravel-rich fans and ramps are not discussed.

carrying capacity (competence) of the fluid phase, consisting of fine silt and clay dispersed in the water. The density and turbulence of that muddy fluid dictate the size and quantity of the larger grains that can be transported. Studies on the tectonic setting of ancient systems may indicate which type of system seems most logical between the fine-grained and coarse-grained for a given area. Therefore those two size types can be considered as end members, and are characterized as such in Figures 1a and 1b. Table 1 presents some of the main differences between the two end members.

The sand-rich turbidite complex is commonly characterized by sand-rich shales that separate the individual systems and are much thinner than the individual sand-rich systems. Each individual fan gradually progrades into the basin, and becomes finer grained and thinner in the downflow direction. Many examples are referenced in some recent books (e.g. Weimer *et al.* 1994, 2000; Pickering *et al.* 1995; Bouma & Stone 2000; Elliot 2000).

The fine-grained, mud-rich complex consists of fine-grained, sand-rich fans that are separated from each other by thick sections of mudstones/shales (Bouma & Wickens 1994; Bouma 2000a,b). They are normally fed by major fluvial-deltaic systems that require a significant sea-level lowering to bring that active deltaic distributary to the shelfbreak. The long overland transport, with commonly periods of burial of sediment, will leave the coarser sand fractions behind, allowing for more diagenetic attack on feldspars, and consequently maturing the sediment.

Deposition of a fine-grained fan may start at the base-of-slope with the formation of broad, shallow channels in the mouth of the wide (4–12 km) opening of the fan valley (lower canyon) (Figs. 1b, 3). Deposits in these channels often have a slightly higher mud content than the other sands in the fan system as fully developed turbulent flow has not yet developed and therefore flows are unable to move much mud from the head to the body and tail of the density current. As soon as the transporting flow reaches the basin floor it starts to form a leveed channel with extensive overbank deposits. Gradually the channel becomes narrower. Most of the time only one distributary channel is active (Weimer 1989). When the levees become too low, the head of a turbidity current will overflow, and sheet sands are deposited. Although the flow changes from confined to unconfined, the flow seems to continue without a significant widening. Oblong sheet sands will onlap onto each other, often with sand-on-sand contact. In the fan fringe area the sheets separate into individual tongues (Fig. 3).

Mississippi Fan

This turbidite system is a fine-sand, mud-rich fan, located on continental crust, belonging to the medium-large size class. The thickness of the entire complex is unknown; therefore the following discussion is restricted to the Upper Pleistocene system. Its maximum thickness of 400 m is located between 60 and 140 km

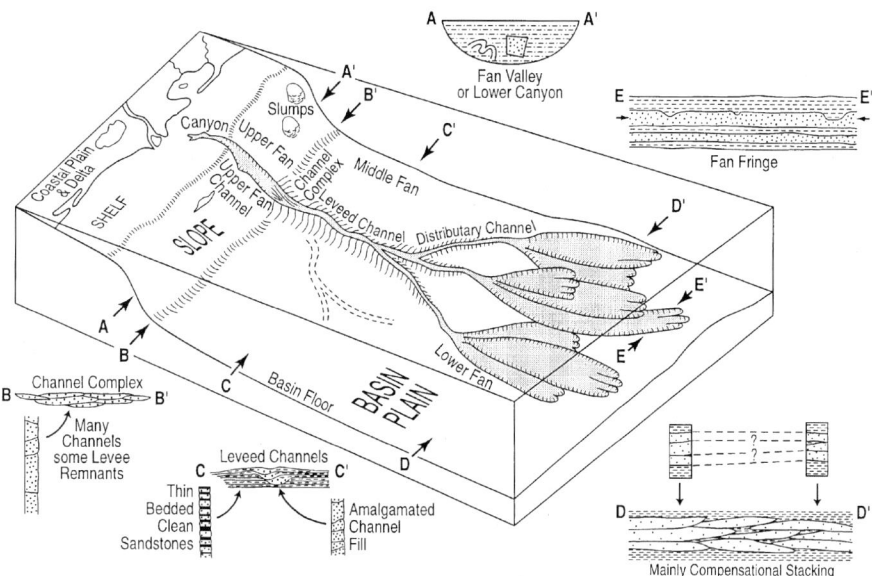

Fig. 3. Block diagram showing the model for the fine-grained turbidite system with schematic cross-sections and cores from four areas. The three depositional end components are: (1) channel complex at the base-of-slope comprising shallow, wide-channel fills, each smaller than the opening of the upper fan channel (canyon mouth); (2) leveed channel and leveed distributary channels with amalgamated channel fill and bedded levee sandstone and shale deposits – the largest part of the mid-fan is covered by mud-rich overbank deposits; (3) sheet sands with mainly compensational stacking and fingered fan fringe. The upper fan channel (cross-section A–A′) is filled with mudstone, some slumps and olistoliths (slide blocks of shelf sand). Not to scale; modified from Bouma (2000a,b; 2001).

basinward of the base-of-slope. For detail see papers in Bouma *et al.* (1985, 1986).

The upper fan consists of the entire canyon and the base-of-slope. The canyon fill consists of mud, whereas it is estimated that the base-of-slope contains about 6–12% of all the sand within the fan system. The mid-fan is characterized by one or more leveed channels and the leveed parts of the distributary channels. Sand occurs only as thick channel fills and bedded levee deposits, which have a sandstone/mudstone ratio of 60–90%. Although the individual channels can be 1 km or much wider, they cover only a small percentage of the mid-fan surface area. The largest surficial area of the mid-fan is covered by overbank deposits, starting with a high sandstone/mudstone ratio of 70% near the outer base of the levee, gradually becoming 0% far away from the channel. Where the fine sand gives way to silt as the largest component is unknown. The percentage of sand in the channelized (channel and levee) mid-fan ranges from 15 to 25% of all the sand entering the entire fan system. The lower fan consists of non-leveed distributaries, sheet sands (depositional lobes) and overbank/fan fringe silty muds. This part of the fan contains about 60–90% of all the fine sand that enters the fan system through the canyon (Bouma 2000a,b; Fig. 3). The distributary channels do not contain much sand, about 3–5% of the entire fan.

Tanqua Karoo

This Permian turbidite complex is located about 200 km northeast of Cape Town, South Africa. The setting was a wide, rather shallow (500–800 m) foreland basin within the Cape fold/thrust belt. It can be classified as an unconfined turbidite basin. The complex is fine-grained and mud-rich, consisting of five fans that are separated by thick shales. The present tectonic setting is unique because there is no tectonic tilt in the N–S direction, and only 1–3° tilt to the east. Excellent outcrops exist along the west side, along the northern and southern boundaries, and along an E–W running river valley in the centre. From one fan to its successor, a progradation to the north can be observed with some topographic compensation to the east. At present the entire complex is about 34 km from north to south and covers an area of approximately

650 km^2 (Bouma & Wickens 1994; Morris et al. 2000; Wickens & Bouma 2000).

Fan 3 is a complete turbidite system, showing base-of-slope, leveed channels, distributary channels, and depositional lobes (sheet sands). Its thickness ranges from about 45 m in the south to zero in the north; average palaeocurrent is towards the NNE (Wickens & Bouma 2000).

Fan 3 was used to calculate the distribution of sand. The data may have an error of 10–15% because it is not known what percentage of this fan has eroded away. The base-of-slope is characterized by sand-rich channel fills, 0.2–2 km in width, and a width/thickness ratio of 40–80/1 (Bouma 2000a). Establishing accurate boundaries between upper, middle and outer fan was difficult, but the excellent outcrops made it possible. Calculations suggest that base-of-slope area stored 7–16% of all the sand that constructed Fan 3, the leveed channels 15–35%, the distributary channels 6–9%, and the outer sheet sands 50–70%.

Channel sands can be very massive, and the lack of graded bedding in fine-grained turbidite sands obscures most of the depositional breaks. Individual levee sandstones are the thickest on the crest of a levee, and can be up to 40–60 cm thick. These sands are the cleanest of the entire fan, where current ripples and climbing ripples indicate that much of the overflowing sand moved as traction currents, thereby winnowing out muds. These levee sandstones are called *low-contrast, low-resistivity, thin-bedded sandstones* by Darling & Sneider (1992), and often have the highest porosity and permeability in a fan system. Because the muds have radioactive isotopes adhering to the clay minerals, micas, and organic matter, those sand/mud couplets are identified on well logs as shaley sandstones because of lack of tool resolution, and thus are often overlooked for reservoir consideration.

The observations from the Mississippi Fan and the Tanqua Karoo suggest that the individual lower fan sheet deposits can be $\frac{1}{2}$–4 km wide and up to 8–10 m thick. There are amalgamated contacts present in the Tanqua Karoo, some being visible where lower Bouma Sequence divisions overlay higher ones. Studies suggest that the various sized, oblong sheets are arranged in a compensational stacking pattern (Fig. 3), rather than in a vertical or singled pattern (Bouma & Rozman 2000).

Peïra Cava Sandstone

The eastern French Maritime Alps are separated from the large Annot area by the N–S oriented sections of the Var and the Tinée Rivers. The major turbidite outcrop locations of the eastern part are Menton, Piena, Sospel, Contes, Peïra Cava, and Tournairet (Fig. 4). All are separated from each other by N–S trending highs of Jurassic limestone, later covered by Middle and Upper Cretaceous formations. The Upper Cretaceous deposits are either of medium-bedded dense limestone or medium to thick marly carbonates. The series has been intensely folded and faulted. Discordantly, the Upper Cretaceous in the Peïra Cava area is overlain by Upper Eocene nummulitic limestone, followed by bluish marls that grade into the brownish-gray Peïra Cava Sandstone (Bouma 1962; Bouma & Coleman 1985).

The Tertiary sediments have undergone at least three orogenic movements. The N–S trending Jurassic ridges were formed near the Middle and Upper Eocene boundary. The second and third movements took place during the Lower to Middle Oligocene and between Upper Miocene and Lower Pliocene, respectively. The late Pliocene and Pleistocene uplift was the last tectonic activity.

The general palaeocurrent direction is to the north. The Piena, Sospel and Contes locations show directions 10–20° to the west (Fig. 4). If this reveals original directions or the result of tectonic turning it is not known to the author. The issue is important because later tectonic turning could have destroyed connections between sub-basins, such as between Contes and Peïra Cava. Significant tectonics, uplifting and erosion are probably responsible for the removal of the other connections.

Common field observations show a south-to-north decrease of the completeness of the Bouma Sequence, together with a thinning of the layers, a fining of the grain size, and a decrease of the sandstone/shale ratio. The original sediment source was a mountain chain located where the northern Mediterranean is now. Although one can observe change from south to north, no geological evidence is left pertaining to connections between the sediment source area and the sub-basins. A connection between Sospel and Peïra Cava may have existed, and became the victim of later erosion and possibly tectonic turning. The Sospel area shows large leveed channels, and the Peïra Cava area consists mainly of distributary channels, small sheet sands, and a mud-rich, thin-bedded fan fringe. Tectonic tilting during deposition can be observed along the southeastern contact between the marls and the turbidites of the Peïra Cava area (Fig. 5).

The Peïra Cava area, together with the surrounding turbidite locations east of the Var-Tinée Rivers, is an excellent example of confined

Fig. 4. Map of the eastern part of the French Maritime Alps with locations of the turbidite basins and measured palaeocurrents as observed at the present.

basins, and of the influences of the tectonic forces before, during, and after deposition of the turbidites. The gradual progradation of sediment into the sub-basins rather than bypassing, the gradual decrease in grain size to the north, together with thinning of the sandstone layers, a decrease in the completeness of the turbidite sequence and a lowering of the sandstone/shale ratio, all point to a major marine area with confined sub-basins – filled with sand-rich turbidite systems.

Diapiric intraslope basins

Diapirism, involving salt or shale, is common all over the world. It can result in simple small or large basins, or in very complicated patterns (Fig. 6). A simple fan complex can be formed downdip from an area of diapirism, and therefore should not be classified as a diapiric basin (Fig. 6a). However, loading of deep buried salt can result in diapirism with the breaking up of fans (Fig. 6). The northern Gulf of Mexico continental slope offshore of Texas and Louisiana presents many examples where active diapers and diapiric ridges prevented transport to deeper water until vertical movement and/or scouring opened an avenue for transport (Fig. 6b; e.g. Martin 1978; Apps *et al.* 1994; Beaubouef *et al.* 2000; Booth *et al.* 2000).

Diapiric ridges can produce a set of sub-basins more or less similar to Figure 2. A canyon can cut through one or more ridges, permitting transport and deposition into a set of sub-basins. Similarly an open passage may have existed that later became blocked by diapers (Sutterfield & Behrens 1990; Apps *et al.* 1994; Pirmez *et al.* 2000).

These mini-basins can be classified as confined basins that require a somewhat different approach from the structurally controlled basins. The Gulf of Mexico is a passive margin, and the deepwater sands are fine-grained. However, the input of sediment might be large for the basin size, with the result that piling-up

Fig. 5. Southeast–northwest oriented photograph of the southern boundary of the Peïra Cava area. On the right-hand bottom one observes onlapping to the NW of the bedded Peïra Cava Sandstones onto the underlying blue marls, suggesting minor downward syndepositional tilting of the western side of the basin. The thicker layers, visible in the photograph, are not entirely parallel. This provides the idea of a later minor post-depositional upward tilt of the western side of the sub-basin.

takes place in the downdip area. As a consequence the characteristics of a fine-grained turbidite system may be difficult to recognize.

Delaware Basin turbidite complex

The descriptive term *fine-grained, mud-rich* does not work for all fine-grained turbidites. An example is the Permian Delaware Basin Brushy Canyon Formation turbidite system in west Texas–New Mexico (Gardner & Borer 2000; Carr & Gardner 2000, and references therein). The turbidites are fine-grained but lack significant amounts of finer-grained siltstones in between. This is an example of aeolian transport of sand to the coast and across the shelf to an outer-shelf reef zone. Final transport took place through reef openings toward the deeper basin. Lack of clay-sized material, and of major channels across the shelf, probably placed these deposits more in a fine-grained/sand-rich architecture.

Conclusions

Tectonics, climate, sedimentary processes and characteristics, and sea-level fluctuations are the main controls in the generation of sedimentary particles, transport over land and the ocean bottom, and final deposition to form a turbidite fan. The controls interact with one another at varying intensities and time schedules.

Tectonics is the major agent that influences the location and height of the sediment-producing mountains, the distance of the mountains to the shoreline, the type of coastal area and the width of the shelf, and the location and other characteristics of the turbidite basin.

Climate controls the freezing–thawing cycle, weathering characteristics, the amount of water available for a sub-aerial sediment transport, the grain size distribution when the material reaches the coastline, and much of the sea-level fluctuation.

Sedimentary characteristics involve composition, grain size range, sandstone/shale ratio,

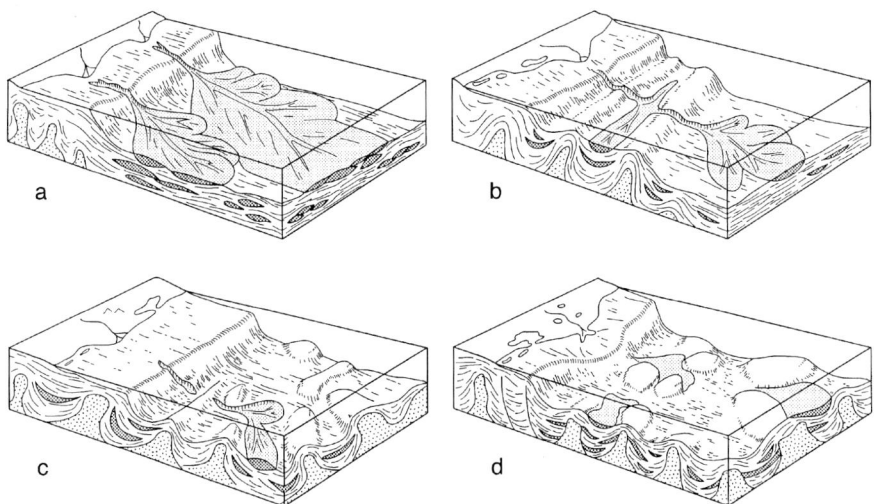

Fig. 6. Selected schematic presentations of the interaction between diapirism and fan deposition: (**a**) no direct interaction between the diapers and the shelf and the turbidite fan; (**b**) diapiric ridges form troughs that have to be filled first (two directional fills) before overflow to a next trough can take place; (**c**) continuous diapiric activity, likely during a high-stand systems tract, breaks a turbidite system into two parts; (**d**) breaking apart of a fan by diapiric activity. The sediment source at the delta front was either not active when the diapers moved upwards, or the delta switched to another location. (**c**) and (**d**) can happen when the fan is still exposed, but it is more common when they are buried. Redrafted from Bouma (1982).

maturity of the sediment, and the types of transport and distribution of sand on the submarine fan.

Sea-level fluctuations dictate when sediment accumulation concentrates on the sub-aerial land or the deepwater area. A major exception is formed by gravel-rich deposits that require rather steep slopes and water for transport to make a 'stubby' fan. Those fans can be constructed during high systems tracts.

There are a large number of submarine fan types. The most common are sand–clay mixtures. Two end members, coarse-grained/sand-rich and fine-grained/mud-rich, bound any combination of those two types. The coarse-grained/sand-rich type is small to medium in size, most common in active margin settings, has a grain size that consists mainly of medium sand and coarser particles, has rather thin sand-rich shales separating the sand-rich turbidite systems, progrades gradually into the basin, thins and fines in downcurrent direction. The fine-grained/mud-rich turbidite complex is characterized by thick clay-fine silt deposits that separate the individual fine-grained/sand-rich turbidite systems. These fans normally are medium to large in size, prograde rapidly into the basin, and transport 60–90% of all of the incoming sand to the distal sheet-sand area.

Turbidite basins can have all types, sizes and shapes, ranging from confined (more typical for active margins) to unconfined (more typical for passive margins, except for diapiric basins that are confined). Most of the confined basins are tectonically controlled. Syn- and post-depositional changes are common and depend on tectonic style and activity. Unconfined basins normally are large, and their sediment source is far from the coastal zone. Confined basins commonly are coarse-grained and may have significant amounts of gravel, as well as a low percentage of fine-grained mud.

In conclusion, it is extremely important to first analyse whether the sediment in the turbidite fan is coarse-grained or fine-grained. Simultaneously, one should search for the tectonic history of the basin. The third step is to determine whether it was a confined or unconfined basin during the time when the turbidites were deposited. Having a better understanding of those issues, the investigator is better prepared for outcrop-sized seismic facies analyses (Prather *et al.* 2000) and reservoir characteristics.

R. H. Kirschner, E. D. Scott, C. E. Stelting, C. G. Stone and J. B. Thomas made constructive remarks on the first draft. T. Elliot and an unknown reviewer provided in-depth reviews that improved the manuscript considerably. C. Wittmann and A. Delery are thanked for typing the various rounds of the manuscript. The invitation of the editors, S. Lomas and P. Joseph, to submit this manuscript is highly appreciated.

References

APPS, G. M., PEEL, F. J., TRAVIS, C. J. & YIELDING, C. A. 1994. Structural controls on Tertiary deep water deposition in the northern Gulf of Mexico. *In*: WEIMER, P., BOUMA, A. H. & PERKINS, B. F. (eds) *Submarine Fans and Turbidite Systems: Sequence Stratigraphy, Reservoir Architecture and Production Characteristics; Gulf of Mexico and International*. Proceedings Gulf Coast Section Society of Economic Paleontologists and Mineralogists Foundation 15th Annual Research Conference, 1–7.

BEAUBOUEF, R. T. & FRIEDMAN, S. J. 2000. High resolution seismic/sequence stratigraphic framework for the evolution of Pleistocene intra-slope basins, Western Gulf of Mexico: depositional models and reservoir analogs. *In*: WEIMER, P., SLATT, R. M. *et al.* (eds) *Deep-Water Reservoirs of the World*. Proceedings Gulf Coast Section Society of Economic Paleontologists and Mineralogists Foundation 20th Annual Bob F. Perkins Research Conference, CD-ROM, 40–60.

BOOTH, J. R., DUVERNAY III, A. E., PFEIFFER, D. S. & STYZEN, M. J. 2000. Sequence stratigraphic framework, depositional models, and stacking patterns of ponded and slope fan systems in the Auger Basin: central Gulf of Mexico slope. *In*: WEIMER, P., SLATT, R. M. *et al.* (eds) *Deep-Water Reservoirs of the World*. Proceedings Gulf Coast Section Society of Economic Paleontologists and Mineralogists Foundation 20th Annual Bob F. Perkins Research Conference, CD-ROM, 82–103.

BOUMA, A. H. 1962. *Sedimentology of Some Flysch Deposits: A Graphic Approach to Facies Interpretation*. Elsevier Publishing Company, Amsterdam.

BOUMA, A. H. 1982. Intraslope basins in northwest Gulf of Mexico: a key to ancient submarine canyons and fans. *In*: WATKINS, J. S. & DRAKE, C. L. (eds) *Studies in Continental Margin Geology*. American Association of Petroleum Geologists Memoir, **34**, 567–581.

BOUMA, A. H. 2000a. Fine-grained, mud-rich turbidite systems: model and comparison with coarse-grained, sand-rich systems. *In*: BOUMA, A. H. & STONE, C. G. (eds) *Fine-Grained Turbidite Systems*. American Association of Petroleum Geologists Memoir, **72**/ Society for Sedimentary Geology Special Publication, **68**, 9–20.

BOUMA, A. H. 2000b. Coarse-grained and fine-grained turbidite systems as end member models: applicability and dangers. *Marine and Petroleum Geology*, **17**, 137–143.

BOUMA, A. H. 2001. Fine-grained submarine fans as possible recorders of long- and short-term climatic changes. *Global Planetary Change*, **28**, 85–91.

BOUMA, A. H., BARNES, N. E. & NORMARK, W. R. 1985. *Submarine Fans and Related Turbidite Sequences*. Springer, New York.

BOUMA, A. H. & COLEMAN, J. M. 1985. Piera-Cava turbidite system, France. *In*: BOUMA, A. H., NORMARK, W. R. & BARNES, N. E. (eds) *Submarine Fans and Related Turbidite Systems*. Springer, New York, 217–222.

BOUMA, A. H. & ROZMAN, D. J. 2000. Characteristics of fine-grained outer fan fringe turbidite systems. *In*: BOUMA, A. H. & STONE, C. G. (eds) *Fine-Grained Turbidite Systems*. American Association of Petroleum Geologists Memoir, **72**/Society for Sedimentary Geology Special Publication, **68**, 291–298.

BOUMA, A. H. & STONE, C. G. (eds) 2000. *Fine-Grained Turbidite Systems*. American Association of Petroleum Geologists Memoir, **72**/Society for Sedimentary Geology Special Publication, **68**, 291–298.

BOUMA, A. H. & WICKENS, H. DE V. 1994. Tanqua Karoo, ancient analog for fine-grained submarine fans. *In*: WEIMER, P., BOUMA, A. H. & PERKINS, B. F. (eds) *Submarine Fans and Turbidite Systems: Sequence Stratigraphy, Reservoir Architecture and Production Characteristics; Gulf of Mexico and International*. Proceedings Gulf Coast Section Society of Economic Paleontologists and Mineralogists Foundation 15th Annual Research Conference, 23–34.

BOUMA, A. H., COLEMAN, J. M. ET AL. 1986. *Initial Reports of the Deep Sea Drilling Project 96*. US Government Printing Office, Washington, DC.

CARR, M. & GARDNER, M. H. 2000. Portrait of a basin-floor fan for sandy deepwater systems, Permian Lower Brushy Canyon Formation, West Texas. *In*: BOUMA, A. H. & STONE, C. G. (eds) *Fine-Grained Turbidite Systems*. American Association of Petroleum Geologists Memoir **72**/Society for Sedimentary Geology (SEPM) Special Publication, **68**, 215–232.

CHIKITA, K. 1990. Sedimentation by river-induced turbidity current: field measurements and integration. *Sedimentology*, **37**, 891–905.

DARLING, H. L. & SNEIDER, R. M. 1992. Production of low resistivity, low-contrast reservoirs, offshore Gulf of Mexico basin. *Gulf Coast Association of Geological Societies Transactions*, **42**, 73–88.

DICKINSON, W. R. 1980. Plate tectonics and key petrologic associations. *In*: STRONGWAY, D. W. (ed.) *The Continental Crust and its Mineral Deposits*. Geological Association of Canada Special Paper, **20**, 341–360.

ELLIOT, T. 2000. Depositional architecture of a sand-rich, channelized turbidite system: the upper Carboniferous Ross Sandstone formation, western Ireland. *In*: WEIMER, P., SLATT, R. M. *et al.* (eds) *Deep-Water Reservoirs of the World*. Proceedings Gulf Coast Section Society Economic Paleontologists and Mineralogists Foundation 20th Annual Bob F. Perkins Research Conference, CD-ROM, 342–373.

GARDNER, M. H. & BORER, J. M. 2000. Submarine channel architecture along a slope to basin profile, Brushy Canyon Formation, West Texas. *In*: BOUMA, A. H. & STONE, C. G. (eds) *Fine-Grained Turbidite Systems*. American Association of Petroleum Geologists Memoir, **72**/Society for Sedimentary Geology Special Publication, **68**, 195–214.

HAMPTON, M. A. 1972. The role of subaqueous flow in generating turbidity currents. *Journal of Sedimentary Petrology*, **42**, 775–793.

MAHAFFIE, M. J. 1994. Reservoir characterization for turbidite intervals at the Mars discovery, Mississippi Canyon 807, Gulf of Mexico. *In*: WEIMER, P., BOUMA, A. H. & PERKINS, B. F. (eds) *Submarine Fans and Turbidite Systems: Sequence Stratigraphy, Reservoir Architecture and Production Characteristics; Gulf of Mexico and International.* Proceedings Gulf Coast Section Society of Economic Paleontologists and Mineralogists Foundation 15th Annual Research Conference, CD-ROM, 233–244.

MARTIN, R. M. 1978. Northern and eastern Gulf of Mexico continental margin: Stratigraphy and structural framework. *In*: BOUMA, A. H., MOORE, G. T. & COLEMAN, J. M. (eds) *Framework Facies, and Oil Trapping Characteristics of the Upper Continental Margin.* American Association of Petroleum Geologists Studies in Geology, 7, 21–42.

MORRIS, W. R., SCHEIHING, M. H., WICKENS, DE V. & BOUMA, A. H. 2000. Reservoir architecture of deepwater sandstones: examples from Skoorsteenberg Formation, Tanqua Karoo sub-basin, South Africa. *In*: WEIMER, P., SLATT, R. M. *et al.* (eds) *Deep-Water Reservoirs of the World.* Proceedings Gulf Coast Section Society of Economic Paleontologists and Mineralogists Foundation 20th Annual Bob F. Perkins Research Conference, 629–667.

MUTTI, E. 1985. Turbidite systems and their relations to depositional sequences. *In*: ZUFFA, G. G. (ed.) *Provenance of Arenites.* Reidel Publishing Co., Dordrecht, 65–93.

MUTTI, E. & NORMARK, W. R. 1987. Comparing examples of modern and ancient turbidite systems: problems and concepts. *In*: LEGGET, J. K. & ZUFFA, G. C. (eds) *Marine Clastic Sedimentology.* Graham and Trotman, London, 1–38.

MUTTI, E. & NORMARK, W. R. 1991. An integrated approach to the study of turbidite systems. *In*: WEIMER, P. & LINK, M. H. (eds) *Seismic Facies and Sedimentary Processes of Submarine Fans and Turbidite Systems.* Springer, New York, 75–106.

MUTTI, E. & RICCI LUCCHI, F. 1972. Le torbiditi dell'Appennino settentrionale: introduzione all'analisi de facies. *Memorie della Societa Geologica Italiana*, 11, 161–199.

NORMARK, W. R. 1970. Growth patterns of deep-sea fans. *American Association of Petroleum Geologists Bulletin*, 54, 2170–7195.

PICKERING, K. T., HISCOTT, R. N. & HEIN, F. J. 1989. *Deep Marine Environments: Clastic Sedimentation and Tectonics.* Unwin Hyman Ltd, London.

PICKERING, K. T., HISCOTT, R. N., KENYON, N. H., RICCI LUCCHI, F. & SMITH, R. D. A. (eds) 1995. *Atlas of Deep Water Environments.* Chapman & Hall, London.

PICKERING, K. T., STOW, D. A. V., WATSON, M. P. & HISCOTT, R. N. 1986. Deep-water facies, processes and models: a review and classification scheme for modern and ancient sediments. *Earth-Science Reviews*, 23, 75–174.

PIRMEZ, C., BEAUBOUEF, R. T., FRIEDMAN, S. J. & MORIG, D. C. 2000. Equilibrium profile and baselevel in submarine channels: examples from late Pleistocene systems and implications for the architecture of deepwater reservoirs. *In*: WEIMER, P., SLATT, R. M. *et al.* (eds) *Deepwater Reservoirs of the World.* Proceedings Gulf Coast Section Society of Economic Paleontologists and Mineralogists Foundation 20th Annual Bob F. Perkins Research Conference, CD-ROM, 782–805.

PRATHER, B. E., KELLER, F. B. & CHAPIN, M. A. 2000. Hierarchy of deepwater architectural elements with reference to seismic resolution: Implication for reservoir prediction and modeling. *In*: WEIMER, P., SLATT, R. M. *et al.* (eds) *Deepwater Reservoirs of the World.* Proceedings Gulf Coast Section Society of Economic Paleontologists and Mineralogists Foundation 20th Annual Bob F. Perkins Research Conference, CD-ROM, 817–835.

READING, H. G. & RICHARDS, M. 1994. Turbidite systems in deepwater basin margins classified by grain size and feeder system. *American Association of Petroleum Geologists Bulletin*, 78, 192–822.

REVELLE, R. (ed.) 1990. *Sea Level Changes.* National Research Council, Studies in Geophysics. National Academy Press, Washington, DC.

SCOTT, E. D., BOUMA, A. H. & WICKINS, H. DE V. 2000. Influence of tectonics on submarine fan deposition, Tanqua and Laingsburg subbasins, South Africa. *In*: BOUMA, A. H. & STONE, C. G. (eds) *Fine-Grained Turbidite Systems.* American Association of Petroleum Geologists Memoir, 72/Society for Sedimentary Geology Special Publication, 68, 47–56.

SHANMUGAM, G. 2000. 50 years of the turbidite paradigm (1950s–1990s): deep-water processes and facies models: a critical perspective. *Marine and Petroleum Geology*, 17, 285–342.

STOW, D. A. V., HOWELL, D. G. & NELSON, C. H. 1985. Sedimentary, tectonic, and sea-level controls. *In*: BOUMA, A. H., NORMARK, W. R. & BARNES, N. E. (eds) *Submarine Fans and Related Turbidite Systems.* Springer, New York, 215–222.

SUTTERFIELD, W. M. & BEHRENS, E. W. 1990. A late Quaternary canyon/channel system, Northeast Gulf of Mexico continental slope. *Marine Geology*, 92, 51–67.

VAN DER PLUIJM, B. A. & MARSHAK, S. 1997. *Earth Structure: An Introduction to Structural Geology and Tectonics.* WCB/McGraw-Hill, New York.

WALKER, R. G. 1978. Deep-water sandstone facies and ancient submarine fans: models for exploration for stratigraphic traps. *American Association of Petroleum Geologists Bulletin*, 62, 932–966.

WEIMER, P. 1989. Sequence stratigraphy of the Mississippi Fan (Plio-Pleistocene), Gulf of Mexico. *Geo-Marine Letters*, 9, 185–272.

WEIMER, P., BOUMA, A. H. & PERKINS, B. F. (eds) 1994. *Submarine Fans and Turbidite Systems: Sequence Stratigraphy, Reservoir Architecture and Production Characteristics; Gulf of Mexico and International.* Proceedings Gulf Coast Section Society of Economic Paleontologist and Mineralogists Foundation 15th Annual Research Conference, Houston, Texas.

WEIMER, P., SLATT, R. M. *et al.* (eds) 2000. *Deep-Water Reservoirs of the World*. Proceedings Gulf Coast Section Society of Economic Paleontologist and Mineralogists Foundation 20th Annual Bob F. Perkins Research Conference, Houston, Texas, CD-ROM.

WICKENS, H. DE V. & BOUMA, A. H. 1995. The Tanqua basin floor fans, Permian Ecca Group, Western Karoo Basin, South Africa. *In*: PICKERING, K. T., HISCOTT, R. N., KENYON, D. H., RICCI LUCCHI, F. & SMITH, R. D. A. (eds) *Atlas of Deep Water Environments, Architectural Style in Turbidite Systems*. Chapman & Hall, London, 317–322.

WOOD, L. J. & ROBERTS, C. 2001. Opportunity in a world-class hydrocarbon basin: Trinidad and Tobago's eastern offshore marine province. *Houston Geologic Society Bulletin*, June, 37–45.

WRIGHT, L. D., YANG, Z. S., BORNHOLD, B. D., KELLER, G. H., PRIOR, D. B. & WISEMAN, W. J. 1986. Hypercynal plumes and plume fronts over the Huanghe (Yellow river) delta front. *Geo-Marine Letters*, **6**, 97–105.

Silled sub-basins to connected tortuous corridors: sediment distribution systems on topographically complex sub-aqueous slopes

RU SMITH

A/S Norske Shell, PO Box 40, 4098 Tananger, Norway
Present address: Shell International E & P, Volmerlaar 6, PO Box 60,
2280 AB Zijswijk ZH, Netherlands (e-mail: Ru.Smith@shell.com)

Abstract: Two end-member classes of sediment distribution systems on topographically complex slopes are distinguished here: (a) *cascades of silled sub-basins*, and (b) *connected tortuous corridors*. In the first scenario a process of filling and spilling of successive silled sub-basins down a slope occurs. For each sub-basin a sill tends to hinder further downslope flow of at least the basal sandy portions of sediment gravity flows until deposition reduces the relief sufficiently to allow spill down-slope. Spill is associated with incision in the sill. In the *connected tortuous corridors* scenario, flows avoid bathymetric obstacles, but follow a (laterally confined) continuous tortuous path down the slope. Without complete three-dimensional imaging of slope architecture it can be possible to incorrectly infer from two-dimensional profiles a *cascade of silled sub-basins* model. Thus flow paths in adjacent apparent sub-basins can be connected out of the plane of section. Convergent thinning and convergent baselap stratal patterns occur in both scenarios, but only in the silled sub-basin case do such patterns occur against closing frontal slopes. For a given complex slope morphology, dominant controls on fill patterns and reservoir architecture are (a) the history of sediment supply character, and (b) rates of structure growth relative to rates of smoothing of topography by erosional and depositional processes. Two particularly important aspects of sediment supply are (i) flow volumes relative to scales of receiving spaces, and (ii) flow properties (in particular, transported grain size distribution, flow thickness and flow concentration), these controlling depositional gradients and the equilibrium profiles to which slopes tend to grade.

Considerable attention has been given to the problems of predicting stratigraphic architectures and patterns of sand deposition on topographically complex sub-aqueous slopes, in both marine (Prather *et al.* 1998; Booth *et al.* 2000; Demyttenaere *et al.* 2000; Winker & Booth 2000) and deep lacustrine (Scholz *et al.* 1990; Smith 1995a) settings. Early published examples documenting the influence of topography on patterns of deposition and erosion include:

(1) California Borderland fill and spill models (Douglas & Heitman 1979);
(2) the Monterey Fan (Normark *et al.* 1984), where channels cut headward into the pirated Ascension Valley when the fan extended south of the Chumash Fracture Zone;
(3) forearc slope basins (e.g. Belderson *et al.* 1984; Stevens & Moore 1985); and
(4) the Welsh Basin uppermost Ordovician to Lower Silurian (James & James 1969; Smith 1987a), where it was inferred that filling of an inboard base of slope depression allowed basinward stepping of a younger channel-fed lobe system.

Intensive study of the Gulf of Mexico slope, which is affected by numerous salt-withdrawal sub-basins (Fig. 1), resulted in detailed models for sequential filling and down-slope 'spilling' of such 'minibasins' (Satterfield & Behrens 1990; Winker 1996; Badalini *et al.* 2000; Beaubouef & Friedman 2000; Winker & Booth 2000). Elsewhere, intraslope basins may result from deformation caused by buried thrust faults, often associated with mud diapirism (e.g. Demyttenaere *et al.* 2000; Pirmez *et al.* 2000). In extensional and transtensional settings perched basins may be associated with tilted basement fault-blocks (e.g. Ravnås & Steele 1998).

Whereas structurally induced bathymetry may form closed depressions on slopes, as in large parts of the Gulf of Mexico slope, in many other cases (e.g. Biegert *et al.* in press, and examples presented here from offshore Brazil and Angola) flow paths extend in a connected, variably tortuous route down complex slopes. This under-appreciated distinction is emphasized here and used to separate two end member types of sediment distribution systems down topographically complex slopes: (a) *cascades of silled sub-basins*, and (b) *connected tortuous corridors*. The latter scenario may prove to be a better descriptor for a number of subsurface and outcrop cases than the well-known fill and spill concept that is a key feature of the cascade of silled sub-basins model. The paper develops two main themes: (a) the interaction between varying types of sediment input character and types of

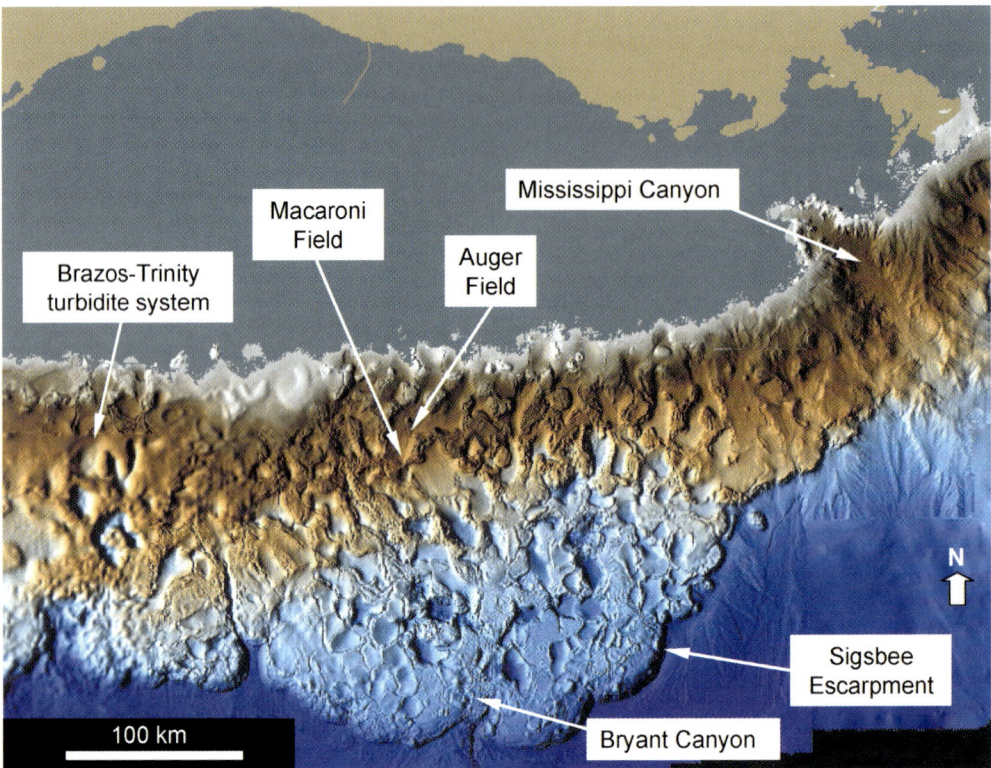

Fig. 1. Rendered seafloor image of the Gulf of Mexico salt-based slope. This is the archetypal example of a complex of silled sub-basins. Note the circular to elliptical salt-withdrawal intraslope basins with diameters ranging between approximately 5 and 20 km. Locations of systems discussed in the text are shown.

slope topography (with and without sills) in governing the depositional and erosional patterns that occur on topographically complex sub-aqueous slopes; and (b) an exploration of the distinction between a basinward downstepping cascade of silled sub-basins and a non-silled tortuous flow path model.

The paper is structured in five parts:

(1) an overview of the main controls on patterns of deposition and erosion on topographically complex slopes (relevant to both silled and non-silled cases);
(2) a discussion of the cascade of silled sub-basin model;
(3) a discussion of the connected tortuous corridors model;
(4) a review of processes and depositional products in silled sub-basins and connected tortuous corridors;
(5) a summary of guidelines for distinguishing between the two end member types of system.

The emphasis is on conceptualizing the controls on the stratigraphic evolution of complex slopes. Much of the supporting evidence has either already been published in some detail (e.g. the extensively published Brazos–Trinity system in the Gulf of Mexico) or else is currently in preparation for future publication (e.g. Biegert 2003 in press).

Explanation of terms

As some of the terminology used in the current literature on complex slope systems has become confusing (for example, see the discussion of the terms *bypass facies assemblage* and *ponded facies assemblage* in Winker & Booth 2000), some terms used in this paper are explained below.

Topographically complex slope. This phrase is used here to describe slopes that exhibit local topographic depressions and highs induced by deformation in the substrate, as a result of faulting, folding, salt tectonics or mud diapirism.

Silled sub-basin. A silled sub-basin is a closed topographic depression on a slope that has a topographic barrier at its downdip margin. Deposition in such a depression need not be ponded. Ponding occurs when sediment gravity flows are large in volume relative to the scale of the receiving depression. Deposition from smaller-volume, non-ponded, flows can result in downflow tapering beds that terminate prior to reaching the bounding slopes.

Connected tortuous corridor. A connected tortuous corridor is an elongate, variably tortuous, laterally confined depression on a topographically complex slope. The word 'connected' is used to indicate that the flow path is not broken into separate discrete sub-basins (even though with incomplete imaging that may appear to be the case). The word 'tortuous' is used to emphasize the common departure from a linear planform pathway. Slopes along the axes of connected tortuous corridors typically vary, with steeper segments alternating with flatter segments ('steps').

Ponding. This term is used (as in Van Andel & Komar 1969) to indicate a situation in which sediment gravity flows are sufficiently large in volume to be contained by the closed bounding slopes of a silled sub-basin. Whereas such bounding slopes may trap a majority of flows during a given stage in the evolution of a sub-basin, unusually thick or energetic flows may overtop the downdip barrier, or lose their upper portions across the barrier. Distal confinement of flows can be expected to result in relatively flat deposits in the distal parts of a silled sub-basin, albeit with draping deposits on adjacent slopes reaching heights equivalent to flow thickness or runup elevations. However, there is no implication that coarser-grained deposits at the updip (input) end of a sub-basin should be flat. They will dip basinward at slopes related to transported size distribution, flow thickness, flow concentration and rates of lateral spreading. As a range of downflow depositional dips can be observed in the *ponded accommodation* of Prather *et al.* (1998), distinctions in the current literature between deposits filling flat-topped 'ponded accommodation' and dipping-topped 'healed slope accommodation' are de-emphasized here.

Distributary and tributary channel patterns. Distributary channel patterns are those in which channels distribute sediment across much of the width of a fan or *depositional lobe* or *distributary channel-lobe complex* (e.g. Beaubouef & Friedman 2000; Pirmez *et al.* 2000) downflow from an input point source. Conversely, tributary channel patterns are those where channels converge as they feed towards a single main erosional conduit, commonly at the outlet of a sub-basin once the sill has been breached.

Convergent baselap and convergent thinning. Convergent baselap describes a stratal geometry in which surfaces terminate in abrupt onlap or downlap against a baselap surface. The word 'baselap' is used to include both onlaps and downlaps, as tectonic rotations may result in onlaps appearing to be downlaps and *vice versa*. Convergent thinning describes a stratal geometry in which surfaces converge more or less progressively, commonly the case where muddy drapes are deposited adjacent (laterally or frontally) to basin floor deposits. When these terms are applied to seismic facies, it must be remembered that the frequency content of the seismic information will usually be insufficient to resolve thin slope drapes, and hence apparently abrupt baselap terminations may in fact be associated with subseismic draping facies adjacent to apparent pinchouts.

Healed slope. A healed slope is the result of the infill of depressions present on an earlier topographically complex slope. Such 'healing' is achieved by emplacement of mud-rich mass transport complexes, deposition from turbidity currents and hemipelagic drape. The process of slope-healing is counteracted by tectonic deformation processes acting to reform the topography.

Topography, flow character and rates of structure growth: some general considerations

This section addresses:

(1) the types of structurally induced topography to be found on topographically complex subaquous slopes;
(2) the role of flow character in influencing the equilibrium profiles towards which systems tend to grade; and
(3) the rates of growth of structure relative to rates of depositional and erosional smoothing of topography.

Types of topography

Three broad classes of complex slope topography can be identified:

(1) silled sub-basins (i.e. closed depressions);
(2) partially silled basins with lateral escape paths;
(3) tectonically induced bounding slopes that guide, but do not block, flow paths.

In the third class flow paths vary from highly tortuous to close to linear and commonly exhibit

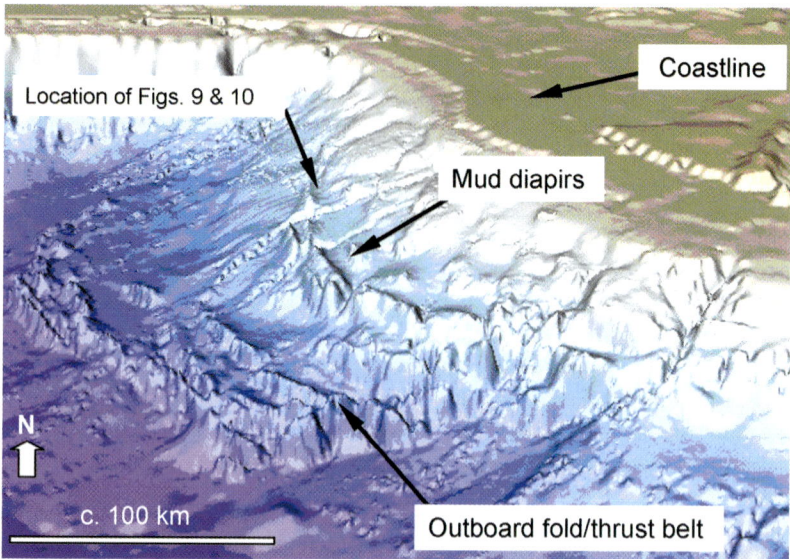

Fig. 2. Thrust-cored ridges and associated intraslope basins, Western Niger Delta slope. The location of the intraslope basin illustrated in Figures 9 and 10 is shown.

segments of lower ('steps') and higher (between 'steps') gradients. Examples of these types of topography are illustrated in Figures 1–4.

The archetypal example of silled sub-basins is the complex of salt-withdrawal minibasins of the Plio-Pleistocene Gulf of Mexico slope (Diegel et al. 1995; Liu & Bryant 2000; Fig. 1). Further examples occur on the western Niger Delta slope (Fig. 2). Examples of partially silled basins with lateral escape paths include the topography associated with the Chumash Fracture Zone (Normark et al. 1984) and the

Fig. 3. Stepped topography on the northwest Borneo slope with tortuous lateral escape paths, between shale-cored ridges, for flows descending from one step to the next (see also Demyttenaere et al. 2000). View towards southwest. Petroleum Geo-Services are gratefully acknowledged for permission to use these data.

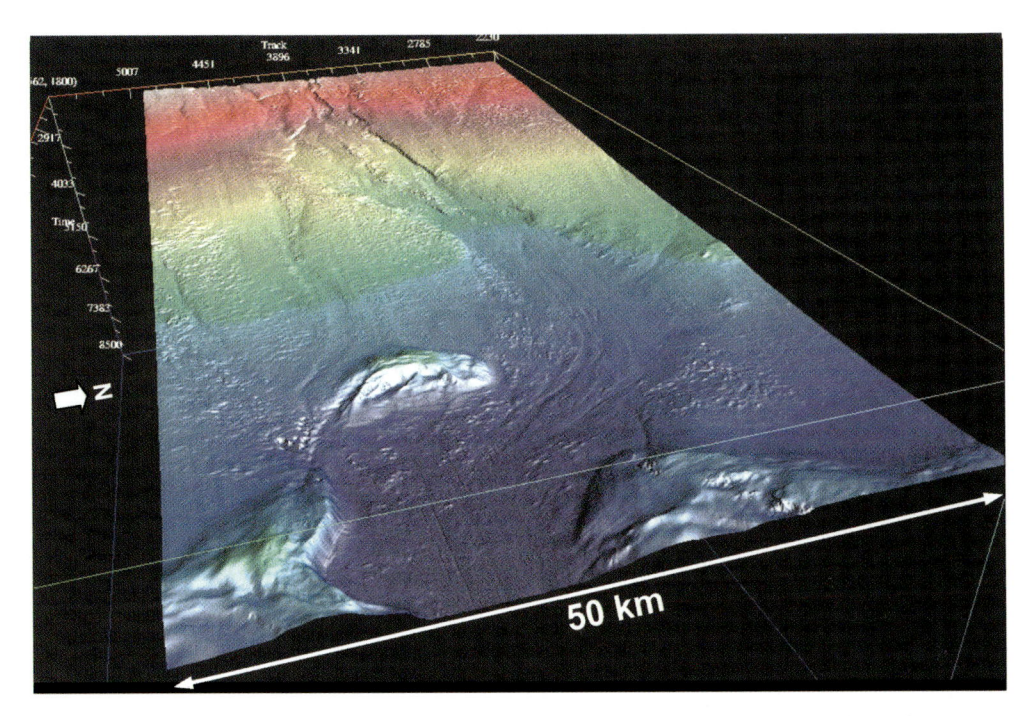

Fig. 4. A seafloor example of flow diversion into a tortuous path around salt-induced obstacles on the Brazilian slope. Water depth ranges from 1500 to 2300 m. View is towards the west. Image courtesy of Peter Diebold and Shell Brazil.

physiography present on the Brunei slope (Demyttenaere *et al.* 2000). Slopes affected by obstacles around which flow paths are continuous include the present day Borneo slope (Fig. 3) and the Campos slope, offshore Brazil (Fig. 4). Different types of complex slope topography may co-exist laterally on a given slope (e.g. Biegert *et al.* in review).

Flow character and equilibrium profiles

Flow volumes relative to areas of receiving depressions. Flow volumes as well as flow runout characteristics for a given volume (related to transported grain size, proportion of fine-grained sediment in a flow and concentration: e.g. Van Andel & Komar 1969; Middleton & Neal 1989; Alexander & Mulder 2002) determine the extent to which potential flow path topography present on a complex slope will be experienced by a flow. Thus flow dimensions that are small relative to receiving basin area (or potential flow path in a tortuous corridor case) result in relatively small systems (or stages in systems; Mutti & Normark 1987) little affected by bounding and opposing slopes. An example would be the youngest deposits of Basin IV in the Brazos-Trinity Turbidite System (Winker 1996; Badalini *et al.* 2000; Winker & Booth 2000). Conversely, flows that are large relative to receiving basin, or potential tortuous corridor flow path, are strongly influenced by all bounding slopes. Examples of this are provided by the early stages of Basin II and Basin IV filling in the Brazos-Trinity system (Winker & Booth 2000) and numerous reservoir-forming ponded fills in the deeper subsurface (Prather *et al.* 1998).

When flow volumes are small relative to the scale of the receiving basin, systematic proximal to distal and axial to lateral changes in properties such as percentage sand, percentage amalgamation and bed thickness are commonly to be expected with distal transitions into fine-grained sections (see, for example, data published by Smith 1987b; di Toro 1995; Smith & Møller 2003 in press). Middleton & Neal (1989) demonstrated the link between bed geometries and both the grain size and concentration of transported sediment, with larger grain size and higher concentration resulting in shorter, thicker beds (for equal volumes of sediment). However, where individual flows were large relative to receiving basin area, and hence confined by opposing slopes, such trends may be strongly truncated as a result of abrupt distal onlaps of sandstone bodies. Bursik & Woods (2000) discussed, using experimental evidence and theoretical arguments, how beds can be expected to thin exponentially along flow paths unless depositing flows are blocked to produce a bore.

Relations between flow character and equilibrium profile. Given a topographically complex slope morphology as described above, the history of deposition and erosion on that slope will be determined by (a) the interaction of sediment gravity flows (of varying character) with the topography and (b) the associated hemipelagic deposition and mass-wasting processes active on the slope. Any sediment gravity flow encountering slopes that are steeper than a theoretical equilibrium profile appropriate for its character and the erosional strength of the bed (and banks, if flow is channelized) will act to adjust the slope along the flow path. This can be achieved either by downcutting (e.g. Satterfield & Behrens 1990; O'Connell *et al.* 1991; Pirmez & Flood 1995), or by increasing the sinuosity of the flow path (e.g. Pirmez *et al.* 2000). Conversely, a sediment gravity flow encountering slopes that are more gently dipping than a theoretical equilibrium profile appropriate for its character may act to change the encountered slope, either by depositing, or by reducing the sinuosity of its flow path. A blocking opposing slope will cause rapid deposition and formation of upstream-propagating bores (e.g. Bursik & Woods 2000). Truncation of the normal tapering bed thickness patterns and ponding will result in a flattening of the depositional surface (sediment that would otherwise have been deposited in the distal parts of tapering wedges must now be deposited updip of the blocking slope).

The physical controls on the ideal equilibrium profiles of sediment gravity flow systems are poorly understood as yet. In a simple two-dimensional model assuming that erosion/deposition is controlled by flow competence (neglecting capacity controls and the erosional strength of bed and banks), at equilibrium (no erosion and no deposition) flow shear velocity at a point along the flow path can be equated to the settling velocity of the coarsest grains being transported. Equilibrium slope at that point in the profile can then be expressed in terms of settling velocity, flow density and thickness (Ben Kneller pers. comm. 2000; Samuel *et al.* 2003):

$$\sin(\text{equilibrium slope}) = \frac{U_s^2(\rho + \Delta\rho)}{g\Delta\rho h} \quad (1)$$

where U_s is the settling velocity of the coarsest grains being transported, ρ is the ambient fluid density, $\Delta\rho$ is the density difference between the sediment gravity flow and the ambient medium, and h is the current thickness.

Thus, as far as this expression is applicable in the real world, decreasing the transported grain size, increasing the flow density and increasing the flow thickness all lead to more gently dipping equilibrium slopes. As noted above, adjustments to achieve such an equilibrium gradient can take the form of either changes in planform sinuosity or incision and deposition. In addition, channel depths, widths and roughness also adjust in response to changes in discharge, grain size of transported sediment and gradients encountered (Leopold & Bull 1979; Pirmez *et al.* 2000).

In river long profiles, increasing water discharge downstream and variations in the erosional resistance of bed and banks have been emphasized (e.g. Allen 1985, p. 85), along with trends of grain-size fining down flowpath. In deepwater depositional systems a trend of downflow grain-size fining generally does apply, but fluid discharge will not usually increase in the same way as seen in rivers with their cumulative contributions from tributaries in a downflow direction. Allen (1985, p. 234) emphasized the overloading of currents by high sediment loads as a major control on the dissipation of turbidity currents through sediment deposition (discussed in terms of a loading factor, the ratio of actual sediment load to theoretically supportable load), and this degree of overloading may be an important control on depositional gradients. The loading factor will usually decrease in a downflow direction (and through time during deposition from a flow), with deposition becoming a result of insufficient flow competence rather than insufficient flow capacity (see also Hiscott 1994). However, addition of new material through erosion (for example, during incision of a breached sill or acceleration down a steeper segment of a connected tortuous flow path) will locally increase the loading factor. A further control is related to the degree of lateral constriction of flows, with rates of deposit thinning increasing as degree of lateral constriction decreases (e.g. Bursik & Woods 2000).

Evidence for the tendency of submarine channels to regrade towards an equilibrium profile following breaks in equilibrium (for example, encountering a new deeper base level following fill to spill and/or avulsion) has been presented by Satterfield & Behrens (1990), O'Connell *et al.* (1991) and Pirmez *et al.* (2000). Ross *et al.* (1995) and Prather *et al.* (1998) argued for the long-term tendencies of basin slopes to grade to a regional

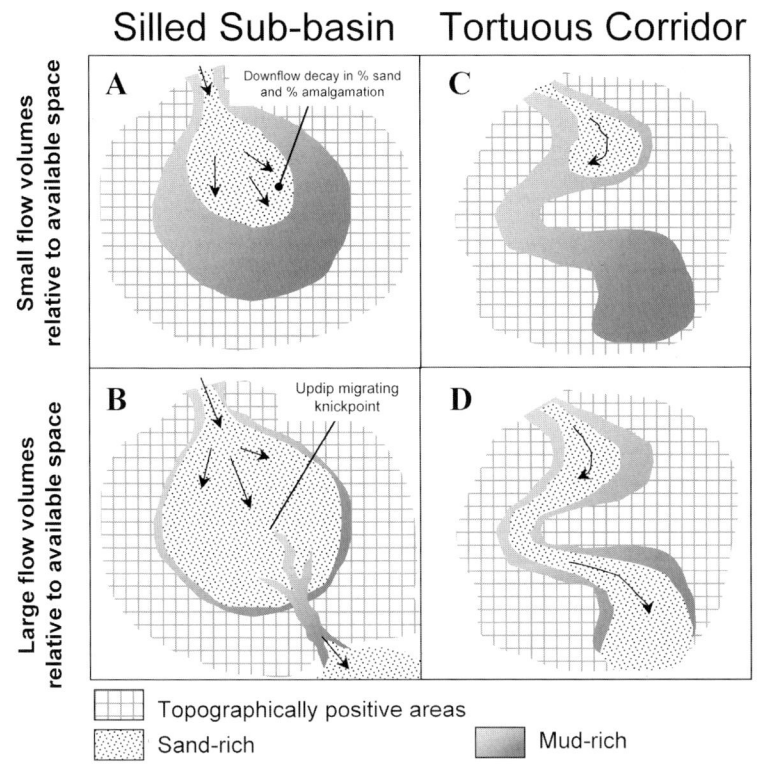

Fig. 5. Schematic diagrams illustrating the importance of the areal extent of sediment gravity flows relative to the areas of receiving depressions. (**A**) Silled sub-basin in which sand-transporting flows are small in volume relative to the scale of the receiving space. (**B**) Silled sub-basin in which sand-transporting flows are large in volume relative to the scale of the receiving space. The diagram shows spill to the next sub-basin downslope with associated incision and bypass in the upper sub-basin. (**C**) Connected tortuous corridor in which sand-transporting flows are small in volume relative to the potential flow path. A possible example is shown in figure 8 of Demyttenaere *et al.* (2000). (**D**) Connected tortuous corridor in which sand-transporting flows are large in volume relative to the potential flow path. Examples from northwest Borneo, Brazil and Angola are shown here in Figures 3, 4, 14 and 15.

equilibrium profile. Agents affecting such grading include deposition and erosion beneath sediment gravity flows and also mass-wasting processes. Volumes of unfilled space beneath assumed long-term time-averaged equilibrium profiles have been divided by Prather (2000, modifying Prather *et al.* 1998) into *ponded accommodation*, defined by a flat top drawn landwards from sill crestal positions, and *healed slope* (fill and blanketing of topographic depressions) and *slope* accommodation between the top of the healed slope and the assumed regional graded profile. An important point to emphasize here is that the slopes towards which sand-transporting flows tend to grade will be different from (and gentler than) those of regional muddy basin slopes.

An issue widely neglected is that of the timescale over which equilibrium profiles can be averaged. A common assumption is that in a given area there is a single average equilibrium profile towards which slopes will tend to grade (e.g. Ross *et al.* 1995; Prather *et al.* 1998; Beaubouef & Friedman 2000). This viewpoint may have some validity if very long periods of time are being considered. However, at the other extreme of timescale, every flow is different and will tend, in detail, to erode and deposit in different areas along a flowpath. This point is illustrated by a dramatic sub-aerial example from the Kota Fan of southern Iceland (Fig. 6) (Thompson & Jones 1986), where deposition from highly concentrated coarse-grained flows (jökulhlaups) associated with eruption of the adjacent volcano in 1727 caused aggradation and steepening of the fan surface. The subsequent return to normal dilute meltwater flows (that is, reduced ratio of sediment load to

Fig. 6. Changes in profile due to changes in flow character: the Kota Fan, southern Iceland. The steeper, upper surface of the fan developed during rapid aggradation resulting from debris-laden *jökulhlaup* flows during the 1727 eruption of the adjacent volcano (Thompson & Jones 1986). The subsequent return to the normal meltwater stream, with far lower sediment load/discharge ratio, resulted in incision (reaching 28.5 m at the fan head) to achieve lower gradients. Similar changes in downflow profiles in response to variations through time in flow character can also be anticipated in deep-marine environments.

discharge, or 'undersaturated' flows) led to stepwise downcutting (with an approximately exponential decay in rate of downcutting through time) in order to achieve a lower-gradient, in-balance, profile. This is an excellent demonstration of deep incision and change of surface gradient due to changes in sediment supply character, but with no change in base level. At intermediate timescales, for example related to segments of a relative sea-level cycle, a succession of flows of similar character will result in grading towards an appropriate average equilibrium profile.

Rates of structure growth relative to depositional smoothing

The rates of structure growth relative to the rate of depositional smoothing of topography determines the longevity of topographic control on patterns of deposition and erosion. Thus Cazzola et al. (1985) and Smith (1987a, 1995a) document cases in which structurally induced topographic depressions were only temporary and, once filled, were replaced by a smooth bathymetry. At the other extreme, topography produced by salt withdrawal on mobile salt substrates is rapidly reformed by salt movement and typically outstrips the effects of depositional and erosional smoothing (Prather et al. 1998; Booth et al. 2000; Winker & Booth 2000). Subsidence rates in intra-slope salt-withdrawal basins can be in the order of several km/Ma (Prather 2000; Winker & Booth 2000).

One scenario may evolve into the other through time. For example, Anderson et al. (2000) describe a progression on the Angola slope from low sedimentation rates relative to fault displacement rates in the early Miocene to high sedimentation rates relative to fault displacement rates in the latest Miocene with resultant rapid filling of hanging wall accommodation. Conversely, a slope lacking closed depressions could evolve into a cascade of silled sub-basins given sufficiently high relative rates of structure growth. An example of a slope-crossing continuous conduit being deformed in such a fashion is provided by the Bryant Canyon in the *Western Deposystem* of the Gulf of Mexico slope (Winker & Booth 2000).

Cascade of silled sub-basins model

The essential features of the well-known fill and spill model for a cascade of silled sub-basins can be summarized as follows (Figs 7 & 8). For each successive sub-basin in a basinward-downstepping 'cascade' of sub-basins (or *minibasins*), a sill tends to prevent further downslope flow of turbidity currents (or at least their basal sandy portions) until deposition reduces the relief sufficiently to allow spill downslope. As the topographic relief at the sill decreases, there is an increasing tendency for the upper portions of flows to overflow the sill (*flow-stripping* of Piper & Normark 1983; Winker & Booth 2000). Spill is eventually associated with headward migrating incision in the sill and updip basin-fill

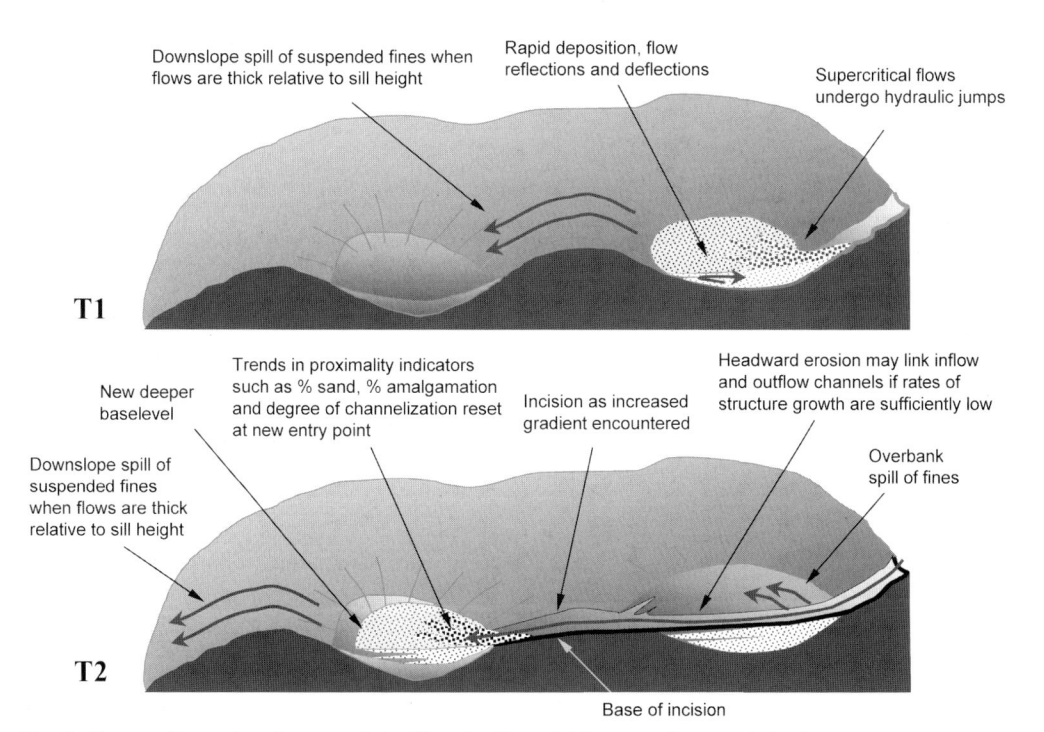

Fig. 7. Cartoon illustrating elements of the fill and spill model for two adjacent sub-basins.

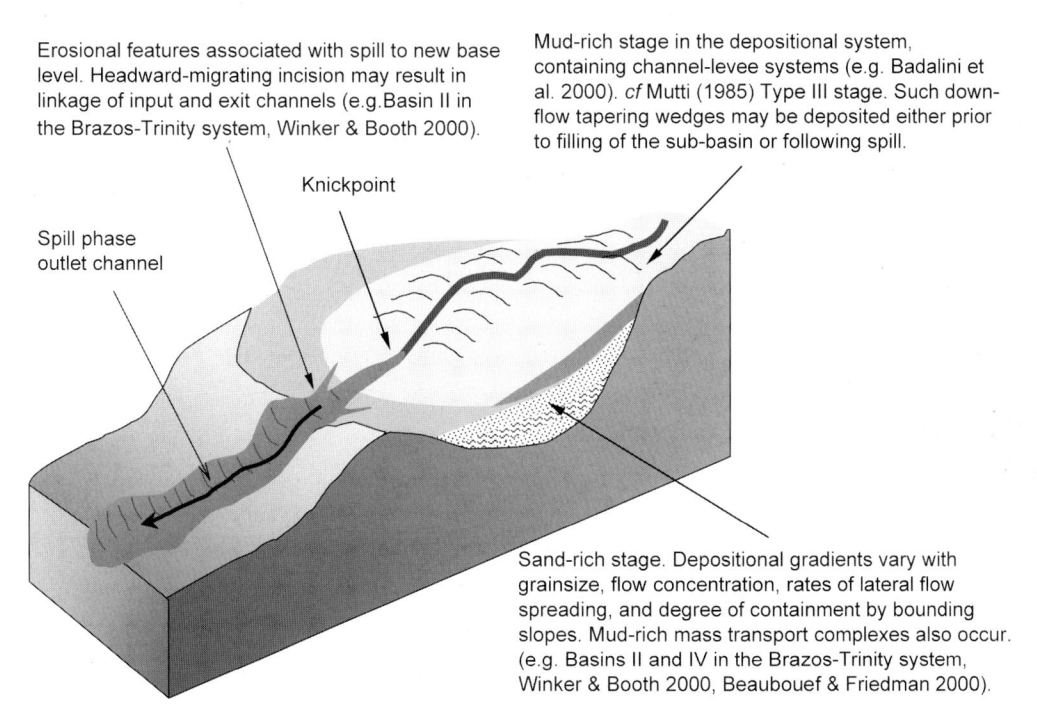

Fig. 8. Cartoon illustrating sand-rich and mud-rich stages of system development in a silled sub-basin. Controls on depositional gradients and erosional features associated with spill to a new deeper base level are noted.

(the *wave of dissection* of Schumm *et al.* 1984), adjusting the gradient of the flow profile such that it tends towards a graded profile appropriate for the nature of flows travelling down the slope, with a new, deeper base level (the deepest point in the basin reached by sediment gravity flows). Such a profile, connecting previously isolated sub-basins, will be achieved and maintained only if rates of structure growth are low relative to the rate of action of slope-modifying processes.

Research and exploration efforts on the Gulf of Mexico slope resulted in refined models supported by (a) high-resolution surface bathymetry (Satterfield & Behrens 1990), and (b) three-dimensional seismic data, both close to the sediment surface and at prospective levels for hydrocarbons in the underlying section (Winker 1996; Prather *et al.* 1998; Badalini *et al.* 2000; Winker & Booth 2000). Prather (2000) presented quantitative forward stratigraphic models (using the Shell-proprietary package STRATAGEM) that successfully simulate the process of sequential fill and spill downslope. These models illustrate the downslope younging of the fills of successive intraslope basins. Further work on the Gulf of Mexico intraslope basins, with the benefit of increased density of well calibration, has yielded more complex models in which basin-fill successions can contain the deposits of more than one turbidite system, associated with different input points (Booth *et al.* 2000; Winker & Booth 2000), or in which successive basin fills may partially overlap in time (Badalini *et al.* 2000).

Winker & Booth (2000) recognized three end-member scenarios for proximal areas in silled sub-basins:

(1) sand-rich fanlobe (named *distributary channel-lobe complex* by Beaubouef & Friedman 2000), exemplified by the shallow deposits of Brazos-Trinity Basin IV and reservoir in the Auger Field;
(2) leveed channel complex (e.g. Brazos-Trinity Basin II);
(3) abundant mud-rich mass transport complexes as seen in the deeper deposits of Brazos-Trinity Basin IV.

Beaubouef & Friedman (2000) emphasized the vertical stacking of these cases into an ideal sequence motif with mass transport complexes at the base overlain by sand-rich distributary channel/lobe complexes and capped by channel–levee complexes and hemipelagic drapes. Winker & Booth (2000) also recognized three end-member cases for the distal portions of silled sub-basins:

(1) a high sill, in which case all flows large enough to encounter bounding slopes were contained;
(2) a low sill, which would favour removal of the upper mud-rich portions of turbidity currents downstream across the sill, hence enhancing the percentage sand of the deposits updip of the sill;

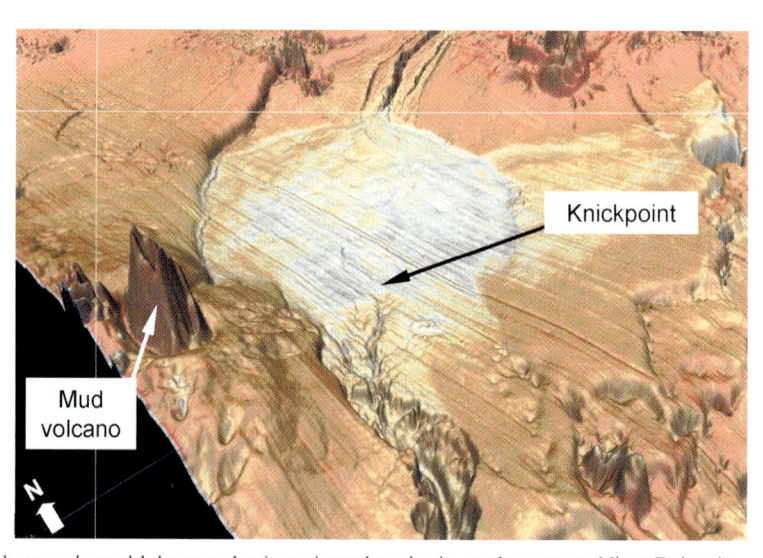

Fig. 9. Distributary channel-lobe complex in an intraslope basin on the western Niger Delta slope. The channel-lobe complex has a diameter of approximately 10 km and a surface gradient of about 1.2°. Note the tributary pattern of erosional channels in the outlet area. Water depth in the field of view ranges from approximately 1300 to 1900 m.

(3) an incised channel case, where channels near exit points may be mud-filled (e.g. the shallow Macaroni Field at the downdip end of the Auger Basin) or contain sand (e.g. the deep Macaroni Field).

Sand deposition in outlet channel fills may be related to local gradient reductions.

The example shown here is from the western Niger Delta slope (Figs 2, 9 & 10). Here, an incised channel (*channel X* in Pirmez *et al.* 2000) is the conduit for a depositional system that occupies an intraslope depression with a diameter of approximately 10 km (Fig. 9). The gradient at the surface of the distributary channel-lobe complex is approximately 1.2°. Seismic profiles show baselapping reflector geometries indicating an early phase of distal confinement; however, the present depositional surface grades to the spillpoint of the sub-basin. A tributary system of erosional channels occurs in the vicinity of the outlet to the sub-basin (comparable with those seen at outlets of the updip basins in the Brazos–Trinity system in the Gulf of Mexico; e.g. Winker & Booth 2000). These feed into the incised exit channel. Points to note are the following:

(1) Gradient reduction together with lateral spreading led to rapid deposition in front of

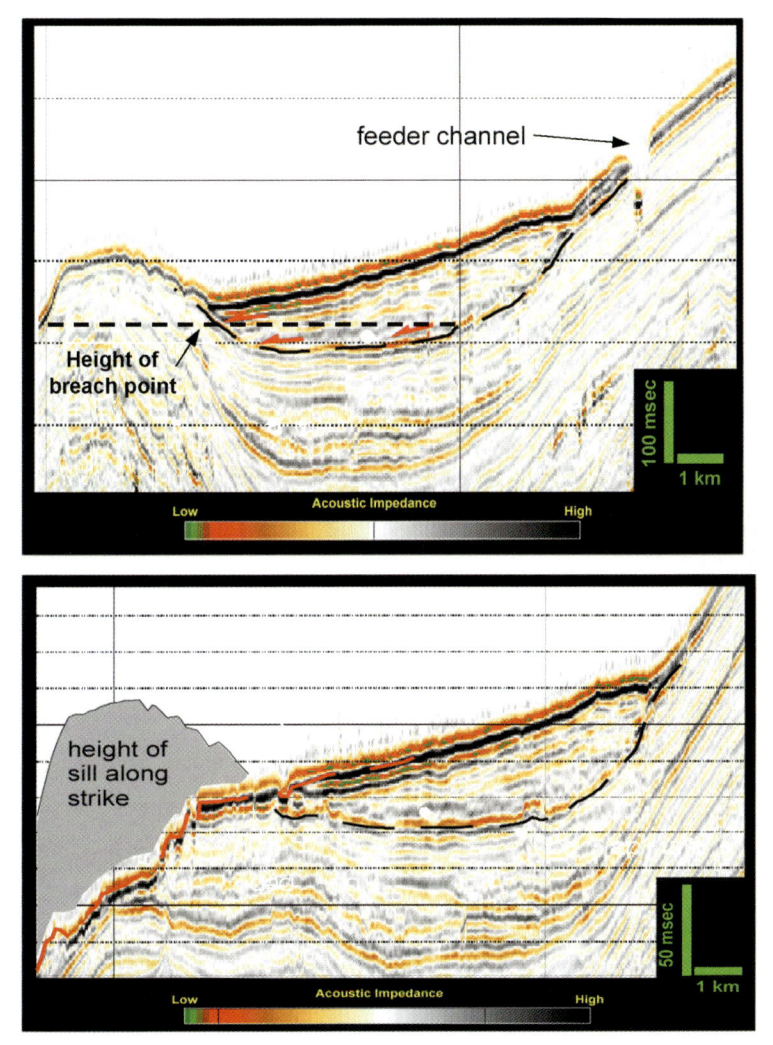

Fig. 10. Seismic sections through intraslope basin shown in Figure 9, showing (a) downlap/onlap terminations against confining topography, and (b) grade to spill point and incipient dissection by the outlet channel in the vicinity of the breach.

the incised slope channel that feeds the sub-basin.
(2) Aggradation and progradation led to the toe of the depositional system reaching the spill point of the sub-basin.
(3) The presence of a steeper gradient downslope of the spill point has resulted in incision of an outlet channel beneath turbidity currents that are large enough to travel beyond the sub-basin.
(4) The overall profile of turbidity current flow paths is being modified over time by incision on the steeper segments above and below the sub-basin and by deposition on the lower gradients in the area of the sub-basin.

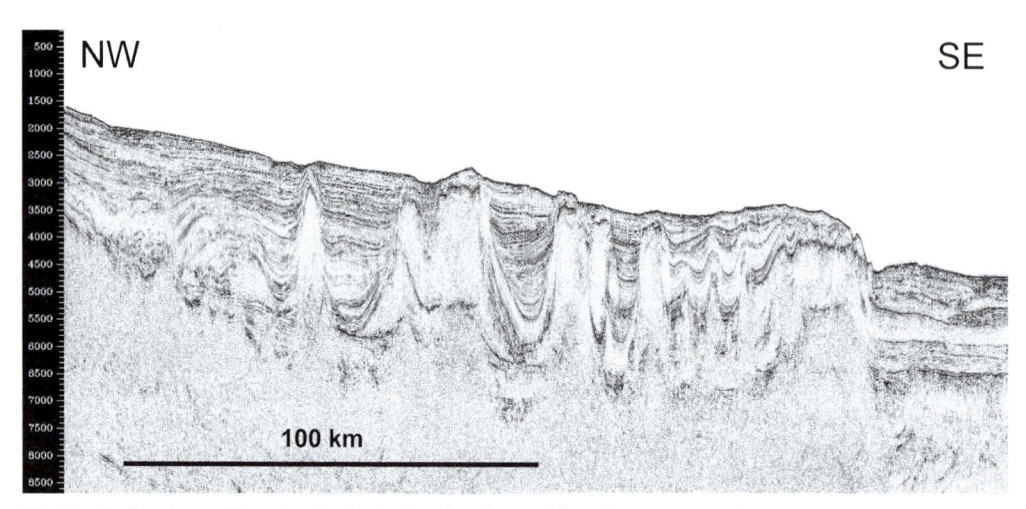

Fig. 11. Regional two-dimensional seismic line showing mobile salt structures and convergent geometries adjacent to salt highs. Apparent 'minibasins' range in length from about 4 to 20 km. Note how the pattern of stratal convergence towards salt-cored highs could easily lead to the inference of a cascade of silled basins model. Image courtesy of Peter Mullin and Shell Brazil.

Fig. 12. Three-dimensional seismic image showing continuity of sediment distribution paths around salt highs. The image shows a stratigraphy-parallel slice (at top Cretaceous level) through a three-dimensional seismic volume (3500 to 5000 ms two-way-time) with RMS amplitude overlay. Blue colours reveal tortuous flow paths (associated with sand deposition) around salt-cored highs. View is towards the northeast and width of image (north–south) is 50 km. Image courtesy of Peter Diebold and Shell Brazil.

The future history of this sub-basin fill depends on the rate at which local subsidence tends to reform the intraslope depression relative to the rate at which the outlet-incised channel cuts headward towards linkage with the input channel.

Connected tortuous corridor model

Two-dimensional sections of continental margins in which mobile salt has produced complex slope topographies commonly appear similar to sections through the Gulf of Mexico Plio-Pleistocene slope deposits. Excellent examples of this can be seen in the South Atlantic continental margins of Brazil and West Africa (e.g. Fig. 11). Thus, as in the silled sub-basin case, seismic facies exhibit convergent thinning and convergent baselap geometries adjacent to syn-depositional salt-related highs. Two-dimensional seismic data are commonly all that is available during the early phases of exploration. However, well penetrations have shown that in many cases the pattern seen in the Gulf of Mexico of the coarse-grained fill components of successive (apparent) sub-basins being progressively younger in a downslope direction is not the case elsewhere. In addition, three-dimensional seismic data, where available, commonly show that apparently separate silled sub-basins are in fact connected in the third dimension (Figs 12 and 13). This type of sediment-receiving depression on a topographically complex slope is here termed a *connected tortuous corridor* to distinguish it from the *cascade of silled sub-basins* scenario (contrast Figs 13 and 7). Changes in gradient will occur along such connected pathways: thus segments of substrate erosion can be expected to alternate with segments of enhanced deposition (e.g. Demyttenaere *et al.* 2000).

Such flow paths traversing topographically complex slopes have been documented on slopes in West Africa and Brazil, and have been inferred for the Brunei slope (Demyttenaere *et al.* 2000). In the Brunei case, prospective reservoirs occur in the form of *fan lobes* deposited in subtle depressions or at local breaks in the slope. In addition, erosional conduits cutting across shale ridges contain mud-rich channel–levee complexes. Moraes *et al.* (2000) discussed the reservoir development of the Cretaceous Carapeba system in the Campos Basin. Here,

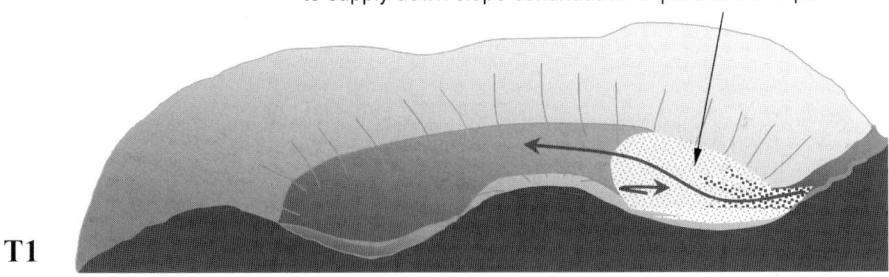

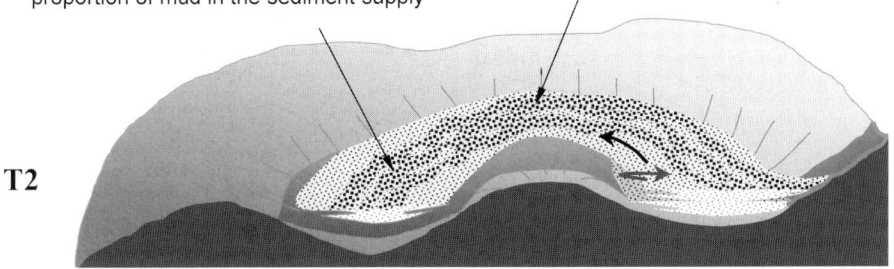

Fig. 13. Summary cartoon of the connected tortuous corridor model showing a case with small volume sand-transporting flows (as in Fig. 5c) and large volume sand-transporting flows (as in Fig. 5d).

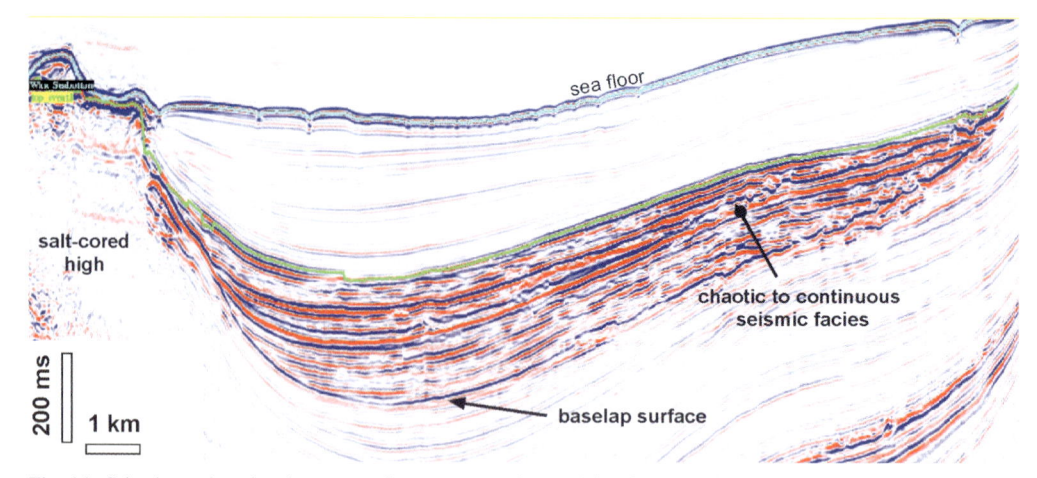

Fig. 14. Seismic section showing external convergence, internal baselap and high reflectivity in a near-seafloor confined turbidite system, offshore Angola. Sonangol are gratefully acknowledged for permission to use these data.

amalgamated sand-rich turbidite channels occur in erosionally modified fault-controlled troughs, comparable in form and dimensions with those of the Annot Sandstone confined turbidite system (SE France). Further examples have been documented on the present-day Gulf of Mexico slope (e.g. Biegert *et al.* in review). Examples illustrated here come from the Campos Basin offshore Brazil, both at present-day seafloor and in the subsurface (Fig. 12) and the

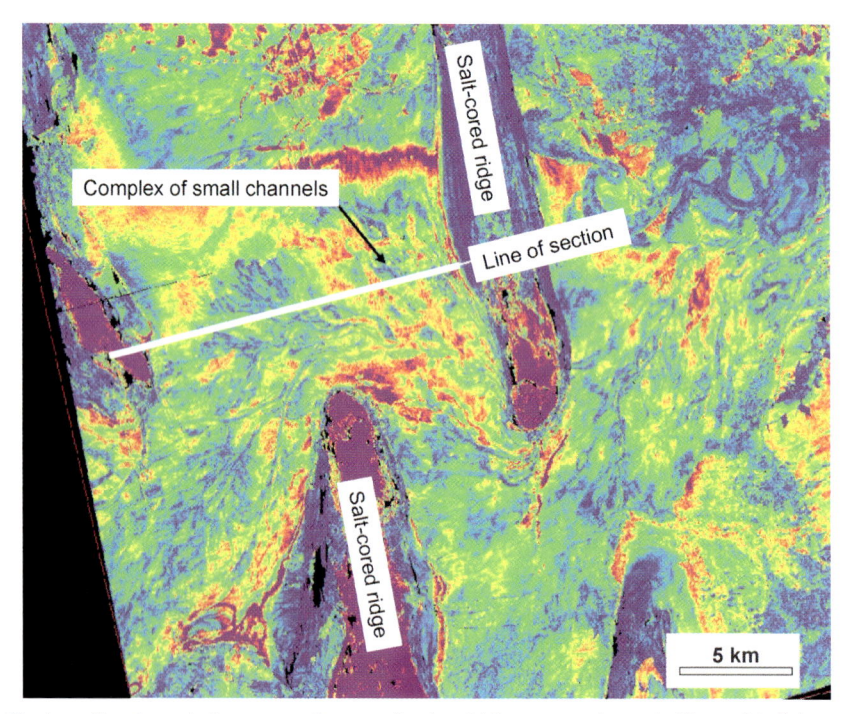

Fig. 15. Horizon slice through the near-seafloor confined turbidite system shown in Figure 14. Colours correspond to maximum amplitudes. Note the tortuous flow paths around discontinuous seafloor ridges as recorded in the orientations of small channel elements. The white line indicates the location of the sections shown in Figure 14. Sonangol are gratefully acknowledged for permission to use these data.

Lower Congo Basin, offshore Angola (Figs 14 & 15). Note how in Figure 12 the seismic facies characteristics of this amalgamated channel and sheet complex show external convergence, internal baselap and high reflectivity, easily confused with closed sub-basins similar to those seen in the Gulf of Mexico. In plan view the complex of channels experiences deflections to the south then north, followed by a further deflection to the south, in response to the discontinuous salt-cored ridges.

Facies architectures and implications for reservoir development

In the sections below the processes acting at entry points and bounding slopes are discussed, and implications for potential reservoir architecture are highlighted.

Processes and products at entry points

Strongly channelized reservoir architectures are to be expected close to entry points of topographic depressions as a result of high velocities acquired along high-gradient laterally constricted input channels. The expression of such channelization varies with the sediment input characteristics, ranging from numerous sand-on-sand erosional contacts in a mud-poor system (e.g. Moraes *et al.* 2000) to channel–overbank complexes when larger proportions of mud are supplied (e.g. see discussion in Mutti 1985; Reading & Richards 1994; Badalini *et al.* 2000). In addition, many supercritical turbidity currents and debris flows will undergo a hydraulic jump at the base of the slope (Menard 1964), especially if there is a distinct break in that slope and if that is also where a flow is able to expand laterally. This scenario can be contrasted with the case in which a channel continues on progressively lower gradients away from the base of slope (e.g. Wierich 1989). The rate of spatial decay in velocity (related to magnitude of slope reduction and magnitude of lateral expansion) will probably have much to do with how far basinward channels extend beyond the base of slope. Rapid deposition occurs downstream of a hydraulic jump (e.g. Wierich 1989; Garcia 1993). Intense scour of the bed is a more speculative consequence (Mutti & Normark 1987), but likely to occur in many cases where changes in gradient and/or lateral constriction are abrupt. Large flute-shaped scours appear to be commonly produced in areas of flow expansion at channel mouths, perhaps associated with hydraulic jumps. Many such scours have been documented from regions between dominantly channelized and dominantly non-channelized areas of turbidite systems, both modern and ancient (summary in Mutti & Normark 1987). However, there are also several good examples from within channel-fills and others from overbank areas (summary in Normark & Piper 1991). Bypassed, mud-draped scours imply that the flow in the region of flow expansion was able to maintain all of its coarser load in suspension (i.e. competence and capacity maintained, hence fast and not overloaded flows) until some position further downstream (see also Komar 1971). As Mutti & Normark (1987) pointed out, flows that are unable to maintain coarse material in suspension in the jump zone, because of the dramatic velocity decrease, will deposit rapidly, filling any scours that have been cut at the bed. This was the interpretation applied to a suite of large-scale amalgamated conglomerate and sand-filled scours in the Pysgotwr Formation of the Welsh Basin (Smith 1987b, 1995b). An alternative interpretation of large filled scours in the region between distinct channelized and non-channelized elements is that they are elongate, downstream-flaring scours comparable with that documented by Sturm & Matter (1978) for turbidite TA-1 in Lake Brienz. Filled scours clearly act to increase vertical sandstone body connectivity.

Sand-rich backfill of the inter-basin erosional conduits could conceivably occur (as suggested by Sinclair 2000) in some cases if the elevation difference between the sill of the updip basin and the level to which basin-fill deposits can aggrade is small. In other cases their fill is expected to include a high proportion of mudstone (as illustrated in Prather *et al.* 1998; Booth *et al.* 2000).

Processes and products at bounding slopes

There are several possible scenarios for processes and reservoir architecture developed in response to the presence of lateral and frontal bounding slopes. Passive downlap will occur if the areal extents of sand-transporting flows are small relative to the receiving space and sand is deposited before they reach the bounding slopes. This will be the case for relatively low sediment supply volumes, or small distances of flow runout relative to volume (*short and thick* versus *long and thin*: see Middleton & Neal 1989). In such a scenario reservoir architecture will be dominated by proportions of mud, sand and gravel in the supply mix and by autocyclic effects such as offset stacking of channel–overbank complexes and depositional lobes (e.g. Gervais *et al.* this volume).

When flows are large relative to receiving spaces, interaction with bounding slopes will occur, potentially with slumping and triggering of debris flows and turbidity currents off opposing slopes (Haughton 1994; Kneller & McCaffrey 1999). An opposing slope often results in deflections and reflections (Van Andel & Komar 1969; Pickering & Hiscott 1985; Kneller *et al.* 1991; Smith & Anketell 1992; Kneller & McCaffrey 1999). Lateral deflections of flows are known from Swiss lakes (Siegenthaler *et al.* 1984), African rift lakes, where sub-lacustrine channels turn to run parallel to small faults (Scholz *et al.* 1990), and are often inferred from palaeocurrent patterns (e.g. Cazzola *et al.* 1985; Smith 1987a; Kneller & McCaffrey 1999). Rapid deposition of suspended load may occur as a result of flow deceleration caused by the opposing slope, although convergent flowlines in an area of active flow deflection may yield distinctive laminated facies interpreted as a record of spatial flow acceleration (Kneller & McCaffrey 1999). Alexander & Morris (1994) and Kneller & McCaffrey (1995) experimentally demonstrated enhanced sand deposition directly updip of tilt block and domal obstacles. A flow that is thick relative to an obstacle in its path may thin and accelerate, leading to erosion and bypass (Komar 1983). Kneller & McCaffrey (1999) reviewed recent literature incorporating the effects of Froude number and density stratification of flow (expressed in the *internal Froude number*). They also provided a general expression for runup height (following Allen 1985).

Stratal architectures at bounding slopes depend strongly on the ratios of aggradation rates on bounding slopes to aggradation rates in the adjacent basin, which in turn depend to a large degree on the thickness and grain-size stratification of the depositing flows (Smith 2004, this volume; Smith & Joseph 2004). Hypothetical examples, modelled in two dimensions by varying the ratio of slope to basinal aggradation rates, are shown in Figure 16. Cases in which slope to basinal aggradation rates are high correspond (if available seismic resolution is sufficiently high) to the convergent thinning seismic facies (Prather *et al.* 1998), whereas low ratios of slope to basinal aggradation rates correspond to the convergent baselap seismic facies. Excellent high resolution seismic images of these stratal styles can be seen in figure 7 of Badalini *et al.* (2000). Convergent thinning facies are usually associated with a low sandstone percentage, although if slope to basinal aggradation rates

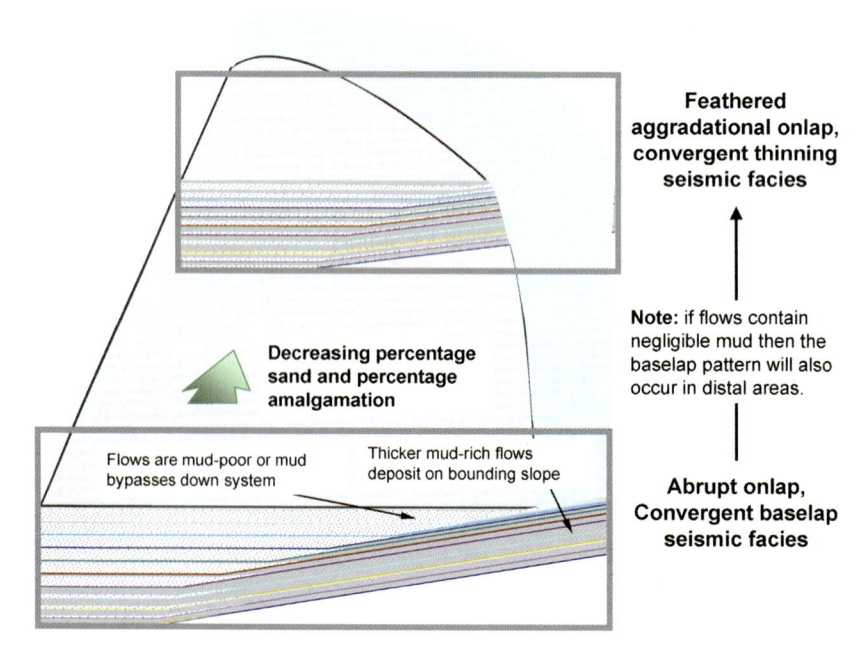

Fig. 16. Illustration of potential downflow change from lateral abrupt (convergent baselap) to aggradational (convergent thinning) onlap style. The sections have been generated using the geometric model described by Smith & Joseph (2004).

vary progressively along a concave-up slope profile, high percentage sandstone bodies may occur in the basin-fill (Smith 2003, this volume). Convergent baselap facies are usually associated with high percentage sandstone unless the basinal succession contains a large proportion of muddy debrites and slides (Smith & Joseph 2003). This facies, in the ponded phases of minibasin fills in the Gulf of Mexico, has been found to correspond to reservoir with excellent connectivity characteristics and consequently high production rate, high ultimate well delivery (Prather *et al*. 1998). Figure 16 illustrates possible downflow transitions from abrupt onlaps (convergent baselap) to aggradational onlaps (convergent thinning), in the case where the fine-grained portion of the sediment load tends to bypass proximal positions and deposit further downflow (see examples in Wynn *et al*. 2002 of downflow patterns of sand and mud deposition for individual turbidites).

In the connected tortuous corridor case, fills of channels guided by topography may contain high-quality reservoir if the sediment source is poor in mud and contains coarser-grade sand and fine gravel, which enable deposition rather than bypass on steeper slopes (e.g. Moraes *et al*. 2000). Coarser grain sizes and the presence of gradient reductions (e.g. 'flats' in a stepped flow-path profile) will enhance deposition of reservoir-forming sands. More mud-rich sediment supply systems result in channelized sandstone bodies enclosed in a mudstone background phase with consequent reservoir connectivity problems (e.g. Reading & Richards 1994; Demyttenaere *et al*. 2000). As with the cascade of silled basins model, mud-poor deposits exhibit abrupt onlap relations with bounding slopes, whereas mud-rich sections exhibit convergent-thinning, slope-mantling wedges.

Recognition criteria

In both the cascades of silled basins and connected tortuous corridor models, the patterns of deposition and erosion along flowpaths are governed by the interactions of successive flows (each differing to some degree in their properties) and the gradients they encounter. The major difference is the presence of sills in the former, providing closed depressions with frontally opposing slopes to sediment gravity flows. Convergent thinning and convergent baselap stratal patterns occur in both classes of slope, but only in the silled sub-basin case do such patterns occur against closing frontal slopes. This downslope separation into discrete sub-basins leads to a pattern of resetting of proximal to distal facies trends in successive sub-basins. Along-flowpath facies variations will also occur in connected tortuous corridor cases in response to changes in gradient, but will be less marked. In both scenarios, areal extents of individual flows relative to the scale of the receiving spaces, fractions of mud, sand and gravel in the sediment supply, and rates of structure growth relative to rates of gradient smoothing by depositional and erosional processes are important controls on the distribution of erosional surfaces and depositional facies.

Discrimination of cascade of silled sub-basins from connected tortuous corridors cases may depend upon demonstration of the following features:

(1) Fills of successive sub-basins will tend to young in the down-palaeoflow direction (Prather *et al*. 1998; Prather 2000). In the subsurface documenting this will require careful mapping in three dimensions, together with biostratigraphic calibration of available well and seismic data.
(2) Proximal to distal facies trends should be reset in each successive sub-basin (e.g. Winker 1996; Beaubouef & Friedman 2000).
(3) There should be evidence for incision, prolonged bypass and commonly fine-grained fill of channels in the vicinity of, and updip of, breached palaeosills (Prather *et al*. 1998; Winker & Booth 2000). Sand-rich backfill into the upper basin incision could be a possibility in the case of there being only a small difference between the height to which the downdip basin fill can aggrade and the elevation of the sill.
(4) Bed-scale evidence for flow reflection in the vicinity of a candidate sill can be supportive of a silled sub-basin model, although such reflections could also be recorded in a configuration with lateral escape routes.
(5) The presence of thick mud caps to individual turbidites may indicate full ponding of flows, although absence of thick mud caps could also be a result of a very mud-poor sediment supply.

Conclusions

(1) Unravelling the depositional and erosional evolution of topographically complex slopes requires an understanding of: (a) the geometric character of the slope topography at different times; (b) the nature of sediment supply through time in terms of flow

volumes, grain size distribution and flow concentrations; and (c) rates of growth in structurally induced topography relative to the erosional and depositional processes acting to 'heal' the slope.
(2) Geometric classes of slope topography include (a) completely closed depressions (or silled sub-basins), (b) partially silled sub-basins with lateral escape paths, and (c) flowpaths, with varying degrees of tortuosity, guided by, but not blocked by, structurally induced bounding slopes. Different classes of slope topography may coexist laterally on a given slope.
(3) Interaction of sediment gravity flows with slope topography results in modifications to the topography, ranging from downcutting (in cases of slopes being steeper than theoretical equilibrium profiles for a given flow type) to upuilding through deposition (in cases of slopes being gentler than theoretical equilibrium profiles for a given flow type). Depositional gradients vary, in particular, with transported grain-size distribution, flow density (concentration), thickness, rates of lateral flow expansion, and degree of containment by bounding slopes.
(4) Two end-member scenarios for deposition and erosion on topographically complex slopes have been identified: (a) a cascade of silled sub-basins model, and (b) a connected tortuous corridor model. The key distinction between the two models is that, in the cascade of silled sub-basins model, topographic barriers between sub-basins are effective in blocking at least the basal sand-rich portions of flows until a sufficient degree of fill is achieved for substantial portions of flows to travel beyond the former barrier. When this happens, flow acceleration on the steep slope downdip of the barrier leads to downcutting and successive flows cut back into the fill of the updip sub-basin.
(5) Distinctions in the current literature between deposits of flat-topped 'ponded accommodation' and dipping-topped 'healed slope accommodation' in the cascade of silled sub-basins model are de-emphasized here. This is because turbidite systems affected by topographic barriers in their distal reaches have variable downflow dips in the vicinity of entry points to a sub-basin, related (as noted above) to flow character (grain-size distribution, thickness and concentration) and rates of lateral expansion. Deposits of flows that are sufficiently large in scale to be contained by the confining topography will, however, tend to have tops that are close to flat.
(6) Distinguishing between the two end-member models can be difficult if knowledge of three-dimensional geometries is incomplete. This can be the case if, for example, only two-dimensional seismic profiles are available, or if a system is only partially preserved in outcrop. Convergent thinning and convergent baselap stratal patterns occur in both scenarios, and, at a bed scale, evidence for flow interactions with bounding slopes can be expected in both cases.
(7) Criteria for distinguishing cascades of silled sub-basins from connected confining corridors (when complete three-dimensional imaging is not available) have been suggested. These are: (a) the tendency for the fills of successive downslope sub-basins to be younger than those of their updip neighbours; (b) the resetting of proximal to distal trends in facies architecture at the entry points to each sub-basin; (c) a tendency for spill-phase incisions in updip basin-fills to be mudstone-filled rather than backfilled with sandstone; (d) bed-scale evidence for flow reflections updip of a candidate palaeosill; and (e) the presence of thick mudstone caps to individual turbidites that can indicate full ponding of flows.

Ideas on the variability of equilibrium profiles with flow character were stimulated by a field season on the Kota alluvial fan of southern Iceland in 1988 (presented at the 1989 BSRG meeting), through discussions with S. O'Connell (at COMFAN II, Parma), and more recently with B. Kneller (California 2000). Ideas on the value of distinguishing between 'cascades of silled sub-basins' and 'connected tortuous corridors' stem from a spell working with Shell Gabon exploration in the mid 1990s, and this paper is based on part of an internal report written at that time. The following people and organizations are thanked for kindly providing images and discussions: B. Prather, J. Booth, P. Diebold, P. Mullin, G. Smith, B. Farrer and J. Lobao. Very helpful reviews from B. Kneller and B. Galloway are gratefully acknowledged. Shell International is gratefully acknowledged for permission to publish this paper.

References

ALEXANDER, J. & MORRIS, S. 1994. Observations on experimental, non-channelised, high-concentration turbidity currents and variations in deposits around obstacles. *Journal of Sedimentary Research*, **A64**, 899–909.

ALEXANDER, J. & MULDER, T. 2002. Experimental quasi-steady density currents. *Marine Geology*, **186**, 195–210.

ALLEN, J. R. L. 1985. *Principles of Physical Sedimentology*. Allen and Unwin, London.

ANDERSON, J. E., CARTWRIGHT, J., DRYSDALL, S. J. & VIVIAN, N. 2000. Controls on turbidite sand deposition during gravity-driven extension of a passive margin: examples from Miocene sediments in Block 4, Angola. *Marine and Petroleum Geology*, **17**, 1165–1203.

BADALINI, G., KNELLER, B. & WINKER, C. D. 2000. Architecture and processes in the Late Pleistocene Brazos–Trinity turbidite system, Gulf of Mexico Continental slope. *GCSSEPM Foundation 20th Annual Research Conference, Deep-Water Reservoirs of the World, 3–6 December 2000*, 16–33.

BEAUBOUEF, R. T. & FRIEDMAN, S. J. 2000. High resolution seismic/sequence stratigraphic framework for the evolution of Pleistocene intra slope basins, western Gulf of Mexico: depositional models and reservoir analogues. *GCSSEPM Foundation 20th Annual Research Conference. Deep-Water Reservoirs of the World, 3–6 December 2000*, 40–103.

BELDERSON, R. H., KENYON, N. H., STRIDE, A. H. & PELTON, C. D. 1984. A 'braided' distributary system on the Orinoco deep-sea fan. *Marine Geology*, **56**, 195–206.

BIEGERT, E. K., STEFFENS, G. S., SUMNER, H. S. & BIRD, D. 2003 in press. Back to basics: a comparative analysis on deepwater settings and associated fan systems. *Marine and Petroleum Geology*.

BOOTH, J. R., DUVERNAY III, A. E., PFEIFFER, D. S. & STYZEN, M. J. 2000. Sequence stratigraphic framework, depositional models and stacking patterns of ponded and slope fan systems in the Auger Basin: Central Gulf of Mexico slope. *GCSSEPM Foundation 20th Annual Research Conference. Deep-Water Reservoirs of the World, 3–6 December 2000*, 82–103.

BURSIK, M. I. & WOODS, A. W. 2000. The effects of topography on sedimentation from particle-laden turbulent density currents. *Journal of Sedimentary Research*, **70**, 53–63.

CAZZOLA, C., MUTTI, E. & VIGNA, B. 1985. Cengio Turbidite System, Italy. *In*: BOUMA, A. H., NORMARK, W. R. & BARNES, N. E. (eds) *Submarine Fans and Related Turbidite Systems*. Springer-Verlag, New York, 179–183.

DI TORO, G. A. E. 1995. Angel Formation turbidites in the Wanaea Field area, Dampier Sub-basin, North-West Shelf, Australia. *In*: PICKERING, K. T., HISCOTT, R. N., KENYON, N. H., RICCI LUCCHI, F. & SMITH, R. D. A. (eds) *An Atlas of Deep-Water Environments: Architectural Style in Turbidite Systems*. Chapman & Hall, London, 260–266.

DEMYTTENAERE, R., TROMP, J. P., IBRAHIM, A., ALLMAN-WARD, P. & MECKEL, T. 2000. Brunei deep water exploration: from sea floor images and shallow seismic analogues to depositional models in a slope turbidite settings. *GCSSEPM Foundation 20th Annual Research Conference. Deep-Water Reservoirs of the World, 3–6 December 2000*, 304–317.

DIEGEL, F. A., KARLO, J. F., SCHUSTER, D. C., SHOUP, R. C. & TAVERS, P. R. 1995. Cenozoic structural evolution and tectonostratigraphic framework of the northern Gulf Coast continental margin. *In*: JACKSON, M. P. A., ROBERTS, D. G. & SNELSON, S. (eds) *Salt Tectonics: A Global Perspective*. American Association of Petroleum Geologists Memoir, **65**, 109–151.

DOUGLAS, R. G. & HEITMAN, H. L. 1979. Slope and basin benthic formanifera of the Californian Borderland. *In*: DOYLE, L. J. & PILKEY, O. H. (eds) *Geology of Continental Slopes*. SEPM Special Publications, **27**, 231–246.

GARCIA, M. H. 1993. Hydraulic jumps in sediment-driven bottom currents. *Journal of Hydraulic Engineering*, **119**, 1094–1117.

GERVAIS, A., SAVOYE, B., PIPER, D. J. W., MULDER, T., CREMER, M. & PICHEVIN, L. 2004. Present morphology and depositional architecture of a sandy submarine system: Golo turbidite system (eastern margin of Corsica). Geological Society, London, Special Publications. 222,

HAUGHTON, P. D. W. 1994. Deposits of deflected and ponded turbidity currents, Sorbas Basin, southeast Spain. *Journal of Sedimentary Research*, **A64**, 233–246.

HISCOTT, R. N. 1994. Loss of capacity, not competence, as the fundamental process governing deposition from turbidity currents. *Journal of Sedimentary Research*, **64**, 209–214.

JAMES, D. M. D. & JAMES, J. 1969. The influence of deep fractures on some areas of Ashgillian-Llandoverian sedimentation in Wales. *Geological Magazine*, **106**, 562–582.

KNELLER, B. C. & MCCAFFREY, W. D. 1995. Modelling the effects of salt-induced topography on deposition from turbidity currents. *Society of Economic Palaeontologists and Mineralogists, Gulf Coast Section*, 137–145.

KNELLER, B. C. & MCCAFFREY, W. D. 1999. Depositional effects of flow non-uniformity and stratification within turbidity currents approaching a bounding slope: deflection, reflection and facies variation. *Journal of Sedimentary Research*, **69**, 980–991.

KNELLER, B., EDWARDS, D., MCCAFFREY, W. & MOORE, R. 1991. Oblique reflection of turbidity currents. *Geology*, **19**, 250–252.

KOMAR, P. D. 1971. Hydraulic jumps in turbidity currents. *Geological Society of America Bulletin*, **82**, 1477–1488.

KOMAR, P. D. 1983. Shapes of streamlined islands on Earth and Mars: experiments and analyses of the minimum-drag form. *Geology*, **11**, 651–654.

LEOPOLD, L. B. & BULL, W. B. 1979. Base level, aggradation and grade. *Proceedings of the American Philosophical Society*, **123**, 168–202.

LIU, J. Y. & BRYANT, W. R. 2000. Sea floor morphology and sediment paths of the northern Gulf of Mexico deepwater. *In*: BOUMA, A. H. & STONE, C. G. (eds) *Fine-Grained Turbidite Systems*. American Association of Petroleum Geologists Memoir 72/SEPM Special Publication, **68**, 33–46.

MENARD, H. W. 1964. *Marine Geology of the Pacific.* McGraw-Hill, New York.

MIDDLETON, G. V. & NEAL, W. J. 1989. Experiments on the thickness of beds deposited by turbidity currents. *Journal of Sedimentary Petrology*, **59**, 297–307.

MORAES, M. A., BECKER, M. R. & MONTEIRO, M. C. & NETTO, S. L. A. 2000. Using outcrop analogs to improve 3D heterogeneity modelling of Brazilian sand-rich turbidite reservoirs. *GCSSEPM Foundation 20th Annual Research Conference. Deep-Water Reservoirs of the World, 3–6 December 2000*, 587–605.

MUTTI, E. 1985. Turbidite systems and their relations to depositional sequences. *In*: ZUFFA, G. G. (ed.) *Provenance of Arenites.* NATO-ASI Series, Reidel Publishing Company, 65–93.

MUTTI, E. & NORMARK, W. R. 1987. Comparing examples of modern and ancient turbidite systems: problems and concepts. *In*: LEGGETT, J. K. & ZUFFA, G. G. (eds) *Marine Clastic Sedimentology: Concepts and Case Studies.* Graham & Trotman, London, 1–38.

NORMARK, W. R. & PIPER, D. J. W. 1991. Initiation processes and flow evolution of turbidity currents: implications for the depositional record. *In*: OSBORNE, R. H. (ed.) *From Shoreline to Abyss.* SEPM Special Publications, **46**, 207–230.

NORMARK, W. R., GUTMACHER, C. E., CHASE, T. E. & WILDE, P. 1984. Monterey Fan: growth pattern control by basin morphology and changing sea levels. *Geo-Marine Letters*, **3**, 681–694.

O'CONNELL, S., NORMARK, W. R., RYAN, W. B. F. & KENYON, N. H. 1991. An entrenched thalweg channel on the Rhone Fan: interpretation from a seabeam and seamark I survey. *In*: OSBORNE, R. H. (ed.) *From Shoreline to Abyss.* SEPM Special Publications, **46**, 259–269.

PICKERING, K. T. & HISCOTT, R. N. 1985. Contained (reflected) turbidity currents from the Middle Ordovician Cloridorme Formation, Quebec: an alternative to the antidune hypothesis. *Sedimentology*, **32**, 373–394.

PIPER, D. J. W. & NORMARK, W. R. 1983. Turbidite-deposits, patterns and flow characteristics, Navy Submarine Fan, California borderland. *Sedimentology*, **30**, 681–694.

PIRMEZ, C. & FLOOD, R. D. 1995. Morphology and structure of Amazon Channel. *In*: FLOOD, R. D., PIPER, D. J. W. et al. (eds) *Proceedings of the ODP, Initial Reports*, **155**, Ocean Drilling Program, College Station, TX, 23–45.

PIRMEZ, C., BEAUBOUEF, R. T., FRIEDMANN, S. J. & MOHRIG, D. C. 2000. Equilibrium profile and baselevel in submarine channels: examples from late Pleistocene systems and implications for the architecture of deepwater reservoirs. *GCSSEPM Foundation 20th Annual Research Conference. Deep-Water Reservoirs of the World, 3–6 December 2000*, 782–805.

PRATHER, B. E. 2000. Calibration and visualisation of depositional process models for above-grade slopes: a case study from the Gulf of Mexico. *Marine and Petroleum Geology*, **17**, 619–638.

PRATHER, B. E., BOOTH, J. R., STEFFENS, G. S. & CRAIG, P. A. 1998. Classification, lithologic calibration, and stratigraphic succession of seismic facies of intraslope basins, deep-water Gulf of Mexico. *American Association of Petroleum Geologists Bulletin*, **82**, 701–728.

RAVNÅS, R. & STEEL, R. J. 1998. Architecture of marine rift-basin successions. *American Association of Petroleum Geologists Bulletin*, **82**, 110–146.

READING, H. G. & RICHARDS, M. 1994. Turbidite systems in deep-water basin margins classified by grain size and feeder system. *American Association of Petroleum Geologists Bulletin*, **78**, 792–822.

ROSS, W. C., WATTS, D. E. & MAY, J. A. 1995. Insights from stratigraphic modeling: mud-limited versus sand-limited depositional systems. *American Association of Petroleum Geologists Bulletin*, **79**, 231–258.

SAMUEL, A., KNELLER, B., RASLAN, S., SHARP, A. & PARSONS, C. 2003. Prolific deep marine slope channels of the Nile Delta, Egypt. *AAPG Bulletin*, **87**, 541–560.

SATTERFIELD, W. M. & BEHRENS, W. E. 1990. A late Quaternary canyon/channel system, northwest Gulf of Mexico continental slope. *Marine Geology*, **29**, 51–67.

SCHOLZ, C.A., ROSENDAHL, B. R. & SCOTT, D. K. 1990. Development of coarse-grained facies in lacustrine rift basins: examples from East Africa. *Geology*, **18**, 140–144.

SCHUMM, S. A., HARVEY, M. D. & WATSON, C. C. 1984. *Incised Channels: Morphology, Dynamics and Control.* Water Resources Publication, Littleton, Colorado.

SIEGENTHALER, C., HSU, K. J. & KLEBOTH, P. 1984. Longitudinal transport of turbidity currents: a model study of the Horgen events. *Sedimentology*, **31**, 187–193.

SINCLAIR, H. D. 2000. Delta-fed turbidites infilling topographically complex basins: a new depositional model for the Annot Sandstones, SE France. *Journal of Sedimentary Research*, **70**, 504–519.

SMITH, R. D. A. 1987a. Structure and deformation history of the Central Wales Synclinorium, NE Dyfed: evidence for a long-lived basement structure. *Geological Journal*, **22**, 183–198.

SMITH, R. D. A. 1987b. The *griestoniensis* Zone Turbidite System, Welsh Basin. *In*: LEGGETT, J. K. & ZUFFA, G. G. (eds) *Marine Clastic Sedimentology: Concepts and Case Studies.* Graham & Trotman, London, 89–107.

SMITH, R. D. A. 1995a. Reservoir architecture of lacustrine turbidite systems, Lower Cretaceous, offshore Gabon. *In*: LAMBIASE, J. (ed.) *Hydrocarbon Habitat in Rift Basins.* Geological Society, London, Special Publications, **80**, 197–210.

SMITH, R. D. A. 1995b. Architecture of channelized and non-channelized sediment bodies in a Welsh Basin turbidite system, Cwm Twrch. *In*: PICKERING, K. T., HISCOTT, R. N., KENYON, N. H., RICCI LUCCHI, F. & SMITH, R. D. A. (eds) *An Atlas of Deep-Water Environments: Architectural Style in Turbidite Systems.* Chapman & Hall, London, 250–254.

SMITH, R. D. A. 2004. Turbidite systems influenced by topography in the multi-sourced Welsh Basin. *In*: LOMAS, S. & JOSEPH, P. (eds) *Confined Turbidite Systems*. Geological Society, London, Special Publications, **222**, 209–228.

SMITH, R. D. A. & ANKETELL, J. M. 1992. Welsh Basin 'contourites' reinterpreted as fine-grained turbidites: the Grogal Sandstones. *Geological Magazine*, **129**, 609–614.

SMITH, R. D. A. & JOSEPH, P. 2004. Onlap stratal architectures in the Grès d'Annot: Geometric models and controlling factors. *In*: JOSEPH, P. & LOMAS, S. (eds) *Deep-Water Sedimentation in the Alpine Foreland Basin of SE France: New Perspectives on the Grès d'Annot and Related Systems*. Geological Society, London, Special Publications, **221**, 389–399.

SMITH, R. D. A. & MØLLER, N. 2003 in press. Sedimentology and reservoir modelling of the giant Ormen Lange Field, Mid Norway. *Marine and Petroleum Geology*, **20**(6–8), 601–613.

STEVENS, S. H. & MOORE, G. F. 1985. Deformational and sedimentary processes in trench slope basins of the western Sunda Arc, Indonesia. *Marine Geology*, **69**, 93–112.

STURM, M. & MATTER, A. 1978. Turbidites and varves in Lake Brienz (Switzerland): deposition of clastic detritus by turbidity currents. International Association of Sedimentologists, Special Publications, **2**, 147–168.

THOMPSON, A. & JONES, A. 1986. Rates and causes of proglacial river terrace formation in southeast Iceland: an application of lichonometric dating techniques. *Boreas*, **15**, 231–246.

VAN ANDEL, TJ. H. & KOMAR, P. D. 1969. Ponded sediments of the Mid-Atlantic Ridge between 22° and 23° north latitude. *Bulletin of the Geological Society of America*, **80**, 1163–1190.

WIERICH, F. H. 1989. The generation of turbidity currents by subaerial debris flows, California. *Geological Society of America Bulletin*, **101**, 278–291.

WINKER, C. D. 1996. High-resolution seismic stratigraphy of a late Pleistocene submarine fan ponded by salt-withdrawal mini-basins on the Gulf of Mexico continental slope. *Offshore Technology Conference Proceedings*, 619–628.

WINKER, C. D. & BOOTH, J. R. 2000. Sedimentary dynamics of salt-dominated continental slope, Gulf of Mexico: integration of observations from the seafloor, near-surface, and deep subsurface. *GCSSEPM Foundation 20th Annual Research Conference. Deep-Water Reservoirs of the World, 3–6 December 2000*, 304–317.

WYNN, R. B., WEAVER, P. P. E., MASSON, D. G. & STOW, D. A. V. 2002. Turbidite depositional architecture across three interconnected deep-water basins on the north-west African margin. *Sedimentology*, **49**, 645–667.

Factors influencing the deposit geometry of experimental turbidity currents: implications for sand-body architecture in confined basins

OMAR S. AL JA'AIDI[1], WILLIAM D. McCAFFREY[2] & BENJAMIN C. KNELLER[3]

[1] *College of Science, Sultan Qaboos University, PO Box 36, Al-Khod, Oman*
[2] *School of Earth Sciences, University of Leeds, Leeds LS2 9JT, UK (e-mail: mccaffrey@earth.leeds.ac.uk)*
[3] *Institute for Crustal Studies, Girvetz Hall, University of California Santa Barbara, CA 93106, USA*

Abstract: Two sets of scaled laboratory experiments were performed to examine the effect of flow volume, flow density and grain-size distribution on the transport efficiency of turbidity currents. The experiments employed two sediment analogues (ballotini and silica flour) intended to model medium- to coarse-grained sand and mud respectively. In the first set of experiments each parameter was varied to examine its effect upon deposit geometry. Increases in the initial flow density, volume and proportion of fines had the effect of increasing the amount of sediment that was transferred to the floor of the experimental tank by the turbidity currents. Increase of each of these parameters has a characteristic effect on the three-dimensional geometry of the deposit: the deposits of large-volume flows are elongate, and those of fines-rich flows are broad. Increase of flow density increases the initial potential energy of the flow, thus increasing the runout distance; increase of the initial density beyond a sediment concentration of 13% by mass results, however, in a reverse of the geometrical trend of deposit elongation, possibly because of turbulence suppression at high densities. Increase of flow volume also increases the initial potential energy, and reduces the rate of velocity decrease due to gravitational spreading. Increase in the proportion of fines leads to maintenance of negative buoyancy, as the fine fraction remains suspended until the flow has virtually come to rest; it also decreases the settling velocity of the coarser fraction and thus delays its sedimentation. The second set of experiments was performed to investigate the influence of flow efficiency on the interaction of turbidity currents with topography. A single arcuate obstacle was placed in the path of the flows. In successive experiments flow efficiency was increased by progressively increasing the proportion of fines (silica flour). Both the proportion of sediment reaching the obstructing topography and the proportion of it able to surmount the topography increased as flow efficiency increased. Thus flow efficiency may determine whether or not an enclosed basin hosts deposits whose geometry has been affected by the confinement, and may also determine the relative effectiveness of the topography in confining inbound turbidity currents, and thus trapping their sediment load.

Flow efficiency can be defined as the ability of the flow to carry sand in a basinward direction (Mutti 1979, 1992). It thus plays a major role in governing the location and geometry of sand deposited from turbidity currents. The factors determining the efficiency of turbidity currents have been discussed, explicitly or implicitly, by several authors (e.g. Normark 1978; Mutti 1979, 1992; Mutti & Normark 1987; Laval *et al.* 1988; Normark & Piper 1991; Bonnecaze *et al.* 1993; Nilsen, *et al.* 1994; Gladstone *et al.* 1998; Bouma 2000). On the basis of field observations, Mutti (1979, 1992) concluded that the final geometry of individual beds or groups of beds is determined not only by the slope and basin configuration but also by the flow efficiency, as controlled by the grain-size composition of the suspended sediment load and the flow volume.

In this interpretation, high efficiency flows are relatively large-volume flows and/or carry substantial amounts of fines, whereas poorly efficient flows are relatively small-volume flows and/or are loaded mainly with coarse-grained sediment (Mutti 1992). These concepts were incorporated by Reading and Richards (1994) in their classification of turbidite systems based on grain size and type of feeder system, in which stratal correlation patterns are largely a function of turbidity current volume and textural composition. Mutti (1992) also noted that in detail the deposition both of lobes and of their component individual sandstone beds is influenced both by the volume of individual flows and by the relief of the basin floor. He suggested that relatively thick (large volume) flows are less affected by the relief of the basin floor than are thinner ones. Mutti did not, however, explicitly consider the impact of flow efficiency upon sediment deposition in partially or fully confined settings.

The aim of the work detailed here was twofold. The first aim was to evaluate systematically the effects of flow volume and grain-size distribution, and also flow density, upon the deposits of effectively unconfined, laboratory-scale turbidity currents. The second aim was to assess the influence of bathymetry upon the deposits of flows as a function of flow efficiency, focusing upon the case in which the parental flows were partially confined. It becomes clear from this work that (1) several factors contribute to flow efficiency, and (2) there are several ways of defining the influence of flow efficiency in terms of deposit. In the light of this, the definition and use of the term *flow efficiency* can be re-evaluated.

Experiments on flow efficiency

Previous experimental work

To date, relatively little experimental work has been published on the flow efficiency of turbidity currents (e.g. Laval *et al.* 1988; Bonnecaze *et al.* 1993; Gladstone *et al.* 1998). Bonnecaze *et al.* (1993) performed several experiments in which both the density and the grain-size distribution of turbulent suspensions were varied. They showed that flows of higher initial density or of finer particle size travelled further downstream. Similar experimental results were reported by Gladstone *et al.* (1998), who demonstrated that the propagation and sedimentation patterns of particle-laden gravity currents were strongly influenced by the size distribution of suspended particles. They showed that the addition of small amounts of fine-grained sediment to a coarse-grained density current had a larger influence on flow velocity and runout distance than adding a small amount of coarse sediment to a fine-grained density current. The experiments performed by Bonnecaze *et al.* (1993) and Gladstone *et al.* (1998) show that the presence of fine-grained sediment reduces the rate of momentum loss because the fine particles remained suspended, thus maintaining an excess current density for a longer period of time. A similar result was obtained numerically by Imran & Parker (1999).

Experimental method

The experiments carried out in the present study were performed in a square tank of dimensions

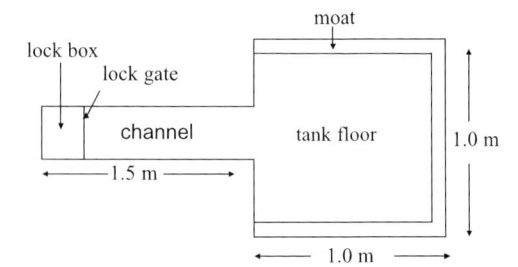

Fig. 1. Schematic plan view diagram of the square flume (T-tank) used in the experiments. The moat is some 7 cm wide, and 3.5 cm deep. Water depth over the main tank floor was 11.5 cm.

1 m × 1 m and 0.2 m depth, with a 1.5 m long channel centred on one side through which the currents entered the tank (Fig. 1). The floors of both the channel and the main body of the tank were horizontal. The currents were released from a lock box at the end of the channel. They were generated by agitating a mixture of sediment and tap water in the lock box with vigorous vertical motions of a grid. Once the sediment was in suspension, the grid was removed simultaneously with the release of the lock gate, and the flow travelled along the horizontal channel, eventually surging onto the main body of the tank, where it spread radially. Part of the sediment was deposited within the lock box and channel, with differing amounts reaching the main body of the tank depending on the properties of the flow. The sediment was then allowed to settle for 12 hours before it was measured. The thickness distribution of the deposit across the tank floor was measured on a 10 mm grid using an ultrasonic probe of nominal precision ±10 μm (technique modified after Best & Ashworth 1994 and Kneller 1995). The thickness of the deposits of these experimental flows ranged up to a few hundred microns. These values were used to calculate the total volume of sediment deposited onto the tank floor. The flow was recorded from above and the side using two video cameras, whose recordings were used to calculate flow head velocities. Two sediment analogues (both of density $2650\,\mathrm{kg\,m^{-3}}$) were used in the experiments: 10–50 μm diameter, approximately spherical silica grains (ballotini) with a mean grain size of 34.5 μm; and 0.1–30 μm silica flour with a mean grain size of 7.65 μm. The coarser grains were intended to model medium- to coarse-grained sand, whereas the silica flour was introduced to model the fine-grained sediment component.

Scaling

Laboratory experiments are often planned as scale models, which attempt to replicate the key features of a natural-scale system (the prototype). Such models should incorporate the principles of similarity theory, in which the key variables that fully characterize the physical system are arranged into dimensionless combinations (e.g. Reynolds and Froude numbers) whose value is the same in the model and the prototype (Buckingham 1915; David & Noelle 1982). Such perfect scaling is generally not possible for the majority of experiments where water is used for both the model and prototype (e.g. Peakall et al. 1996). However, through scaling the most influential variables, the scale model may preserve the essential features of the prototype. In these experiments a Froude scaling approach was adopted, in which the subcritical state of the prototype was honoured, but in which control over the flow Reynolds number was relaxed (albeit with the proviso that the Reynolds number be above the laminar to turbulent transition). Reynolds numbers were calculated to be of the order $Re \approx 7 \times 10^3$, using the head velocity as the characteristic velocity and assuming the viscosity was equal to that of water. However, because body velocities may be 40–50% greater than head velocities (e.g. Kneller et al. 1997, 1999) this value is conservative. Flow is generally considered fully turbulent at flow Reynolds numbers of 2000 or above (e.g. Leeder 1982 and references therein). However, other workers have concluded that Reynolds numbers of at least $Re \approx 2 \times 10^5$ are required if the energy cascade is to be preserved across all scales of turbulence (e.g. Parsons & Garcia 1998 and references therein), with the implication that in this case the sediment carried in suspension may not have been fully supported by the turbulence. These experiments may, therefore, more accurately model the depositional regimes of natural-scale currents, rather than those in which significant bypass occurs. The precise geometries of the experimental deposits may not reflect those produced at the natural scale. However, the analysis presented below is focused upon *trends* in deposit geometry produced by varying the flow properties. Such trends are more likely to be scale-independent, as the effect of turbulence intensity variations at any one scale is likely to be small. Grain-size scaling was achieved by scaling settling velocity with the head velocity of the flow, the rationale being that sedimentation from suspension is dependent on the shear velocity of the flow, which (assuming constant bed slope and roughness) is a linear function of the head velocity. This assumption is commonly applied in depth-averaged theoretical models of turbidity currents (e.g. Parker et al. 1986).

Experimental programme: unobstructed flows

Three series of experiments were performed to investigate, independently, the factors influencing flow efficiency of turbidity currents and their effect on depositional geometries. These factors included:

(1) the initial flow density;
(2) the initial volume of the suspension; and
(3) the proportion of fine-grained sediment within the flow.

In each set of experiments one parameter was varied from a base case, which was common to each series. The starting conditions of the three sets of experiments are presented in Table 1.

In the first series of experiments, the efficiency of the flow was investigated by varying the initial flow density (i.e. mass fraction of suspended sediment) of the gravity current. In each of the six experiments, 10–50 μm ballotini were used to create the excess density. The initial flow density was varied from $1060 \, \text{kg m}^{-3}$ to $1160 \, \text{kg m}^{-3}$ (Flows 1–6), whereas the flow volume was kept constant. In the second set of experiments, the suspension volume was varied while keeping the initial suspension density and grain-size distribution constant. This was achieved by introducing a variable-volume lock box. The lock gate remained in a fixed position, and an adjustable baffle was inserted within the lock box. The position of the baffle was then altered to vary the lock volume. Table 1 illustrates the four different flow volumes (Flows 7–10). In nature, this may be caused by variation in starting conditions such as the size of mass failure on a slope. In the third set of experiments the grain-size distribution of the suspension was varied. In each of the four runs (Flows 11–14), the initial flow density was $1080 \, \text{kg m}^{-3}$ (yielding an excess density over the ambient of 8%), and the initial volume of the suspension in each case was $4.59 \times 10^{-3} \, \text{m}^3$. However, in this case the proportion of fine-grained material in suspension was systematically varied by substituting fine-grained silica flour for ballotini in increasing proportion (Table 1).

The relationship between the proportional increase in the initial mass of suspended sediment versus the proportional increase in sediment mass deposited on the basin floor is shown in

Table 1. *Experimental starting conditions*

Series	Flow	Initial flow density Density (kg m^{-3})	Density excess (%)	Initial flow volume Volume (m^3)	Volume excess (%)	Initial sediment composition Ballotini (%)	Silica flour (%)
1	1	1060	6	4.59×10^{-3}	0	100	0
1	2	1080	8	4.59×10^{-3}	0	100	0
1	3	1100	11	4.59×10^{-3}	0	100	0
1	4	1130	13	4.59×10^{-3}	0	100	0
1	5	1150	15	4.59×10^{-3}	0	100	0
1	6	1160	16	4.59×10^{-3}	0	100	0
2	7	1080	8	4.59×10^{-3}	0	100	0
2	8	1080	8	5.049×10^{-3}	10	100	0
2	9	1080	8	5.508×10^{-3}	20	100	0
2	10	1080	8	5.967×10^{-3}	30	100	0
3	11	1080	8	4.59×10^{-3}	0	100	0
3	12	1080	8	4.59×10^{-3}	0	90	10
3	13	1080	8	4.59×10^{-3}	0	80	20
3	14	1080	8	4.59×10^{-3}	0	70	30

Flows 1–14 comprise the unobstructed experimental turbidity currents of Series 1, 2 and 3. The ambient fluid was tap water (density ~1000 kg m^{-3}). Note: flow 2 is the 'standard flow'. The starting conditions for the series of obstructed experimental turbidity currents in which the proportion of fines was progressively increased were the same as for flows 12, 13 and 14, respectively, of Series 3. In this case, however, a single arcuate obstacle 15 mm high and of radius 115 cm was placed in the path of the flows, 370 mm from the channel mouth.

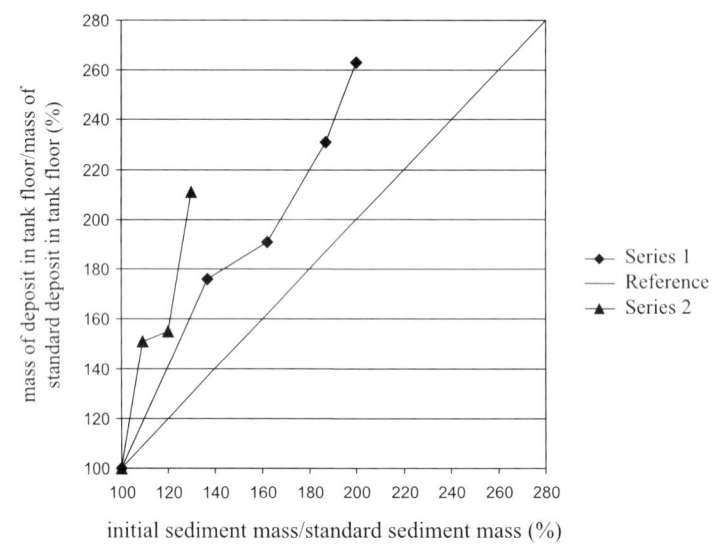

Fig. 2. Relationship between the proportional increase in mass of sediment in the initial flow versus the proportional increase in sediment mass deposited on the basin floor, with reference to a 'standard flow'. Flow density and volume were progressively increased in Series 1 and 2 experiments respectively (Table 1). The reference line indicates the hypothetical case in which these two values increase in direct proportion to one another. The data indicate that proportionally more of the sediment budget is deposited on the tank floor with increasing initial sediment mass, suggesting that the flow efficiency was progressively increased. Note: only the sediment reaching the tank floor was measured. However, because no sediment was lost from the system, the data do indicate changes in the relative distribution of the total sediment budget in each case. 'Standard' flow conditions are given in Table 1.

Figure 2. Proportionally more of the sediment budget is deposited on the tank floor with increasing initial sediment mass (due either to increased density or to volume), suggesting that the flow efficiency was progressively enhanced (particularly in Series 2, in which flow volume was progressively increased). The head velocities in each series increased with flow density, flow volume and the proportion of fines (Fig. 3).

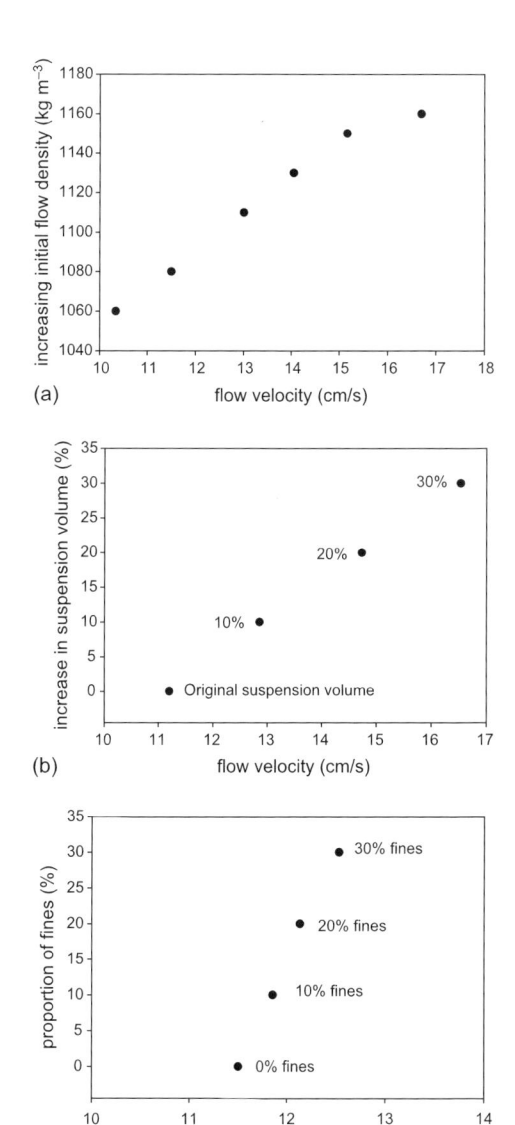

Fig. 3. Relationship between flow head velocity and (a) increasing initial flow density (Series 1), (b) increasing suspension volume (Series 2), and (c) increasing proportion of fines (Series 3).

Deposit geometries

To facilitate comparison between the deposits of flows of differing character (in terms of suspension density, volume and grain-size distribution), aspect ratio diagrams were constructed in which the ratios length/width (L/W) and width/height (W/H) were co-plotted. Such diagrams allow the representation of three-dimensional shapes on a two-dimensional diagram. To construct the plots, a thickness contour was chosen (0.4 mm) that occurred in all the plots; the length and maximum width of the area contained within the contour was measured, and the deposit thickness was determined at the intersection point of the transects along which length and width were measured (Fig. 4). This was repeated for the 0.5 mm contour to establish the repeatability of the trends. Figure 5 illustrates the results of this analysis. In Series 1 the deposit gradually becomes relatively longer and thinner with increasing initial flow density, up to a critical density of 1130 kg m^{-3}, above which the trend changes, and the deposits of denser flows tend to show decreasing length-to-width ratios. In the experiments of Series 2 the deposit becomes relatively longer and thinner with increasing initial suspension volume. However, unlike the Series 1 experiments, there appears to be no threshold above which the trend in deposit shape change is altered. The experiments in which the suspended grain-size distribution was varied (Series 3) show a different characteristic evolution of deposit geometry, in which the deposit becomes progressively relatively wider and thinner as the proportion of fine-grained particles in the flow is increased. Thus the deposits of flows carrying a relatively larger proportion of fines tend to be more areally extensive than flows with relatively fewer fines, and to be more equidimensional in plan view.

Interpretation of single flow deposit geometries

The driving force behind spreading gravity currents and those flowing down slopes is their potential energy. For currents on a horizontal surface, assuming an initially uniform vertical sediment distribution, the potential energy per unit volume is given by

$$E = \frac{1}{2}\rho_f g \left(\frac{\rho_f - \rho}{\rho}\right) d \qquad (1)$$

where E is potential energy per unit volume, ρ_f is the density of the suspension, g is the acceleration due to gravity, ρ is the density of the ambient

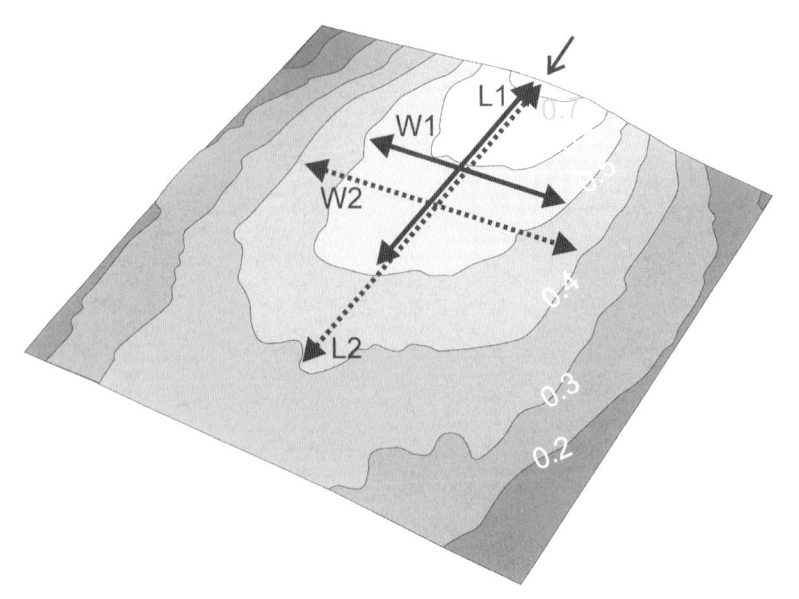

Fig. 4. A representative isopach map illustrating the length (*L*) and width (*W*) measurements derived from the 0.4 and 0.5 mm contours; height measurements (*H*) were taken at the intersection of the *L* and *W* measurement traces. Contour unit is millimetres. Arrow indicates flow direction.

fluid, and *d* is the flow thickness. Gravity-surge collapse and subsequent motion across a horizontal surface are understood to be controlled primarily by the pressure gradient existing at the current head (e.g. Huppert 1980; Middleton & Neal 1989). Considering the conversion of potential energy to kinetic energy, the velocity of the head of a turbidity current is given by

$$U_0 = Fr(g'h)^{1/2} \qquad (2)$$

where *h* is the height of the head of the current, and *Fr* is a Froude number that is principally a function of the cross-sectional geometry of the current, and its fractional depth with respect to the ambient fluid, which together largely control the drag (Keulegan 1957; Middleton 1966; Simpson 1982, 1997). The reduced gravity depends upon the flow density (which depends in turn on both the density of the sediment and its concentration). Thus increasing the density or thickness of the flow both have the effect of increasing U_0 (e.g. Fig. 3). Increasing the current velocity also increases the shear velocity (U^*):

$$U^* = \sqrt{\frac{f_b}{8}} u \qquad (3)$$

where *u* is the mean velocity of the flow below the height of the velocity maximum, and f_b is the basal friction factor. As long as the shear velocity of the flow remains above the suspension threshold of the suspended grains, the flow will continue to suspend its load. The increase in velocity thus has the effect of maintaining sediment in suspension for longer, and extending the deposit.

Increasing the suspension density. The foregoing analysis, in which increasing the flow density causes U_0 to increase (e.g. equation (2), Fig. 3), assumes fully turbulent suspension. At higher sediment concentrations, however, turbulence may be suppressed (Lowe 1982; Middleton & Hampton 1976 and references therein). Flow Reynolds numbers for varying density can be calculated using a modified Chezy equation to estimate the effect of fractional grain concentration on mean flow velocity, and the empirical relationship given by Davidson *et al.* (1977) to estimate apparent viscosity:

$$\frac{\mu_s}{\mu} = \left(1 - \frac{C}{C_{max}}\right)^{-2.5 C_{max}} \qquad (4)$$

where μ_s is apparent viscosity, μ is the viscosity of the ambient fluid and *C* is the fractional grain concentration. This suggests that, regardless of the absolute value of Reynolds number, flow Reynolds numbers for a flow of a given thickness, flowing down a slope of a given angle, may reach a maximum at a concentration of around 5% solids by volume, or 13% by mass assuming a grain density of 2650 kg m^{-3}. The change in the geometric trend with increasing density shown in Figure 5 occurs at a similar

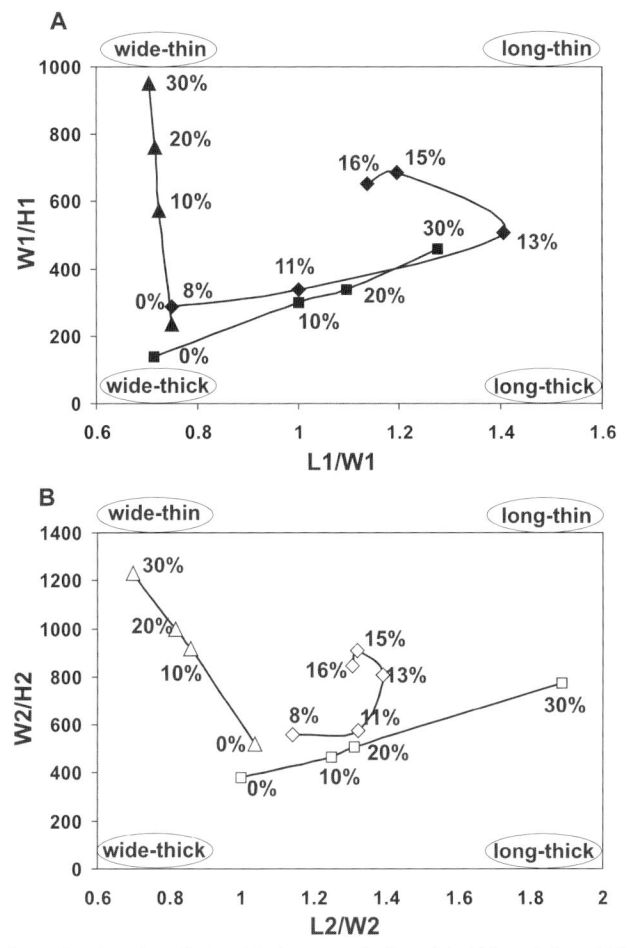

Fig. 5. Aspect ratio plots, showing the relationship between the length/width and the width/height of the deposit; end-member relative deposit characters are indicated at the plot margins. Diamond symbols show Series 1 data, with the density excess of the initial flow over the ambient expressed as a percentage. Square symbols show Series 2 data, with the volume excess of the initial flow over a standard flow expressed as a percentage. Triangular symbols show Series 3 data, with the proportion of the fine-grained sediment analogue in the initial sediment mix expressed as a percentage. (**A**) Closed symbols indicate data measured from the 0.5 mm deposit thickness contour. (**B**) Open symbols indicate data measured from the 0.4 mm contour. See Table 1 for experimental conditions.

threshold value of about 13%, and, at the experimental scale at least, it is possible that this is a consequence of progressive turbulence suppression at higher concentrations (although the 13% value relates to the initial concentration, and it is not known how density evolves during flow propagation). However, it is not clear at what concentration turbulence suppression in natural-scale turbidity currents may begin to affect sediment suspension.

Increasing the suspension volume. In the two-dimensional case, the effect of varying initial flow volume can be considered through increasing current length or current height, or both. The consequences of increasing flow length can be illustrated using box models of flow propagation (see Huppert 1998 for review), in which the flow is regarded as a collapsing column of dense fluid that, neglecting sedimentation, can be represented by a series of rectangles of equal area. The rate of spreading (that is, the propagation velocity of the front) is a function of the flow thickness (equation (2)). For rectangles of greater initial length/height, a given propagation distance will result in smaller decreases in flow thickness, and thus smaller decreases in front velocity. It follows that where the flow volume is increased by increasing the flow's length rather than its height (see above, for example), the velocity, and hence shear velocity, decline relatively more slowly than for shorter currents of the same initial height. Thus they maintain their suspended loads over longer distances. Once deposition begins, loss of mass leads to a decrease in momentum, which accelerates the decay of the current. Nevertheless, this reasoning may be applied to explain *trends* in deposit geometry produced by predominantly depositional currents, such as the experimental currents described above. Increasing flow volume by increasing the flow height will increase potential energy, and thus velocity (equations (1) & (2)). The concomitant increase in shear velocity will extend the bypass phase of the current, thus increasing flow efficiency by promoting greater runout lengths.

Increasing the proportion of fines. The distance travelled by grains suspended in a turbidity current before coming to rest on the bed depends upon the settling velocity of the particles (a function of the fluid viscosity and the size, shape and density of the particles) and the rate of decline of shear velocity exerted by the current (Bagnold 1956; Dietrich 1982). Particles with very small settling velocities remain in suspension until the shear velocity of the flow approaches zero: that is, when the flow has almost come to rest. The rate of negative buoyancy loss through sedimentation while the flow is moving is thus reduced compared with flows with a suspended load exclusively of faster-settling grains, and flow velocity (and thus shear velocity) does not decay nearly as rapidly as in a current predominantly composed of coarser particles. This allows a flow with a modest amount of fines to preserve its coarser-grained load in suspension for longer (e.g. Imran and Parker 1999).

An additional factor contributing to the efficiency of fines-rich flows is the buoyancy effect of dispersed fines. The presence of significant amounts of suspended fine material reduces the density difference between the coarse-grains and interstitial 'fluid' that consists of a dense mixture of fines and water: this reduces the settling velocities of the coarser grains (Lowe 1982; Middleton & Southard 1984; Gladstone *et al.* 1998), reducing the suspension threshold, and thus delaying sedimentation from suspension. A further factor is the reduction of the drag coefficient associated with clay-bearing flows in salt water (Best & Leeder 1993). Smaller initial rates of propagation associated with low rates of momentum loss in fines-rich currents lead to greater lateral spreading, producing the typical equidimensional deposits of fines-rich flows (e.g. Fig. 3).

These results are broadly in agreement with the flow efficiency concepts summarized by Mutti (1992), although the effects of flow density were not considered in that scheme. Note, however, that with the experimental configuration used here it was not possible to produce experimental flows that bypassed proximal regions to produce detached facies tracts *sensu* Mutti (1992).

Theoretical analysis. Dade & Huppert (1995) performed a box-model-type analysis to determine the governing equations of a radially spreading, deposit-forming gravity current. It should be borne in mind that the effects of channelized input are not considered in this analysis, and in particular the impact on deposit aspect ratios of inherited momentum in a direction parallel to the channel axis. Nevertheless, the analysis shows the area A inundated by the deposit varies as

$$A \sim \left(\frac{gV^3C}{W^2}\right)^{1/4} \qquad (5)$$

where V is the initial flow volume, C is the initial sediment concentration, W is a characteristic particle settling velocity, and $\sim$ implies dimensionally consistent proportionality. It follows that

$$\frac{W}{H} \sim \left(\frac{V}{W^6 C^5}\right)^{1/8} \qquad (6)$$

where W/H is the cross-sectional aspect ratio (equivalent to the vertical axis of Fig. 5). Thus the deposit aspect ratio should increase with increasing flow volume, diminishing particle fall speed or decreasing concentration. These anticipated trends are borne out in the Series 2 and 3 data (relating to increasing flow volume and proportion of fines respectively; see Fig. 5), but not in the Series 1 data (relating to increasing flow concentration). This may be an effect of hindered settling in which lower particle fall rates may prolong the transport phase before the effects of turbulence suppression dominate (see above).

The axial deposit aspect ratio, L/W (the horizontal axis of Fig. 5), is not incorporated into the analysis of Dade & Huppert (1995). Along the axis L/W, as the flows become more 'efficient' (that is, denser, of larger volume or finer grained), opposite trends are seen in the deposits of progressively finer flows compared with those of progressively larger flows. Also, the trend of increasing L/W seen in the deposits of increasingly dense flows reverses at a value of initial concentration 13% by mass (c. 5% by volume). Because no general pattern is produced, it follows that these trends are unlikely to be an artefact of the experimental set-up. Rather it suggests that the starting conditions are the key control.

Investigating the effect of partial blocking of turbidity currents of variable efficiency

A further set of experiments was performed to investigate the effect of partial blocking on flows of different efficiency. A single arcuate obstacle 15 mm high and of radius of 1.15 m was placed in the path of the flows, 0.37 m from the channel mouth. In successive experiments flow efficiency was progressively increased by systematically increasing the proportion of fines (silica flour). In the first experiment the flow consisted of 90% ballotini and 10% silica flour, whereas silica flour made up 20% and 30% respectively of the total mass in the second and third experiments (see Table 1 for starting conditions). The resultant sediment distribution maps of the deposits are presented in Figure 6. The arcuate obstacle divided the tank floor into two zones (A and B). The total volume of the sediment was calculated in both zones and plotted against the silica flour content (Fig. 7). Sediment samples from zone B were obtained along a tank centreline profile at Sites 1 and 2, located some 0.50 m and 0.80 m from the channel mouth respectively. Sample grain-size distributions were analysed using a Malvern Mastersizer Plus laser diffraction grain-size analyser (Figs 8a & 8b). The amount of sediment increased both upstream and downstream of the obstacle as the proportion of fines was increased (Figs 6 & 7). In addition, flows initially comprising 30% fines were able to transport a larger fraction of coarser grains downstream than flows that were initially coarser (Fig. 8).

Interpretation

Relatively more sediment was introduced into the depositional area as the initial proportion of fines in each parental flow was increased (as was observed in the earlier, unobstructed experiments of Flows 12, 13 and 14). This trend can be related to the runout length of turbidity currents increasing with flow efficiency (*sensu* Mutti 1992). As well as introducing more sediment into the upstream 'sub-basin' (Zone A), the volume of that sediment that was able to surmount the obstacle and escape into the

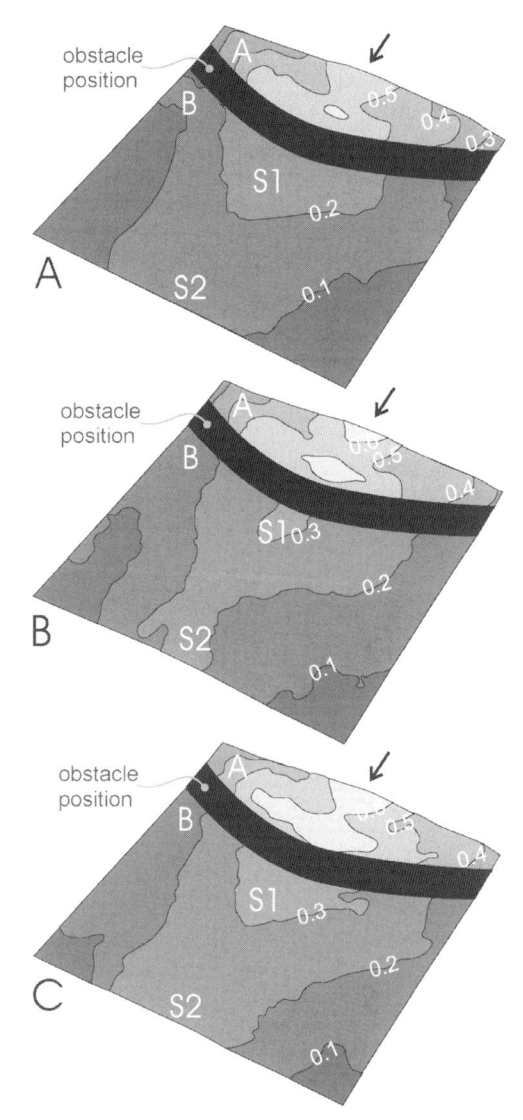

Fig. 6. Three sediment distribution maps produced by partially obstructed flows of increasing fines content. The contour interval is 0.1 mm. Arrow indicates flow direction: (**A**) 10% fines; (**B**) 20% fines; (**C**) 30% fines. The division of the tank floor into Zone A, upstream of the obstacle, and Zone B, downstream is indicated, as is the location of the sampling locations for grain-size analysis, Sites 1 and 2, located some 0.50 m and 0.80 m downstream of the channel mouth respectively.

downstream 'sub-basin' (Zone B) also increased with the proportion of fines in the initial flow. This indicates a flow efficiency control on the effectiveness of confining bathymetry to trap inbound turbidity currents, with more efficient currents better able to surmount obstructing

% of fines "vs" Deposit volume

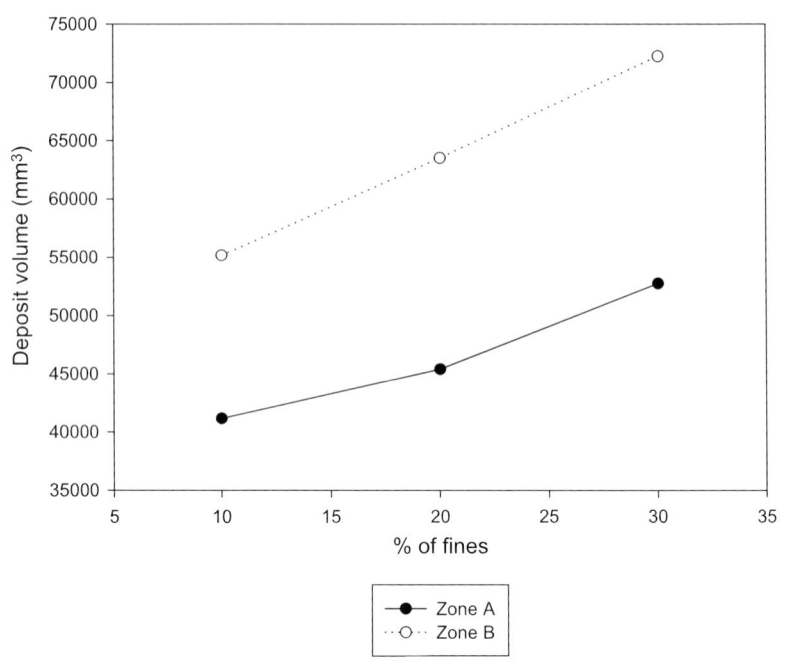

Fig. 7. Volume of sediment deposited in zones A and B versus the proportion of fines (silica flour) in the parental flow.

bathymetry. This may be because the relative velocity of flows as they impinge upon the bathymetry is likely to be greater for more efficient flows. Consideration of kinetic to potential energy conversion indicates that the effective runup height will be greater for relatively more efficient flows (see discussion in Kneller & McCaffrey 1999). Such flows are therefore relatively more likely to surmount a bathymetric barrier of a given height than are their 'less efficient' counterparts. Also, flow behaviour during interaction with a bounding slope may in part be determined by the vertical flow stratification (Kneller & McCaffrey 1999), with flow deflection (and therefore containment) more likely for flows with a pronounced density stratification. Density stratification is related to vertical gradients in suspended-sediment concentrations, which may in turn be related to shear velocity (U^*) via the Rouse number (e.g. Middleton & Southard 1984); intensity of stratification increases with decreasing shear velocity, as more efficient flows are likely to be characterized by higher shear velocities, and are thus likely to be less well stratified. Therefore they are less likely to be deflected by confining bathymetry, but more likely to surmount it – escaping to downstream basins, if the runup height exceeds the height of the confinement.

Both the absolute volume and the proportion of coarser material deposited in distal areas of the tank increased as the overall proportion of fines was increased in parental flows (which were of the same initial density and volume). The observation that coarser particles are transported greater distances in currents containing larger amounts of fines is in agreement with the conclusions of Bonnecaze et al. (1993) and Gladstone et al. (1998) detailed above.

Discussion

Turbidity current interactions with confining topography may be complex, producing a wide range of effects in the geometry and internal character of the resulting deposits. These interactions include flow deflection and reflection, and partial to complete confinement (Kneller et al. 1991; Haughton 1994; Alexander & Morris 1994; Kneller & McCaffrey 1999). In the deposits, this may result in changes in event bed character approaching the onlap slope that may be either progressive (e.g. Haughton 1994;

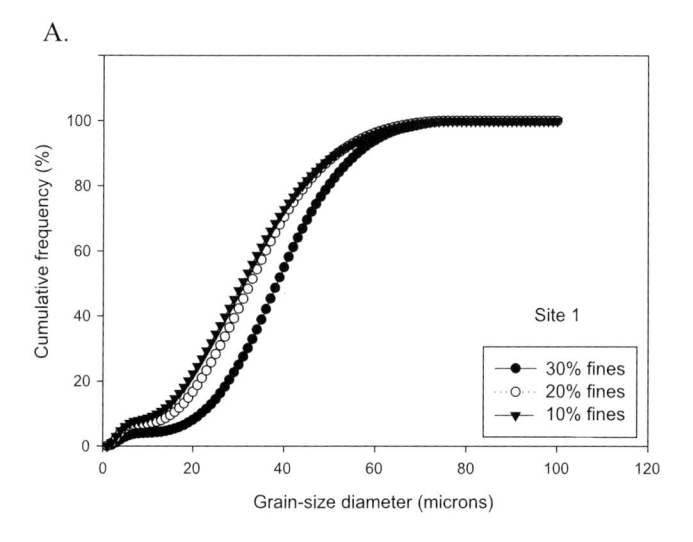

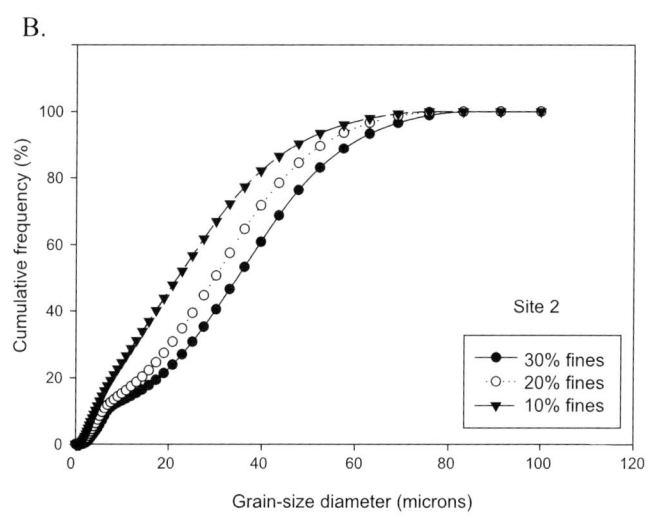

Fig. 8. Cumulative volume percentage versus grain size of samples taken from (**A**) Site 1 and (**B**) Site 2, which were located some 0.50 and 0.80 m downstream of the channel mouth, respectively.

Hurst et al. 1999; Kneller & McCaffrey 1999) or abrupt (e.g. McCaffrey & Kneller 2001), and may dictate whether sediment is confined within the local basin, or partially escapes into downstream basins (see the experimental work detailed above). Within composite sediment accumulations systematic changes in sedimentary architecture, such as the degree of vertical connectivity, may be related to slope proximity (e.g. Hurst et al. 1999; Amy 2000).

It is self-evident that the range of effects outlined above may occur only when the inbound turbidity currents are effectively confined, at least to some degree. It is also clear that the existence of confining bathymetry is a necessary, but not sufficient, condition for confinement effects to be produced. This is because the inbound flows may decay through sedimentation before reaching the basin margins. The Series 1–3 experiments on unconfined flows indicate that the effective runout length is directly influenced by the flow efficiency, as controlled via flow volume, density and grain-size distribution, confirming the analysis of Mutti (1979, 1992). Thus, all other things being equal, larger-volume, denser or more fines-rich flows, being more efficient, are more likely to reach basin margins, where present. Within confined basins, flow efficiency will therefore dictate whether the flows interact with the confining bathymetry

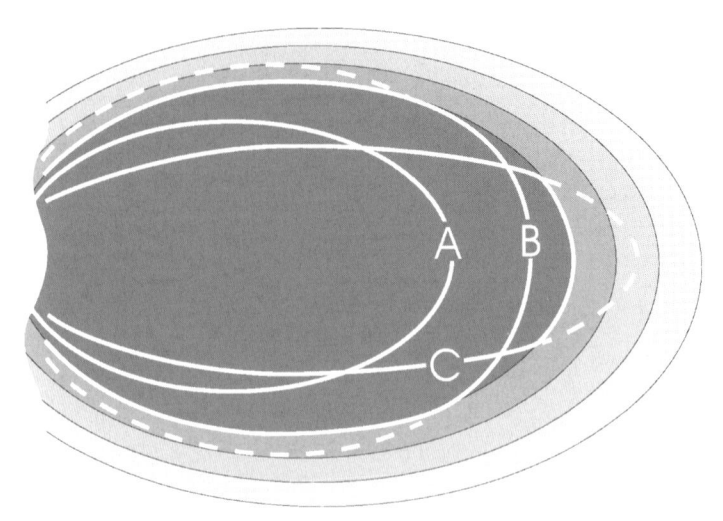

Fig. 9. Schematic summary diagram illustrating the anticipated control of flow efficiency upon the degree of confinement experienced by sand-bearing turbidity currents entering enclosed basins. Solid lines indicate depositional limit of the sandy part of the deposit within the confined basin; dashed lines indicate hypothetical sand limit in the unconfined case. Deposit A is that of a relatively low-efficiency flow, which did not reach the basin margins, and therefore has an essentially unconfined geometry. Deposit B is that of a flow that was relatively efficient by virtue of carrying an elevated proportion of fines within its sediment budget, and was therefore relatively equant. The flow interacted with the confining bathymetry at its lateral margins. Deposit C is that of a flow that was relatively efficient by virtue of being high volume, and was therefore relatively elongate. The flow interacted with the confining bathymetry at its distal margins. Darker shading indicates deeper areas. Not to scale.

and thus whether or not the basin hosts deposits whose geometry has been affected by the confinement. The Series 1–3 experiments also showed that trends in deposit shape were different depending upon whether flow volume, density or grain-size distribution was being varied. In particular, the deposits of flows of progressively greater volume were increasingly elongate, whereas the deposits of progressively more fines-rich flows were increasingly equant. Thus the specific factors controlling flow efficiency may dictate whether the lateral or distal margins of a basin effectively confine the inbound flows. These conclusions are summarized in Figure 9. It is worth bearing in mind that as a confined basin is infilled by gravity flow deposits, the effective surface area of the basin floor will increase. Thus, if flow efficiency remains unchanged, it is possible that an upward evolution from confined to effectively unconfined deposits may be produced.

Conclusions

The experiments performed in this study support the view of Mutti (1979, 1992) that flow volume and grain-size distribution have a major effect on the transport efficiency of turbidity currents and the shape of the resulting deposits. Moreover, flow density appears to have an analogous effect. However, the use of *flow efficiency* as a single measure of flow behaviour masks the diversity of effects that variations in these flow parameters may produce individually or when operating together. Each control has a characteristically different effect upon the depositional geometry (Fig. 5). Flows of higher 'efficiency'— that is, those of higher density, larger volume or higher fines content—were found to produce deposits of higher aspect ratio (Fig. 5). The experimental deposits of high-density and high-volume turbidity currents formed elongate deposits, whereas turbidity flows with a high proportion of fines produced widely dispersed but equant deposits (Fig. 5). When trying to interpret and understand the development of deepwater clastic deposits, a key set of variables to be considered would therefore appear to be changes in the initial flow parameters: that is, flow volume, flow density and type of sediment supplied to the system. Most other factors can be considered secondary and will tend only to modify the flow processes, and hence the geometry, of the deposit. For example, enclosed basin floor bathymetry will fundamentally affect the flow direction and velocity of the turbidity currents, which may modify the resultant depositional geometry of the turbidite. However, the degree of flow deflection and confinement will depend mainly on the flows' efficiency, which will be controlled by the amount and type of sediment supplied to a particular system.

The paper is based in part upon the doctoral work of O. Al-Ja'aidi, funded by Petroleum Development Oman. This study was also part-supported by the Leeds Turbidites Research Group (TRG) Phases 2 and 3 Consortia. We thank our TRG colleagues for discussion and logistical help, and in particular we thank M. Felix for discussions regarding scaling. We thank E. Mutti for stimulating discussions in the field. We thank J. Parsons and T. Mulder for insightful and constructive comments on an earlier version of this manuscript. We thank reviewers B. Dade and S. Morris, who provided helpful reviews on this submission. In particular, we express our thanks to B. Dade, whose suggestions for the geometrical analysis of the experimental deposits were incorporated into the Theoretical analysis section of the text.

References

ALEXANDER, J. & MORRIS, S. 1994. Observations on experimental, nonchannelized, high-concentration turbidity currents and variations in deposits around obstacles. *Journal of Sedimentary Research, Section A: Sedimentary Petrology and Processes*, **64**, 899–909.

AMY, L. A. 2000. *Architectural analysis of a sand-rich confined turbidite basin: the Grès de Peïra Cava, South-East France.* Ph.D. thesis, Leeds University, UK.

BAGNOLD, R. A. 1956. The flow of cohesionless grains in fluids. *Philosophical Transactions of the Royal Society of London*, **A249**, 235–297.

BEST, J. L. & ASHWORTH, P. 1994. A high resolution ultrasonic bed profiler for use in laboratory flumes. *Journal of Sedimentary Research*, **64**, 674–675.

BEST, J. L. & LEEDER, M. R. 1993. Drag reduction in turbulent muddy seawater flows and some sedimentary consequences. *Sedimentology*, **40**, 1129–1137.

BONNECAZE, R. T., HUPPERT, H. E. & LISTER, J. R. 1993. Particle-driven gravity currents. *Journal of Fluid Mechanics*, **250**, 339–369.

BOUMA, A. H. 2000. Coarse-grained and fine-grained turbidite systems as end-member models: applicability and dangers. *Marine and Petroleum Geology*, **17**, 137–143.

BUCKINGHAM, E. 1915. Model experiments and the forms of empirical equations. *Transactions of the American Society of Mechanical Engineers*, **37**, 263–292.

DADE, W. B. & HUPPERT, H. E. 1995. Runout and fine-sediment deposits of axisymmetrical turbidity currents. *Journal of Geophysical Research*, **100**, 18597–18609.

DAVID, F. W. & NOELLE, H. 1982. *Experimental Modelling in Engineering*. Butterworths, London.

DAVIDSON, J. F., HARRISON, D. & CARVALHO, G. D. 1977. On the liquid like behaviour of fluidised beds. *Annual Review of Fluid Mechanics*, **9**, 55–86.

DIETRICH, W. E. 1982. Settling velocities of natural particles. *Water Resources Research*, **18**, 1615–1626.

GLADSTONE, C., PHILLIPS, J. C. & SPARKS, R. S. J. 1998. Experiments on bidisperse, constant-volume gravity currents: propagation and sediment deposition. *Sedimentology*, **45**, 833–843.

HAUGHTON, P. 1994. Deposits of deflected and ponded turbidity currents, Sorbas basin, Southeast Spain. *Journal of Sedimentary Research*, **64**, 233–246.

HUPPERT, H. E. 1980. The slumping of gravity currents. *Journal of Fluid Mechanics*, **99**, 785–799.

HUPPERT, H. E. 1998. Quantitative modelling of granular suspension flows. *Philosophical Transactions of the Royal Society of London Series A*, **356**, 2471–2496.

HURST, A., VERSTRALEN, I., CRONIN, B. & HARTLEY, A. 1999. Sand-rich fairways in deep-water clastic reservoirs: genetic units, capturing uncertainty and a new approach to reservoir modeling. *American Association of Petroleum Geologists Bulletin*, **83**, 1096–1118.

IMRAN, J. & PARKER, G. 1999. On the role of mud in keeping sand in suspension in a turbidity current. *American Association of Petroleum Geologists 1999 Annual Meeting, San Antonio, Expanded Abstracts*, 66.

KEULEGAN, G. H. 1957. Twelfth progress report on model laws for density currents: An experimental study of the motion of saline water from locks into fresh water channels. *US National Bureau of Standards Reports*, **5168**, 21.

KNELLER, B. 1995. *Beyond the Turbidite Paradigm: Physical Models for Deposition of Turbidites and Their Implications for Reservoir Prediction.* Geological Society, London, Special Publications, **94**, 31–49.

KNELLER, B. & MCCAFFREY, W. 1999. Depositional effects of flow nonuniformity and stratification within turbidity currents approaching a bounding slope: deflection, reflection and facies variation. *Journal of Sedimentary Research*, **69**, 980–991.

KNELLER, B. C., BENNETT, S. J. & MCCAFFREY, W. D. 1997. Velocity and turbulence structure of density currents and internal solitary waves: potential sediment transport and the formation of wave ripples in deep water. *Sedimentary Geology*, **112** (3–4), 235–250.

KNELLER, B. C., BENNETT, S. J. & MCCAFFREY, W. D. 1999. Velocity structure, turbulence and fluid stresses in experimental turbulent density currents. *Journal of Geophysical Research*, **104**, 5381–5391.

KNELLER, B. C., EDWARDS, D., MCCAFFREY, W. & MOORE, R. 1991. Oblique reflection of turbidity currents. *Geology*, **19**, 250–252.

LAVAL, A., CREMER, M., BEGHIN, P. & RAVENNE, C. 1988. Density surges: two-dimensional experiments. *Sedimentology*, **35**, 73–84.

LEEDER, M. R. 1982. *Sedimentology*. George Allen & Unwin, London.

LOWE, D. R. 1982. Sediment gravity flows: II. Depositional models with special reference to the deposits of high-density turbidity currents. *Journal of Sedimentary Petrology*, **52**, 279–297.

MCCAFFREY, W. & KNELLER, B. 2001. Process controls on the development of stratigraphic trap potential on the margins of confined turbidite systems and aids to reservoir evaluation. *American Association of Petroleum Geologists Bulletin*, **85**, 971–988.

MIDDLETON, G. V. 1966. Experiments on density and turbidity currents: [Part] 1, Motion of the head. *Canadian Journal of Earth Sciences*, **3**, 523–546.

MIDDLETON, G. V. & HAMPTON, M. A. 1976. Subaqueous sediment transport and deposition by sediment gravity flows. *In*: STANLEY, D. J. & SWIFT, D. J. P. (eds) *Marine Sediment Transport and Environmental Management*. Wiley, New York, 197–218.

MIDDLETON, G. V. & NEAL, W. J. 1989. Experiments on the thickness of beds deposited by turbidity currents. *Journal of Sedimentary Petrology*, **59**, 297–307.

MIDDLETON, G. V. & SOUTHARD, J. B. 1984. *Mechanics of Sediment Movement*. Society of Economic Palaeontologists and Mineralogists, Short Courses, **3**.

MUTTI, E. 1979. Turbidites et cones sous-marins profonds. *In*: HOMEWOOD, P. (ed.) *Sedimentation detritique (fluvial, littoral et marine)*. Institut de Géologie, Universite de Fribourg, Suisse, 353–419.

MUTTI, E. 1992. *Turbidite Sandstones*. Agip, San Donato Milanese, 275.

MUTTI, E. & NORMARK, W. R. 1987. Comparing examples of modern and ancient turbidite systems; problems and concepts. *In*: LEGGETT, J. K. & ZUFFA, G. G. (eds). *Marine Clastic Sedimentology; Concepts and Case Studies*. Graham and Trotman, London, 1–38.

NILSEN, T. H., IMPERATO, D. P. & MOORE, D. W. 1994. Reservoir geometry and architecture of productive upper Cretaceous mud-rich and sand-rich submarine-fan systems, Sacramento Basin, California. *In*: WEIMER, P., BOUMA, A. H. & PERKINS, B. F. (eds) *Submarine Fans and Turbidite Systems*. GCSSEPM Foundation, Houston, 269–280.

NORMARK, W. R. 1978. Fan valleys, channels, and depositional lobes on modern submarine fans; characters for recognition of sandy turbidite environments. *American Association of Petroleum Geologists Bulletin*, **62**, 912–931.

NORMARK, W. R. & PIPER, D. J. W. 1991. Initiation processes and flow evolution of turbidity currents; implications for the depositional record. *In*: OSBORNE, R. H. (ed.) *From Shoreline To Abyss; Contributions in Marine Geology in Honor of Francis Parker Shepard*. Society of Economic Paleontologists and Mineralogists, Tulsa, Special Publications, **46**, 207–230.

PARKER, G., FUKUSHIMA, Y. & PANTIN, H. M. 1985. Self accelerating turbidity currents. *Journal of Fluid Mechanics*, **171**, 145–181.

PARSONS, J. D. & GARCÍA, M. H. 1998. Similarity of gravity current fronts. *Physics of Fluids*, **10**, 3209–3213.

PEAKALL, J., ASHWORTH, P. & BEST, J. L. 1996. Physical modelling in fluvial geomorphology: principles, applications and unresolved issues. *In*: RHOADS, B. L. & THRON, C. E. (eds) *The Scientific Nature of Geomorphology: Proceedings of the 27th Binghamton Symposium in Geomorphology*, 221–253.

READING, H. G. & RICHARDS, M. 1994. Turbidite systems in deep-water basin margins classified by grain size and feeder system. *American Association of Petroleum Geologists Bulletin*, **78**, 792–822.

SIMPSON, J. E. 1982. Gravity currents in the laboratory, atmosphere and ocean. *Annual Review of Fluid Mechanics*, **14**, 213–234.

SIMPSON, J. E. 1997. *Gravity Currents in the Environment and the Laboratory*, 2nd edn. Cambridge University Press, Cambridge.

Present morphology and depositional architecture of a sandy confined submarine system: the Golo turbidite system (eastern margin of Corsica)

ANNE GERVAIS[1], BRUNO SAVOYE[2], DAVID J. W. PIPER[3], THIERRY MULDER[1], MICHEL CREMER[1] & LAETITIA PICHEVIN[1]

[1] *Université Bordeaux I, Département de Géologie et Océanographie, UMR 5805 EPOC, 33405 Talence Cedex, France*
[2] *IFREMER, DRO/GM, Laboratoire Environnements Sédimentaires, BP70, 29280 Plouzané Cedex, France*
[3] *Geological Survey of Canada (Atlantic), Bedford Institute of Oceanography, PO Box 1006, Dartmouth, Nova Scotia, B2Y 4A2 Canada*

Abstract: The modern sandy Golo turbidite system (500 km^2) is located in a confined basin on the eastern margin of Corsica. The Golo turbidite system is fed by a single river, which supplies coarse sand derived from active weathering of the neighbouring mountains. The late Quaternary deposits have been imaged using a closely spaced grid of 1000 km of sparker seismic-reflection profiles (line spacing close to 1.6 km, vertical resolution of 2 m). The turbidite system is composed of four non-coalescent fans that were at times active simultaneously and of two small deposits onto the slope. The resulting sedimentation pattern is characterized by stacked turbidite deposits. At a regional scale, there is a continuum of fan morphologies and geometries from south to north. The use of both seismic and sedimentary facies, together with mapped seismic geometry of sedimentary bodies, allowed definition of four architectural elements: (1) submarine valley (canyon and gully), (2) sandy channel, (3) muddy levee, and (4) sandy lobe. Some of these architectural elements can be recognized at a scale that is comparable to outcrop examples. Features such as progressive lateral migration and avulsion, or complex longitudinal evolution (progradation and retrogradation), can also be accurately described. Despite the active tectonics along the studied margin, the main variations in sedimentation appear to be controlled by eustatic changes, pre-existing seafloor topography, and sediment source characteristics. The general pattern of sedimentation is controlled by the influence of a confining slope, leading to the predominance of aggradation and to specific morphology and architecture of sedimentary bodies.

Deepwater sand bodies, both modern and ancient, have interested geoscientists and petroleum geologists for a long time. Such sands are potential reservoirs for oil and gas, and consequently have a large economic interest (Shanmugam & Moiola 1988). However, among numerous articles dealing with modern deep-sea systems (see review by Stow & Mayall 2000), relatively few are relevant to smaller turbidite systems in which channel and lobe sands are volumetrically important. The nature, variety and geometry of such sandy systems have been poorly documented, and consequently they remain an interesting problem in deepwater sedimentation (Kneller 1995).

The understanding of turbidite systems still depends on the resolution of the survey techniques (Piper *et al.* 1999). At present, the processes that transport and deposit sand in deepwater systems are disputed (Shanmugam 2000), and it is uncertain whether or not individual seismically recognized lobe deposits represent single beds (Piper *et al.* 1999). The relative importance of hyperpycnal flow from rivers is unknown (Piper & Normark 2001), and the general architectural evolution of sandy systems is described for only a few case studies. Research on modern deep-sea systems using high-resolution bathymetry, seismic reflection and sediment cores is now able to provide information about reservoir characteristics at a scale and order of magnitude comparable to that of industry three-dimensional seismic in the near-seafloor systems and superior in the deeper subsurface.

The present study is focused on a small sandy turbidite system that is located in a small, narrow and shallow basin along the eastern Corsican coast (Mediterranean Sea). The large dataset available on this system (multibeam echosounder data and high-resolution seismic-reflection profiles) has been analysed in order to address two main topics: (1) the overall morphology of the turbidite system; (2) the geometry and architecture of the sedimentary bodies.

From: LOMAS, S. A. & JOSEPH, P. (eds) 2004. *Confined Turbidite Systems*. Geological Society, London, Special Publications, **222**, 59–89. 0305-8719/04/$15.00 © The Geological Society of London 2004.

Regional setting

Previous studies

Segre & Debrazzi (1960), Gabin (1972) and Stanley *et al.* (1980) carried out the first studies on the eastern margin of Corsica (Mediterranean Sea) in terms of the bathymetric and general processes occurring in the Corsican trough. Bellaiche *et al.* (1993) regionally mapped the eastern margin of Corsica and discovered a series of small sandy turbidite fans along the whole margin (Fig. 1). After this initial work, IFREMER ran several cruises over the area using a large range of instruments (Corstage in 1997 and 1998, Corfan 1 in 1998, and Cork in 1998). The eastern margin of Corsica was selected because of the sandy lithology of the turbidite systems (their small size allowing detailed analysis of the whole system from canyon head to distal lobe), and the shallow water depth, which permits the use of surface-towed high-resolution and very-high-resolution seismic systems. The present study is focused on the largest turbidite system of the margin: the Golo system.

General morphology

In the area of the Golo system, the continental slope deepens eastwards until it merges with the N-S trough named the Corsican Trough or 'Canal de Corse' at a water depth of 700-900 m (Fig. 1). The Corsican Trough actively trapped sediments during the Tertiary (Mauffret *et al.* 1999). Active sedimentation from the Pliocene to the Quaternary is demonstrated by the large thickness (1000-1500 m) of unconsolidated deposits in this area (Viaris de Lesegno 1978) overlying Messinian evaporites or an erosional surface. Stanley *et al.* (1980) showed that this trough is asymmetrical, with major progradation on the western Corsican flank and gradual uplift of the eastern flank, the Pianosa Ridge (Figs 1 & 2). The slope of the Corsican trough deepens southward (Stanley *et al.* 1980; Fig. 1). Sills isolate small basins such as the 850-900 m deep Golo Basin in the south of the study area (Fig. 1).

Cenozoic geological history: tectonic control

The western Mediterranean Sea has been interpreted as a marginal oceanic basin created by the north-northwest subduction of the Africa-Apulian Plates beneath the European Plate (Auzende *et al.* 1973). Tectonic and sedimentary evidence of the Oligo-Miocene rifting episode, which resulted in stretching and thinning of the continental margin and in the creation of the Pianosa ridge, is well preserved on margins and on land. The western Corsica margin is

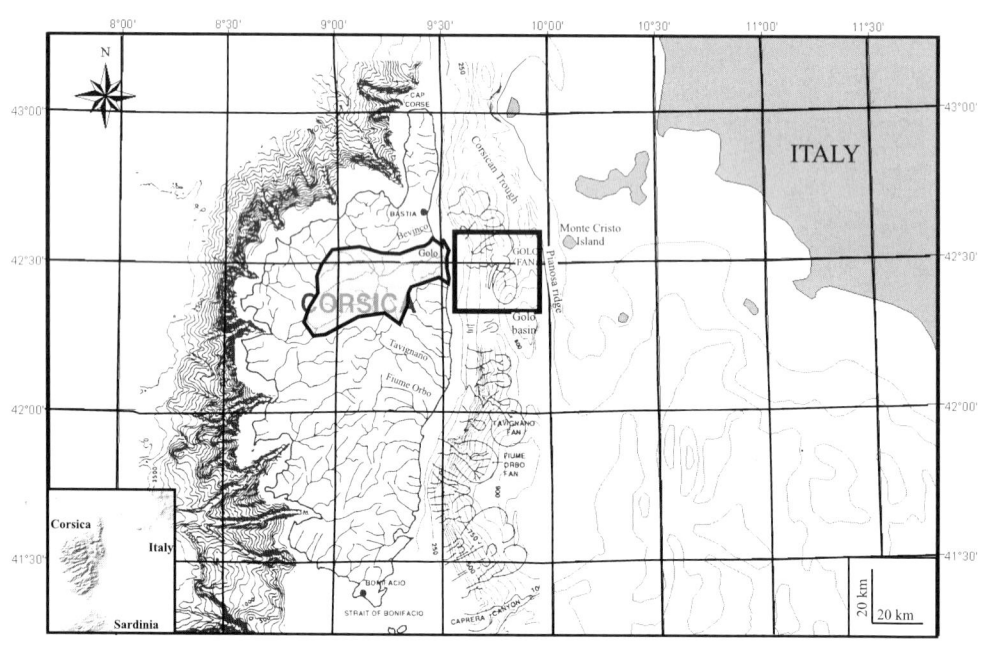

Fig. 1. General location map of the East Corsica margin showing the series of fans described by Bellaiche *et al.* (1993), the drainage basin of the Golo river, and the location of the study area. Contours are in metres.

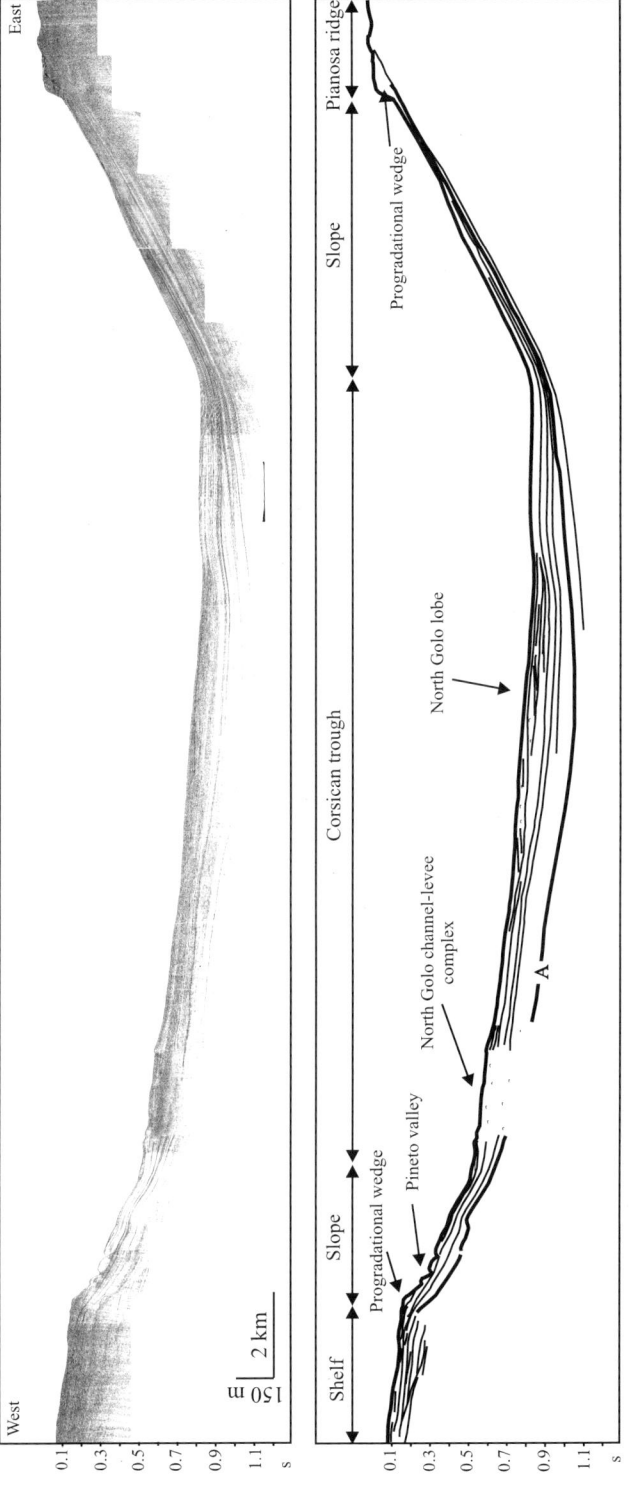

Fig. 2. Structural cross-section of the East Corsica margin (profile 24; see Fig. 3 for location) and location of the base of the study (i.e. maximum seismic penetration corresponding to the reflector 'A'). Depth is in seconds two-way travel time (TWTT) below sea-level.

characterized by numerous deep canyons developed on steep slopes without a continental shelf, whereas the eastern Corsica margin shows a moderately developed continental shelf with shallower canyons on more gentle slopes (Fig. 1). The eastern margin of Corsica is seismically active, whereas the western margin is inactive (Ferrandini et al. 1994). Such seismicity could trigger sediment failures on the continental slope and in canyon heads, but this has yet to be demonstrated.

On the eastern margin the Corsican trough is a half graben, with the master fault along the east coast of Corsica largely obscured by delta progradation. Uplift of the footwall of the east Corsican fault is still active and has led to the formation of high relief of eastern Corsica (up to 2800 m at Monte Cinto). The basinal configuration east of the Golo river is due to tectonic control, and the uplift of the adjacent alpine Corsica and Pianosa ridge (Mauffret et al. 1999). Uplift and correlative rejuvenation of graben-bounding faults have enhanced erosion and are suggested to explain several observations within the Golo drainage area (Ottman 1958; Conchon 1975; Rehault et al. 1984).

Hydrologic and hydrodynamic context

Previous studies carried out on this half-graben faulted margin suggest that sedimentation is sensitive to climatic-eustatic variations.

The rivers in eastern Corsica flow down the fault-line scarp and are short with very high gradients. The resulting high sediment supply has created many kilometres of deltaic progradation through the Pliocene and Quaternary. The granitic hinterland (Mulder & Maneux 1999) and widespread Miocene sandstones (Bellaiche et al. 1994) result in abundant sand, which is distributed on beaches, delta plains, and in large Quaternary alluvial cones along the coast (Ottman 1958; Conchon 1977).

The Golo river is the largest drainage basin of the island, with an area of about 1080 km^2 and a maximum altitude of about 2700 m. During the Quaternary, the Golo river was very active and built a large alluvial delta (Ottman 1958). Several recent studies show that, except in the coastal plain, the Golo river has a highly erosional profile (Mulder & Maneux 1999). At the present time, concentrations of suspended matter measured in the Golo river are low. This results from the effects of human activities on land lowering the concentration of suspended matter through the last century (Mulder & Maneux 1999). Without this effect, floods in the river could have been higher (Gauthier & Prone 1980; Mulder & Maneux 1999) and sufficient to produce hyperpycnal turbidity currents due to the existence of alluvial deposits and easily erodable sedimentary formations along the river bed (Mulder & Maneux 1999). Extreme rainfall and snow melt events are described during the last century for the Golo river (Conchon 1975; Conchon 1984; Mulder & Maneux 1999), which may have led to increased supply of sediments to the sea. Large river floods could have moved boulders of more than 1 m in size (Conchon & Gauthier 1985). During periods of major discharge of sediments or during storms, the narrowness of the continental shelf will favour supply into canyon heads. The present shoreline is microtidal (<100 cm mean tidal range) and wave fetch from the east is broken by Monte Cristo island only 40 km from the Corsican coast. Some sediment accumulations could be of littoral drift origin (Orszag-Sperber & Pilot 1976; Conchon 1999). North-flowing littoral currents have produced, for example, five sandy shoals on the continental shelf fed by the Tavignano river (Gauthier & Prone 1980). Thus direct river sedimentation, shelf storms and longshore drift may all play a role in filling submarine canyon heads with sediment. Based on seismic profile, the effect of bottom currents in the northern Strait of Corsica trough has been inferred (Marani et al. 1993), but no bottom current activity has been observed in the 'Canal de Corse' itself.

According to Gauthier (1981) and to Mulder & Maneux (1999), the seaward transport of sediment is more important during regressive periods. Based on pollen analysis, the climate was more wet, and humid during glacial periods (Ottman 1958). Evidences of previous glaciations can be seen in the eastern Golo valley, with moraines of Younger Dryas age at 1700 m altitude and of Younger Würm at 1100 m altitude (Conchon 1975). Based on continental fossils, during interglacial periods climate was similar to present conditions and vegetation was more abundant leading to a reduction in mechanical erosion (Conchon 1975).

The exact origin of marine sediments is difficult to determine petrographically because of:

(1) the similarity between the alluvial deposits of rivers (Conchon 1975; Gauthier & Prone 1980; Mulder & Maneux 1999);
(2) the change of river courses during the Quaternary (Conchon 1975); and
(3) redistribution of sediments on the shelf by littoral drift (Gauthier 1981).

Data and methods

The bathymetry (Fig. 3) was collected using a multibeam echosounder (SIMRAD EM12) during the MESIM survey (Bellaiche et al. 1993). No usable backscatter data were obtained during this cruise because of the high speed at which the survey was run. The bathymetric map (Fig. 3) and three-dimensional view (Fig. 4) have been made using Caraibe software (IFREMER). About 30 high-resolution seismic-reflection lines parallel or perpendicular to the coast were collected during recent IFREMER cruises (Fig. 3) using a 300J multitip sparker, in the 200–800 Hz frequency range, giving vertical resolution around 2–3 m. Line spacing is about 1.6 km. Seismic-reflection data were located by DGPS (resolution around 2 m). This survey provided more than 1000 km of profiles over an area of about 500 km^2 extending from north of Pianosa Island (about 42°40'N lat.) to south of the Golo basin (about 42°20'N lat.): that is, covering the whole Golo system.

Seismic profiles were processed with SITHERE software (IFREMER). Each profile has been interpreted in terms of seismic facies according to the methodology of Nely et al. (1985). Sparker resolution allows detailed study of the upper 200 m of sedimentation, the base being marked by the 'A' reflector in Figure 2. Regional key reflectors were distinguished according to major changes in acoustic facies or to local unconformities on each seismic profile. Each interval has then been mapped in terms of palaeofacies based on lithoseismic correlation and in terms of thickness of deposits: that is, isopach maps using SURFER software.

Morphology of the Golo margin

Offshore from the Golo river, the *continental shelf* reaches its maximum width and forms a well-developed bulge (Figs 3 & 4). The shelf width varies from 9 km (in the southern part of the study area) to a maximum of 12 km (in the

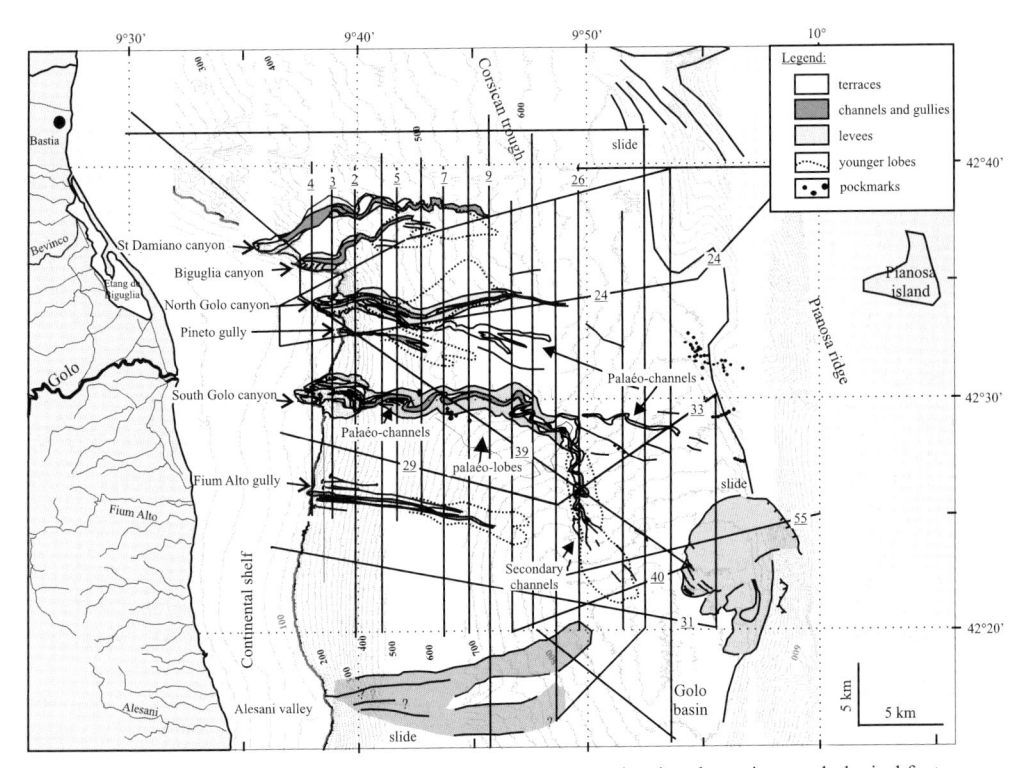

Fig. 3. Morpho-sedimentary map of the Golo depositional system showing the main morphological features and sparker seismic lines locations. Contours of the bathymetric map are in metres. White areas correspond to areas with no data available. Contours on the shelf are 'Service Hydrographique et Océanographique de la Marine' data. Modern lobes characterize stage K–L defined in Figure 11. Underlined letters correspond to seismic profiles illustrated in this paper. Locations of Stanley et al. (1980) cores (55, 56 and 57) are indicated.

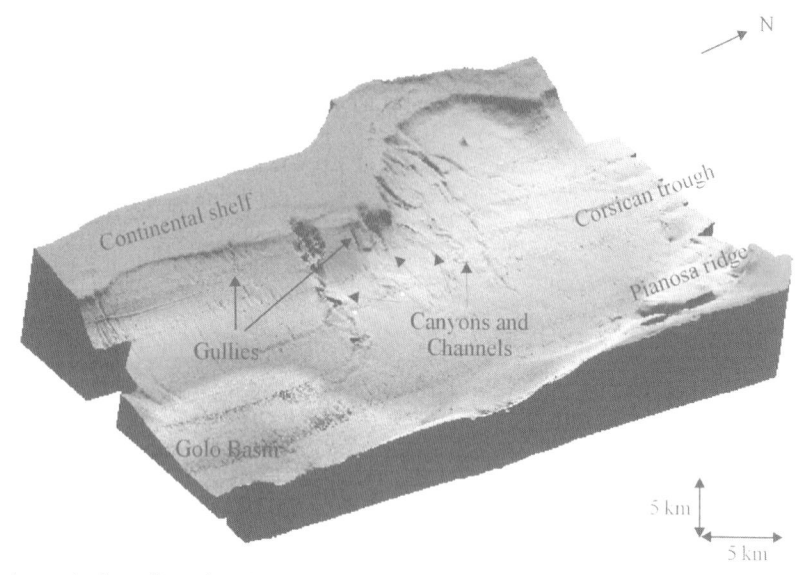

Fig. 4. Bathymetric three-dimensional view of the Golo depositional system.

northern part of the area). The shelf break is located at a water depth of about 110 m. In the south of the margin, a slide from the edge of the continental shelf is observed (Fig. 3).

The shelf is incised by several *submarine valleys* (Fig. 3). Only six of these valleys, composing the Golo system, have been recorded in detail on seismic-reflection data. From south to north these are Alesani, Fium'Alto, South Golo, Pineto, North Golo, Biguglia and St Damiano (Fig. 3). Each valley has been named using the topographic map of Corsica. Most of the valley names correspond to the river directly onshore. *Two types* are observed: large deep canyons (South Golo, North Golo, St Damiano and Biguglia) that deeply incised the shelf, and small narrow gullies (Pineto and Fium Alto) that slightly incised the shelf break (Fig. 3). The two best-developed canyon and channel-levee-lobe complexes that are directly connected to the Golo river have been named North Golo and South Golo respectively. Numerous small gullies are observed along the whole margin. Two of these gullies, the Pineto and the Fium Alto gullies, are also located close to the Golo river and thus have been studied in detail. In the sector of the Golo system, the *slope* deepens eastwards with an average gradient of 1:30 (2°) (profile 39; Fig. 5).

The *South Golo canyon* (Fig. 3) exhibits a V-shaped sinuous trough (sinuosity of 0.8) (profile 3 in Fig. 5). It is 5 km long. It is the only canyon showing a well-developed canyon head. Canyon depth rapidly increases from 100 to 170 m less than 2 km downslope, associated with an increase of width (from 1500 to 3500 m). Downslope (450 m water depth), width decreases. Terraces are observed (profiles 3 and 2 in Fig. 5). The upstream section of the canyon is characterized by furrows oriented parallel to the canyon (E–W; Fig. 3). The *North Golo, Biguglia* and *St Damiano canyons* are relatively straight (sinuosity of 1) and V-shaped (Fig. 3 and profile 3 in Fig. 5). The Biguglia and St Damiano canyons parallel the canyons to the south, and are quite oblique to the local slope, perhaps as a result of tectonic control (Fig. 3). The mean depth of the North Golo canyon is about 120 m, of the Biguglia canyon about 70 m, and of the St Damiano canyon less than 40 m. The North Golo, Biguglia and St Damiano canyons are respectively 3 km, 4 km and 2 km long, and 1000 m wide (Fig. 3 and profile 3 in Fig. 5). Terraces are observed along the steep walls of the canyons, but are less pronounced and numerous than those observed in the South Golo canyon. Transverse furrows with N–S trends are observed (Fig. 3). They indicate lateral supply of sediment from the shelf.

The change from canyons to *channel–levee complexes*, defined by the onset of levee deposition, occurs at different water depths. The *South Golo channel* extends from 450 to 700 m water depth (Fig. 3). The channel is 1.5 km wide and 19 km long. It shows an E–W trend

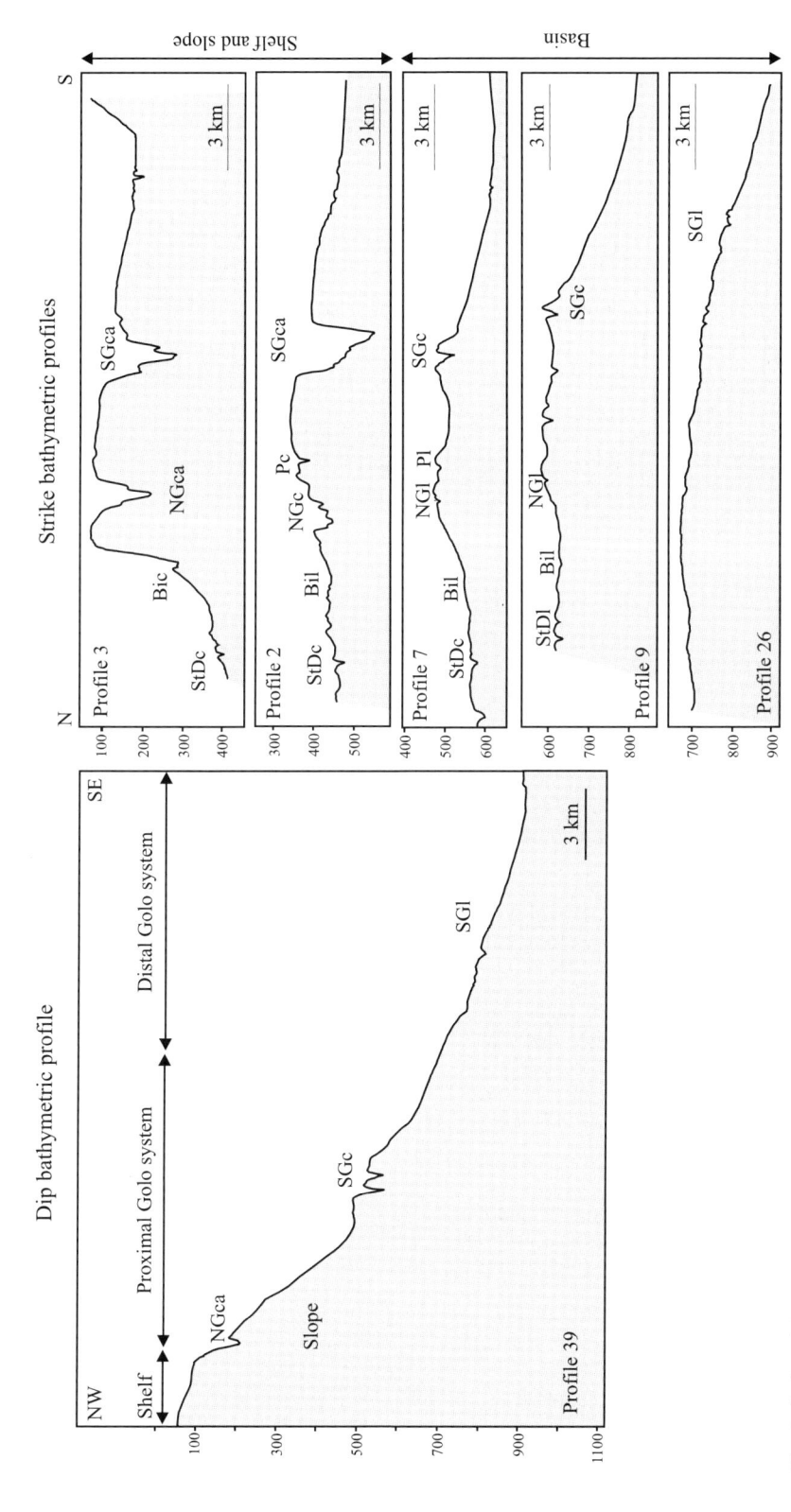

Fig. 5. Series of dip and strike bathymetric profiles across the Golo depositional system: SG, South Golo; P, Pineto; NG, North Golo; Bi, Biguglia; StD, St Damiano; ca, canyon; c, channel; l, lobe. The location of these profiles is indicated in Figure 3. All these profiles are N–S oriented except the NW–SE profile 39. Depths are in metres below sea-level.

before 42°30'N of latitude and then turns suddenly N–S following the 9°50'E longitude southward (Fig. 3). The number of meanders and terraces increases basinward separating a relatively low sinous (0.7) upstream channel from a sinuous (0.5) downstream channel with terraces (Fig. 3). Slope, measured in the talweg of the channel, is strong, and varies from 0.85° in the upstream section of the channel to 0.68° in the downstream section of the channel. It is bounded by lateral asymmetrical levees. Upstream, levees are greater than 55 m high and 1 km wide (Fig. 3 and profile 7 in Fig. 5). Basinward, the channel becomes narrower, until it can be recognized only by the levee bulge (Fig. 3 and profile 9 in Fig. 5). The levees become less prominent.

The *North Golo canyon* rapidly develops a large channel–levee complex from 250 to 400 m water depth (profile 2 in Fig. 5). The North Golo channel is less sinuous than the South Golo channel (0.8). Slope is around 1°. It shows well-developed levees (about 35 m high; profile 2 in Fig. 5). The lengths of the *St Damiano* and *Biguglia channels* are 9 km and 5 km respectively, and they are located between 200 and 500 m. These channels are less sinuous and steeper in their upstream section than the South Golo channel (sinuosity of 0.8 and slope of 1.4°). The levees become rapidly less prominent basinward. Maximum height is observed near the shelf break (35 m high; profile 3 in Fig. 5). Less than 1 km downstream, no levees can be seen (profile 2 in Fig. 5).

Downslope from the major channels, the channel–levee complexes pass into *channel-termination lobes* (Fig. 3). The South Golo channel, down to the 800 m isobath, passes into a prominent lobe on the northern edge of the Golo basin (Fig. 3 and profile 26 in Fig. 5). Slope remains strong (0.68°). The channel then develops secondary channels (Fig. 3). Secondary channels are shallow (10 m maximum depth) and about 100 m wide (profile 26 in Fig. 5).

The *Pineto* and *Fium Alto gullies* are small, narrow, straight (sinuosity of 0.9), and developed from the shelf break to the middle or the base of the slope (from 120 to 300 m isobath; Fig. 3). Length is about 1 km, and depth varies from 50 to 10 m basinward. Slope is stronger than those observed in canyons and channels (from 4.3 to 3.4°). No levees are visible (profile 2 in Fig. 5). Associated gullies onto the slope are observed closed to the Fium Alto gully (Fig. 3).

The *Pianosa ridge* is characterized by a steep slope (6°; Fig. 2). Two slumps and numerous pockmarks are observed on the Pianosa slope (Fig. 3).

Seismic facies analysis and architectural elements

Seismic-reflection facies

The seismic-character mapping, the seismic facies and geometry, and the interpretation integrate multi-beam echosounder data (bathymetry). In the study area, five main types of seismic-facies have been recognized (Fig. 6):

(1) continuous bedded reflective facies;
(2) continuous bedded more transparent facies;
(3) chaotic facies;
(4) hyperbolic facies;
(5) transparent facies.

Continuous bedded facies is characterized by numerous parallel reflections or by convergent reflections. Continuous bedded more transparent facies is characterized by numerous low-amplitude parallel reflections. Chaotic facies is characterized by short reflection segments of variable dip and generally high amplitude.

Internal geometry of sedimentary bodies

Seismic data show the limits of sedimentary bodies and characterize their facies and geometries. Recognition of different bodies on sparker seismic data is aided by easy recognition of sub-bottom levee facies with divergent reflections, chaotic facies, or inclination of reflectors due to channel shape. Canyons are erosive whereas channels exhibit constructional character. Lateral bulges are made of parallel continuous divergent reflectors interpreted as levee deposits. Lenticular, bulge structures at the termination of channels are composed of chaotic facies and are interpreted as sand prone lobe deposits (Stanley *et al.* 1980; Piper *et al.* 1999).

Onshore from the Golo turbidite system, the *continental shelf* is characterized by bedded continuous facies with parallel reflections of high amplitude and frequency (Fig. 7a).

Filled incised valleys are located on the continental shelf upslope of canyons, except the area upslope from the Pineto and the Fium Alto gullies (Fig. 7b). They correspond to a wide valley (minimum width of 500 m) that incises the continental shelf to a depth of at least 35 m. In these valleys, numerous reflectors of high amplitude define several successive palaeo-valleys and suggest progressive infill by deltaic sediments.

The *shelf break* zone shows well-developed *progradational wedges* (Fig. 7c). They are composed of seaward-dipping progradational

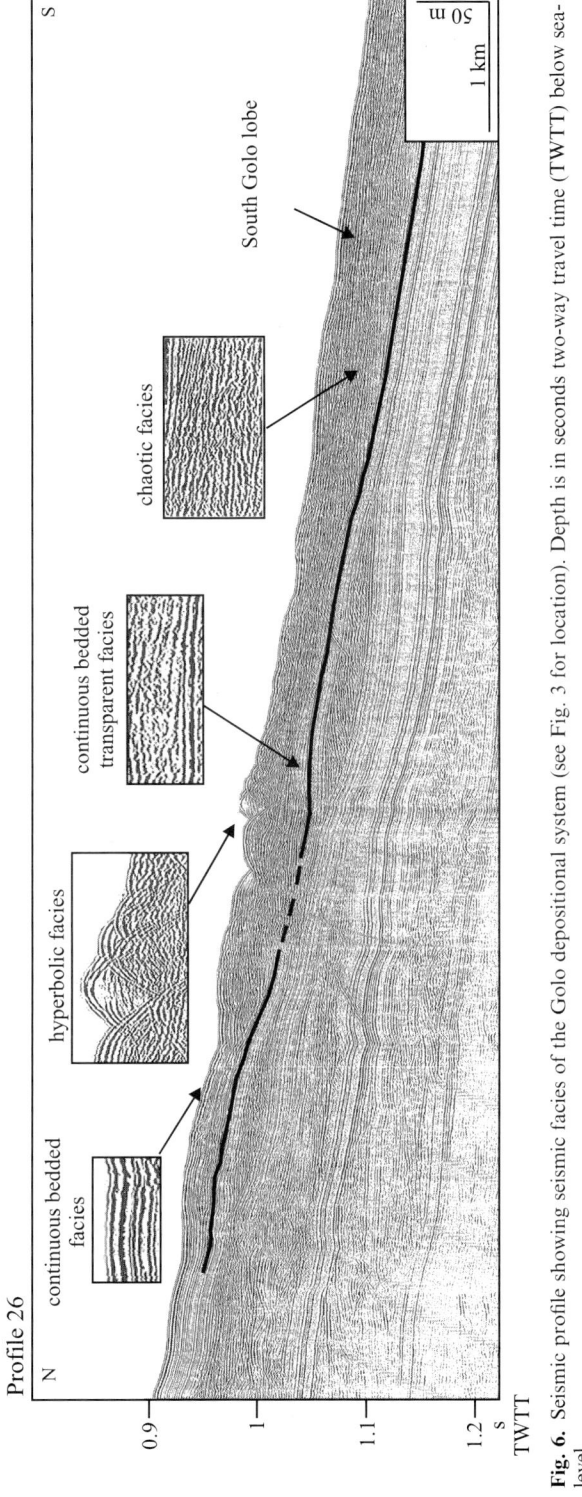

Fig. 6. Seismic profile showing seismic facies of the Golo depositional system (see Fig. 3 for location). Depth is in seconds two-way travel time (TWTT) below sea-level.

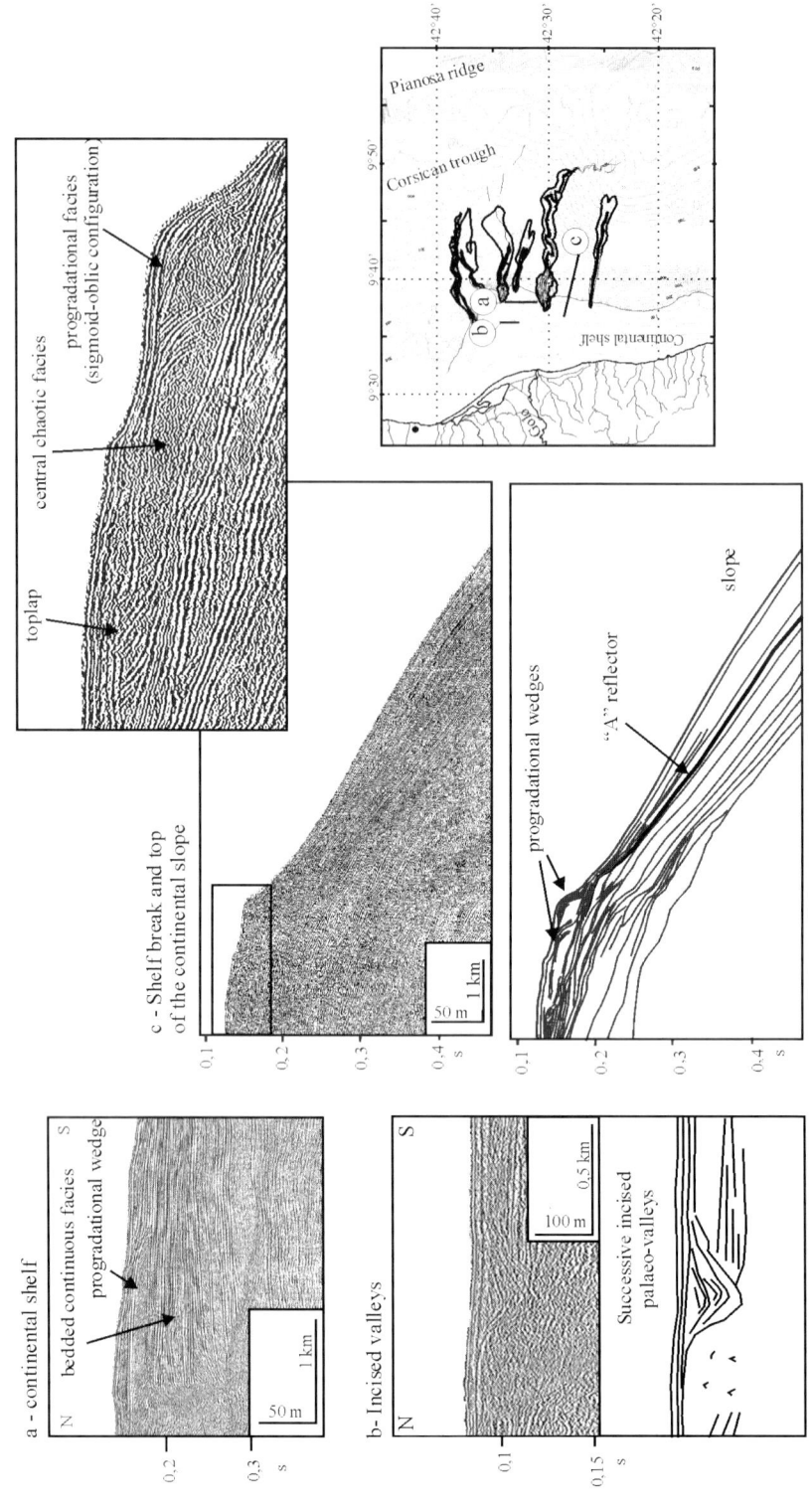

Fig. 7. Selected parts of seismic profiles showing the continental shelf, incised valleys and continental shelf break (profiles 4, 24 and 29). Depth is in seconds two-way travel time (TWTT) below sea-level. The 'A' reflector is defined in Figure 11.

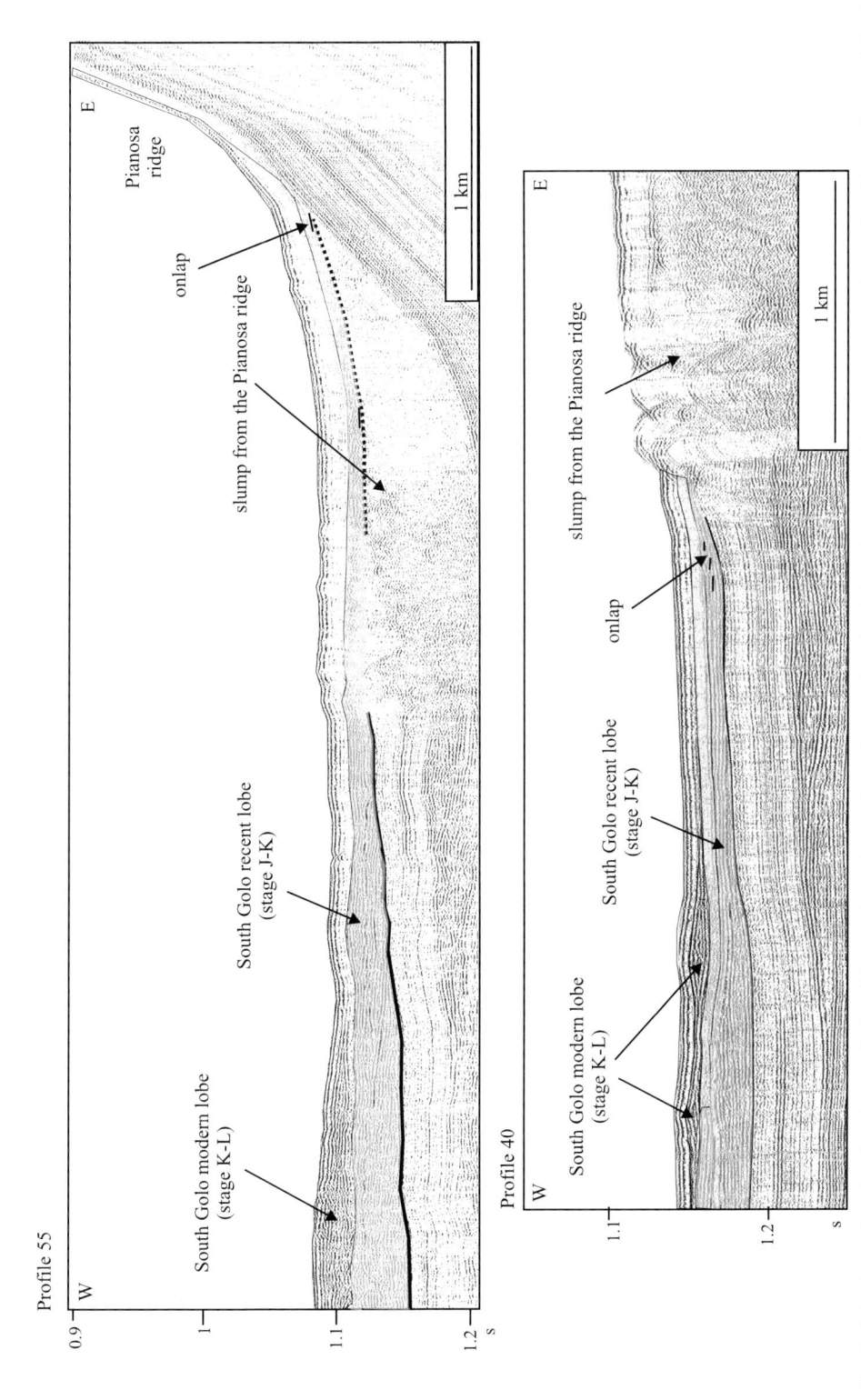

Fig. 8. Seismic profiles showing characteristics of the southern slump of the Pianosa slope (see Fig. 3 for location). Letters correspond to regional key reflectors defined in Figure 11. Depth is in seconds two-way travel time (TWTT) below sea-level.

clinoforms. The lower boundary is a downlap surface, the upper boundary is a toplap surface, and internal downlap surfaces are common. Seaward-dipping layers converge toward the basin, and generally display a transparent to weak acoustic response. Two seismic-reflection internal structures are differentiated: (1) a lower and upper progradational facies with a sigmoid-oblique configuration, becoming transparent in a seaward direction, and (2) a central chaotic facies (Fig. 7c).

Areas of the slope between canyons/channels are characterized by numerous onlap surfaces (Fig. 7c). Continuous bedded reflectors diverge seaward from 20 ms (16 m) on the shelf break to more than 200 ms (160 m) in the basin. The base of the interval studied by seismic-reflection is marked by the 'A' reflector, and it is located just under the first major progradational wedge on the shelf (Fig. 7c). East of the studied area, continuous bedded reflectors converge toward the Pianosa slope. The 'A' reflector is located just beneath the outer shelf progradational wedge on the Pianosa ridge (Fig. 2). Slumps of the continental slope and of the Pianosa slope (Fig. 3) are characterized by transparent facies (Fig. 8).

Canyons are characterized by chaotic facies on their floor (Fig. 9). Parallel sub-bottom reflectors are truncated in the vicinity of canyons. Canyon

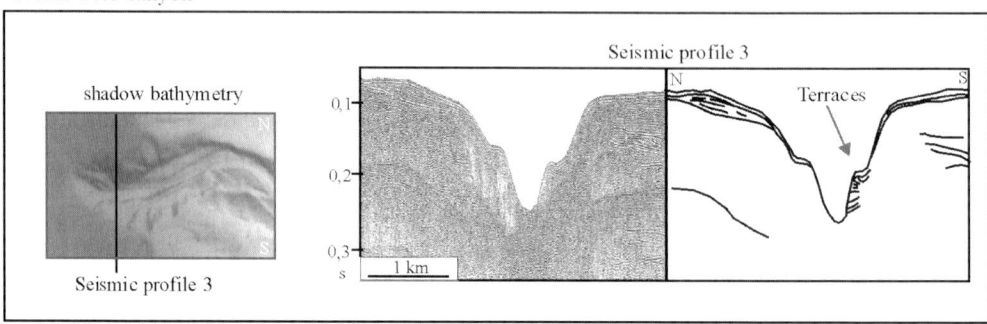

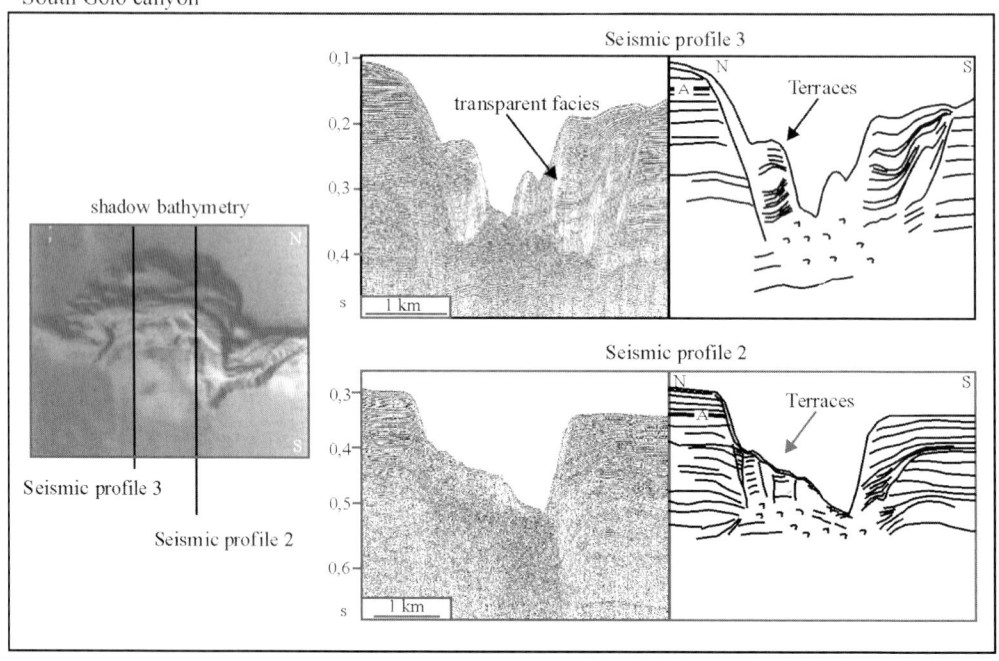

Fig. 9. Selected part of seismic profiles 2 and 3 (see Fig. 3 for location) showing characteristics and geometries of the North and South Golo canyons. Depth is in seconds two-way travel time (TWTT) below sea-level.

walls are characterized by deep erosion. The location of canyons is relatively stable over time. Indications of faults are visible close to the canyons (Fig. 9). Terraces are visible along their steep flanks (Fig. 9). They can be characterized by continuous bedded parallel or convergent reflections. Reflectors cannot be traced from intercanyon areas to terraces.

In the Golo system, two types of *channel–levee complex* are observed: (1) a major well-developed deep channel with well-developed levees (Figs 10b & d), and (2) a shallow channel with no or low levees (Fig. 10e).

The first type is characterized by chaotic facies at the channel floor. Their shapes are difficult to define because of the vertical stacking pattern of channel–levee complexes (presence of the chaotic facies over several stages of deposition; Figs 10d, 10g & 11d). Levees are well developed. Erosion and aggradation characterize the floor and the flanks of the channels respectively. High-amplitude reflection packets (HARPs) are observed at the base of some channel–levee complex floors (Fig. 10). Terraces can be observed on channel flanks (Figs 10d & g). Based on boomer profiles, most of these terraces are characterized by continuous divergent bedded facies, and are interpreted as confined levees (Pichevin 2000).

The second type is characterized by continuous bedded facies with numerous parallel sub-bottom reflections and commonly occur on channel floors and flanks (Figs 10e, 11a & c). Some channels are filled by chaotic facies due to lobe deposition above them (Figs 10a & 11a). Some channels correspond to an earlier erosional channel that became inactive and was passively filled by continuous bedded facies (Figs 10e, 11b & 12a).

Chanel-lobe transition is characterized by chaotic facies and small secondary channels observed on bathymetric map and seismic profile (Figs 3 & 12a). Active, not filled small secondary channels (type 2; Figs 11c & 12a) may be present at the top. Filled secondary channels can be observed at the top (Figs 11a & b) or at the base (Fig. 12a).

Lobes are characterized by chaotic seismic facies and continuous bedded facies with downlap terminations on basal surface and toplap terminations on top surfaces. We have observed two lobe deposit types: thick radial lobes in the basin (up to 70 km^2 area and usually more than 50 m thick; Fig. 12b) and thinner elongated lobes at the base of the slope (up to 10–15 km^2 area and usually less than 20 m thick; Figs 10f (stage K–L) & 11a).

All these lobes show a *proximal convex-up* portion (Fig. 12b). Numerous onlap, downlap and toplap surfaces are observed. Some continuous reflectors are visible within the chaotic facies defining several internal units. These units are more visible in large lobes (compare Figs 10f & 12b). These units vary from 10 m to 45 m in thickness. They suggest several stages of deposition. For smaller lobes, this convex-up chaotic configuration represents the greater part of the lobe deposit. Lobes pass abruptly laterally, with onlap terminations, to bedded continuous facies or to a more transparent bedded facies (Fig. 12b). Numerous secondary channels are observed within lobe deposits (type 2 defined above; Fig. 12b).

This proximal convex-up part of the lobes passes down-fan to *flat distal lobe areas* (Fig. 12c). These lobes have numerous continuous bedded reflectors with a predominance of onlap terminations. Scattered chaotic facies and small secondary channels are sometimes observed. The lateral transition to the bedded continuous or continuous more transparent facies is more progressive compared with the lateral passage observed in their proximal part (Fig. 12c).

At the surface, small channels observed in the channel–lobe transition and lobe deposit correspond to small secondary channels as observed on the bathymetric map (Fig. 3). However, some of these features within channel–lobe transition and lobe deposit could perhaps correspond to scours, as observed by Normark & Piper (1991) and Wynn *et al.* (2002). Steep slope described in the Golo system would favour these processes.

The *small Pineto* and *Fium Alto gullies* (Fig. 13a) are characterized by the absence of chaotic facies or terraces. The floor of the gully is characterized by numerous parallel reflectors. The feeder thalweg shows a lower degree of erosion than canyon and channel fans.

The *Pineto* and *Fium Alto lobes* are characterized by chaotic facies (Figs 10c & 13b). They are elongated thin lobes (15 m thick) of 9 km^2 area. Seismic configuration seems to indicate a deposit that migrated laterally from south to north and not basinward, with a sigmoidal configuration dipping toward the north.

Lateral and longitudinal evolution of sedimentary bodies

The upper fan can be divided into two sub-segments: (1) a relatively short section with stable well-developed channel–levee complexes that evolve rapidly downslope into (2) laterally migrating channel–levee complexes. Stable

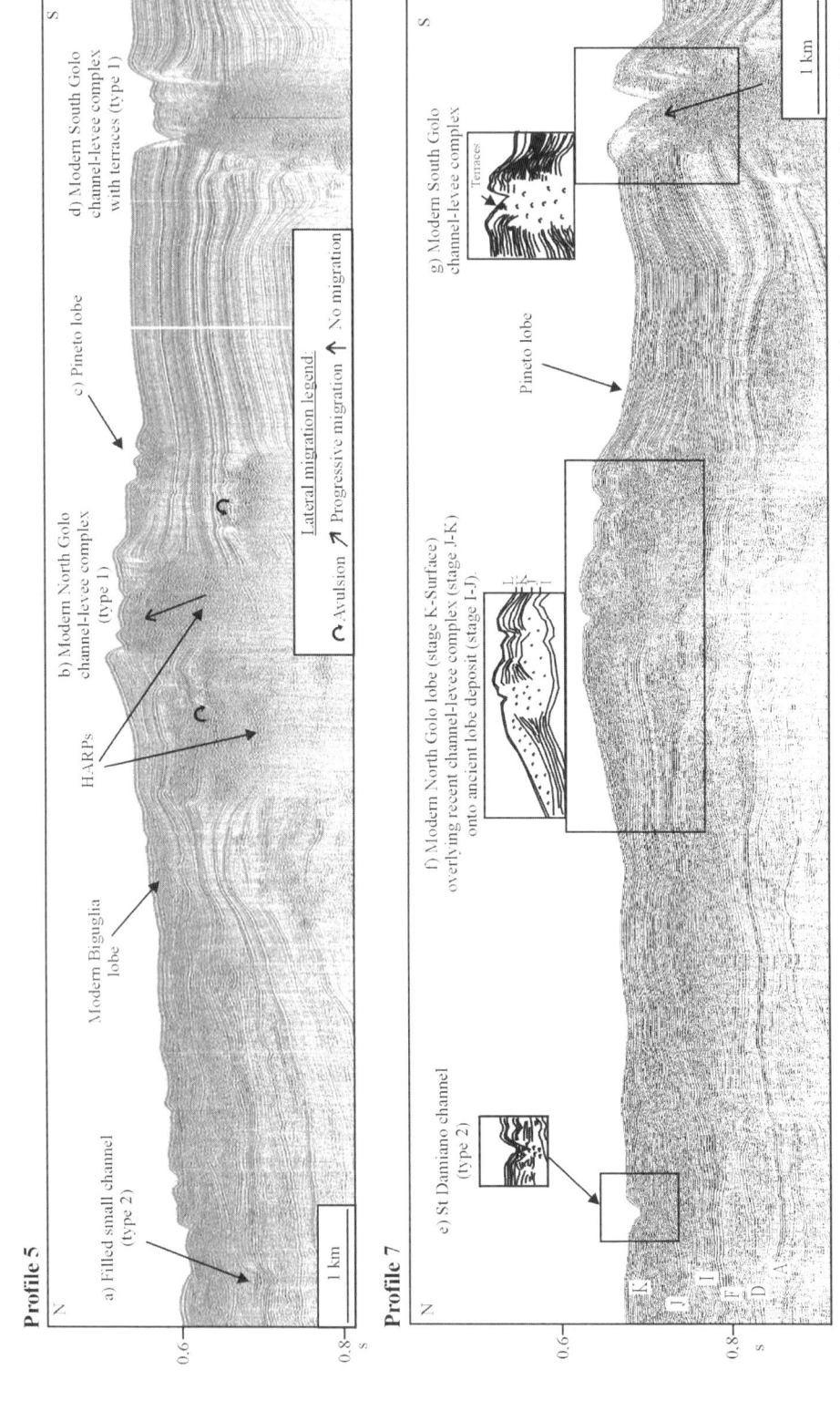

Fig. 10. Geometries of sedimentary bodies and type of lateral migration observed in the Golo depositional system based on interpreted seismic profiles 5 and 7 (see Fig. 3 for location). Letters correspond to regional key reflectors defined in Figure 12. Depth is in seconds two-way travel time (TWTT) below sea-level.

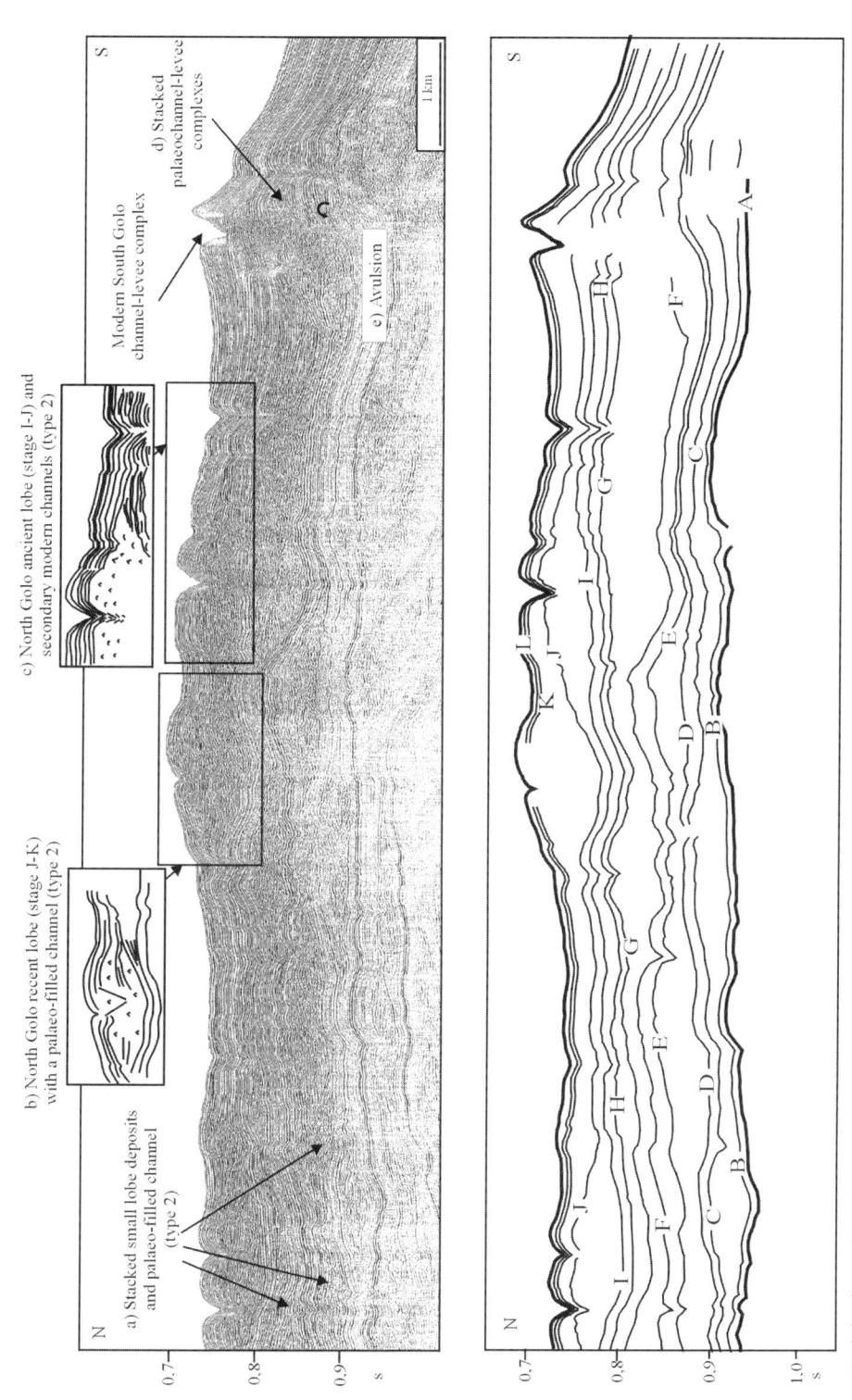

Fig. 11. Seismic profile 9 (see Fig. 3 for location) showing geometries of sedimentary bodies of the Golo system and lines that represent regional key reflections (A to surface) defining 12 sedimentary stages in Figures 17 and 18. The L reflector is located between the K reflector and the surface. Depth is in seconds two-way travel time (TWTT) below sea-level.

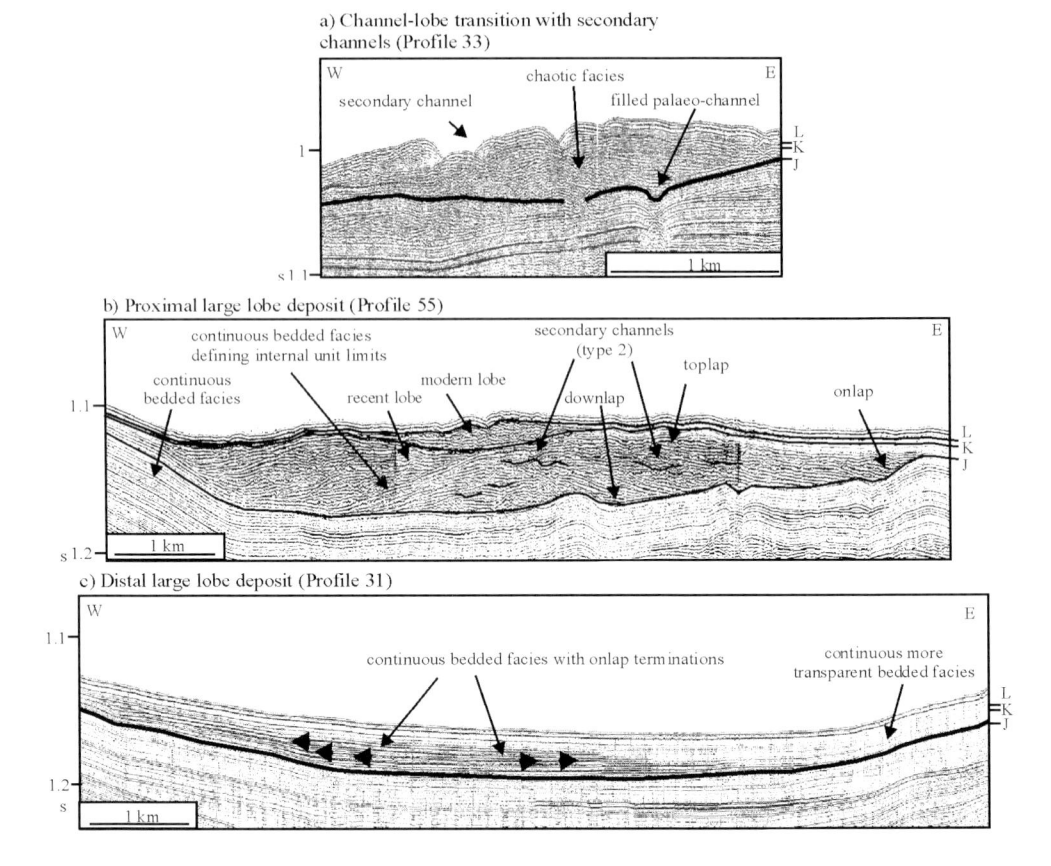

Fig. 12. Selected parts of seismic profiles (see Fig. 3 for location) showing seismic characteristics and geometries of large lobes. Letters correspond to regional key reflectors defined in Figure 11. Depth is in seconds two-way travel time (TWTT) below sea-level.

channel–levee complexes are characterized by stacked complexes, where it is difficult to define the base of the channel because of the vertical stacking of chaotic facies (Fig. 10d). Lateral migration of channel–levee complex occurred either progressively or abruptly by avulsion (Flood et al. 1991).

Progressive migration is characterized by stacked channel–levee complexes that are slightly shifted laterally with respect to their predecessors. Vertical continuity of the chaotic facies characterizing the floor of the channel is observed (Figs 10b & g). In the South Golo channel–levee complex, a northward lateral

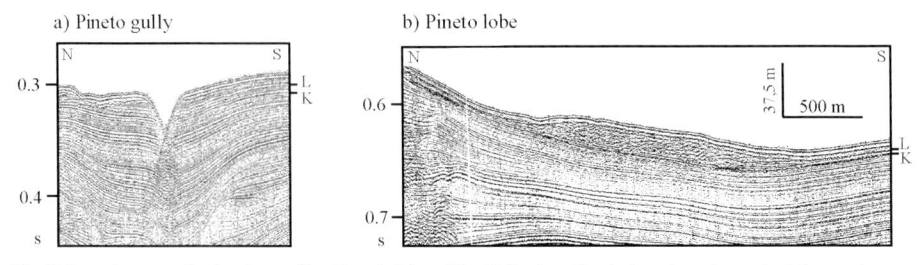

Fig. 13. Selected parts of seismic profiles 2 and 6 (see Fig. 3 for location) showing characteristics and geometries of the Pineto gully and lobe. Depth is in seconds two-way travel time (TWTT) below sea-level. Letters correspond to regional key reflectors defined in Figure 11.

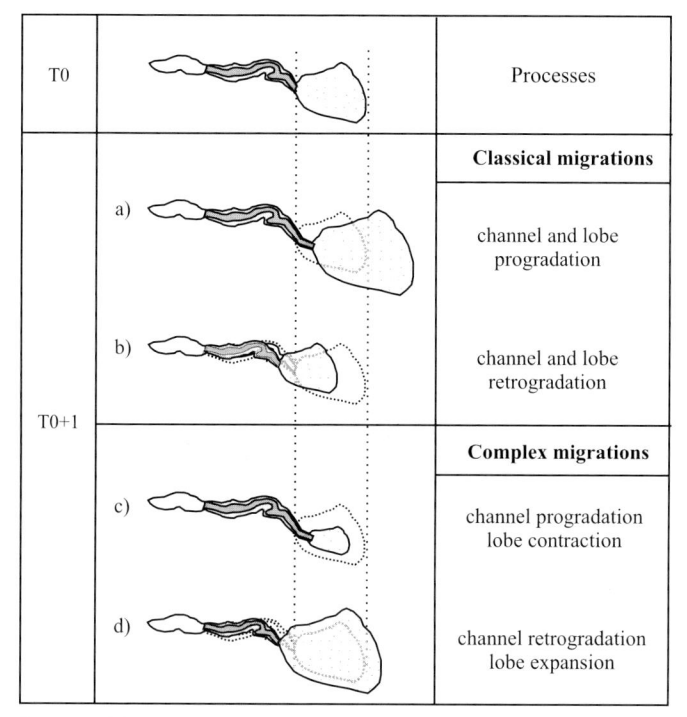

Fig. 14. Synthetic diagram of the main longitudinal migrations observed in the Golo turbidite system.

progressive migration is observed (Fig. 3), and less than 2 km downstream progressive migration is observed toward the south.

Avulsion of channel–levee complexes is also observed (Figs 10b & 11e). In this case, there is no continuity between the HARPs and chaotic facies characterizing the floor of the channel before and after the migration. During the period of time studied, each fan presents several abrupt changes in the location of the main channel. In general, each fan has one channel active during a sedimentary stage. However, two channels were active simultaneously during a short period of time for the North Golo fan during the F–G stage because of an avulsion (Fig. 10b).

During deposition of the studied interval (from the 'A' reflector to the surface), *longitudinal migration* (progradation or retrogradation) has occurred. Examples are shown in Figures 10f and 11a. Progradational configurations are characterized by chaotic lobe deposits overlain by a new channel (Figs 11a & b). Retrogradational configurations are characterized by the infilling of a channel by a lobe deposit (Fig. 10f) or a hemipelagic drape (Fig. 11b). Care must be taken in interpreting the seismic configurations in this general way. The infill by hemipelagic drape could mean retrogradation but also inactivity of the fan, or lateral migration of the whole fan resulting in the abandonment of the active channel. Moreover, this lateral migration can be associated with progradation or retrogradation of the main channel or of the lobe deposit. In the case of Figure 10f, the prograding channel is associated with expansion of the lobe deposit. Thus we have observed and defined in Figure 14 several types of longitudinal migration. The time $T0$ corresponds to a pre-existing fan configuration (extension of the channel–levee complex and the lobe deposit). At time $T0 + 1$, we observe different type of longitudinal migration in the four fans composing the Golo system. The extent of the channel–levee complexes and lobe deposits has changed.

Distribution of sedimentary bodies and variations in fan characteristics

Based on morphological and seismical data, a new map of the recent sedimentation of the Golo turbidite system has been made (Fig. 15). The Golo depositional system is composed of four individual fans and two small deposits onto the slope. In detail, the South Golo canyon is the only fan of the Golo system with a well-developed canyon head (Fig. 16a). Its channel–levee complex is the best developed

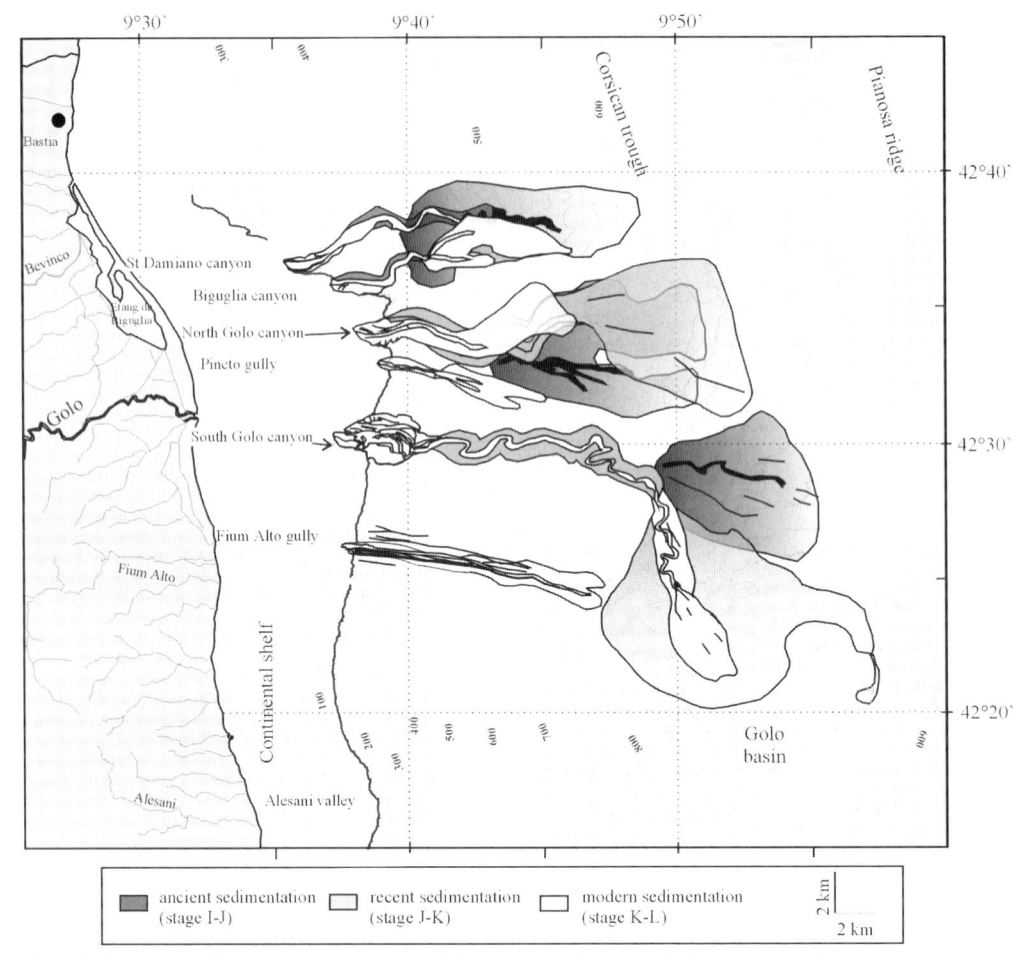

Fig. 15. New map of the Golo system: four individual fans and two small deposits onto the slope. Letters correspond to regional key reflectors defined in Figure 11. Compare this new map with Bellaiche *et al.* (1993) in Figure 1.

(Fig. 16b). It is longer and more sinuous. The North Golo fan is represented at the surface by a short relatively straight channel–levee complex with numerous secondary smaller channels and a lobe deposit onto the slope. However, during the period of time studied, the North Golo fan can be compared to the South Golo fan because its channel–levee complex also has been long and sinuous (Fig. 15), although less so than the South Golo. Moreover, it has deposited large lobes comparable to those of the South Golo fan (Figs 16c & d). Avulsion and progressive migrations are described within these fans, and longitudinal migrations are of small extent with lobes located mainly in the basin (Fig. 16). Biguglia and St Damiano canyons are less deep and canyon heads are not well developed. These fans consist of channel–levee complexes, well developed in their upstream part (type 1 defined above), that pass rapidly to smaller channels with no or weak levees (type 2 defined above; Fig. 16). Only avulsion has occurred, and longitudinal migration is of large extent, with small lobe deposits onto the slope and lobe deposits in the basin (Fig. 16). A recent study, based on boomer data, also shows a clear difference in boomer facies between coarser levees of minor channels and thinner levees of major channels (Pichevin 2000).

Whatever the size of lobe, lobe deposits are not only vertically stacked but horizontally offset stacked (Figs 11, 16c & d).

To summarize, there is a *continuum* from south to north, from the major South Golo fan to the minor Biguglia and St Damiano fans (Figs 15 & 16), with:

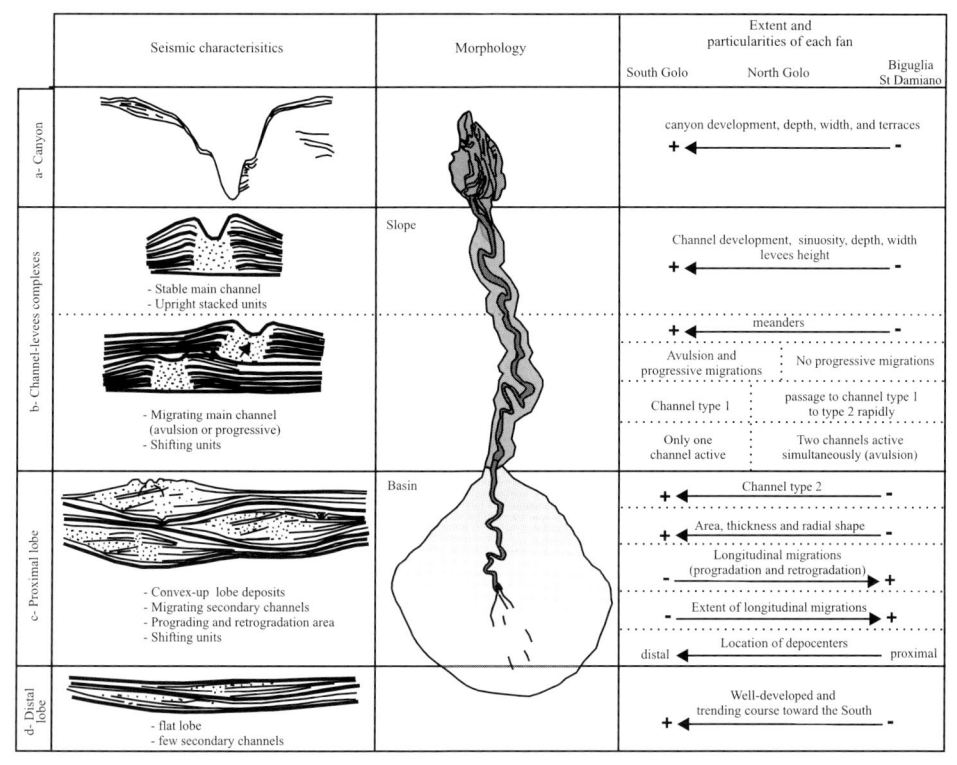

Fig. 16. Schematic diagram showing relative size, shapes and mode of assemblage of sedimentary bodies according to their emplacement on fan, and evolution of seismic configuration.

(1) a decrease of canyon depth, width, sinuosity, incision into the shelf and slope, and of height of terraces;
(2) a decrease of channel sinuosity, depth, and width;
(3) a decrease of levee height, length and of lateral extent associated with a increased of sand content of levees; and
(4) a decrease of lobe size and thickness, change to radial shape to more elongated shape, and change from basinal to more proximal location of lobe deposits.

However, Biguglia and St Damiano fans are grouped together in term of minor fans and separated from the South and North Golo fans named major fans.

Stratigraphic and palaeogeographic evolution

Sedimentary evolution

Sedimentation is characterized by numerous stacked sedimentary bodies separated by bedded continuous facies (Fig. 11). Because of the complexity and multiplicity of stacked units, we have distinguished 12 regional seismic markers (Fig. 11). These are labelled 'A' to 'Surface', with 'A' being the deepest regional seismic markers. The L-Surface stage represents present sedimentation, and the others are late Quaternary in age. Each stage has been then mapped in terms of palaeo-facies and in terms of deposit thickness, i.e. isopach maps (Figs 17 & 18).

Four depositional models for lobe deposits are distinguished in the Golo system (Fig. 17). One corresponds to a phase of hemipelagic sedimentation (Fig. 17a). The second model corresponds to true lobe depositional phases. The lobes are elongated, of small extent, and thin (15 m, 10–15 km^2), and are located mainly in the proximal part of the basin. These phases affect all fans at the same time. They characterize periods of moderate sediment gravity flow activity in the system (Fig. 17b). The third model corresponds to large and thicker lobe deposition. The lobes are located in the middle part of the basin floor and were deposited in all fans. They characterize periods of intermediate sediment gravity flow activity in the system (Fig. 17c). The fourth

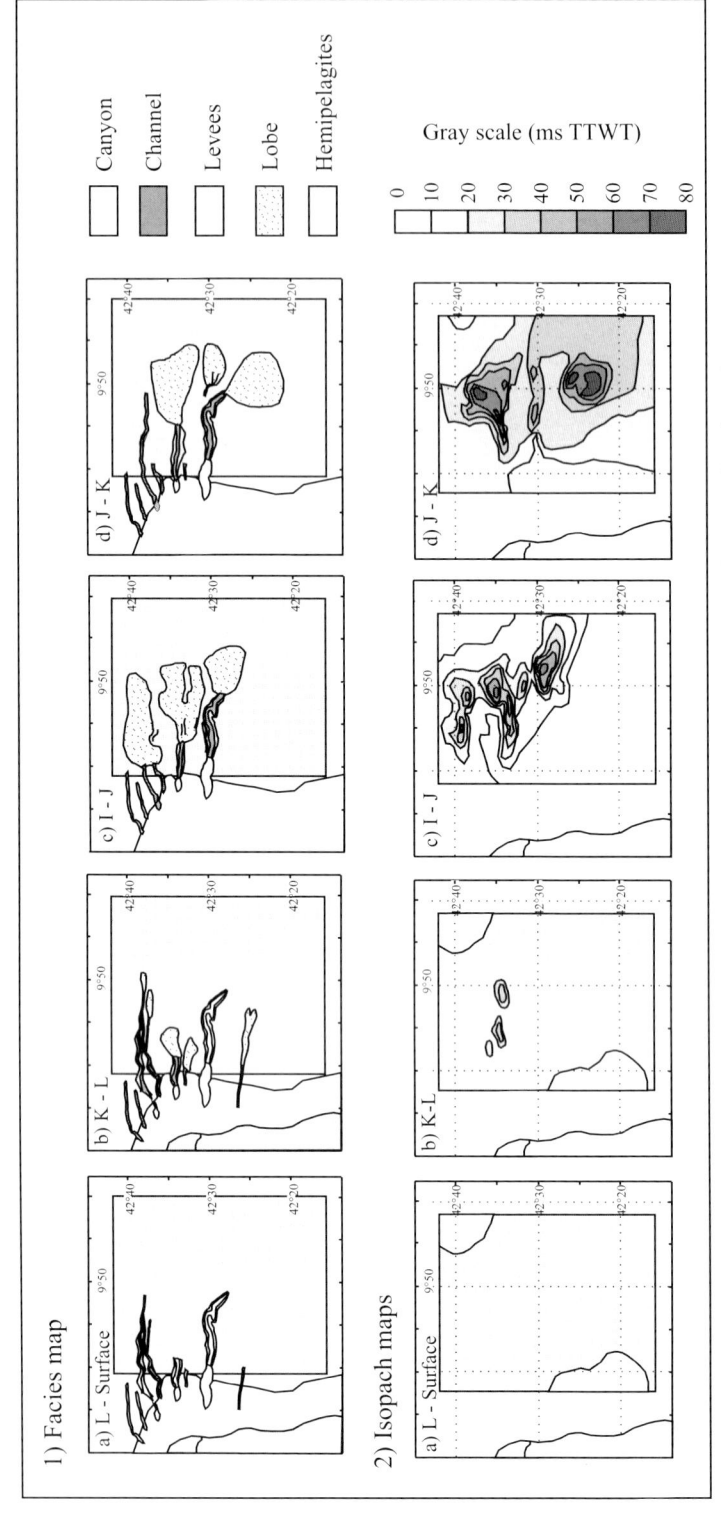

Fig. 17. Four types of sedimentary facies and distribution at the stages defined in Figure 11: (1) maps of facies and (2) isopach maps.

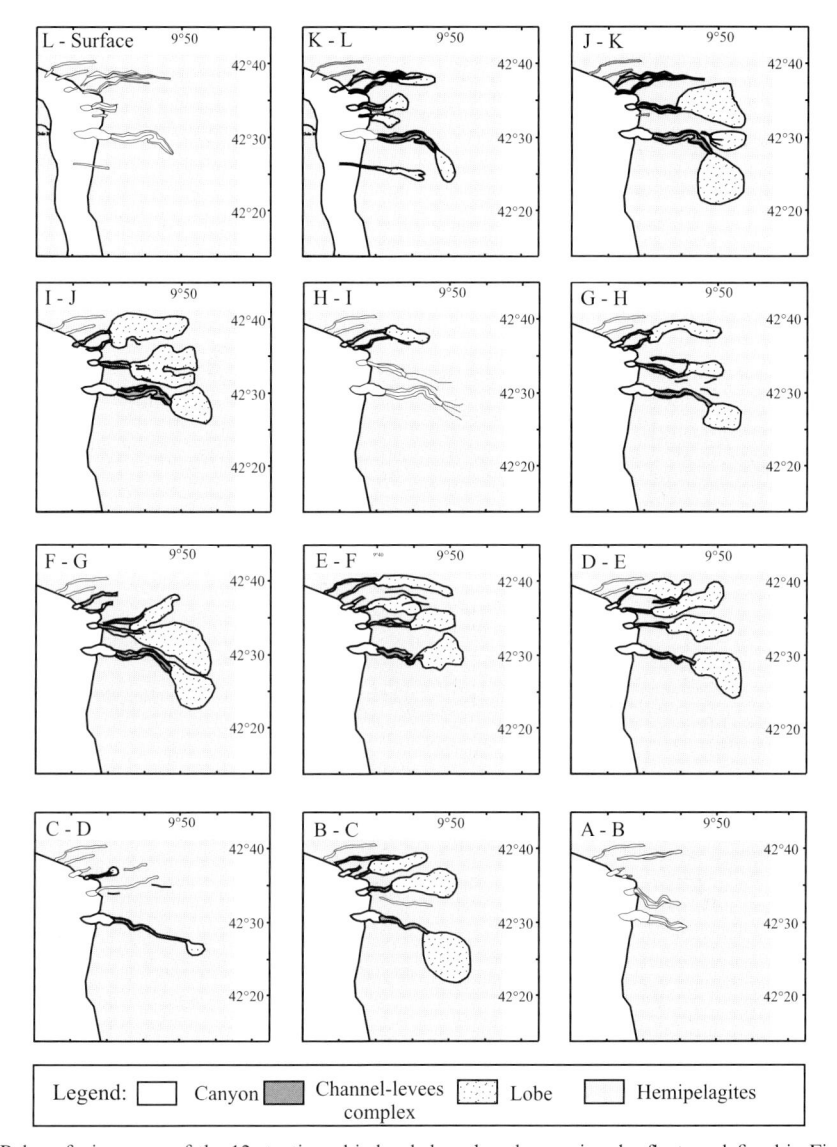

Fig. 18. Palaeo-facies maps of the 12 stratigraphic levels based on key regional reflectors defined in Figure 12.

model corresponds to large radial thick lobe deposition (50 m and 70 km^2). The lobes are located in the distal part of the basin floor and are observed only within the major South and North Golo fans. They correspond to periods of high sediment gravity flow activity in the system (Fig. 17d). For the whole phase of sedimentation studied, there is an alternation of these four turbidite models (Fig. 18).

Key reflectors were chosen to separate the maximum stages of the growth pattern of the depositional system. However, a stage could correspond to several episodes of sedimentation (stage B–C in Fig. 18). Lobe deposits mapped are not necessarily simultaneous. However, large lobe deposits characterizing high turbiditic periods are simultaneous.

During periods of high sediment gravity flow activity (fourth model defined above and in Fig. 17d), major fans (North and South Golo fans) prograde (Fig. 14a). During periods of intermediate sediment gravity flow activity (third model defined above and in Fig. 17c), all fans are supplied and longitudinal migrations are very variable. For major lobe deposits, contraction is observed, associated with channel retrogradation (Fig. 14b) or slight development (Fig. 14c). Lobes of major fans are still located in the basin but in proximal areas. For minor fans, these periods characterized the maximum of sedimentation with lobe deposits. Progradation or expansion of lobes are observed (Fig. 14a). During periods of moderate sediment gravity flow activity (second model defined above and in Fig. 17b), we can observe retrogradation of the minor fans with lobe deposits at the base of the slope (Fig. 14b). Major fans are usually characterized by contraction of the lobe that is still located in the basin, and no development of levees is observed. However, for stage K–L, a global retrogradation is observed for the North Golo fan with lobe deposition onto the previous channel–levee complex (Figs 10f & 14b).

Stratigraphic context

High sedimentation rate characterizes the Corsican Trough (Corsica Basin is 8.5 km deep and filled by Tertiary sediments), with up to 1.5–2 km of post-Messinian (5.2 Ma) sediments originating from Corsica (Mauffret et al. 1999). Plio-Quaternary hemipelagic sediments cover the whole margin, as a result of the ubiquitous post-Messinian transgression. If we assume a constant rate of sedimentation for the whole sedimentation period, then 0.2 km of the Corsica basin fill would be equivalent to 500 000–700 000 years: that is, the Late Quaternary (middle and late Pleistocene–Holocene). However, sedimentation is unlikely to have been steady throughout the Pliocene and Quaternary and particularly for the Golo basin, where active sedimentation restarted after the post-Messinian transgression. The 'A' reflector passes just beneath the youngest major progradational wegde on the shelf or perhaps in the middle of this wedge. Seven similar wedges have been recognized on the shelf break, and we infer that they could correspond to major variations of sea-level, with a periodicity of about 100 000 years. If this is the case, the reflector 'A' would represent some time younger than 130 000 years BP. Radiocarbon dates (Stanley et al. 1980) show that the most recent large South Golo sandy lobe is of last glacial maximum period in age (dates around 20 ka at 4 m on cores 56 and 57; Fig. 3). The L–Surface stage characterized by hemipelagic sedimentation would represent the Holocene drape.

Discussion

Source of sediments

Regionally, the morphology of the shelf (with a seaward bulge) reflects the fact that the Golo river is the main source of sediment on the eastern margin of Corsica. Conchon (1977) has shown that the Golo river was flowing in a more northerly direction in middle Würm times, in the direction of the Biguglia and St Damiano canyons. The secondary source of sediment is the Bevinco river (Fig. 3). Alluvial deposits are present all along the course of this river, but at present sediments are trapped in the coastal Biguglia lagoon. In the past, the Bevinco river could have fed several canyons. However, the trends of the canyons suggest that it was probably not the major source of sediments for the basin. The littoral drift on the shelf is presently S to N, preventing supply of sediment to the canyons by the Bevinco river. This river could have been a sediment source for canyons located towards the north (Gauthier and Prone 1980). Thus, based on size of rivers, drainage areas, and the morphology of the shelf, we assume that terrigenous sedimentation in the Golo submarine depositional system was principally supplied by the Golo river. This river provides coarse sand derived from active weathering of the mountainous hinterland.

Factors controlling fan growth

Sea-level variations. Deposition in the Golo depositional system has probably been controlled by relative sea-level variations. This control is suggested by the presence of stacked progradational wedges and palaeo-filled incised valleys on the shelf (Vail et al. 1977). This control is also suggested by radiocarbon dating of Stanley et al. (1980). The recent large thick lobe deposit of the South Golo fan between J and K seems to correspond to the Last Glacial Maximum period (blue sedimentation in Fig. 15). Small lobe deposits during the stage K–L would perhaps correspond to the rising sea-level just after the last glacial period (red sedimentation in Fig. 15). Even if we lack precise dating of the different observed stages, the alternation of four sedimentary models (Fig. 17) seems to be plausibly linked to the alternation of glacial and interglacial periods. It seems that we have to consider four conditions:

(1) *full interglacial highstands* (marine isotopic stage 1), when sea-level was close to present

level and sediment was trapped in deltas and on the shelf. Widespread hemipelagic sedimentation occurred in the basin (stage L–Surface in Fig. 17a);

(2) *times of rapidly rising sea-level* when sea-level was at least 50 m below present sea-level, but was either quite high or rising, so that sediment tended to be trapped on the shelf or in retrograding deltas, explaining limited coarse sediment in the basin. A stillstand may permit a progradation if sediment supply outstrips rate of space creation, as for example during the Younger Dryas (stage K–L in Fig. 17b);

(3) *glacial lowstands of sea-level*, with emerged continental shelf (marine isotopic stages 2, 4 and 6) (sea-level was −100 m according to Lambeck & Bard 2000) so that deltas could be located at the present shelf break and prograde and supply a large quantity of coarse sediment to the basin (last glacial period with stage J–K, and F–G in Figs 17c & 18);

(4) *times of falling sea-level* when sea-level was at least 50 m below present sea-level, but was either quite low or falling, so that deltas were prograding and supplying coarse sediment to the basin (stage I–J in Fig. 17d).

Because of changes in sediment supply and in the 'efficiency' of sand transport by turbidity currents, depositional system facies periodically shifted landward or basinward. The updip–downdip longitudinal shifts of fan depocentres were probably caused by cyclic changes in sediment load, which resulted in variations of the activity of basinward transport in relation to sea-level changes.

Pre-existing morphology control. The influence of pre-existing seafloor topography on the growth and morphology of the Golo depositional system is visible over several spatial scales.

(1) Regionally, in the Golo system, the position of the major feeder valleys and the basinward limit of the system do not appear to have changed significantly during the late Quaternary. We have also seen large longitudinal migrations of fan depocentres in small fans (Fig. 18). In addition, the thickness of sediment deposited above reflector 'A' is similar across the basin area (Fig. 11). These observations suggest that growth of the Golo system has been characterized by long-term *aggradation*, not progradation. The system is confined within this active half-graben and the Pianosa ridge, thus enhancing aggradation compared with progradation. Aggradational processes also characterized the rapidly subsiding shelf, and, as a result, the slope is also aggradational, with a progressive landward retreat of shelf-edge deltas through time. The distal parts of large lobe deposits are deflected towards the south by the Pianosa ridge and by the plunge of the Corsican trough (Fig. 15). Where smaller lobe deposits are located near the base of the slope, i.e. in the proximal part of the basin, they are not constrained and are not deviated toward the south. Sedimentation is regionally redirected by the general slope of the Corsican trough, inherited from the opening of the Tyrrhenian Sea to the south and the presence of the Pianosa ridge and its large slumps. In detail, the South Golo lobe is confined by the continental slope and Pianosa ridge slumps and by the topographic low created by previous lobe deposits. Resulting lobe terminations are characterized by onlap directly onto the Pianosa slope or its large slumps (Fig. 8). Flows are constrained laterally and longitudinally. This *prevents progradation towards the east and favours aggradation of the lobe deposit*. Shapes of sedimentary bodies are influenced by the bounding slope (Pianosa ridge) only when supply volumes are sufficiently large. Moreover, large lobe deposits developed in the distal part of the basin show a radial form. Smaller lobe deposits located near the base of the slope, on steeper gradients, have elongated forms (Fig. 15). The slope gradient thus has a strong influence on the morphology of lobe deposits. These processes are accentuated by the present rising of alpine Corsica and of the Pianosa ridge.

(2) At the scale of a single fan, we have seen that each fan shows avulsion of the main channel in its upstream part. The location of the new channel–levee complex is controlled by the available space created by the previous complex (Fig. 19-1). The degree of turbidity current confinement within channels probably has an impact on the existence of channel progressive migrations (Fig. 19-2). These progressive migrations are observed only in the upstream part of major channels where the relative stability of the channel (levee height) probably favours progressive migration rather than avulsion. Basinward, the channel is less stable and flows are less confined. No progressive migration is observed, and avulsion of the channel is more frequent. Moreover, we have seen that during the last stage, K–L, only the South Golo fan deposed a lobe in the basin. The confinement of flows in this best-developed channel of the margin allows a sand prone lobe deposit in the basin by increasing flow efficiency to transport sand.

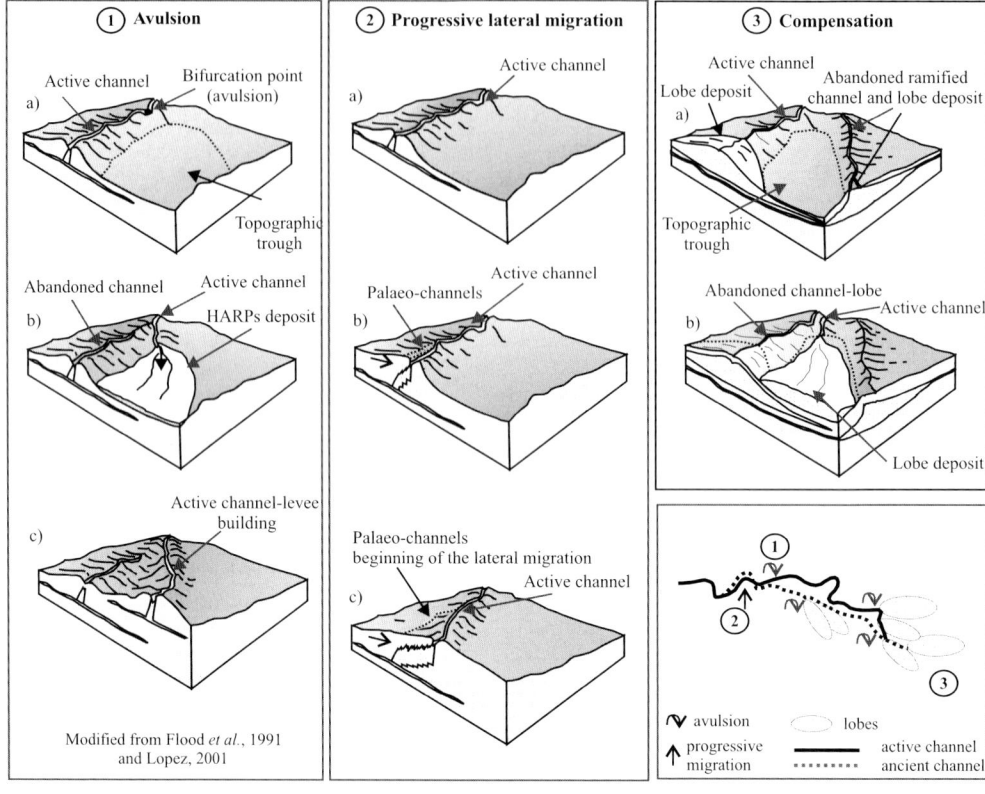

Fig. 19. Schematic diagrams showing processes and resulting geometries of channel–levee complex lateral migration by (1) avulsion and (2) progressive migration, and associated lobe deposit processes by compensation (3) (see text for explanation).

Based on isopach maps of lobe deposits, we have thus observed that depocentres are not simply vertically stacked (Fig. 11) but are shifted laterally. Lobe deposition follows the same way (Fig. 19-3a). The new lobe deposit will occupy and be confined within the low created by the previous lobe deposit (Fig. 19-3b). Lobe sedimentation thus occurs by compensation (Mutti & Sonnino 1981). This control is also well expressed in the proximal part of lobe deposits, where sharp lateral changes are observed from onlap terminations to hemipelagic sedimentation (Fig. 12b) or to slumps or slopes of Pianosa ridge or the continental slope (Fig. 8). Therefore the present morphology is the heritage of the previous sedimentary history of the depositional system.

The dispersal and accumulation pattern of sediment and the resulting morphology and architecture of the Golo system have thus been strongly influenced by seafloor topography at different levels.

Local tectonic control of sedimentation. Although the Golo depositional system is located in an active tectonic area, it is very difficult to extract the true influence of tectonics on Quaternary sedimentation and on the orientation and location of canyons. It seems that the major sedimentation changes occurring during the Late Quaternary are influenced mainly by eustacy and pre-existing topography.

Growth pattern

Lateral and longitudinal migration processes. The absence of *progressive migration* on minor fans could be due to the lack of seismic data in their upstream portions or to the nature of turbidity current processes. Channel–levee complexes of major fans are better developed and more sinuous. Their channels have terraces interpreted as confined levees (Pichevin *et al.* 2000). According to Pichevin *et al.* (2000), the confined levees are interpreted as the result of low-energy flows allowing vertical accretion. Terraces are located on the inside of meander bends. Erosional processes of the outside of meander bends are involved to explain the lateral stacking pattern of terraces. Similar flows could also explain progressive lateral migration of the thalweg observed only in major fan meanders (Fig. 19-2).

Avulsion (Flood *et al.* 1991) is observed in the four fans. This abrupt migration can be due to (1) infilling of the active channel by hemipelagic drape (inactivity) or by a lobe deposit (retrogradation) and then branching and growth of a new channel in another place, or (2) development of a new channel due to levee breach or large overflow, leading to the abandonment of the ancient active channel and then infilling. No retrogradational lobe deposits are observed under the lateral migration above the F reflector. As the North Golo fan shows activity of the main channel and of the new main channel after the abrupt migration above the 'F' reflector (Fig. 9b), the migration seems not to have occurred as a result of the inactivity or infill of the previous main channel, but rather because of levee breach. Avulsion in the upstream part of channels is observed during particularly intense turbiditic flows (stages F–G, I–J and J–K). It is probable that large sediment supply induced an increase of the pressure exerted by the turbiditic column on the levee walls. This causes inner-levee slope failure (Fig. 19-1a; Lopez 2001). However, levees of the Golo system are still of slight height compared with large muddy systems, and could have favoured large overflow of turbidity current leading to an avulsion, without breaking of the levees. The turbidity current rapidly spreads out across the adjacent interchannel area to form sheet-like, sand-rich bodies (Fig. 19-1b). These deposits, reducing the slope, would favour the progressive progradation of a new channel–levee complex over the sheet sands (Fig. 19-1c). In the Golo system, avulsions have occurred toward the south and the north. In contrast to the Danube system (Popescu *et al.* 2001), the Golo avulsions seem not be influenced by Coriolis force.

The downstream parts of channel–levee complexes are relatively complex because of *longitudinal migration*. We have seen that several types of longitudinal migration have been observed in the Golo system (Fig. 14). Different mud/silt/sand ratios within turbidity currents are involved. An increase in silt and mud will favour channel–levee complex growth and progradation. Associated with (1) an increase in sand content, lobe expansion will occur; (2) with a decrease of sand content, lobe contraction occurs.

Lobes of minor fans have shifted landward or basinward frequently, whereas major fans have deposited lobes in the distal part of the basin. Channels of minor fans have a relatively unstable downstream part with smaller channels (type 1 defined above) compared with major fan channel–levee complexes (type 2 defined above). A major sand/mud ratio could explain the instability of minor channel–levee complexes, allowing more or less landward or basinward shifting of lobe deposits.

Over the entire margin, major fan patterns of longitudinal evolution are synchronous. However, minor fans have very variable configurations, suggesting that they are fed by a multiplicity of turbidity current processes in a random way, whereas major fans are probably fed mainly by a similar suite of turbidity current processes, leading to more stable and repetitive responses to relative sea-level change.

Conceptual model of the Late Quaternary Golo system growth. Along the Golo margin, four depositional models alternate through time. They are linked to eustasy. The shelf break is composed of progradational stacked wedges. They are interpreted as the product of river discharge during low relative sea-level (Vail *et al.* 1977). Four canyons feed into four fans composing the Golo depositional turbidite system. These canyons have been directly linked to the Golo river. Filled incised valleys upslope of them on the shelf are probably the result of the erosion of the continental shelf by rivers during periods of lowstand sea-level and filling during highstands of sea-level. We consider gravity flow processes as the main mechanism of deposition. The mud/silt/sand ratio variations involved to explain lateral and longitudinal migrations are probably linked to variations in turbidity current processes linked to climatic-eustasy variations and to the position of canyons with regard to the main source. Based on these interpretations, we suggest the following growth pattern for the Golo depositional system.

During *glacial lowstands of sea-level* (defined above), rivers incised the continental shelf and flowed directly into canyon heads, with the exception of the Pineto and Fium Alto gullies. During the Quaternary, changes of the Golo river course may explain the existence of several canyons fed by the same river (Conchon 1977). However, the synchroneity of large lobe deposition on both the North and South Golo fans suggests that, during relative low sea-level periods, the Golo river built a delta with several distributaries that fed several canyons simultaneously.

During the successive sea-level lowstands (Fig. 20b), canyons may reactivate. River-derived mud is more likely to be deposited in canyons

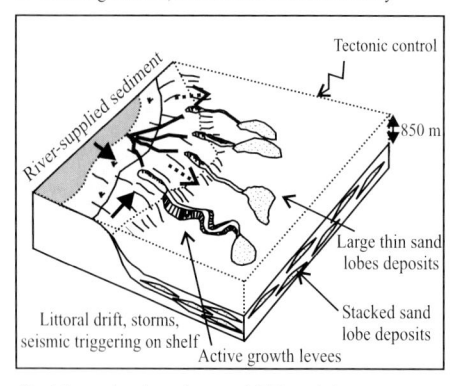

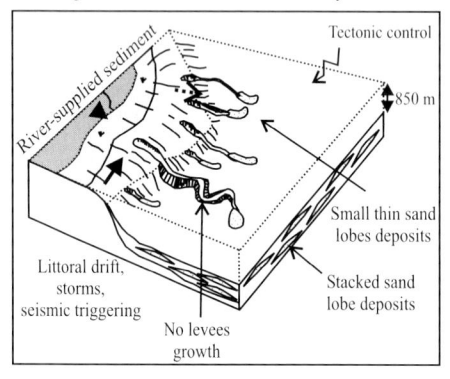

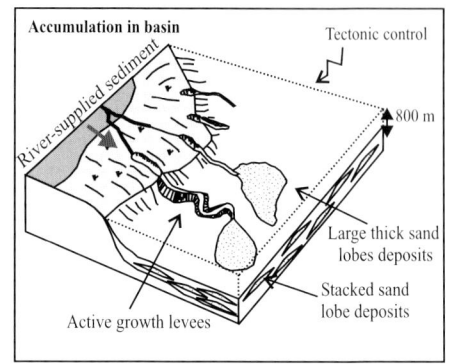

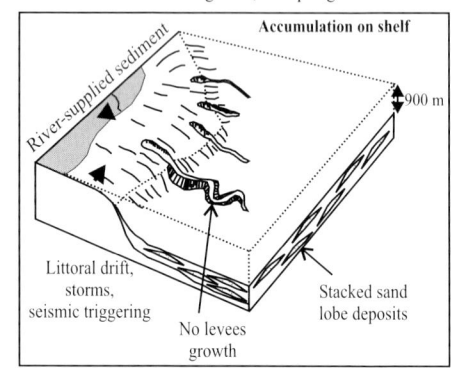

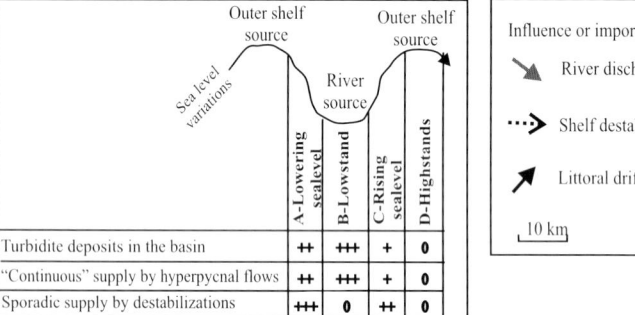

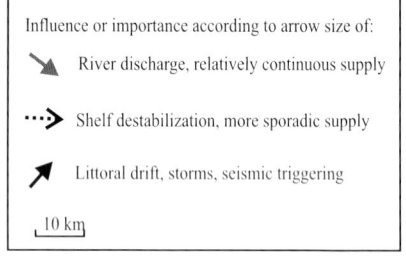

Fig. 20. Conceptual model of the late Quaternary Golo sedimentation according to major change in climatic-eustasy variations associated with dominant sedimentary processes.

near the river mouth, where rapid deposition may lead to failure, generating turbidity currents. River flow directly into canyon heads as channelized mud/sand low-energy relatively continuous flows allows the transport of both fine and coarse material into the basin. These currents could correspond to hyperpycnal turbidity currents as suggested by hydrologic data (Mulder & Maneux 1999). These relatively frequent currents feed only proximal canyons and thus only major fans. They allow a more persistent, relatively mud/sand-rich point source through time. This type of current is suggested to explain the development of stable, well-developed channel–levee complexes, progressive lateral migrations, and small longitudinal migrations and thick extensive lobes in the distal part of the basin characterizing major fans.

During *times of rising or falling sea-level* (defined above; Figs 20a & c), coarse material accumulated on the shelf or delta was probably transported by littoral drift or longshore currents toward the north of the study area. All the canyons of the margin would have then captured some of this remobilized sediment. Sporadic turbidity currents may be initiated by focusing storm waves in canyons (Fukushima *et al.* 1985), by rapid deposition of sediment on the upper steep slopes, or by seismic triggering on steep slopes (Normark & Piper 1991). Any one of these mechanisms might trigger turbidity currents in any of the canyon heads of the margin, and would allow the transport of sand into the basin as relatively channelized sand-laden turbidity currents. Major floods can also generate hyperpycnal currents from delta distributaries (Mulder & Maneux 1999). The supply of coarse sediment will be larger during times of falling sea-level as a result of the general trend of prograding delta during a global fall of sea level explaining the thickness of lobe deposits (third model in Fig. 17).

During *full interglacial highstands* (defined above), coarse detrital sediments were (and are) deposited mainly on the shelf, and the fine particles were transported seaward in suspension to form hemipelagic deposits (Fig. 20d) (Vail *et al.* 1977; Shanmugam & Moiola 1988; Normark *et al.* 1998). Some rare turbidity currents, as described for times of rising or falling sea-level, could be initiated by littoral drift, storm waves or major floods. The littoral drift will favour feeding of northern minor fans rather than the South Golo fan located just in front of the Golo river. The resulting sedimentation will be characterized by a very low sand/mud ratio in the basin. According to currently available dating, no significant turbidity current activity has occurred during the Holocene.

The location of a fan with respect to the river source location will determine the main feeding processes (quasi-continuous or more sporadic currents) and the periods of major supply.

The location of the Pineto and Fium Alto gullies on the shelf break and their architecture revealed by seismic-reflection data suggest that they are less mature and probably more recent. They have never been directly connected to a river. They could correspond to sediment failure near the shelf break. Several small bodies like the Pineto or the Fium Alto bodies are visible all along the eastern margin of Corsica. They could also play a significant role in the transfer of sediment at the scale of the entire eastern margin.

Implications for modern and ancient studies

A useful comparison of modern with ancient, ancient with ancient, and modern with modern turbidite systems can be based only on well-understood and thoroughly mapped systems. The examples selected for comparison must represent depositional systems similar in characteristics such as type of basin, volume and type of sediment source available, physical and temporal scales, and stage of development (Mutti & Normark 1987). What is required is a clear definition of the common characteristics of the bodies being compared, together with an appreciation of the wide range of variables that influence their development. Many fan sedimentation models currently in use do not meet these criteria because they are applied to turbidite systems that do not have a similar geological context as defined above. In the following discussion, we shall compare only small sand-prone turbidite systems developed on active or immature passive margins.

Indications for modern studies. This study provides a modern example of small sand-rich turbidite systems. The Golo depositional system shows numerous similarities with the Navy fan, Hueneme fan and associated fans (Normark *et al.* 1998; Piper *et al.* 1999) and with other small systems in the southern part of the eastern margin of Corsica, such as the Tavignano and the Fium'Orbo systems (Guiomar & Savoye 1998).

In these similar geological environments shelves are relatively narrow, and a single river provides an important sandy source. Sedimentation patterns are always characterized by numerous stacked bodies. Several feeders and associated sedimentary bodies of various morphologies and geometries are usually observed. We recognize major well-developed deposits generally located in front of the main source in addition to small deposits such as the Pineto and the Fium Alto lobes (adjacent to the main source).

The most important factors controlling deposit morphology and geometry, and thus sand-bed geometry, in these systems appear to be the source and textural composition of turbidity currents, in relation to sea-level changes and the pre-existing topography. This influence of pre-existing topography on the morphology and architecture of sandy systems has been observed for the Navy and Monterey fans (Normark 1978). Comparable lateral shifts of channel–levee complexes and of lobe deposits are observed in the Rhone and Indus fans (Droz & Bellaiche 1985; Droz & Bellaiche 1991), but the closest analogue is in the growth of sandy lobes on the Navy fan described by Normark *et al.* (1979). The strong influence of climato-eustasy variations has been argued for the Fium'Orbo fan (Gauthier & Prone 1980; Guiomar & Savoye 1998), for the Hueneme and associated fans by Normark *et al.* (1998) and Piper & Normark (2001), for the Valencia system (Palanques *et al.* 1994), and for the Navy system (Normark & Piper 1985). The Navy fan is active at lowstands and almost inactive at highstands, as observed in the Golo system. For the Hueneme system, active periods of sedimentation during glacial periods (reflector $J = 12$ ka) and during the Holocene have been recognized using radiocarbon dates from an ODP site.

In these areas, similar processes as those deduced from the Golo system have been suggested:

(1) Efficient hyperpycnal turbidity currents with a mixed sediment load, derived directly from river input, via major valleys, deposit prominent sandy lobes in the distal basin that are of large area (around 60 km^2) and several tens of metres thick. Lobe reflections downlap onto older fan surfaces. Proximal depositional system channels associated with these multiple input points appear highly stable as a result of the growth of high-relief levees along active channels. These channels have been persistently reoccupied, although the size of the thalweg channel has varied through time.

(2) Low-efficiency sand-laden turbidity currents, generated by upper slope sediment instability, deposit small elongated lobes in the proximal part of the basin. Cores taken across the entire Golo system will allow us to detail these processes more precisely.

Canyons of the Golo system could be active simultaneously, as in the Monterey system (Normark 1978), or not simultaneously, as in the Hueneme system (Normark *et al.* 1998). In the Fium'Orbo, two canyons have been active simultaneously only during short periods of the Quaternary history of the fan (Guiomar & Savoye 1998), in contrast to the Golo system, where five canyons have been active over a long period of time. In the Golo system, several feeders are active during sea-level lowstands, or times of falling sea-level. However, we have seen that delta distributaries are not equally active. The north and south canyons are fed principally by direct delta distributaries, whereas minor fans probably receive sediment only during major floods. This unequal discharge through delta distributaries has been described in the modern fan deltas in the western Gulf of Corinth (Piper *et al.* 1990). The influence of hydrodynamic conditions in favouring some canyon feeding observed in the Golo system is also observed in the modern fan deltas in the western Gulf of Corinth and in the Hueneme system.

Understanding these variations in sediment feeder type and delivery processes to the basin is fundamental to predicting and evaluating the likely geometry and architecture, facies distribution and reservoir quality of deep-sea clastic systems.

Comparison with ancient rocks case study. Today, numerous fan models can be found in the literature (Mutti & Ricci Lucchi 1972; Normark 1978; Walker 1978; Heller & Dickinson 1985). The use of these models leads to several problems inherent in using modern submarine systems to understand ancient systems and vice versa. Most models address fan geometries and facies relationships based almost exclusively on examination of outcrops and/or modern submarine system morphology (Posamentier *et al.* 1991). A good summary of problems involved is provided in Mutti & Normark (1987) and Shanmugam & Moiola (1988).

In this study, our very high-resolution sparker seismic-reflection survey of the northern part of the eastern margin of Corsica has provided outcrop-scale (better than 2 m vertical and horizontal resolution) seismic-facies and geometry recognition. We have documented a great range in morphology and geometry of sedimentary bodies that have also been recognized in several ancient systems—for example the Upper Oligocene–Lower Miocene Cengio Turbidite System (Cazzola *et al.* 1985), and the Eocene Hecho turbidite system (Mutti 1977).

The present study shows the variability and the complexity of such small areas. It points out several facts that authors should be aware of when studying ancient systems. The sandy material transported to a small basin by one river does not necessarily build one fan with one major valley but could also build several individual fans with several feeders that potentially could be active simultaneously, as for the Golo margin, or not simultaneously, as for the Hueneme fan (Normark *et al.* 1998). The resulting fan sedimentation patterns and morphologies could be very variable although the area is narrow. The recognition in ancient rocks of sand-rich deposits on the slope, such as the Pineto sandy lobe, does not necessarily indicate the end of the sedimentary basin. Lateral and longitudinal extents of sedimentary bodies are relatively variable and complex. So precautions must be taken in ancient systems when studying local outcrops. Channelized lobes are located at the end of the upper fan valley. They are not always well developed in each fan. Moreover, secondary channels on the lobes are very difficult to differentiate from small channel complexes of minor fans, as in the Biguglia and St Damiano fans. On the Golo margin, the presence of small secondary channels does not characterize the proximal or distal section of the system. An accurate knowledge of the present morphology and very high seismic resolution is needed. These could have important implications in recognition of channelized lobes and secondary channels in ancient systems where the recognition of channel deposits is not always possible.

Conclusions

(1) Our high-resolution sparker seismic-reflection survey of the northern part of the eastern margin of Corsica has provided outcrop-scale (better than 2 m vertical and horizontal resolution) seismic facies and geometry recognition in a modern, small, sand-rich continental margin turbidite depositional system in the late Quaternary (at least 100 000 years BP). These seismic facies and geometries provide a good potential analogue for many modern and ancient turbidite deposits.
(2) The study area is characterized by a narrow bulge in the continental shelf that is fed by the Golo river, which supplies coarse sand derived from active weathering of the mountainous hinterland. The transport of the coarse material from the river to the confined basin does not build one fan with a single major valley. The Golo system is composed of four canyons that feed four individual fans and of two gullies that feed two small, sand-prone lobes on the slope. These canyons apparently were active simultaneously. The resulting sedimentation pattern is characterized by numerous stacked turbidite deposits and lateral migration (avulsion or progressive), and complex longitudinal migrations (progradation–retrogradation) are observed. It corresponds to a mud/sand-rich multiple-source ramp as defined by Reading & Richards (1994).
(3) There is a continuum of fan morphologies and geometries from the south to the north of the study area. These differences and those observed in lateral and longitudinal migration processes are linked to climatic-eustatic variations and to canyon location with regard to the main source leading to different canyon feeding processes. (a) Major fans, located in front of the river, are fed mainly by a persistent mud/sand-rich point source during times of lowstands or falling sea-level, leading to the construction of deep, stable, more or less sinuous channels with well-developed levees, avulsion and progressive lateral migrations and large thick lobes in the distal part of the basin. (b) Minor fans, located far from the river, are also fed by these types of current during lowstand periods, but mainly by sandier currents that may be different in origin and generated by shelf storms or failure in canyon heads during times of rising or falling sea-level. This leads to the construction of unstable, relatively straight channels with low, sandier levees. Only avulsions and small thin lobes in the proximal part of the basin are described.
(4) The transfer of sediments into the basin occurs mainly via the two major North and South Golo canyons and secondarily by the minor Biguglia and St Damiano canyons. However, the Pineto and the Fium Alto gullies, associated with similar sedimentary gullies, probably also play an important role in the transfer of sediment to deep-sea at the scale of the margin.
(5) Despite the active margin tectonics, major sedimentation changes can be related to eustatic sea-level changes and pre-existing topography leading to regional and local confined sedimentation.

The authors are grateful to IFREMER for data availability. Multibeam bathymetry and seismic data have been compiled by A. Normand and E. LeDrezen. The authors are also grateful to J. St Paul, D. Poirier, G. Chabaud, G. Floch, R. Kerbrat and R. Apprioual for technical assistance. Special thank to S. Alfonso, J. C. Faugeres, E. Gonthier, J. Lofi, S. Migeon and S. Zaragosi for their discussions and reviews. This is a UMR/EPOC CNRS, No. 5805 contribution No. 1501.

References

AUZENDE, J. M., BONNIN, J. & OLIVET, J. L. 1973. The origin of the western Mediterranean basin. *Journal of the Geological Society*, London, **129**, 607–620.

BELLAICHE, G., DROZ, L., GAULLIER, V. & PAUTOT, G. 1994. Small submarine fans on the eastern margin of Corsica: Sedimentary significance and tectonic implications. *Marine Geology*, **117**(1–4), 177–185.

BELLAICHE, J. P., PAUTOT, G., GAULLIER, V., VANNEY, J. R. & DROZ, L. 1993. Les appareils sédimentaires de la marge orientale de la Corse: interprétation hydrodynamique et implications structurales. *Marine Geology*, **316**(II), 513–517.

CAZZOLA, C., MUTTI, E. & VIGNA, B. 1985. Cengio Turbidite System, Italy. In: BOUMA, A. H., NORMARK, W. R. & BARNES, N. E. (eds) Submarine Fans and Related Turbidite Systems. Frontiers in Sedimentary Geology, Springer, New York, 179–183.

CONCHON, O. 1975. Les formations quaternaires de type continental en Corse orientale. Ph.D. thesis, University of Paris, VI.

CONCHON, O. 1977. Néotectonique en Corse orientale d'après l'étude des formations Quaternaires: comparaison entre la Marana et la plaine d'Aléria. Bulletin de la Société Géologique de France, **7**(3), 631–639.

CONCHON, O. 1984. Corrélations entre la sédimentation fluviale et la sédimentation marine littorale en Corse. Bulletin de l'Association française pour l'étude du Quaternaire, **1–2–3**, 151–156.

CONCHON, O. 1999. Le littoral de Corse (France) au Quaternaire. Quaternaire, **2–3**, 95–105.

CONCHON, O. & GAUTHIER, A. 1985. Phénomènes naturels exceptionnels en Corse: intérêt pour l'étude géologique de la période Quaternaire. Bulletin de la Société des Sciences Historiques et Naturelles Corse, **648**, 141–165.

DROZ, L. & BELLAICHE, G. 1985. Rhône deep-sea fan: morphostructure and growth pattern. American Association of Petroleum Geologists Bulletin, **69**, 460–479.

DROZ, L. & BELLAICHE, G. 1991. Seismic facie sand geologic evolution of the central portion of the Indus Fan. In: WEIMER, P. & LINK, M. H. (eds) Seismic Facies and Sedimentary Processes of Submarines Fans and Turbidite Systems. Springer, New York, 383–312.

FERRANDINI, J. et al. 1994. First study of the microseismicity of Corsica island with data from a regional seismological network and SISBALIG II network. Comptes Rendus de l'Académie des Sciences, **319**(6), 705–712.

FLOOD, R. D., MANLEY, P. L., KOWSMANN, R. O., APPI, C. J. & PIRMEZ, C. 1991. Seismic facies and late Quaternary growth of Amazon submarine fan. In: WEIMER, P. & LINK, M. H. (eds) Seismic Facies and Sedimentary Processes of Submarine Fans and Turbidite Systems. Springer, New York, 415–433.

FUKUSHIMA, Y., PARKER, G. & PANTIN, H. M. 1985. Prediction of ignitive turbidity currents in Scripps submarine canyon. Marine Geology, **67**, 55–81.

GABIN, R. 1972. Résultats d'une étude de sismique réflexion dans le Canal de Corse, et de sondeur de vase dans le Basin de Toscan. Marine Geology, **13**, 267–286.

GAUTHIER, A. 1981. Contribution à l'étude du débit solide et genèse des sédiments au cours de celui-ci: Exemple du Fium'orbo, fleuve de la Corse Orientale. BRGM Internal Reports, **30**, 163–172.

GAUTHIER, A. & PRONE, A. 1980. Histoire des sables dans un basin fluviatile par exoscopie des quartz: exemple du Fium'Orbo (Corse). Compte Rendu Sommaire des Scéances de la Société Géologique de France, **22**(1), 16–20.

GUIOMAR, M. & SAVOYE, B. 1998. Etude des éventails turbiditiques du Fiume Orbo et du Tavignano, Marge Est-Corse. Institut Géologique Albert-de-Lapparent Memoir.

HELLER, P. L. & DICKINSON, W. R. 1985. Submarine ramp facies model for delta-fed, sand-rich turbidite systems. American Association of Petroleum Geologists Bulletin, **69**, 960–976.

KNELLER, B. C. 1995. Beyond the turbidite paradigm: physical models for deposition of turbidites and their implications for reservoir prediction. In: HARTLEY, A. J. & PROSSER, D. J. (eds) Characterization of Deep Marine Clastic Systems. Geological Society, London, Special Publications, **94**, 31–49.

LAMBECK, K. & BARD, E. 2000. Sea level change along the French Mediterranean coast from the past 30 000 years. Earth and Planetary Science Letters, **175**(3–4), 203–222.

LOPEZ, M. 2001. Architecture and depositional pattern of the Quaternary deep-sea fan of the Amazon. Marine and Petroleum Geology, **18**, 479–486.

MANLEY, P. L., PIRMEZ, C., BUSCH, W. & CRAMP, A. 1997. Grain-size characterization of the Amazon Fan deposits and comparison to seismic facies units. In: FLOOD, R. D., PIPER, D. J. W., KLAUS, A. & PETERSON, L. C. (eds) Proceedings of the Ocean Drilling Program, Scientific Results, **155**, 35–52.

MARANI, M., ARGNANI, A., ROVERI, M. & TRINCARDI, F. 1993. Sediment drifts and erosional surfaces in the central Mediterranean: seismic evidence of bottom-current activity. Sedimentary Geology, **82**, 207–220.

MAUFFRET, A., CONTRUCCI, I. & BRUNET, C. 1999. Structural evolution of the Northern Tyrrhenian Sea from new seismic data. Marine and Petroleum Geology, **16**, 381–407.

MULDER, T. & MANEUX, E. 1999. Flux et bilan sédimentaires: impact des apports fluviatiles sur la construction des éventails sous-marins profonds de la marge Est-Corse. University of Bordeaux I Memoir.

MUTTI, E. 1977. Distinctive thin-bedded turbidite facies and related depositional environments in the Eocene Hecho Group (south-central Pyrenees, Spain). Sedimentology, **24**, 107–131.

MUTTI, E. & NORMARK, W. R. 1987. Comparing examples of modern and ancient turbidites systems: problems and concepts. Marine Clastic Sedimentology, 1–38.

MUTTI, E. & SONNINO, M. 1981. Compensation cycles: a diagnostic feature of sandstone lobes. International Association of Sedimentologists, 2nd European Meeting, Bologna, 120–123.

MUTTI, E. & RICCI LUCCHI, F. 1972. Turbidites of the northern Apennimes: introduction to facies analysis (English translation by NILSEN, T. H., 1978). International Geology Review, **20**, 125–166.

NELY, G., COUMES, F. & CREMER, M. 1985. La réussite exemplaire d'une exploration en mer par des méthodes complémentaires: l'éventail sédimentaire du Cap-Ferret (Golfe de Gascogne, France). Bulletin du Centre de Recherche, Exploration et Production, Elf-Aquitaine, Pau, **9**(2), 253–334.

NORMARK, W. R. 1978. Fan valleys, channels, and depositional lobes on modern submarine fans: characters for recognition of sandy turbidite environments. *American Association of Petroleum Geologists Bulletin*, **62**(6), 912–931.

NORMARK, W. R. & PIPER, D. J. W. 1985. Navy Fan, Pacific Ocean. *In*: BOUMA, A. H., BARNES, N.E. & NORMARK, W. R. (eds) *Submarine Fans and Related Turbiditic Systems*. Springer, New York, 87–93.

NORMARK, W. R. & PIPER, D. J. W. 1991. Initiation processes and flow evolution of turbidity currents: implications for the depositional record. *In*: OSBORNE, R. H. (ed.) *Shoreline to Abyss, Contributions in Marine Geology in Honor of Francis Parker Shepard*. Society of Economic Paleontologists and Mineralogists, **46**, 207–230.

NORMARK, W. R., PIPER, D. J. W. & HESS, G. R. 1979. Distributary channels, sand lobes, and mesotopography of Navy submarine fan, California borderland, with applications to ancient fan sediments. *Sedimentology*, **26**, 749–774.

NORMARK, W. R., PIPER, D. J. W. & HISCOTT, R. N. 1998. Sea level controls on the textural characteristics and depositional architecture of the Hueneme and associated submarine fan systems, Santa Monica Basin, California. *Sedimentology*, **45**(1), 53–70.

ORSZAG-SPERBER, F. & PILOT, M. D. 1976. Grands traits Néogène de Corse. *Bulletin de la Société Géologique de France*, **7**(5), 1183–1187.

OTTMAN, F. 1958. Les formations pliocènes et quaternaires sur le littoral corse. *Memoires de la Société Géologique de France*, **37**(84), 176–181.

PALANQUES, A., ALONSO, B. & FARRAN, M. L. 1994. Progradation and retreat of the Valencia fanlobes controlled by sealevel changes during the Plio-Pleistocene (northwestern Mediterranean). *Marine Geology*, **117**, 195–205.

PICHEVIN, L. 2000. *Etude sédimentaire et sismique d'un éventail turbiditiques sableux: le système récent du Golo (Marge Est-Corse)*. University of Bordeaux I Memoir.

PIPER, D. J. W. & NORMARK, W. R. 2001. Sandy fans: from Amazon to Hueneme and beyond. *American Association of Petroleum Geologists Bulletin*, **85**, 1407–1438.

PIPER, D. J. W., KONTOPUOLOS, N., ANAGNOSTOU, C., PANAGOS, A. G. & CHRONIS, G. 1990. Modern fan deltas in the western Gulf of Corinth, Greece. *Geo-Marine Letters*, **10**, 5–12.

PIPER, D. J. W., HISCOTT, R. N. & NORMARK, W. R. 1999. Outcrop-scale acoustic facies analysis and latest Quaternary development of Hueneme and Dume submarine fans, offshore California. *Sedimentology*, **46**(1), 47–78.

POPESCU, I., LERICOLAIS, G., PANIN, N., WONG, H. K. & DROZ, L. 2001. Late Quaternary channel avulsions on the Danube deep-sea fan, Black Sea. *Marine Geology*, **179**, 25–37.

POSAMENTIER, H. W., ERSKINE, R. D. & MITCHUM, R. M. J. 1991. Models for submarine-fan deposition within a sequence-stratigraphic framework. *In*: WEIMER, P. & LINK, M. H. (eds) *Seismic Facies and Sedimentary Processes of Submarine Fans and Turbidite Systems*. Springer, New York, 127–136.

READING, H. G. & RICHARDS, M. 1994. Turbidite systems in deep-water basin margins classified by grain size and feeder system. *American Association of Petroleum Geologists Bulletin*, **78**(5), 792–822.

REHAULT, J. P., BOILLOT, G. & MAUFFRET, A. 1984. The Western Mediterranean Basin geological evolution. *Marine Geology*, **55**, 447–477.

SEGRE, A. & DEBRAZZI, E. 1960. *Carta bathymetrica del Mediterraneo centrale. Mar Ligure e Tirreno settentrionale*. Istituto Idrographico Marina, Genova (chart).

SHANMUGAM, G. 2000. 50 years of the turbidite paradigm (1950s–1990s): deep-water processes and facies models—a critical perspective. *Marine and Petroleum Geology*, **17**, 285–342.

SHANMUGAM, G. & MOIOLA, R. J. 1988. Submarines fans: characteristics, models, classification and reservoir potential. *Earth Science Reviews*, **24**, 383–428.

STANLEY, D. J., REHAULT, J. P. & STUCKENRATH, R. 1980. Turbid-layer by passing model: the Corsican trough, northwestern Mediterranean. *Marine Geology*, **37**, 19–40.

STOW, D. A. V. & MAYALL, M. 2000. Deep-water sedimentary systems: new models for the 21st century. *Marine and Petroleum Geology*, **17**, 125–135.

VAIL, P. R., MITCHUM, R. M. & THOMPSON, S. 1977. Part four: Global cycles of relative changes of sealevel. *In: Seismic Stratigraphy: Implications to Hydrocarbon Exploration*. American Association of Petroleum Geologists Memoirs, **26**, 83–97.

VIARIS DE LESEGNO, L. 1978. *Etude structurale de la Mer Tyrrhénienne septentrionale*. Ph.D. thesis, University of Paris.

WALKER, R. G. 1978. Deep-water sandstones facies and ancient submarine fans: models for stratigraphic traps. *American Association of Petroleum Geologists Bulletin*, **62**, 932–936.

WYNN, R. B., KENYON, N. H., MASSON, D. G., STOW, D. A. & WEAVER, P. E. 2002. Characterization and recognition of deep-water channel-lobe zones. *American Association of Petroleum Geologists Bulletin*, **86**, 1441–1462.

Multiple terraces within the deep incised Zaire Valley (ZaïAngo Project): are they confined levees?

N. BABONNEAU[1], B. SAVOYE[1], M. CREMER[2] & M. BEZ[3]

[1]*IFREMER, Géosciences Marines, Laboratoire Environnements Sédimentaires, BP70, 29280 Plouzané, France (e-mail: nbabonne@ifremer.fr)*
[2]*Université de Bordeaux I, Département de Géologie et d'Océanologie, Avenue des Facultés, 33000 Bordeaux, France*
[3]*Totalfinaelf, 64018 Pau Cedex, France*

Abstract: Terraces have been frequently observed and described along turbidite valleys. Many interpretations have been aimed at determining the origin of these structures, including a tectonic origin, succession of infilling and incision processes, channel-wall slumps, or inner levee aggradation. The Zaire submarine valley presents a complex structure with multiple terraces bordering a deep incised meandering thalweg. The detailed analysis of the morphology, the seismic structure and the recent sedimentation (in cores) along the Zaire upper-fan valley show that terraces are inner levees confined within the incised valley. Many terraces correspond to the infilling of abandoned meanders, and aggrade by deposition of turbidite sequences due to current overflows. The major process affecting the initiation and the development of terraces inside the valley is the vertical incision of the thalweg, simultaneously with meander migration.

The development of 'terrace' features inside submarine valleys has been frequently described on reflection seismic or sonar imagery across numerous canyons and channels in the world (Damuth *et al.* 1988; Hagen *et al.* 1994; Kenyon *et al.* 1995; Masson *et al.* 1995; O'Connell *et al.* 1995; Carter & Carter 1996; Klaucke & Hesse 1996; Hübscher *et al.* 1997; Torres *et al.* 1997; Nakajima *et al.* 1998; Piper & Normark 2001). The origin of terraces in turbidite valleys is often not well identified. In some cases, structural influence seems necessary to generate terraces, e.g. fault or gravity tectonics (Cramez & Jackson 2000; Chow *et al.* 2001). But, in most cases, terraces seem to be generated by sedimentary processes: slump or true depositional unit.

The Zaire turbidite valley is particularly well imaged by bathymetric and seismic data (ZaïAngo data). Multiple terraces have been recognized along the upper-fan valley. They are stepped, and their organization appears very complex. The understanding of these structures has required a detailed morphological analysis (bathymetry) and the identification of their internal architecture (seismic data). Also, the characterization of the sedimentary processes occurring in the valley is obtained by the analyses of several Kullenberg cores collected onto terraces. The main purposes of this paper are:

(1) to identify the origin of the terraces of the Zaire valley;
(2) to understand the processes occurring in this area (sedimentation, flow behaviour); and
(3) to determine the recent evolution of the valley.

Description and interpretation of terraces along submarine valleys

In turbidite environments, the term *terrace* is generally used for topographic flats bordering a channel (with no specific implications for process or origin).

Along most large turbidite valleys studied in the world, terraces have been frequently described inside canyons or channels. Side-scan sonar imagery has been used to describe the detailed morphology of terraces in the canyon of the Rhone Fan (O'Connell *et al.* 1995; Torres *et al.* 1997), the Monterey Fan (McHugh *et al.* 1992; Masson *et al.* 1995), the NAMOC channel (Klaucke & Hesse 1996), and other canyons located on the Japan forearcs (Soh *et al.* 1990; Klaus & Taylor 1991; Nakajima *et al.* 1998) and within the Peru–Chile forearc (Hagen *et al.* 1994). With seismic-reflection data, the term 'terrace' has not always been used, even for features that morphologically resemble terraces. Some are interpreted in terms of *canyon-fill sediment*, e.g. the Indus channels (McHargue & Webb 1986), in terms of *inner levees*, e.g. the Bengal channel (Hübscher *et al.* 1997), or in terms of *slump* or *slide* as for the Kaoping canyon (Liu *et al.* 1993). Some different teams have proposed different origins for the

From: LOMAS, S. A. & JOSEPH, P. (eds) 2004. *Confined Turbidite Systems.* Geological Society, London, Special Publications, **222**, 91–114. 0305-8719/04/$15.00 © The Geological Society of London 2004.

a Entrenchment of a small channel into the valley fill in response to an avulsion

Rhône Fan (O'Connel et al., 1995)

b Rotational slumps of channel walls

Indus Fan (Kenyon et al., 1995)

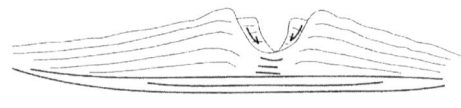

c Slumps induced by active gravity tectonic

Kaoping Canyon (Liu et al., 1993)
Zaire/Congo Canyon (Cramez and Jackson, 2000)

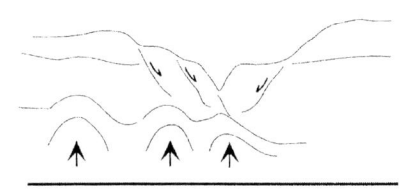

d True inner-levees developed during sea-level rise (decrease of flow volume)

Bengal Fan (Hübscher et al., 1997)
Hueneme Fan (Piper et al. 1999)

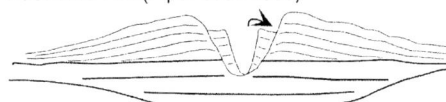

e Aggradation of deposits during thalweg migration or Partial infilling of channel as flow volume decreased

Indus Fan (Mc Hargue et al., 1986)
Amazon Fan (Damuth et al., 1988)
Toyama Channel (Nakajima et al., 1998)

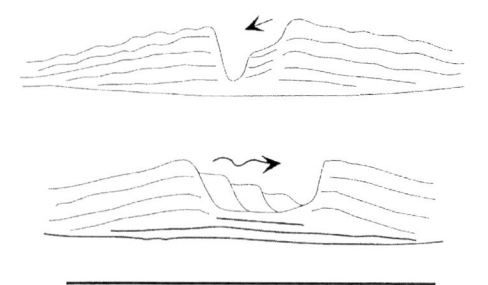

f Development of a small confined channel-levee on valley fill

Rhône Fan (Torres et al., 1997)

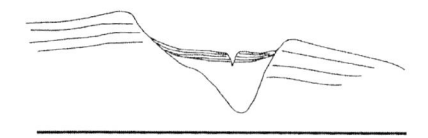

g Meander abandonement and continuous thalweg entrenchment

Arequipa Basin (Hagen et al., 1994)
Indus Canyon (Von Rad and Tahir, 1997)

No seismic data

Fig. 1. Various models for the interpretation of terraces along turbidite channels.

same terraced reach (for example, O'Connell et al. 1995, and Torres et al. 1997, concerning the Rhone valley). There are at least four main processes explaining terrace formation in submarine areas.

Valley incision

The first process is valley incision as for fluvial valleys, where terraces correspond to relic floodplain relief due to incision of a channel. A succession of filling and incision phases could lead to the formation of a terrace complex. Comparable scenarios have been used to explain the formation of some terraces observed along submarine canyons. For example, along the Petit Rhone canyon, terrace formation has been attributed to the incision of the modern thalweg into the fan-valley fill (O'Connell et al. 1995), in response to an avulsion located downslope (Fig. 1a). The first stage was the filling of the fan valley during the present sea-level rise. Then an avulsion occurred along the channel–levee system and induced the entrenchment of a new minor thalweg, to readjust the slope gradient (O'Connell et al. 1995).

Slump or slide

A slump origin has mostly been used to explain the terraces observed within submarine valleys (Stubblefield et al. 1982; Carlson & Karl 1988; Klaus & Taylor 1991; McHargue 1991; Liu et al. 1993; Carter & Carter 1996; Klaucke &

Hesse 1996). As slumps are commonly observed in canyons, many terraces have been interpreted as channel-wall slumps without any particular evidence for this interpretation (Fig. 1b). Along the Monterey channel, arcuate scarps seen on side-scan images and associated with terrace morphology were first interpreted as slump scars on channel walls (Masson et al. 1995). However, they look approximately parallel to the direction of overbank flows, suggesting a relationship with overbank processes (Masson et al. 1995).

In tectonically active areas, slumps could be linked to faulting. For example, along the Kaoping canyon (Liu et al. 1993), slumps of canyon walls are probably linked to gravity tectonics (Fig. 1c), generated by mud diapirs (Chow et al. 2001).

For the Bounty channel, a mixed structural and sedimentary interpretation has been suggested (Carter & Carter 1996). Terraces appear to have originated as sidewall slumps (basal disturbed strata), subsequently draped by overflow sediments.

Inner levees

In other cases, some terrace features have been interpreted as *inner levees* (Fig. 1d), for example along the Bengal channel–levee system (Hübscher et al. 1997) and along the Hueneme channel (Piper et al. 1999; Piper & Normark 2001). During the Holocene, the youngest channel was still active but narrower and constricted (as flow volume decreased). When the outer levee growth stopped, inner levee segments aggraded within the channel.

There is also evidence for inner levee aggradation in many published seismic data from submarine fans (Fig. 1e), for example the Amazon Fan (Damuth et al. 1988), the Toyama channel (Nakajima et al. 1998), and the Indus Fan (McHargue & Webb 1986). The main hypotheses for the origin of these structures are (1) the aggradation of deposits during thalweg migration (McHargue & Webb 1986; Nakajima et al. 1998), and (2) the partial infilling of the channel as flow volume decreases (Nakajima et al. 1998). Without complementary sonar imagery or bathymetric data, it is difficult to discriminate between the two hypotheses.

It is interesting to note that terrace features, which look very similar on seismic reflection profiles, have been interpreted as rotational slumps of channel walls (Fig. 1b) along the Indus channel–levee system (Kenyon et al. 1995) and as inner levees (Fig. 1d) along the Bengal channel–levee system (Hübscher et al. 1997).

On the Petit Rhône canyon, terrace formation was first attributed to the incision of a modern thalweg into a fan-valley fill (O'Connell et al. 1995). Later, this hypothesis was re-evaluated by Torres et al. (1997) in the light of new data (Fig. 1f). Cores showed that the inner thalweg is associated with true overbank deposits. Thus the thalweg and the upper unit of the valley fill are genetically related and represent a small channel–levee system confined within the Petit Rhône valley (Torres et al. 1997). During Quaternary lowstands of sea-level, a large Rhône fan valley was created. During the last sea-level rise, the incision was progressively filled. The turbidity current activity decreased but remained active, leading to the formation of a small channel–levee complex confined inside the large fan valley (Torres et al. 1997).

Abandoned meanders

Finally, small terraces, bordered by arcuate canyon-facing scarps and described along the main submarine canyon of the central part of the East Arequipa basin (Peru–Chile forearc; Hagen et al. 1994), and along the Indus canyon (Von Rad & Tahir 1997), seem to be generated by meander migration. These features were initially interpreted as being the borders of large slumps. Improvements in side-scan techniques led Hagen et al. (1994) to reinterpret these arcuate scarps as the borders of meanders formed and abandoned during the entrenchment of the meandering channel (Fig. 1g). The presence of abandoned channel segments or cut-off meander loops on some of these terraces supports this interpretation. In this case, terrace formation appears linked to simultaneous meander development and thalweg entrenchment (Hagen et al. 1994). Deposition of levee-type sediment over the lowest terraces is suggested, but the internal structure of terrace units is not known (no seismic data).

Thus we see that the origin of terraces observed within the modern submarine canyons or valleys is controversial. The detailed study of the terrace complex located within the Zaïre valley may help to discriminate between these different structural or sedimentary origins.

Geological setting of the studied area

The Zaire deep-sea fan is located on a mature (Fig. 2), passive continental margin resulting from the Early Cretaceous opening of the South Atlantic Ocean (Jansen et al. 1985; Marton et al.

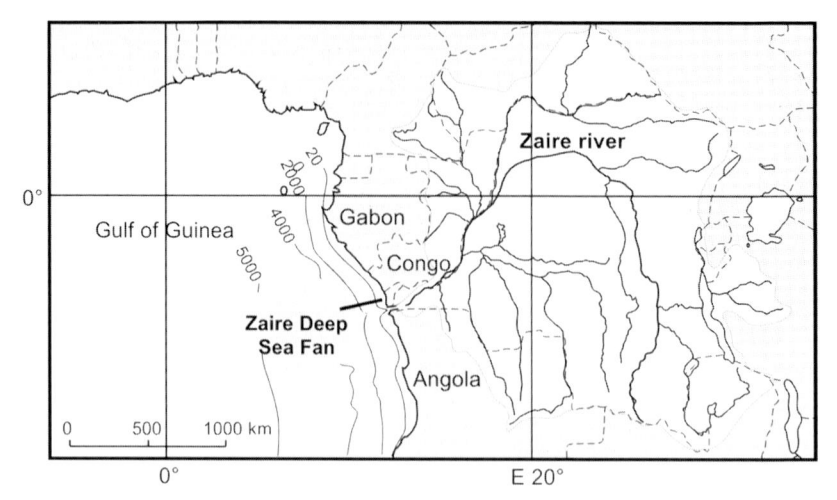

Fig. 2. Location map of the Zaire-Congo deep-sea fan in the Gulf of Guinea. The extension of the Zaire watershed is indicated in light grey.

2000). Since Oligocene time, the Lower Congo basin has been affected by the huge Zaire turbidite system, which constitutes the Zaire deep-sea fan. The sedimentary succession is highly deformed by gravity tectonics: fault, salt creeps and diapiric structures (Fig. 3).

The Zaire fan is one of the largest modern deep-sea fans. It extends from the Congo–Angola continental slope to abyssal depths (>5100 m) (Fig. 3). It is currently still active and fed by a single canyon, which is connected directly with the Zaire River Estuary (Heezen

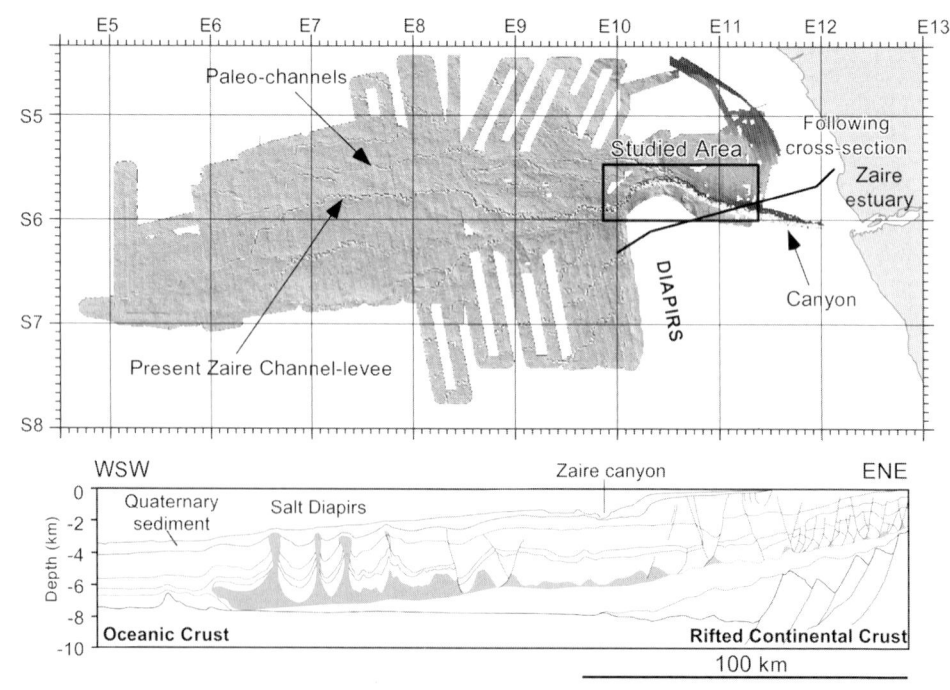

Fig. 3. Shaded bathymetric map acquired during ZAIANGO and GUINESS cruises. The canyon, the present channel and the palaeo-channels are clearly visible in the morphology of the seafloor. Below the map, a schematic cross-section of the margin shows the deep structure of the margin and the influence of salt diapirism deformation (after Marton et al. 2000).

et al. 1964; Van Weering and Van Iperen 1984).

Multibeam bathymetry surveys show that the Quaternary Zaire fan is characterized by a multitude of fossil channel–levee systems (Droz *et al.* 1996; Rigaut 1997; Savoye *et al.* 2000; Droz *et al.* submitted), frequently still visible on the present morphology of the sea floor (Fig. 3). The active channel is located in the axial part of the system. Based on a detailed morphological and architectural analysis of the active channel (Babonneau *et al.* 2002), the turbidite system has been divided in several zones: the canyon, the upper-fan valley, the channel–levee and the distal lobe complex.

The *Zaire canyon* is relatively straight, and deeply incises the continental shelf and the slope (depth of 1300 m at 100 km from the coast). Its morphology (V-shaped profile) and the sedimentary patterns (gullies, distributary canyons) indicate intense erosive processes (Babonneau *et al.* 2002). Across the lower part of the continental slope the Zaire valley is characterized by an incised meandering thalweg bordered by a complex of terraces, and corresponds to the *upper-fan valley*.

Toward the continental rise and the abyssal plain, the morphology and structure of the present Zaire turbidite system are typical of a *channel–levee system*. The channel is highly meandering along its upper part, but becomes less sinuous downstream. The whole present channel floor is incised beneath the base of the levees, and no significant thickness of sediments has been identified in the thalweg, on seismic sections (Babonneau *et al.* 2002), indicating a recent stage of incision. The present Zaire turbidite system terminates at about 760 km from the coast, with a *lobe complex* at 5000 m water depth (Savoye *et al.* 2000).

Data and methods

The results presented in this paper are based on the analysis of bathymetric, seismic and core data acquired during the ZAIANGO 1 and 2 (1998), ZAISAR (2000), ZAICAR (2000) and ZAIROV (2000) cruises. During these cruises, ship navigation was based on DGPS positioning (±3 m). In addition, data collected on the earlier GUINESS cruises (Droz *et al.* 1996; Rigaut 1997) have been incorporated.

Bathymetric data

The bathymetric map is a synthesis of Simrad EM12 dual multibeam data acquired during the GUINESS and ZAIANGO cruises. Complementary data have been collected with the Simrad EM300 dual multibeam during the ZAICAR cruise (2000). EM300 bathymetric data have higher vertical and lateral resolution for water depths shallower than 3500 m. At 3000 m depth, EM12 data provide a terrain numerical grid with 100 m horizontal grid spacing and a vertical resolution around ±10 m, whereas EM300 data provide a grid with 50 m horizontal spacing and a vertical resolution around ±5 m.

The abrupt channel flanks (cliffs) are not always well surveyed by the EM12 tool, because of the large horizontal cover for a single beam (80–100 m). The relatively low horizontal resolution of the EM12 tool induces a high variability of the bathymetric data along the channel axis due to artefacts linked to the influence of channel flanks in the measurement. The EM300 provides a more accurate bathymetric map (Fig. 4). Channel contours and steep morphologies are better defined. The comparison of bathymetric profiles following the present thalweg (obtained with either the EM12 or the EM300) shows that the EM12 profile is very irregular, whereas the EM300 profile is smooth (Fig. 4). Because of the larger horizontal beam cover and the narrowness of the morphology, the EM12 bathymetric data of the channel floor are frequently undervalued (too shallow). Thus the true bathymetric profile of the channel floor is closer to the EM300 profile.

Acoustic and seismic data

3.5 kHz echo-sounder profiles were obtained with the SAR system during the ZAISAR cruise. The SAR is a deep-towed side-scan sonar operating at 170–190 kHz that is towed at 100 m above the seafloor and is equipped with a 3.5 kHz acoustic profiling system. The SAR system provides high-resolution information on seafloor texture and microtopography.

Seismic profiles (locations shown in Fig. 5) were acquired during the ZAIANGO 1 and 2 and ZAISAR cruises. Three different seismic acquisition configurations were used. Profiles Z2-29, Z2-42 and TZ2-81 are high-resolution seismic profiles with air-gun sources (6 GI) and a 2500 m long, 96-channel streamer. Profile ZSAR-37 is a high-resolution profile with air-gun source (mini-GI) and a single-channel streamer. Profiles ZSAR-38 and 40 are PASI-SAR profiles with air-gun source (mini-GI) and a single-channel streamer towed close to the seafloor behind the SAR system.

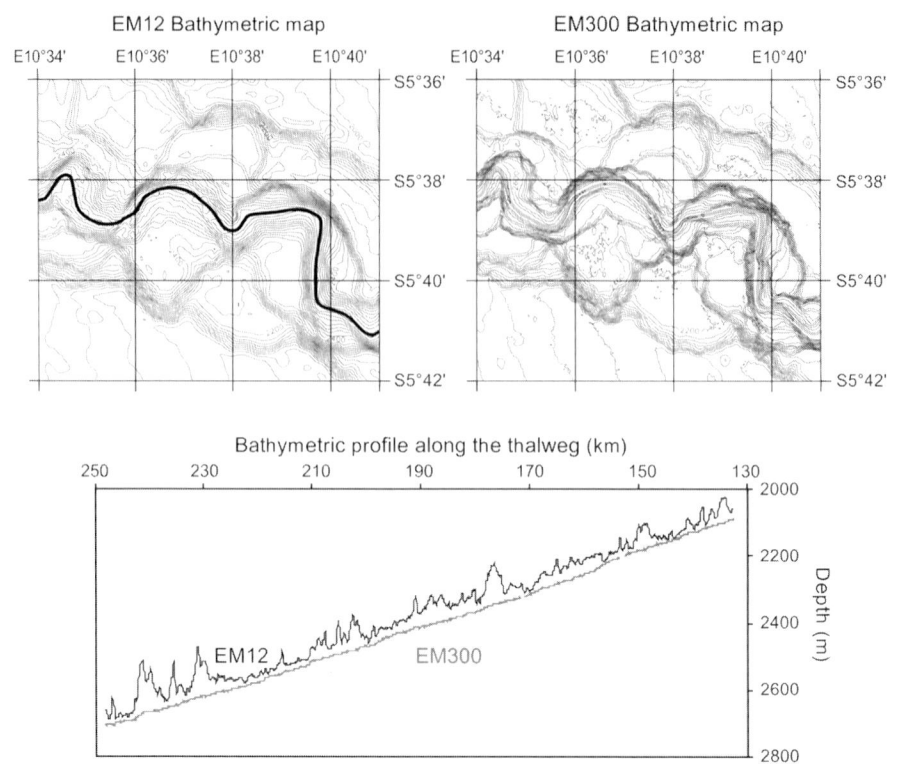

Fig. 4. Comparison of EM12 and EM300 bathymetric data in map form and along bathymetric profiles of the thalweg (contour interval: 20 m).

Throughout our studied area, the seismic signal is perturbed by the presence of gas hydrate and gas in the sediments, and especially by the BSR (bottom simulating reflector). Thus the interpretation of seismic facies in term of sedimentary units requires precise and careful attention.

Cores

Two Kullenberg cores, KZR23 and KZR24, were collected on two terraces during the ZAIROV cruise (2000). They have been analysed using a Multi-Sensor Core Logger (Geotek Ltd), which provides gamma density, magnetic susceptibility and P-wave velocity measurements. Numerous fissures along these cores disturbed the measurements. Thus only magnetic susceptibility curves gave significant results along the two studied cores.

Visual descriptions distinguished several main lithologies (clay, silty clay, silt and fine sand), and vertical successions of sedimentary facies. A series of 1 cm thick sediment slabs were collected for each split core section and X-radiographed using a digital X-ray imaging system SCOPIX (Migeon *et al.* 1999). Detailed images of the sedimentary structures are obtained after X-ray image processing.

Results

The studied area is focused on the upper-fan valley (Fig. 5), which is located across the lower part of the continental slope (between 1500 and 3000 m water depth).

General morphology

Along this area, the Zaire valley evolves from a wide and deep canyon to a channel–levee system. A series of four seismic profiles (Fig. 6) underlines the important changes in shape and dimensions that occur along the section of the Zaire valley.

The upstream part visible on the bathymetric map (Fig. 5) corresponds to the canyon area.

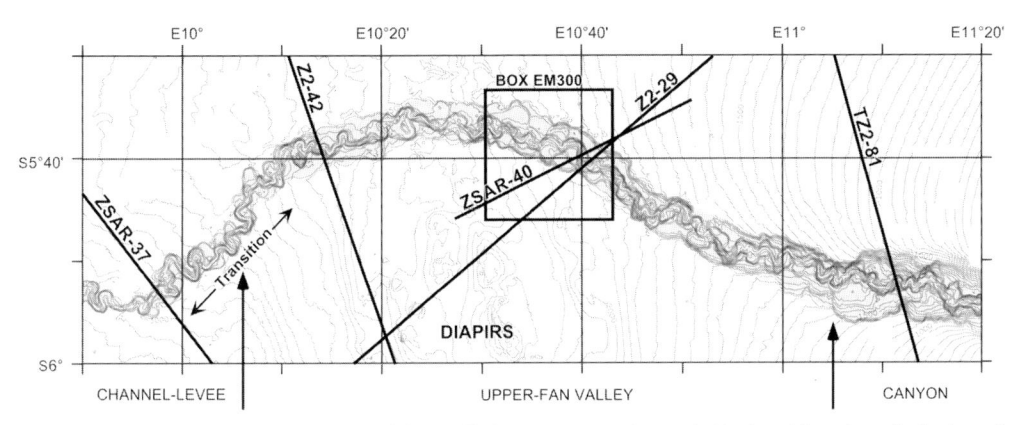

Fig. 5. General EM12 bathymetric map of the studied area (contour interval: 50 m) and location of seismic and EM300 data. Greyed area corresponds to the Zaire turbidite valley.

The Zaire Canyon is 10–12 km wide and more than 800 m deep. The thalweg is 200–250 m wide and moderately sinuous. The morphology of the canyon walls shows large concave patterns, which could suggest slump scars. The seismic profile TZ2-81 (Fig. 6) shows an overall V-shaped profile, typical for erosional canyons, both down the canyon flanks and along the canyon thalweg. The processes occurring along this incised canyon are mainly erosion and transport of sediment. Flow energy along the canyon is powerful enough to maintain a regular and empty thalweg. Several small terraces are visible on the lower part of the valley flanks (see profile TZ2-81).

Downslope, the valley direction changes toward the WNW and then curves to avoid an area to the south uplifted by salt diapirism. Through this area, the valley width is relatively constant around 8 km. The thalweg is highly meandering. The valley relief gradually decreases from 750 m down to 350 m downslope. Valley walls have several levels of terrace. Seismic profiles Z2-29 and Z2-42 (Fig. 6) illustrate the morphology and the structure of the valley along this area, corresponding to the upper-fan valley.

Downslope, the valley direction is SW. The thalweg is highly sinuous and bordered by small isolated terraces (Fig. 5). The valley progressively becomes a channel less than 350 m deep and about 2 km wide. On seismic profile (ZSAR-37, Fig. 6), true levee structures are identified on both sides of the channel. Levee deposits are about 70 m thick for a total channel depth of about 300 m. Thus the channel floor is highly over-incised (about 230 m below the levee base). It is highly meandering and locally bordered by small terraces, which look like abandoned meanders.

The upper-fan valley

The upper-fan valley is characterized by the presence of stepped terraces (Figs 7 & 8). Locally, there are up to four or five levels of terrace on the same section.

Detailed morphological analysis of EM300 data. A detailed morphological analysis of the EM300 bathymetric data has been carried out. These data are available only in a small area of the upper-fan valley (see location in Figs 7 & 8). The morphology of terraces varies according to their position in the valley. We distinguish three types of terrace. First, *type A* corresponds to the highest and largest terraces. They are very flat and commonly elongated (from 5 to 12 km). Second, terraces of *type B* are characterized by an arcuate pattern. They are 2–3 km long and wide, and very flat too. The third type (*type C*) corresponds to the lowest topographic flats inside the valley. They are situated at less than 100 m above the thalweg floor. They are narrow (less than 1 km) and elongated along the thalweg axis. They are generally inclined toward the thalweg, and commonly correspond to local gentle slopes along the flank.

Figure 9 presents a bathymetric map in which the colours correspond to the relative height of terrace according to the thalweg floor (dark blue shows the highest terraces and red the lowest). No preferential levels of terraces can be found along this section of valley. The terrace pattern does not appear organized, each terrace being at a different level. Terraces of type A are mainly dark blue, light blue and green (more than 300 m above the present thalweg floor). Terraces of type B are characterized by yellow and light orange colours (between 150 and 300 m

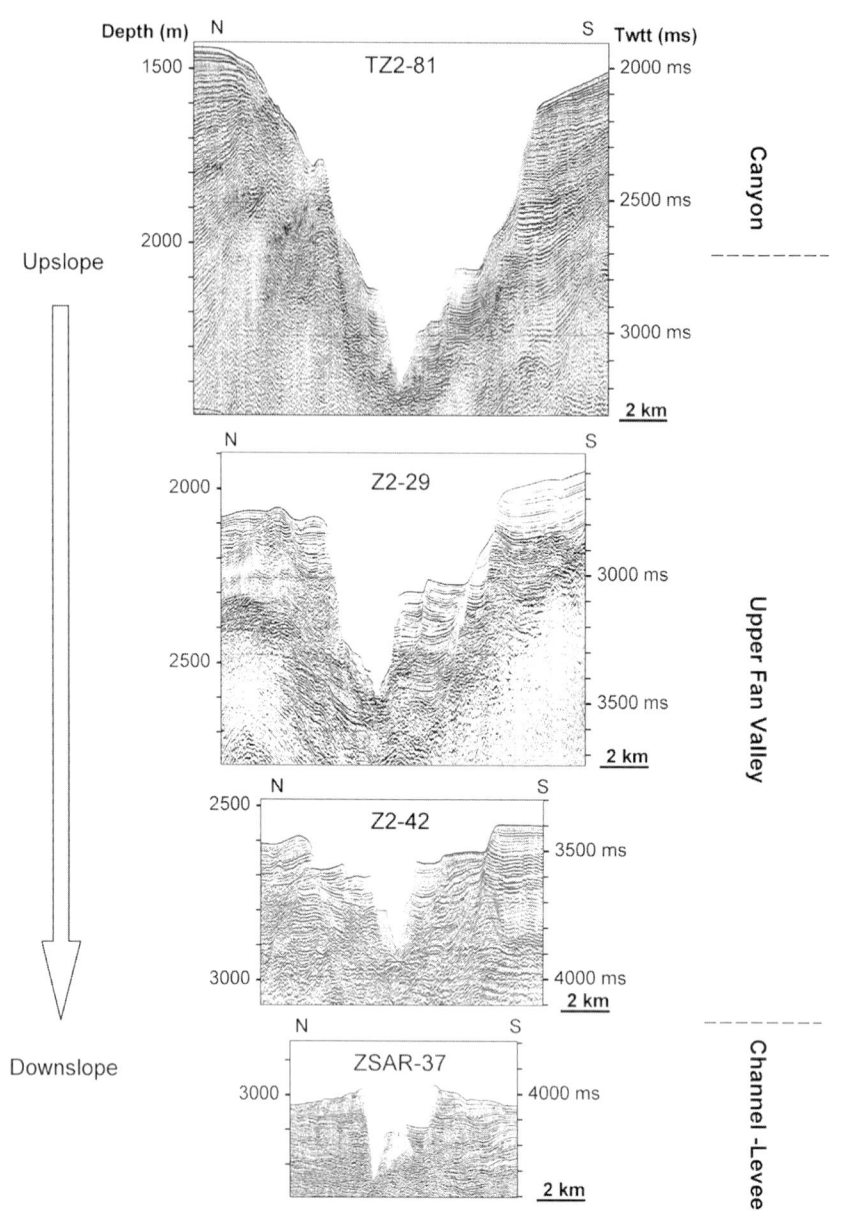

Fig. 6. Series of seismic profiles across the Zaire valley, from upslope to downslope. Three main areas are distinguished: the canyon, the upper-fan valley, and the upper channel–levee.

above the thalweg floor). Terraces of type C are red and dark orange (less than 150 m above the channel floor).

Seismic profiles. Most parts of the continental slope, except for the canyon, are overlain by a continuous and parallel seismic unit (profile Z2-42, Fig. 10). Its seismic facies is characterized by continuous, low-to-moderate amplitude reflectors. Cores indicate that it corresponds to thick hemipelagic sediments, adjacent to the Zaire canyon and upper-fan valley (cores KZR31 and FZ2-11 in Fig. 7). This proves that there is no recent overflow of turbidity currents out of the canyon in this area. In Fig. 10 (Z2-42), the thalweg floor is 375 m below the regional

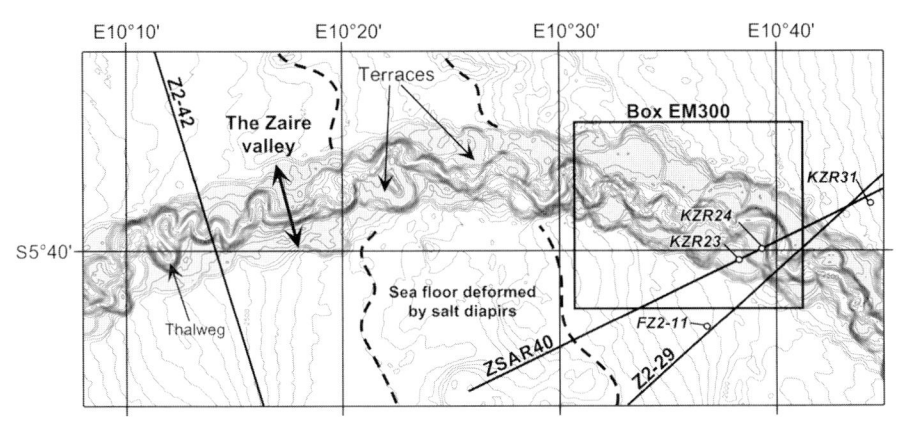

Fig. 7. Detail of the EM12 bathymetric map along the upper-fan valley (contour interval: 25 m). Note location of the cores.

seafloor. On this seismic profile, several terraces are well developed on both sides of the channel, between 325 and 250 m above the channel floor. They seem to be built by depositional units, which are characterized by moderate amplitude and relatively continuous reflectors and by a wedge-shaped architecture that thickens toward the channel.

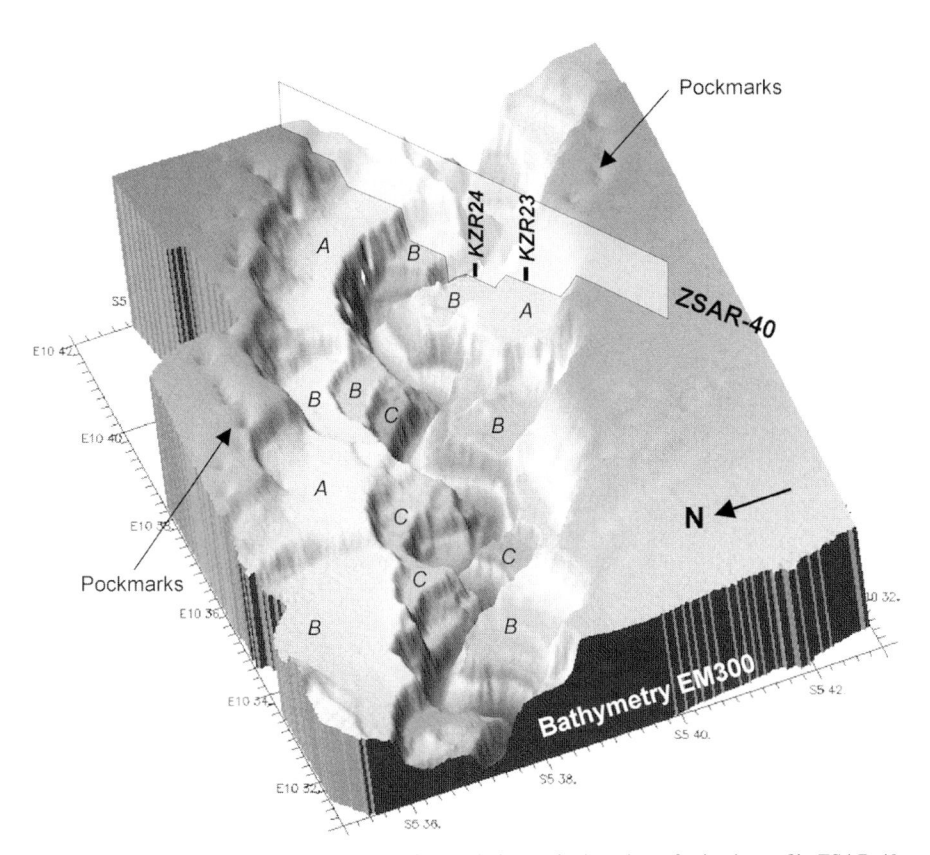

Fig. 8. Three-dimensional view of the EM300 bathymetric box. The location of seismic profile ZSAR 40 and cores KZR23 and KZR24 is indicated.

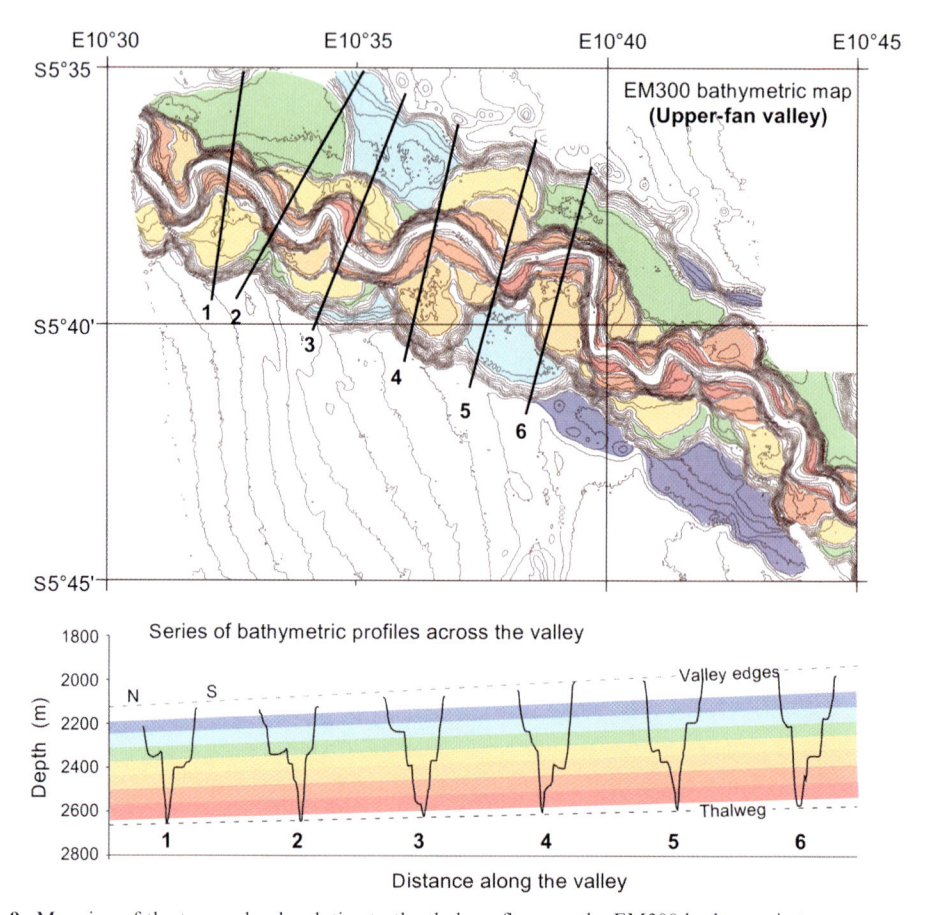

Fig. 9. Mapping of the terrace levels relative to the thalweg floor on the EM300 bathymetric map.

Seismic profile Z2-29 (Fig. 11) is located upstream from Z2-42. The thalweg is deeper with respect to the regional seafloor (550 m below the valley edges). As in the previous seismic line, hemipelagic deposits are identified on the canyon edges. The pre-existing sedimentary successions are highly deformed, especially where dome structures induced by underlying salt diapirs are observed. Two main terraces are visible on the north side of the valley, situated about 350 m above the thalweg. The sedimentary units that create the terraces have a bedded seismic facies and wedge-shaped architecture thickening towards the channel (about 200 m thick). Under these units, seismic facies have higher amplitude, are less well bedded, and probably correspond to the basal units of the terraces.

ZSAR-40 (Fig. 12) is a PASISAR seismic acquisition configuration, located near Z2-29 (Fig. 7). It shows the same seismic sequences as on Z2-29, with a better resolution but a lower penetration. Four main terraces are visible within the Zaire valley (two on each side). The terraces described on this profile appear slightly concave up, rising towards both the canyon wall and the thalweg. This concave shape is also seen in many of the buried reflectors.

3.5 kHz acoustic profile. Figure 13 illustrates the part of the 3.5 kHz SAR acoustic profile corresponding to the seismic line ZSAR-40 that crosses the terraces on the south flank of the valley. Cores have sampled the top of each terrace (see their location in Figs 8 & 13). The lower terrace is located at 230 m above the present thalweg and the upper terrace at 380 m. On this profile, several main acoustic facies are identified. Valley edges are composed of a transparent facies, which corresponds to hemipelagic deposits. The top of the lower terrace is composed of more than 60 m of bedded facies (the estimation of the thickness is limited by the penetration of the 3.5 kHz tool). The upper terrace is composed of an upper unit (from 15

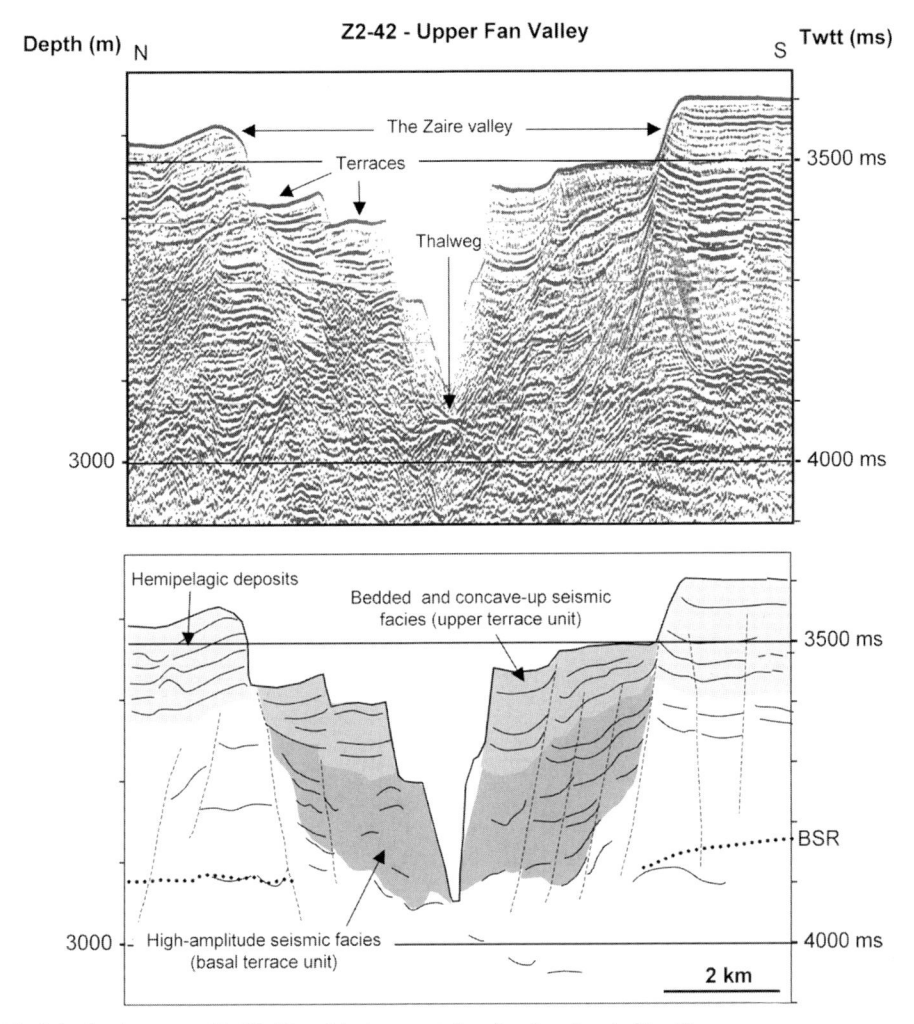

Fig. 10. Seismic air-gun profile Z2-42 and its interpretation (see location in Fig. 7).

to 20 m thick), characterized by a wedge pattern and a quasi-transparent facies with some low amplitude reflectors. The underlying unit is also about 15–20 m thick with a strongly bedded acoustic facies. The 3.5 kHz signal is strongly attenuated by this facies, but it is possible to identify a third unit that resembles the uppermost unit in terms of thickness and facies. It overlies a fourth unit probably comparable to the second. The top of this terrace thus seems to be composed of an alternation of 20 m thick units of bedded facies and quasi-transparent facies.

Core results. Two cores have sampled the top of two terraces in the upper-fan valley: KZR23 and KZR24 (see their location in the acoustic profile ZSAR-40 in Fig. 13). KZR24 sampled 18.50 m of the top of the lower terrace situated at 230 m above the thalweg floor, entirely within the bedded facies. KZR23 sampled 19.59 m of the top of the upper terrace situated at 380 m above the thalweg, entirely within the quasi-transparent facies. Simplified sedimentological logs and magnetic susceptibility curves are presented in Figure 14. These two cores can be compared to the core FZ2-11, which sampled 6.52 m of the top of the hemipelagic succession outside the turbidite valley (see location in Fig. 7).

Core FZ2-11 is composed of homogeneous dark brown-green mud. No structures, lithological changes or bioturbation are visible. The microfacies is characterized by abundant diatoms and radiolaria and a few foraminifera.

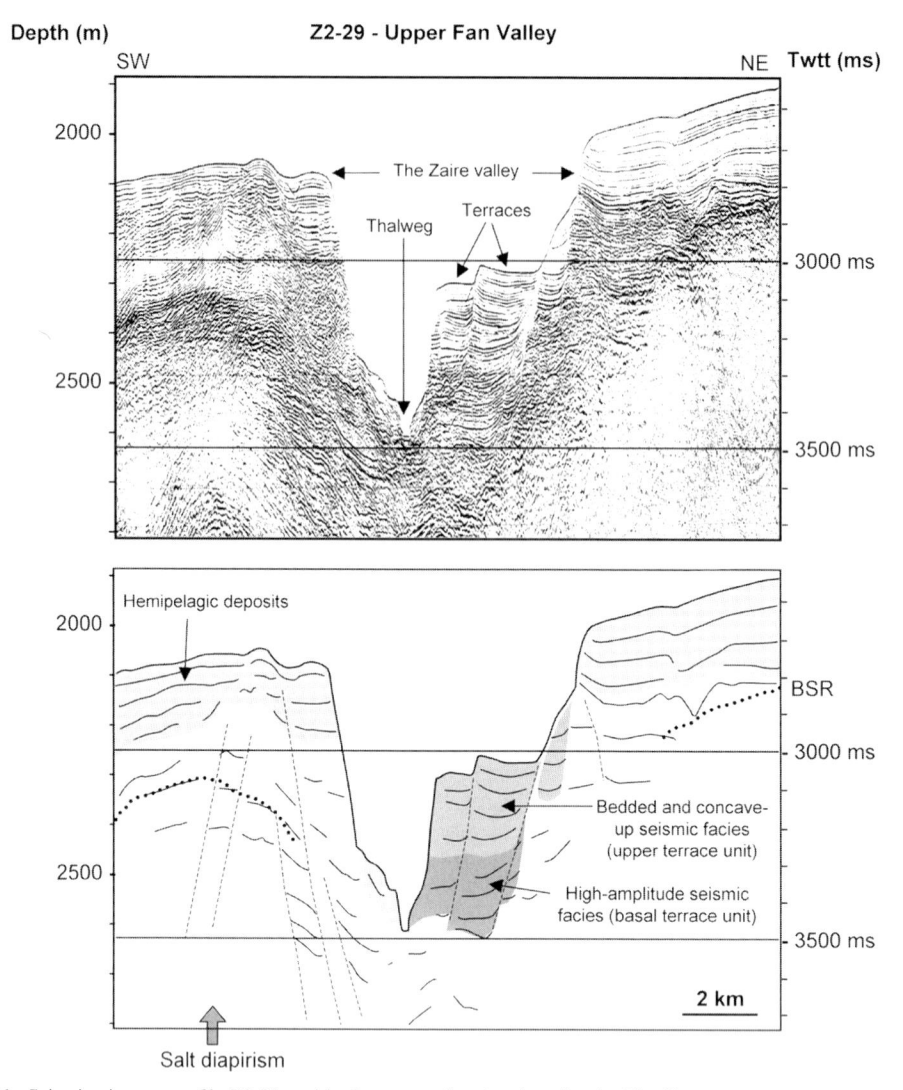

Fig. 11. Seismic air-gun profile Z2-29 and its interpretation (see location in Fig. 7).

In *KZR24*, the first 3 m below the sea floor (Fig. 14) are composed of homogeneous clay. No structures or sequential changes in lithology are visible. Only a gentle bioturbation disturbs the apparent homogeneity of sediments. Deeper in the core, the facies is less homogeneous. Some vague alternations of dark-brown mud and brown-green mud and fine silty lenses are distinguished. X-ray images show that these alternations are true sedimentary structures (Fig. 15). Changes of intensity of X-ray transmission and smear slides sampled along the core suggest that these structures probably correspond to very fine turbidite sequences. Two types of turbidite sequence are determined (Fig. 15). The most common turbidite sequence in core KZR24 is sequence A, characterized by a sharp lower contact of structureless silty clay grading upward into clay. The silty fraction is low and changing. It is generally limited to the lower 5 cm of the sequence. Rare small plant remains are observed. The clayey upper part of sequence A is very homogeneous and often bioturbated. Its thickness varies from 10 cm to 50–60 cm. Diatoms and radiolaria are abundant. Sequence B is characterized by a basal silty laminated unit (Fig. 15). This unit is 1–2 cm thick and shows sharp lower and upper contacts. It is overlaid by structureless silty clay grading upward into homogeneous clay. The coarser granulometry, the better

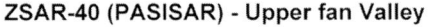

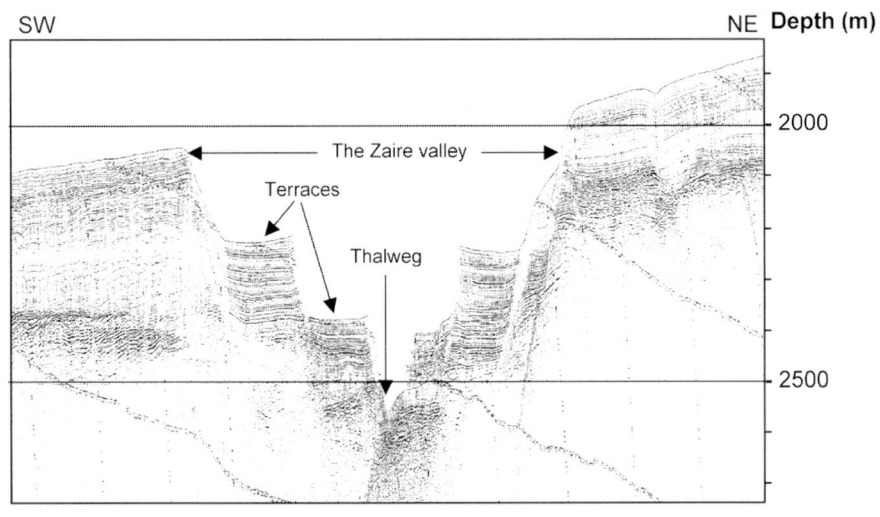

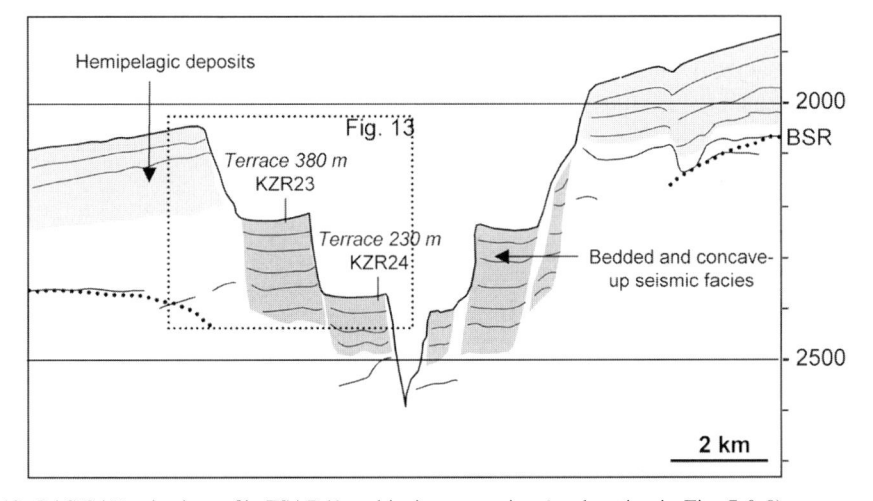

Fig. 12. PASISAR seismic profile ZSAR40 and its interpretation (see location in Figs 7 & 8).

grading and the laminated structure indicate higher current energy than for the sedimentation of sequence A. Only four sequences B are observed along the entire core KZR24.

Core KZR23 is composed of bioturbated facies along its whole length. The upper 5 m are characterized by dark clay with big burrows filled by green clay (Fig. 16a). No sedimentary structures and lithological heterogeneity have been identified by visual description or X-ray image analysis. Smear slides show abundant biogenic material, essentially diatoms, radiolaria and foraminifera. The quantity and diversity of foraminifera are low, and decrease downward along the core. It seems very similar to the foraminifera content of the regional hemipelagic sediment (sampled on the continental slope out of the Zaire canyon). Below 15.5 m of homogeneous clayey sediments, several graded clayey and silty sequences are identified (mainly type A sequences). The most marked turbidite sequence along core KZR23 is shown in Fig. 16b (sequence C). Sequence C is characterized by coarser facies than in sequences A and B, and probably indicates higher current energy. The basal unit of this sequence is 1 cm of black silty clay. The lower contact is sharp, and inverse grading is observed. It is overlaid by 1–2 cm of structureless, very fine sand. Upward, fine laminae of clayey silt and very fine sand are visible. Except for the basal unit, the sequence is normally graded and grades upward into structureless clayey silt and finally

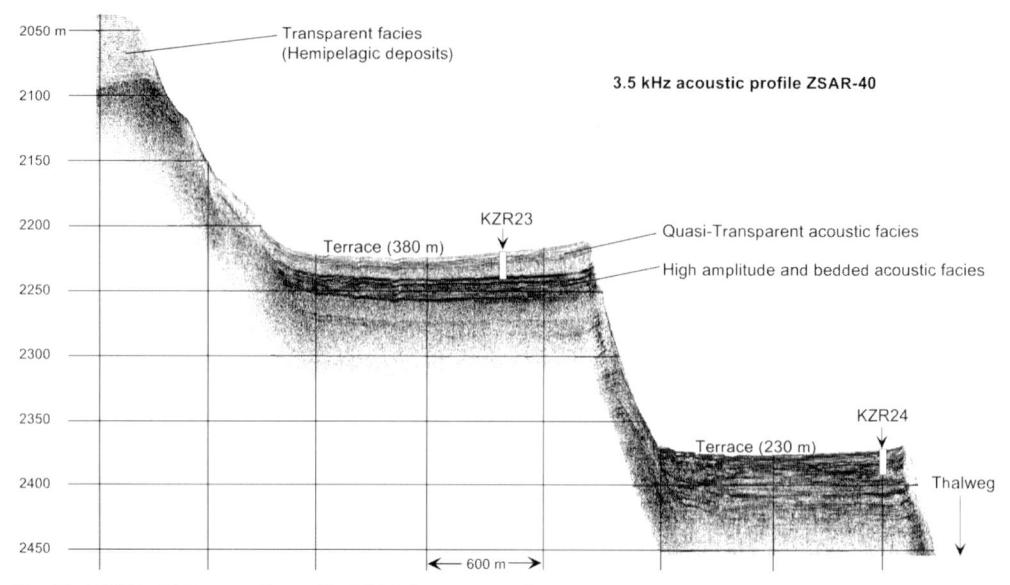

Fig. 13. 3.5 kHz SAR acoustic profile ZSAR40 corresponding to the previous seismic profile and cores location (KZR23 and KZR24).

into clay. Plant remains are relatively frequent in the black clayey-silt facies. Carbon-14 dating provide age estimates of the deposits: 6.5 and 12.5 ka BP respectively at 810 and 1690 cm (KZR23), and an average accumulation rate of about 145 cm/ka.

Significance of the acoustic facies. The integration of core results with the 3.5 kHz profile ZSAR-40 allows correlation of the sedimentological facies with the two main acoustic facies (Fig. 13). The quasi-transparent acoustic facies corresponds to the highly bioturbated facies composing the upper 15 m of KZR23, where big burrows are observed. The deposits are mainly composed of clay. Although the lithological facies seems similar to the regional hemipelagic deposits, the acoustic facies shows that these two sedimentary accumulations are different. The quasi-transparent facies is vaguely bedded and the sedimentary thickness varies, so that it has a wedge-shaped architecture. Moreover, the accumulation rate, estimated at 145 cm/ka for this facies, is very high in comparison with the regional hemipelagic sedimentation calculated at 15–20 cm/ka (Dennielou *et al.* 2001). Thus the origin of this upper sedimentary unit sampled in core KZR23 is interpreted as mixed: from the settling of the upper turbidity plume and from the true hemipelagic sedimentation.

The bedded acoustic facies correlates with the fine-grained turbidite sequences, as in KZR24 and the lower part of KZR23. Clayey and silty sequences are well preserved in spite of bioturbation. Silty layers are more frequent, and rare fine sandy laminae are observed at the base of sequences (sequences A, B and C in Figs 15 & 16).

Transition to the channel–levee system

The transition area from the upper-fan valley to the channel–levee system is located at the continental slope foot, between E9°55′ and E10°10′ (Fig. 5).

Morphology. The channel is characterized by a highly meandering thalweg bordered by small isolated terraces (Fig. 17). Meanders are locally very tight, and most terraces seem to correspond to abandoned meanders. No wider valley can be identified in this area. Most of the terraces described along this area correspond to type B terraces. They are about 2–3 km wide, and are characterized by arcuate morphology toward the thalweg. Some terraces have a topographic high located at their centre and imprints of meander loops.

Seismic data. Profile ZSAR-37 (Fig. 18) shows the structure of the Zaire channel across this area (see location on Fig. 17). On both sides of the channel there are levees. Here, turbidity

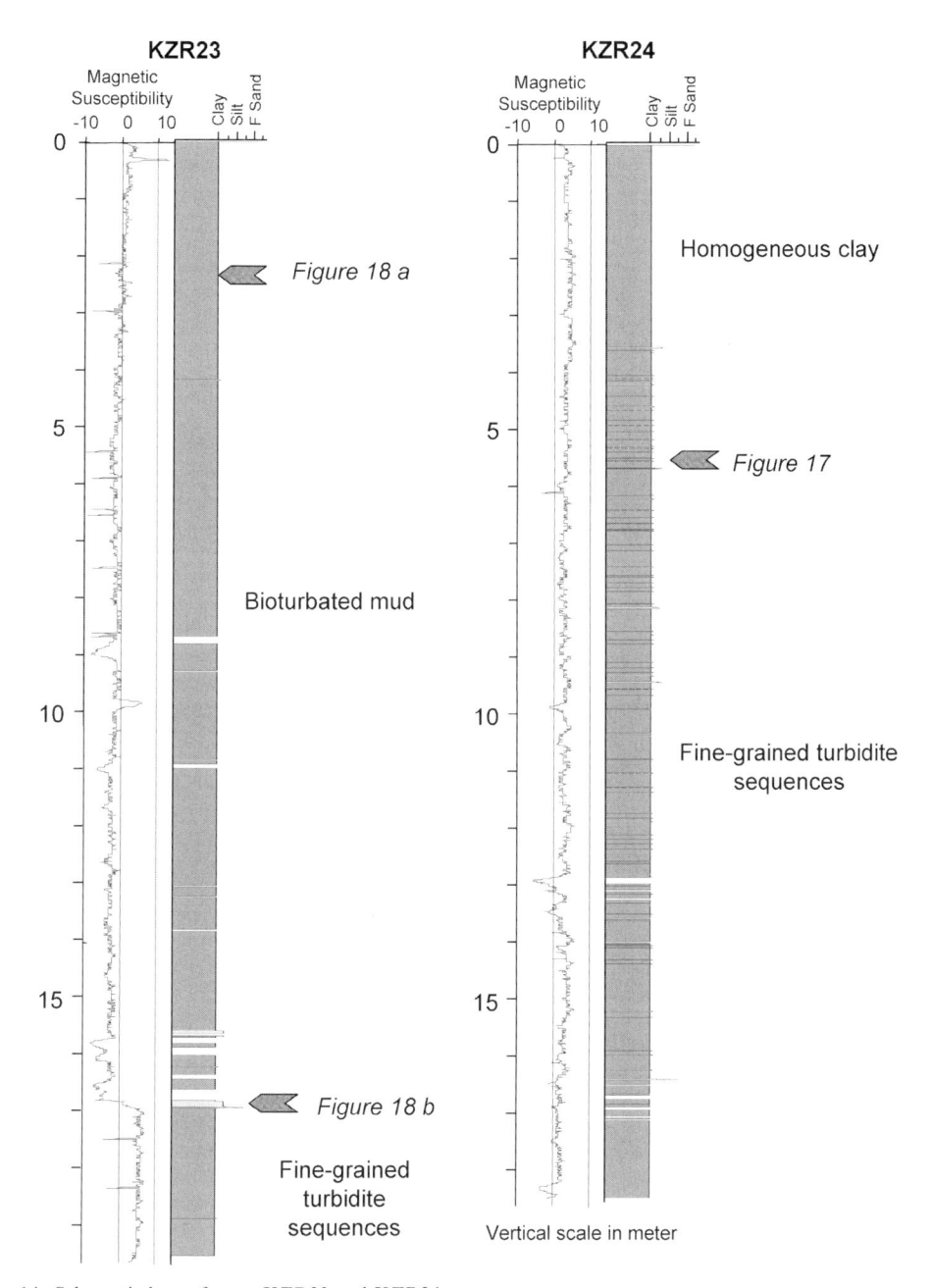

Fig. 14. Schematic logs of cores KZR23 and KZR24.

currents can overflow and build up levees over the channel edges. Profile ZSAR-37 crosses two small terraces. The upper terrace is situated at about 150–160 m above the floor. Its internal structure is characterized by a thick sedimentary unit composed of concave-up reflectors. Their seismic structure and facies seem very similar to the terrace units described upslope.

Discussion

What are the Zaire terraces?

Most terraces described along channel–levee systems (Kenyon et al. 1995; Hübscher et al. 1997; Nakajima et al. 1998) have been interpreted in terms of slump or slide processes, or

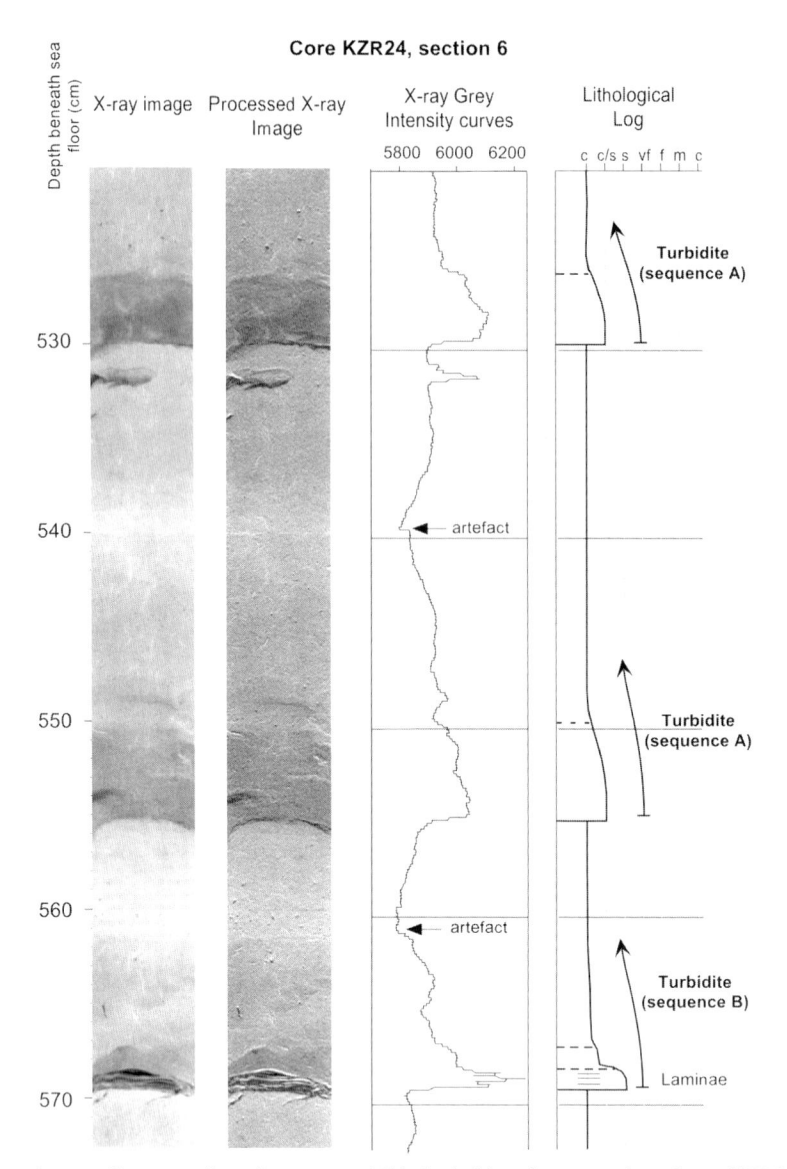

Fig. 15. X-ray images, X-ray grey intensity curve and lithological log along a section of core KZR24.

the aggradation of confined levees. In most of these studied cases the levees are generally well developed and the channel is moderately incised (less than 150 m deep).

In canyons, terrace features have generally been interpreted as the result of the succession of filling and incision stages (as in fluvial environments) or as being caused by tectonic activity (gravity tectonic).

Along the upper-fan valley of the Zaire system, no evidence of a larger and deeper valley has been clearly identified in seismic profiles, and the mapping of relative terrace heights (Fig. 9) has not shown any terrace levels correlatable at the scale of the area. All the terraces appear to be situated at different levels, indicating they are probably initiated by autocyclic or random processes. Thus the terraces of the Zaire valley do not result from a succession of filling and incision stages, as in fluvial settings.

Along the upper-fan valley of the Zaire system the valley is 300–800 m deep, apparently too deep for the development of true external levees.

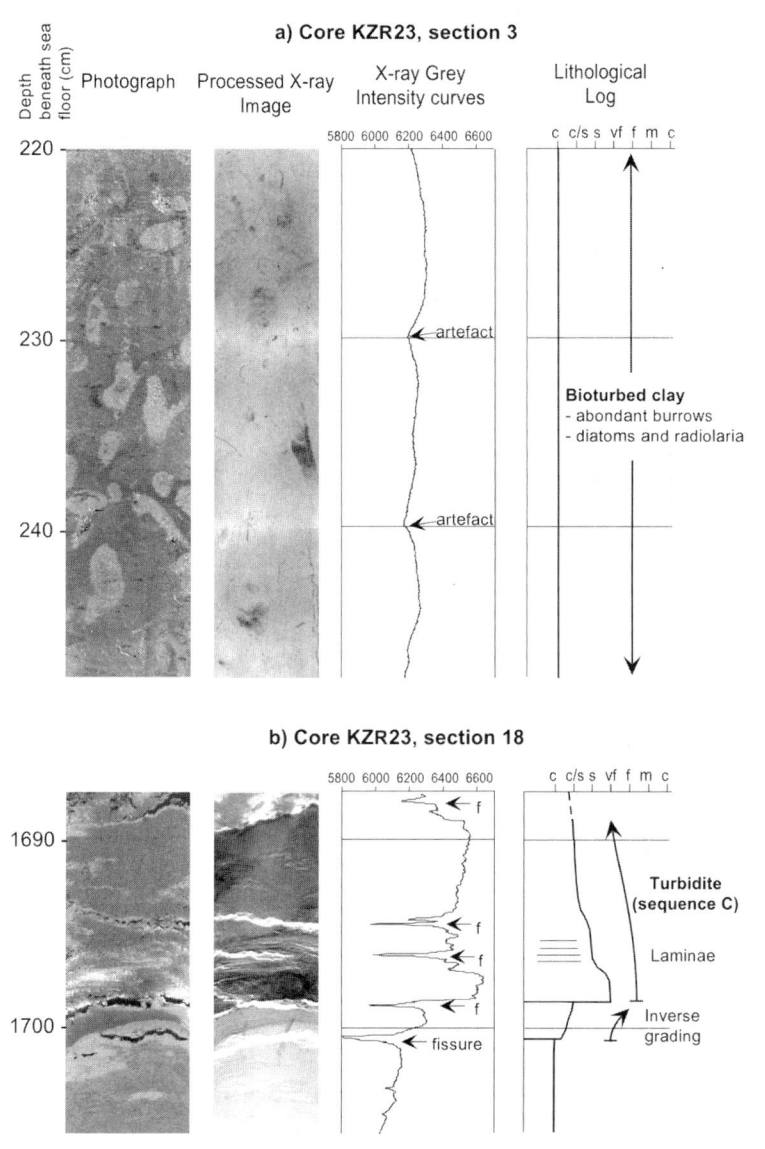

Fig. 16. Two examples of photograph, X-ray image, X-ray grey intensity curve and lithological log along sections of core KZR23.

Detailed analysis of seismic data has shown evidence of vertical aggradation of the terraces and wedge-shaped geometry of the deposits (Figs 11, 12 & 13), which suggests a levee type of sedimentation. Core KZR24 consists of fine-grained turbidites (Fig. 15), typical of overbank deposits. Thus the present sedimentation over the terraces is generated by activity of turbidity currents in the Zaire valley, or more precisely by overflow of turbidity currents (Piper & Normark 1983). Thus the terraces of the Zaire valley currently work as 'inner' or 'confined' levees.

Dynamic of recent turbidite flows along the Zaire valley

Behaviour of the turbidite flows. Figure 19 illustrates the behaviour of turbidity currents which could flow across the Zaire upper-fan valley. (A) shows the possible behaviour of small turbidity currents, which remain confined along the thalweg. No major turbidite deposition occurs over the terraces. In contrast, (B) illustrates the behaviour of larger turbidity currents, which overflow on terraces and deposit confined levees.

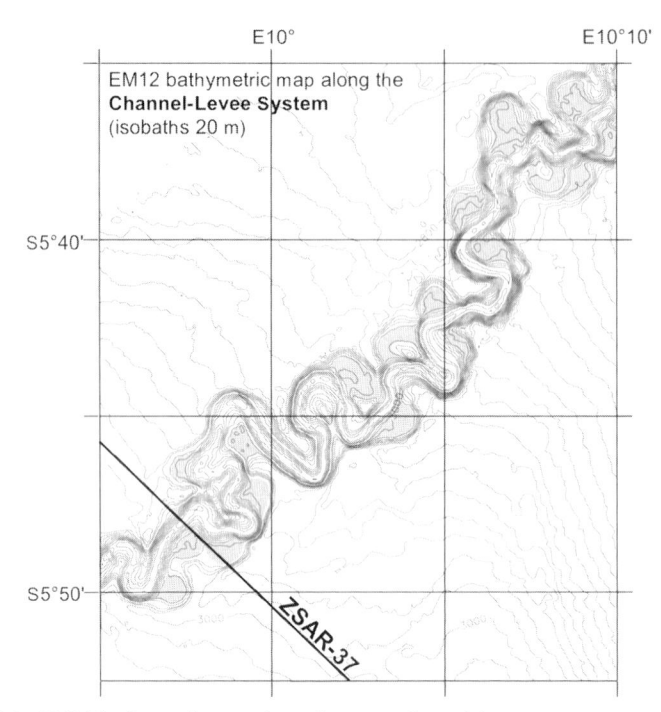

Fig. 17. Detail of the EM12 bathymetric map along the upper channel–levee.

Whether a terrace is fed by overflow sedimentation depends on the relative heights of the current and the terrace. Upper terraces are occasionally affected by major turbidity current overflow. In contrast, lower terraces are more frequently affected by turbidite sedimentation. The variability of facies observed in cores KZR23 and KZR24 (Figs 15 & 18) could be explained by the different terrace height according to the thalweg. The clayey bioturbated facies of KZR23 (quasi-transparent acoustic facies) is not really overflow sedimentation, but consists mainly of settling of turbidite plumes mixed with hemipelagic sediments.

As the grain-size distribution in a turbidity current decreases toward the top of the current (Stacey & Bowen 1988), the upper part of turbidity currents and their plume carry only clay. Sediments, deposited by settling of the uppermost turbidite plume, are difficult to distinguish from true hemipelagic deposits in terms of lithology and facies. The microfauna appears quite similar in smear slides (diatoms and radiolaria), although the sedimentation rate of true hemipelagic deposits should be lower than the sedimentation rate of the mixed deposits (plume settling).

Overflows of turbidity currents are more frequent on lower terraces. True turbidite sequences are identifiable in KZR24 (Fig. 15). Coarser sediments (silt or fine sand) are transported onto terraces. The same facies has been observed in the lower part of KZR23 and corresponds to true overflow deposits (bedded acoustic facies). The sedimentation rate is probably higher than for hemipelagic or mixed deposits, but no carbon-14 dating has been acquired, because of the lack of calcareous microfauna in the turbidite sequences.

The alternation of sedimentary units with quasi-transparent acoustic facies and bedded facies (3.5 kHz acoustic profile; Fig. 13) on the upper terrace shows that the flow regime variation of turbidity currents is probably cyclic. Bedded facies could correspond to periods of high occurrence of large turbidite flow, and transparent facies to periods of low occurrence. The last transition between the two facies is dated at 12 500 years BP, during the last sea-level rise. It could mark the flow regime decreasing during the present highstand phase. Thus the facies alternation may be related to sea-level variations.

Flow thickness. The boundary between the upper-fan valley and the channel–levee system is defined by the development of true external levee structures. Although no cores are available, the acoustic facies and wedge-shaped geometry

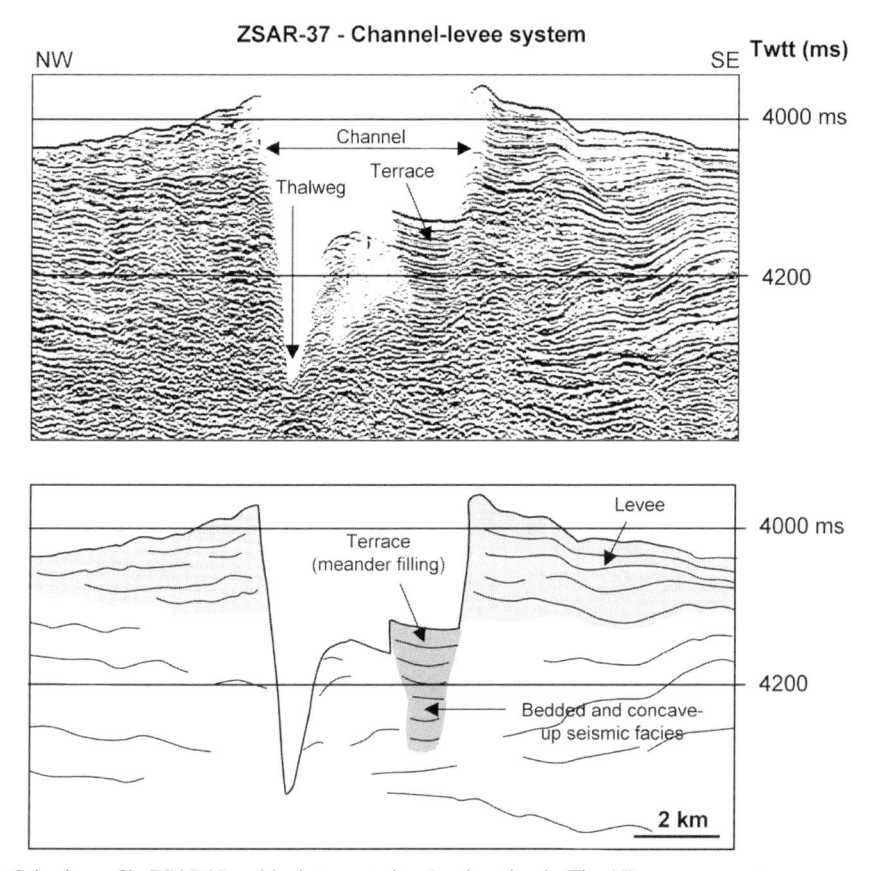

Fig. 18. Seismic profile ZSAR37 and its interpretation (see location in Fig. 17).

of the external levees are evidence of recent turbidity current spillover. Along the upper-fan valley turbidity flows remain confined in the valley, and only hemipelagic sediments accumulate outside the valley. Once the channel depth decreases below 350 m the currents may overflow and build levee-type accumulation. Thus 350 m corresponds to an estimate of the maximum thickness reached by turbidity currents.

A) Small flow volume: turbidity currents remain confined along the thalweg. There is no overflow onto the terraces.

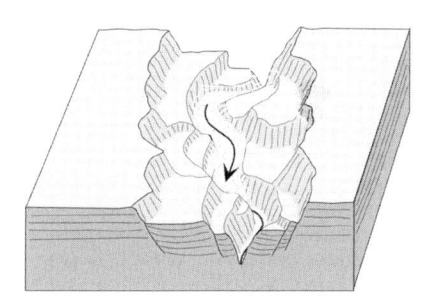

B) Large flow volume: turbidity currents overflow onto terraces and lays down levee sequences (confined levees). However, turbidity currents always remain confined inside the valley.

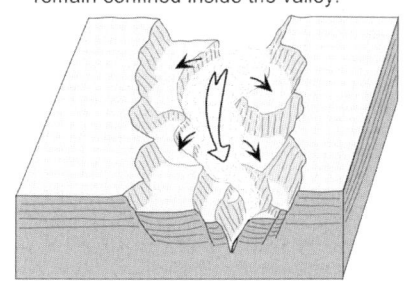

Fig. 19. Schematic three-dimensional blocks illustrating the behaviour of turbidity currents into the Zaïre valley with different flow volumes.

Along the upper-fan valley, the cores taken on terraces themselves indicate that a flow thickness threshold of about 350 m also has a meaning. KZR23 and KZR24 have been collected on terraces at 380 and 230 m respectively above the thalweg. At 380 m sedimentation is currently dominated by hemipelagic and turbidite plume settling. The boundary between turbidite- and settling-dominated sedimentation is situated 15–16 m underneath the top of the terrace. Thus the maximum height of overflows depends on period of turbidite activity and can also reach about 350 m.

At present the terraces situated at more than 350 m above the thalweg are 'abandoned' (green and blue terraces in Fig. 9). Their aggradation is due mainly to hemipelagic sedimentation and plume settling. The change of sedimentation type (from turbidite to settling) is not just due to aggradation on the terrace or to thalweg incision. As we have explained previously, this change could also be caused by decreases of turbidite activity, related to an allocyclic phenomenon (climatic or eustatic).

Dying terraces. The terraces currently situated more than 350 m above the thalweg (green and blue terraces in Fig. 9) are draped by hemipelagic and plume-settling deposits. They correspond to type A terraces and have probably reached a final stage of filling by overflow.

The outer scarp of type A terraces seems to be composed of several arcuate scarps. This suggests that their origin is similar to that of the lower terraces (type B). A type A terrace is thus probably initially composed of small arcuate type B terraces, which have aggraded until they reached 350 m above the thalweg (height threshold of overflows). The topographic compensation of the different small initial terraces could be explained by inner-levee aggradation during periods when the thalweg was not or only weakly in incision. The generation of several levels of type A terrace implies the alternation of active incision and less active incision stages. The present diversity of terrace levels could be explained by the most recent incision stage of the thalweg.

Creation of available space

For several terraces (for example on Z2-29; Fig. 11), the thickness of the sedimentary units associated with the terrace is at least 200 m. Considering that this whole unit is built by vertical aggradation, this implies that there was, before its deposition, available space in the canyon for sedimentation. Several hypotheses can be put forward to create available space.

Destabilization of channel walls. In canyons, available space could be created by destabilization of the valley walls. On the Zaire margin, several processes could destabilize canyon walls and create large slumps: entrenchment of the thalweg, which increases the slope gradient of the walls and enhances instability risks; tectonic activity (salt diapirism and faulting), which induces deformations of buried sedimentary successions; or gas hydrate destabilization and fluid-escape phenomena.

Along the Kaoping canyon (Taiwan), terraces structures appear clearly controlled by diapirism (Liu *et al.* 1993). The structure of the canyon is explained by faulting and slumps along preferential surfaces created along the fault planes. Within our study area, the Zaire margin is also deformed by active diapirism (Fig. 3). On seismic profiles (Figs 10 & 11), terrace borders seem relatively similar to normal faults. Based on regional seismic data, Cramez & Jackson (2000) have previously interpreted the structure of Zaire valley as the result of faulting linked to the salt tectonic.

In light of the ZaïAngo data, no deep fault has been identified in the prolongation of the terrace borders, and no slump or deformed structures have been recognized at the base of terrace sedimentary units. Moreover, the analyses of cores and acoustic profiles show evidence that the sedimentary units building terraces are genetically different from the sediments deposited outside the valley.

Although the thalweg entrenchment is up to 800 m deep along the canyon, and results in high slope gradients along the canyon walls (average slope gradient >30° with local near-vertical cliffs), no sedimentary unit resulting from slump processes has been securely identified in seismic data.

Processes of valley flank destabilization seem relatively minor along the Zaire valley. They affect probably a small sediment volume, and the resulting deposits must be removed by erosion and integrated into turbidity flows.

Channel migration, meander development and abandonment. The identification of meander loop imprints on some terraces located along the channel–levee system clearly suggests a sedimentary origin linked to the meander migration. It appears very similar to the terraces observed along the channel of the Arequipa basin (Hagen *et al.* 1994). Hagen *et al.* (1994) interpret terraces as the result of simultaneous thalweg

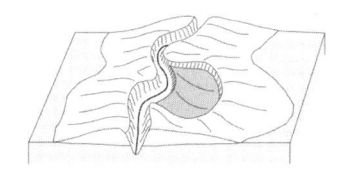

Fig. 20. Schematic sketches showing how a channel abandons a meander and builds inner levees.

1_ Channel migration and meander development

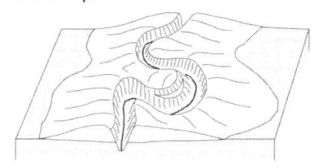

2_ Meander cutoff

3_ Filling of the abandoned meander by sediment supplied by turbidite current

4_ Filling by deposition of inner-levee (terrace morphology)

entrenchment and meander development. Numerous morphological similarities (sinuosity, meander loops) between the two systems lead us to consider that they have a similar behaviour.

The detailed bathymetric data provided by the EM300 tool along the Zaire valley (Fig. 9) have shown that the type B terraces have characteristics typical of filled meander cut-off (arcuate canyon-facing scarps, and sometimes topographic high in the centre). A synthetic sketch of meander abandonment and filling is illustrated in Figure 20. In stage 1 the meander grows. Stage 2 illustrates the cut-off of the meander. Stage 3 corresponds to the first stage of filling by the sediment supplied by turbidity currents and trapped into the topographic depression. The last stage illustrates the filling by overflow of turbidity currents and the aggradation of sediments as inner levees.

During the incision stage of the thalweg, levees and inner levees become relatively higher. Overflows become less frequent as terraces aggrade and thalweg incises. The sedimentation rate depends on the terrace height regarding the thalweg, and on the terrace location relative to the meander geometry. Overflows are higher and more frequent on the terraces located near the external flanks of meanders, because centrifugal forces induce flow-stripping processes (Komar 1973; Piper & Normark 1983).

Thus the terrace height depends on their location but mostly on their age (age of the meander cut-off): the oldest terraces are the highest in the valley, because they have been filled and built by overflow sedimentation for a longer time.

Transition between canyon and channel–levee

Canyons and channel–levee systems are two environments that are different in term of behaviour. Although used as a general term to qualify submarine valleys, the word 'canyon' corresponds to the upper area where the main processes occurring are sediment erosion and transport (Shepard 1981). The upper part of the Zaire turbidite system present typical canyon characteristics.

Several small topographic flats marked by horizontal reflectors are visible on the lower part of the canyon flanks (seismic profile TZ2-81; Fig. 6). They correspond to the first terraces built by aggradation of small inner levees. It shows how difficult it is to identify the limit of the true canyon area, where theoretically no deposition occurs. The lower limit of the canyon has been defined at about E11°05′ (Fig. 5), where the morphology changes: the valley width becomes more regular and true planar terraces become well developed.

Along the channel–levee system, both transport and sedimentation processes occur. Turbidity currents are channelized and transport a large volume of sediments as suspended load. During a turbidity event the channel floor can be in erosion or in sedimentation as flow energy decreases. Channel–levee systems are also responsible for the lateral deposition of large volumes of sediments by overflow, thus leading to an interaction between flow volume and channel section.

In term of behaviour, the upper-fan valley corresponds to a confined channel–levee complex, because there is deposition and aggradation of levees. Although its morphology is clearly different from that of true levees, and the available space for sedimentation is reduced, the sedimentary sequences observed in cores are similar to true levee sequences.

Recent evolution of the present Zaire valley

The present morphology of the Zaire canyon suggests recent incision. As proposed by Hagen et al. (1994), thalweg entrenchment and active meander migration occur simultaneously. The recent evolution of the morphology is illustrated by the schematic sketch in Figure 21.

(1) The first stage corresponds to a sinuous channel–levee system. The channel path is free to migrate and to develop meanders. Sinuosity increases with time, limited only by meander cut-offs. Topographic depressions, induced by the abandoned meanders, are partly filled by trapping of sediments supplied by turbidity currents.

(2) The simultaneous channel-floor incision and acute meander migration induce numerous meander cut-offs. They are progressively filled by turbidite deposits. The fill thickness (or the terrace height) depends on the age of the abandonment and the incision level during cut-off.

(3) The continuous entrenchment of the sinuous thalweg limits overflow outside the valley. Turbidity currents remain confined inside the valley. The sinuosity decreases because the thalweg is trapped within the valley (its migration is less free). Meander cut-offs still occur in the central part of the valley, but the rate of lateral migration of the thalweg is probably reduced. Deposition occurs only inside the valley, filling ancient meander loops and creating stepped terraces.

(4) The strong entrenchment of the thalweg limits its lateral migration. At this stage, the initiation of new terraces is reduced (only small type B and C terraces). Major terraces are features inherited from the previous stages of evolution. They are still fed by turbidity currents and continue their vertical aggradation as inner levees. The final stage of evolution, if the thalweg continues its entrenchment, is the erosion of walls and terraces. It corresponds to the canyon stage, where the morphological profile tends toward a V-shape. Sedimentation processes become ephemeral.

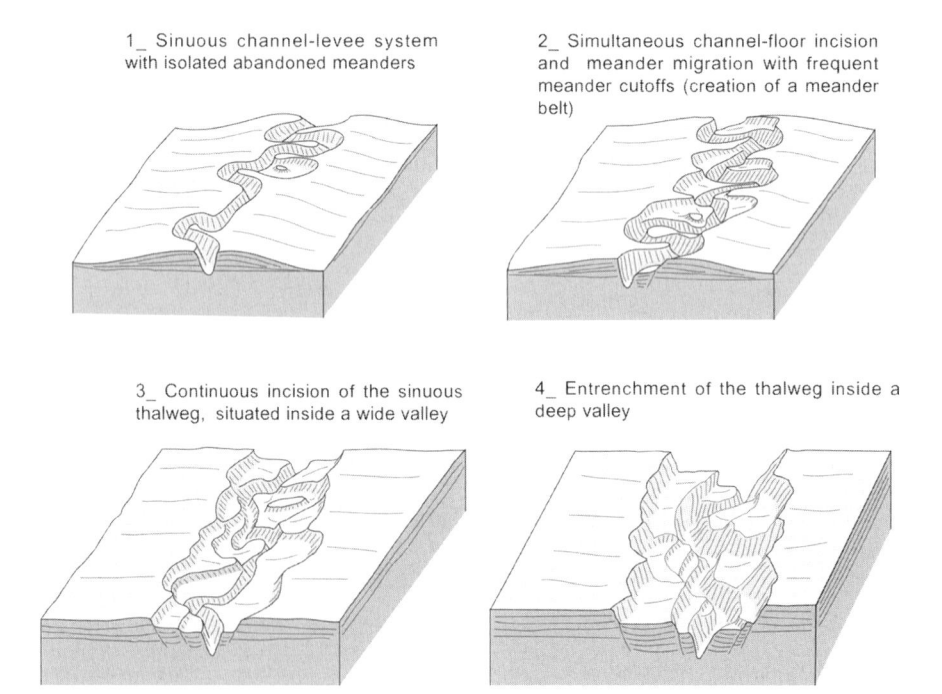

Fig. 21. Schematic sketch of the recent evolution of the Zaire valley: simultaneous meander migration and cut-offs, and thalweg entrenchment.

Conclusions

A complex of terraces has been identified along the Zaire upper-fan valley. The development of these structures is caused mainly by sedimentary processes, generated by turbidity currents flowing inside the valley. The role of slump processes and tectonic activity seems relatively minor.

The valley has been created by incision and lateral migration (meander development and cut-off) of the thalweg; it corresponds to an entrenched meander belt. Available spaces within the valley (created by meander migration) are progressively filled by inner levees, forming a complex of stepped terraces. The Zaire upper-fan valley can be considered as a confined channel–levee system, located at the transition between the canyon and the true channel–levee system (building external levees).

Along the valley, the maximum thickness of turbidity currents is estimated to be 350 m and corresponds to the maximum height of levees (inner or external levees) fed by overflow.

Three main types of sedimentation have been recognized along the upper-fan valley: true turbiditic sequences deposed by turbidity current overflows (levee deposits) over the terraces; hemipelagic deposits outside the upper-fan valley; and mixed deposits composed of turbiditic plume settling and hemipelagic sedimentation over the highest terraces. The distinction between true hemipelagic and mixed sediments is established by acoustic facies, the seismic patterns of the sedimentary bodies and average accumulation rates (the lithological facies and fauna content being quite similar).

An alternation of bedded and quasi-transparent facies is identified in some terraces. It probably relates to flow regime changes, possibly associated with sea-level variations.

N. Babonneau acknowledges IFREMER for the Ph.D. financial support. The ZAIANGO project is conjointly financially supported by TotalFinaElf and IFREMER. The authors thank TotalFinaElf for authorization to publish the results of this study. They also thank the technicians team of Genavir, crew members, and the captains of the R/V *L'Atalante*, R/V *Le Nadir*, and R/V *Le Suroit*, on board which the data presented in this paper were recorded during the ZAIANGO cruises. Processing of data was realized by B. Loubrieu, A. Normand and B. Gueguen (bathymetric data), E. Le Drezen (SAR data), J. P. Le Formal, H. Nouzé and L. Petit de la Villéon (seismic data) from IFREMER. Special thanks go to D. J. W. Piper for his constructive and critical review of the first draft of the manuscript. The manuscript benefited significantly from reviews by D. Masson and R. Holmes.

References

BABONNEAU, N., SAVOYE, B., CREMER, M. & KLEIN, B. 2002. Morphology and architecture of the present canyon and channel system of the Zaire deep-sea fan. *Marine and Petroleum Geology*, **19**, 445–467.

CARLSON, P. R. & KARL, H. A. 1988. Development of large submarine canyons in the Bering Sea, indicated by morphologic, seismic, and sedimentologic characteristics. *Geological Society of America Bulletin*, **100**, 1594–1615.

CARTER, R. M. & CARTER, L. 1996. The abyssal Bounty Fan and lower Bounty channel: evolution of a rifted-margin sedimentary system. *Marine Geology*, **130**, 181–202.

CHOW, J., LEE, J. S., LIU, C. S., LEE, B. D. & WATKINS, J. S. 2001. A submarine canyon as the cause of a mud volcano: Liuchieuyu Island in Taiwan. *Marine Geology*, **176**, 55–63.

CRAMEZ, C. & JACKSON, M. P. A. 2000. Superposed deformation straddling the continental-oceanic transition in deep-water Angola. *Marine and Petroleum Geology*, **17**, 1095–1109.

DAMUTH, J. E., FLOOD, R. D., KOWSMANN, R. O., BELDERSON, R. H. & GORINI, M. A. 1988. Anatomy and growth pattern of Amazon deep-sea fan revealed by long-range side-scan sonar (GLORIA) and high-resolution seismic studies. *American Association of Petroleum Geologists Bulletin*, **72**(8), 885–911.

DENNIELOU, B., JOUANNEAU, J. ET AL. 2001. Holocene Sediment Accumulation Rates on the Zaire Deep-Sea Fan. EUG IX, Strasbourg, France.

DROZ, L., RIGAUT, F., COCHONAT, P. & TOFANI, R. 1996. Morphology and recent evolution of the Zaire turbidite system (Gulf of Guinea). *Geological Society of America Bulletin*, **108**(3), 253–269.

DROZ, L. ET AL. 2003. Architecture of an active mud-rich turbidite system: The Zaire Fan (Congo-Angola margin southern Atlantic). Results from ZaïAngo 1 and 2 cruises. *American Association of Petroleum Geologists Bulletin*, **87**(7), 1145–1168.

HAGEN, R. A., BERGERSEN, D. D., MOBERLY, R. & COULBOURN, W. T. 1994. Morphology of a large meandering submarine canyon system on the Peru–Chile forearc. *Marine Geology*, **119**, 7–38.

HEEZEN, B. C., MENZIES, R. J., SCHNEIDER, E. D., EWING, W. M. & GRANELLI, N. C. L. 1964. Congo Submarine Canyon. *American Association of Petroleum Geologists Bulletin*, **48**(7), 1126–1149.

HÜBSCHER, C., SPIESS, V., BREITZKE, M. & WEBER, M. E. 1997. The youngest channel–levee system of the Bengal Fan: results from digital sediment echosounder data. *Marine Geology*, **141**, 125–145.

JANSEN, J. H. F., VAN WEERING, T. C. E., GIELES, R. & VAN IPRESEN, J. 1985. Middle and late Quaternary oceanography and climatology of the Zaire–Congo fan and the adjacent eastern Angola Basin. *Netherlands Journal of Sea Research*, **17**(2/4), 201–249.

KENYON, N. H., AMIR, A. & CRAMP, A. 1995. Geometry of the younger sediment bodies of the Indus Fan. *In*: SMITH, R. D. A. (ed.) *Atlas of Deep Environments: Architectural Style in Turbidite Systems*. Chapman & Hall, London, 89–93.

KLAUCKE, I. & HESSE, R. 1996. Fluvial features in the deep-sea: new insights from the glacigenic submarine drainage system of the Northwest Atlantic Mid-Ocean Channel in the Labrador Sea. *Sedimentary Geology*, **106**, 223–234.

KLAUS, A. & TAYLOR, B. 1991. Submarine canyon development in the Izu-Bonin forearc: a SeaMARC II and seismic survey of Aoga Shima Canyon. *Marine Geophysical Researches*, **13**, 131–152.

KOMAR, P. D. 1973. Continuity of turbidity current flow and systematic variations in deep-sea channel morphology. *Geological Society of America Bulletin*, **84**, 3329–3338.

LIU, C. S., LUNDBERG, N., REED, D. L. & L., H. Y. 1993. Morphological and seismic characteristics of the Kaoping Submarine Canyon. *Marine Geology*, **111**, 93–108.

MARTON, L. G., TARI, G. C. & LEHMANN, C. T. 2000. Evolution of the Angola passive margin, West Africa, with emphasis on post-salt structural styles. *In*: TALWANI, M. (ed.) *Atlantic Rifts and Continental Margins*. American Geophysical Union, Washington, DC, 129–149.

MASSON, D. G., KENYON, N. H., GARDNER, J. V. & FIELD, M. E. 1995. Monterey Fan: channel and overbank morphology. *In*: SMITH, R. D. A. (ed.) *Atlas of Deep Environments: Architectural Style in Turbidite Systems*. Chapman & Hall, London, 74–79.

MCHARGUE, T. R. 1991. Seismic facies, processes, and evolution of Miocene inner-fan channels, Indus submarine fan. *In*: LINK, M. L. (ed.) *Seismic Facies and Sedimentary Processes of Submarine Fans and Turbidite Systems*. Springer, New York, 403–413.

MCHARGUE, T. R. & WEBB, J. E. 1986. Internal geometry, seismic facies, and petroleum potential of canyons and inner fan channels of the Indus submarine fan. *American Association of Petroleum Geologists Bulletin*, **70**(2), 161–180.

MCHUGH, C., RYAN, W. B. F. & HECKER, B. 1992. Contemporary sedimentary processes in the Monterey canyon–fan system. *Marine Geology*, **107**, 35–50.

MIGEON, S., WEBER, O., FAUGÈRES, J. C. & SAINT-PAUL, J. 1999. SCOPIX: a new X-ray imaging system for core analysis. *Geo-Marine Letters*, **18**, 251–255.

NAKAJIMA, T., SATOH, M. & OKAMURA, Y. 1998. Channel–levee complexes, terminal deep-sea fan and sediment wave fields associated with the Toyama Deep-Sea Channel system in the Japan Sea. *Marine Geology*, **149**, 25–41.

O'CONNELL, S., MCHUGH, C. & RYAN, W. B. F. 1995. Unique fan morphology in an entrenched thalweg channel on the Rhone Fan. *In*: SMITH, R. D. A. (ed.) *Atlas of Deep Environments: Architectural Style in Turbidite Systems*. Chapman & Hall, London, 80–83.

PIPER, D. J. W. & NORMARK, W. R. 1983. Turbidite depositional patterns and flow characteristics, navy submarine fan, california bordeland. *Sedimentology*, **30**, 681–694.

PIPER, D. J. W. & NORMARK, W. R. 2001. Sandy fans—from Amazon to Hueneme and beyond. *American Association of Petroleum Geologists Bulletin*, **85**(8), 1407–1438.

PIPER, D. J. W., HISCOTT, R. N. & NORMARK, W. R. 1999. Outcrop-scale acoustic facies analysis and latest Quaternary development of Huenene and Dume submarine fans, offshore California. *Sedimentology*, **46**, 47–78.

RIGAUT, F. 1997. *Analyse et Evolution Récente d'un Système Turbiditique Méandriforme: l'Eventail Profond du Zaïre*. Université de Bretagne Occidentale, Brest.

SAVOYE, B., COCHONAT, P. *ET AL*. 2000. Structure et évolution récente de l'éventail turbiditique du Zaïre: premiers résultats scientifiques des missions d'exploration ZaïAngo 1 & 2 (Marge Congo-Angola). *Comptes-Rendus de l'Académie des Sciences de la Terre*, **331**, 211–220.

SHEPARD, F. P. 1981. Submarine canyons: multiple causes and long-time persistence. *American Association of Petroleum Geologists Bulletin*, **65**, 1062–1077.

SOH, W., TOKUYAMA, H., FUJIOKA, K., KATO, S. & TAIRA, A. 1990. Morphology and development of a deep-sea meandering canyon (Boso Canyon) on an active plate margin, Sagami Trough, Japan. *Marine Geology*, **91**, 227–241.

STACEY, M. W. & BOWEN, A. J. 1988. The vertical structure of density and turbidity currents: theory and observations. *Journal of Geophysical Research*, **93**(C4), 3528–3542.

STUBBLEFIELD, W. L., MCGREGOR, B. A., FORBE, E. B., LAMBERT, D. N. & MERRILL, G. F. 1982. Reconnaissance in DSRV Alvin of a 'fluvial-like' meander system in Wilmington Canyon and slump features in South Wilmington Canyon. *Geology*, **10**, 31–36.

TORRES, J. *ET AL*. 1997. Deep-sea avulsion and morphosedimentary evolution of the Rhone Fan Valley and Neofan during the Late Quaternary (northwestern Mediterranean Sea). *Sedimentology*, **44**, 457–477.

VAN WEERING, T. C. E. & VAN IPEREN, J. 1984. Fine-grained sediments of the Zaire deep-sea fan, southern Atlantic Ocean. *In*: PIPER, D. J. W. (ed.) *Fine-grained Sediments: Deep-Water Processes and Facies*. Geological Society, London, Special Publications, **15**, 95–113.

VON RAD, U. & TAHIR, M. 1997. Late Quaternary sedimentation on the outer Indus shelf and slope (Pakistan): evidence from high-resolution seismic data and coring. *Marine Geology*, **138**, 193–236.

Factors controlling foredeep turbidite deposition: the case of Northern Apennines (Oligocene–Miocene, Italy)

U. CIBIN[1], A. DI GIULIO[2], L. MARTELLI[1], R. CATANZARITI[3], S. POCCIANTI[1], C. ROSSELLI[1] & F. SANI[4]

[1]*Servizio Geologico, Sismico e dei Suoli, Regione Emilia-Romagna, V. le Silvani 4/3, Bologna, Italy*
[2]*Dipartimento di Scienze della Terra, Via Ferrata 1, Pavia, Italy (e-mail: digiulio@unipv.it)*
[3]*CNR-Centro di Studio per la Geologia Strutturale e Dinamica dell'Appennino, Via S. Maria 53, Pisa, Italy*
[4]*Dipartimento di Scienze della Terra, Via La Pira 4, Firenze, Italy*

Abstract: Three major controlling factors affect turbidite deposition in foredeep basins: tectonics in the source area, tectonics in the belt–basin system, and variations of sea-level (local or global). These factors are expected to have different effects on the volume, grain size, provenance and distribution of clastic sediments during the evolution of the basin. The interplay of these factors is investigated for the latest Oligocene–Middle Miocene Northern Apennines Foredeep turbidite wedges by means of turbidite-system-based lithostratigraphy and field mapping, integrated with nannoplankton biostratigraphy and sedimentary petrography. Almost all recognized turbidite systems, unless tectonically truncated, show an overall stacking pattern formed by a lower sand-rich, thickly bedded stage (*depocentre stage*) passing upward into mud-rich, thinly bedded stages, eventually grading up to mostly mudstone units (*abandonment stage*). This rhythmically repeated pattern is interpreted as the result of the abrupt switching on and off of coarse-grained input, coupled with an alternating increase/decrease of depositional rate recorded in all detected systems. Biostratigraphy makes it possible to distinguish the switching-off of turbidite systems due to depocentre migration (a new system is switched on basinward) from that due to a regional decrease of clastic input. Sandstone petrography records the compositional variation related to tectonically induced source reorganization. In the latest Oligocene–Middle Miocene NAF foredeep wedges, this integrated dataset allows us to recognize: two different phases of source tectonics in the latest Oligocene and the middle Burdigalian; two major episodes of basin tectonics and related depocentre shift in the latest Oligocene and the Langhian, plus a minor middle Aquitanian phase; and three intervals of reduced regional turbidite deposition during the Late Aquitanian, Middle Burdigalian and Early Serravallian, possibly linked to sea-level rises.

Foredeep basins commonly comprise turbidite systems that form in topographically confined settings. The inherited or actively developing topography strongly influences the gross geometry of depositional systems and their architecture in terms of stacking pattern and the distribution of coarse materials and related porous bodies, but this is only one of the controlling factors acting on such depositional systems.

Deep-sea clastic wedges accumulated in foredeep basins all over the world are a major challenge for petroleum exploration (e.g. Weimer & Link 1991). Thus understanding how the different controlling factors acting on deep-sea foredeep deposition interact in determining the thickness, shape, internal stratigraphy and heterogeneity of foredeep turbidite systems and complexes proves particularly relevant. In many cases, this can be a difficult task because of the real or apparent monotony of many turbidite successions.

As a general model, three major controlling factors can be expected to have a role of paramount importance in determining the accumulation of turbidite wedges in foredeep basins:

(1) the tectonics active in the source region that modulates the volume and composition of clastic sediment produced;
(2) the tectonics active in the belt whose load generates the foredeep subsidence that modulates the foredeep migration and controls the geometry of the basin;
(3) relative variations in sea-level, whatever their cause (eustatic or tectonic) and scale (regional or global), which modulate the supply of detritus produced in the source region

From: LOMAS, S. A. & JOSEPH, P. (eds) 2004. *Confined Turbidite Systems.* Geological Society, London, Special Publications, **222**, 115–134. 0305-8719/04/$15.00 © The Geological Society of London 2004.

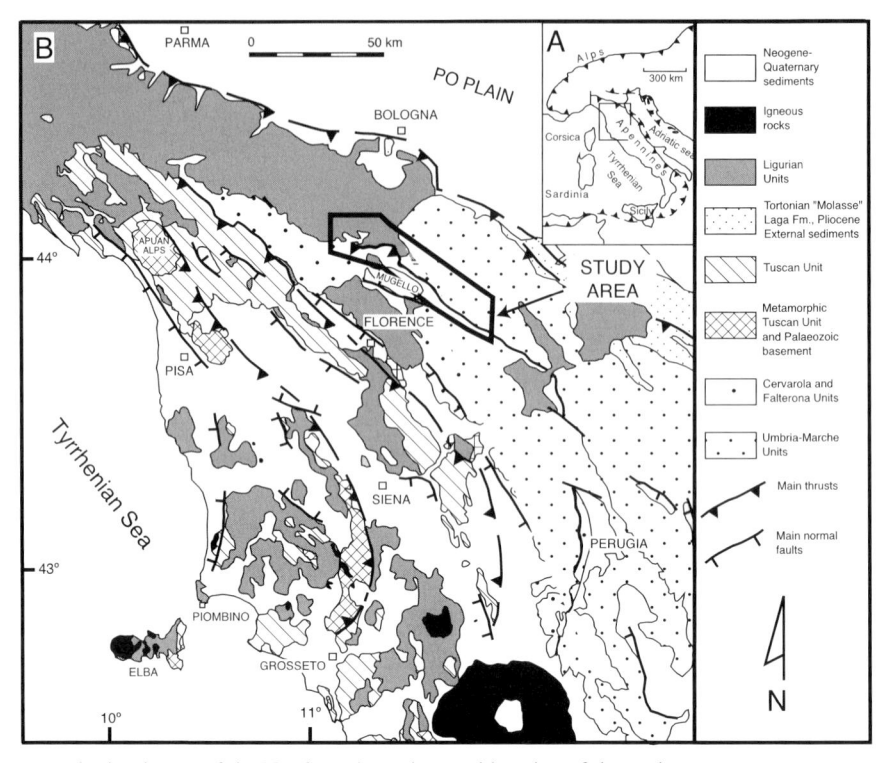

Fig. 1. Tectonic sketch map of the Northern Apennines and location of the study area.

and regulate the deposition of coarse clastics into peripheral areas or deep water environments; in this case, when sea-level change is due to climate, it can have a reinforced effect due to the change of the erosion rate in the source region and the related clastic input in the basin.

This paper discusses how and when each of the cited controlling factors acted on the deposition of the huge turbidite wedges that accumulated in the Northern Apennines Foredeep during the latest Oligocene–Miocene time span, and which form the bulk of the chain cropping out in the Tuscan-Emilia region in the present day (Fig. 1). In addition, we discuss how the influence of each factor can be recognized as a dominant control during the stratigraphic evolution of the turbidite complexes.

For this task, the Northern Apennines Foredeep provides an ideal field laboratory because of its relatively well-known lithostratigraphy, recently improved by means of new detailed field-mapping and biostratigraphical dating for the new Italian Geological Map at 1:50 000 scale (CARG Project), combined with the fact that the major foredeep turbidite systems were fed mainly by the Alps rising at the NW end of the basin and not by the Northern Apennines belt whose load generated the basin (Fig. 2). This particular source–basin configuration gives us the chance to discriminate between the effects of tectonics in the source region and those of tectonics in the belt generating the foredeep.

Geological setting

The Northern Apennines belt (NA) is part of the major system of chains occurring in the central Mediterranean (Fig. 1). It is composed of mostly NE-verging tectonic units piled up first during the Cretaceous–Eocene pre-collisional convergence, and then during the Oligocene–Neogene collisional steps of the NA orogeny.

The top of the nappe stack is composed of Ligurian units, including ophiolites and their sedimentary cover (Jurassic to Eocene) originally deposited in the Ligurian-Piedmont ocean (NW sector of the Tethyan Ocean) and progressively accreted at the convergent margin (e.g. Principi & Treves 1984 and references therein).

The Ligurian units tectonically overlie the Tuscan and Umbria-Marche units. The latter were detached from the Adria passive margin during the Oligocene–Miocene collisional setting,

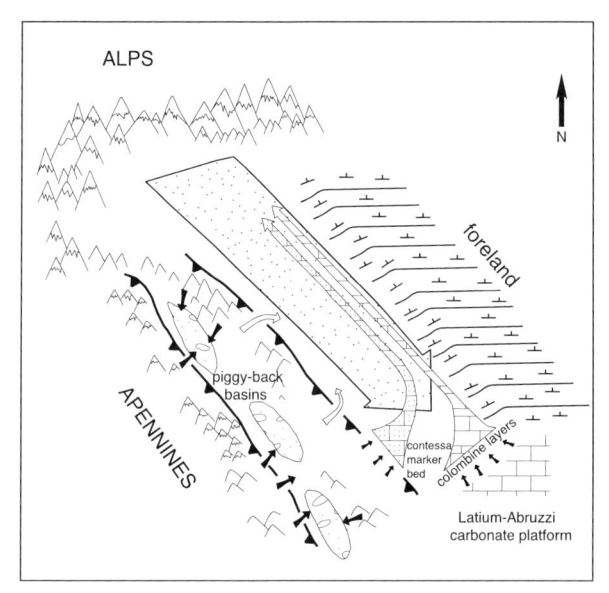

Fig. 2. Very simplified palaeogeographical scheme showing the basic features of the Northern Apennines Foredeep source–basin system during Late Oligocene–Miocene times (redrawn, simplified and modified after Gandolfi et al. 1983).

which developed at the western margin of the Adria Plate contemporaneously with widespread extensional processes developed in the central Mediterranean, leading to the formation of the Ligurian–Balearic Basin during the Aquitanian–Burdigalian, and the Tyrrhenian Basin from the Late Miocene up to the present (Boccaletti & Guazzone 1972, 1974; Barberi et al. 1973; Dewey et al. 1973; Biju-Duval et al. 1977; Burrus 1984; Finetti & Del Ben 1986; Coward & Dietrich 1989).

The Tuscan and Umbria-Marche units consist of a thick stratigraphic succession, comprising 1000–1500 m of basal Triassic evaporites (Burano Formation) overlain by a mainly carbonate Mesozoic–Cenozoic succession. In turn Cenozoic carbonates are topped by 2000–3000 m thick Oligocene–Miocene deep-sea siliciclastic sediments accumulated in the morphologically complex foredeep system that developed at the leading edge of the NA chain. Provenance studies on the foredeep sediments indicate that they were fed mostly from the rapidly uplifting central-western Alps at the northern end of the basin (e.g. Gandolfi et al. 1983; Valloni et al. 1991; Andreozzi & Di Giulio 1994; Di Giulio 1999; Fig. 2).

Currently, these Oligocene–Miocene turbidite complexes are widely exposed in the NA belt even if, northwestward, they are widely tectonically covered by the Ligurian units, which prevent study of their upcurrent part (Fig. 1); they become progressively younger eastwards and downwards in the nappe pile as a result of the eastward migration of the thrust front. From west to east, they are classically named the Macigno, Cervarola-Falterona and Marnoso-Arenacea units (e.g. Ghibaudo, 1980; Ricci Lucchi 1981, 1986).

This paper focuses on the Cervarola-Falterona and Marnoso Arenacea units in an area located along the main NA watershed between Florence and Bologna (Fig. 1).

Rationale and methods

In the last three decades, knowledge of turbidite sedimentology and deep-water systems has increased dramatically, especially concerning the internal lithological heterogeneity, vertical and lateral facies changes, and the hierarchy of turbidite depositional units (see Mutti 1992 for a comprehensive review). In particular, the gross architecture and vertical evolution of turbidite-dominated basin fills were emphasized by Mutti (1985) and Mutti & Normark (1987), who defined a simple model for their stratigraphic evolution based on the evolution observed in many turbidite systems, which records a gradual reduction of mass flow volume through time. This model, and the related hierarchical classification, are summarized in Figure 3.

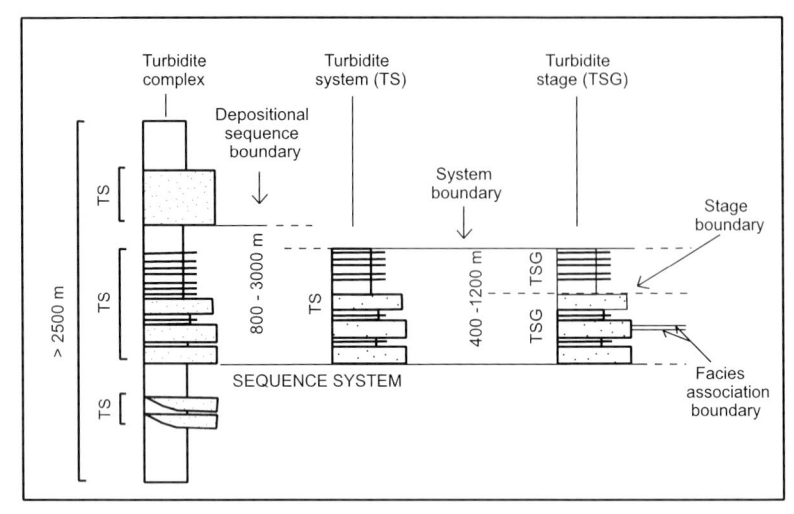

Fig. 3. Hierarchical scheme for classification of turbidite successions based on the physical scale of the various units (redrawn after Mutti & Normark 1987 and Mutti 1992). Note that the scale of the different unit types has been adapted to the Northern Apennine foredeep case.

The infill of a turbidite-dominated basin is considered a *turbidite complex*: it is generally made up of several *turbidite systems* (TS), which are the physical record of individual fans that form stratigraphically continuous units. In turn, each TS is composed of *turbidite stages* (TSG) arranged within the TS, in fining- and thinning-upward sequences (FUTU) reflecting the evolution of the fan. Finally, the substages (TSSG) represent the typical facies association several metres thick, defining single turbidite–fan sub-environments (such as a channel or levee).

In this framework, TS and TSGs are the main building blocks of turbidite complexes suitable for mapping in regional-scale studies (Mutti 1992). In addition, biostratigraphy provides the necessary temporal framework to compare the foredeep evolution with that of the surrounding belts, and sedimentary petrography gives some insights into the geological history of the source area.

For these reasons, this study has been performed by integrating:

(1) detailed field-mapping at 1:10 000 scale, based on the hierarchical classification of turbidite units, made directly in the field according to few key depositional features (sandstone/mudstone ratio, bed thickness), following the TS/TSG stratigraphic approach. This approach was chosen because in the study region turbidite stages form the smallest, relatively homogeneous mappable units, correlatable at a regional scale;

(2) the systematic study of the nannoplankton associations preserved in each recognized TSG; and

(3) the systematic study of sandstone petrography, performed mostly on sand-rich stages occurring at the base of each system.

Collectively, 4 km of stratigraphic sections were analysed. A part (1 km thick as a whole) was studied through 32 high-resolution bed-by-bed stratigraphic logs; the rest (3 km thick as a whole) was studied through low-resolution logs.

A total of 869 samples have been studied for their calcareous nannofossil contents by light microscope technique (normal light and crossed nicols) on smear slides. Preparation of smear slides followed standard procedures; a small amount, 5–10 mm^3, is smeared onto a glass slide, using a drop of water and a flat toothpick. For the time period investigated, the regional calcareous nannofossil biostratigraphical schemes proposed by Fornaciari & Rio (1996) and Fornaciari *et al.* (1996) were adopted; they are reported in Figure 4 together with their correlation with standard schemes. In order to overcome problems concerning reworking or absence of marker species due to poor preservation, a quantitative approach in data collecting was used according to the methods discussed by Backman & Shackleton (1983) and Rio *et al.* (1990). Particularly the following routine was used:

(1) counting the index species, common in the assemblage, in approximately 500 species of the entire assemblage; it has been applied to

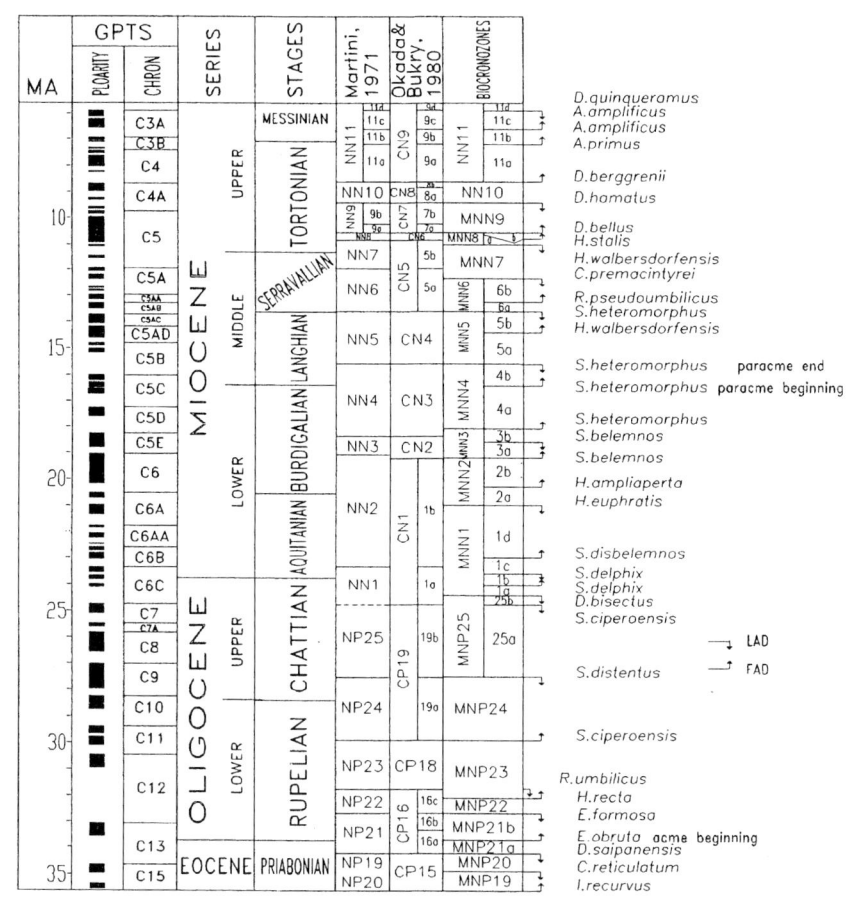

Fig. 4. Adopted calcareous nannofossil biostratigraphic schemes (after Fornaciari & Rio 1996 and Fornaciari *et al.* 1996) and correlation with standard schemes (Martini 1971; Okada & Bukry 1980). Chronostratigraphy after Berggren *et al.* (1995) and geochronology after Cande & Kent (1992, 1995). GPTS, geomagnetic polarity time scale; FAD, first appearance datum; LAD, last appearance datum.

the *Dictyococcites bisectus* and *Calcidiscus premacntyrei* LAD (last appearance datum), and to the *Reticulofenestra pseudoumbilicus* FAD (first appearance datum);

(2) counting the index species relative to a prefixed number of taxonomically related forms. With this respect, FAD and LAD of *S. ciperoensis*, *S. delphix*, *S. disbelemnos*, *S. belemnos*; FAD, LAD, discontinuous and scattered occurrence (paracme) of *S. heteromorphus* were established by counting 100 sphenoliths. FAD, LAD of *H. euphratis, H. carteri, H. ampliaperta* and *H. walbersdorfensis* were established by counting 50 helicoliths.

Sandstone petrography was studied for 108 samples from 24 measured sections. It was investigated by microscope modal analysis using the G-D method (Gazzi 1966; Dickinson 1970; see also Di Giulio & Valloni 1992 for a methodological review). A double analysis was performed on each sample: the first considered all rock constituents (essential and accessory framework grains, matrix, cement; at least 200 essential grains were counted), and the second focused on fine-grained rock fragments (at least 100 lithic grains were counted). These double analyses were used to determine the framework composition of samples (synthetically described by the conventional QFL+C diagram), and the composition of fine-grained rock fragments (including carbonates; synthetically described by the conventional LmLvLs+C diagram), which is particularly useful for highlighting the compositional evolution of Oligocene–Miocene foredeep units of the Northern Apennines (Valloni *et al.* 1991; Andreozzi & Di Giulio 1994; Di Giulio 1999).

Turbidite systems of the Oligocene–Miocene foredeep succession

Detailed field-mapping, integrated with biostratigraphy and sandstone petrography, led to the identification of seven TSs in the study area (Fig. 5). According to the regional nomenclature they are, in stratigraphic order: the Falterona turbidite systems (FAL) part of the Macigno turbidite complex; the Acquerino (ACQ), Torrente Carigiola (TCG), Stagno (STA) and Castiglione dei Pepoli (CDP) turbidite systems, which all belong to the Cervarola turbidite complex; and two superimposed turbidite systems in the Inner Marnoso Arenacea turbidite complex (MAIa and MAIb). The younger (Tortonian) and outer Marnoso Arenacea turbidite complex was not considered in this paper.

Turbidite stages (TSGs), i.e. relatively homogeneous rock bodies, are distinguishable and mappable at the regional scale in all TSs. In the new geological map of Italy (1:50 000 scale) they are generally mapped with the rank of members.

Each TS is made mainly of beds having typical turbidite features, as graded beds with Bouma intervals, sole marks and tabular geometry. Usually, each TS has a thickly bedded, medium- to coarse-grained sandy stage at the base, with high to very high sandstone/mudstone ratio generally related to a sandstone lobe depositional environment, an intermediate stage made up of sandstone/mudstone couplets ranging from sand-dominated to mud-dominated beds traditionally referred to lobe–fringe associations, and an upper stage, usually characterized by thinly bedded mud-dominated turbidites and hemipelagic mudstones, which provide important mappable horizons. Some TSs do not show the complete fining upward-thinning upward (FU-TU) facies trend (not fully developed TS) because they were truncated by sin-depositional emplacement of the Ligurian units.

Relatively common toolmarks (mostly flute and groove casts) provide the reported palaeocurrent data, which support a persistent axial distribution pattern of detritus (Fig. 2), with a dominant siliciclastic source from NW and minor carbonate supply from SE (Ricci Lucchi 1975, 1981; Bruni & Pandeli 1980; Gandolfi et al. 1983).

In this chain sector, downslope and laterally thin and wedge-out are observable, usually toward NE and SE. As the relationship between TSs is actually tectonic, through thrust-faults with NE vergence (Figs 5 & 6), northeastward facies changes are not preserved. Sometimes, fan-fringes facies, like bed closures, increase of low-density turbidites and pelagic marls, are observable downcurrent, toward SE (e.g. in the TCG and MAIa systems).

The structural relationships between the seven TSs and the Ligurian units are shown in the cross-sections of Figure 6. As a whole, the structure comprises a thrust system emplaced in a piggyback progradation (*sensu* Butler 1987) from southwestern sectors towards the NE. This structural setting derived from the progressive NE shift of depocentres of both foredeep and associated piggyback basins, as evidenced by the stratigraphic reconstruction described in detail below. Nevertheless, later compressive tectonics persisted up to recent times, indicated by cross-cutting relationships between thrust faults, as well as thrust faults cutting both fore and back limbs of folds, accounting for the presence of out-of-sequence thrusting throughout the area.

In the following the main lithostratigraphical (bed thickness, sand/shale ratio), biostratigraphical and petrographical features of each TS are briefly described in stratigraphic order from the oldest; bed thickness classes are referred to Campbell (1967). This provides the necessary database from which to discuss the role of basin tectonics, source tectonics and sea-level changes in guiding the depositional/structural evolution of the NA turbidite systems from the latest Oligocene to Middle Miocene.

Falterona turbidite system (FAL)

The Falterona system represents a single fully developed TS, overlying deep-sea pelagic mudstones: it is about 2000 m thick (maximum thickness) and has an age spanning from Chattian up to Aquitanian (Figs 7 & 8).

The Falterona System is split into five TSGs. The lowermost (FAL1) is made up of thickly bedded, frequently amalgamated, medium- to coarse- and granular-grained sandstones, virtually without shales. The middle interval (FAL2) is made up of sand-dominated sandstone/mudstone couplets (sandstone/mudstone ratio between 10 and 2) forming thick to very thick beds with medium-grained sandstone at the base. The overlying stages (FAL3–FAL5) grade from sand-dominated to mud-dominated, thick to thin beds with fine- to very fine-grained sandstone intervals. Palaeo-current data suggest predominant provenance from W–NW, rarely from E–SE.

Falterona TS dates from the Chattian (MNP25a zone) up to the Aquitanian (MNN1d

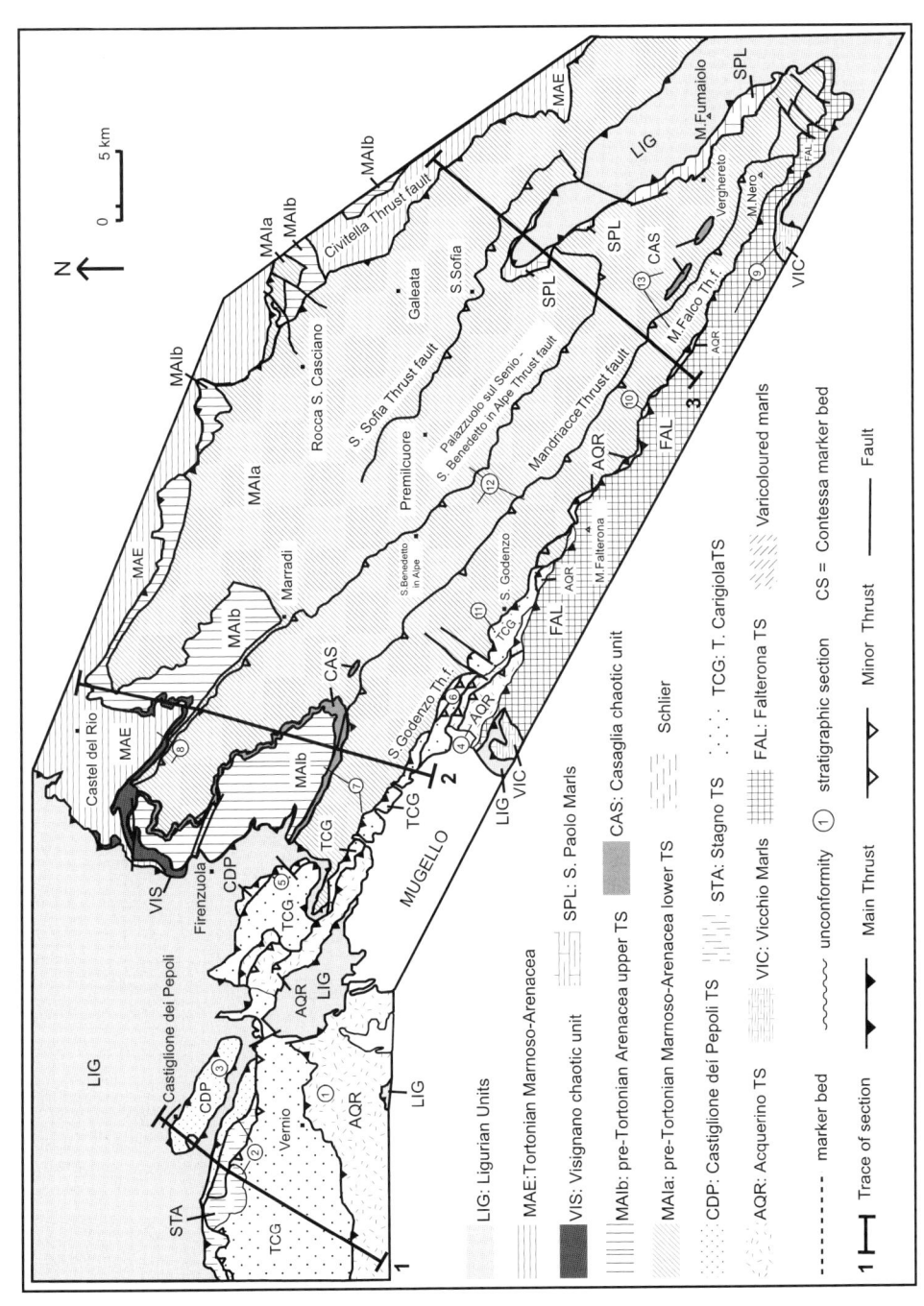

Fig. 5. Tectonic map of the study area reporting the location of the studied stratigraphic sections; thick black lines refer to cross-sections of Figure 6. Symbols also referred to in Figures 6, 7 and 8.

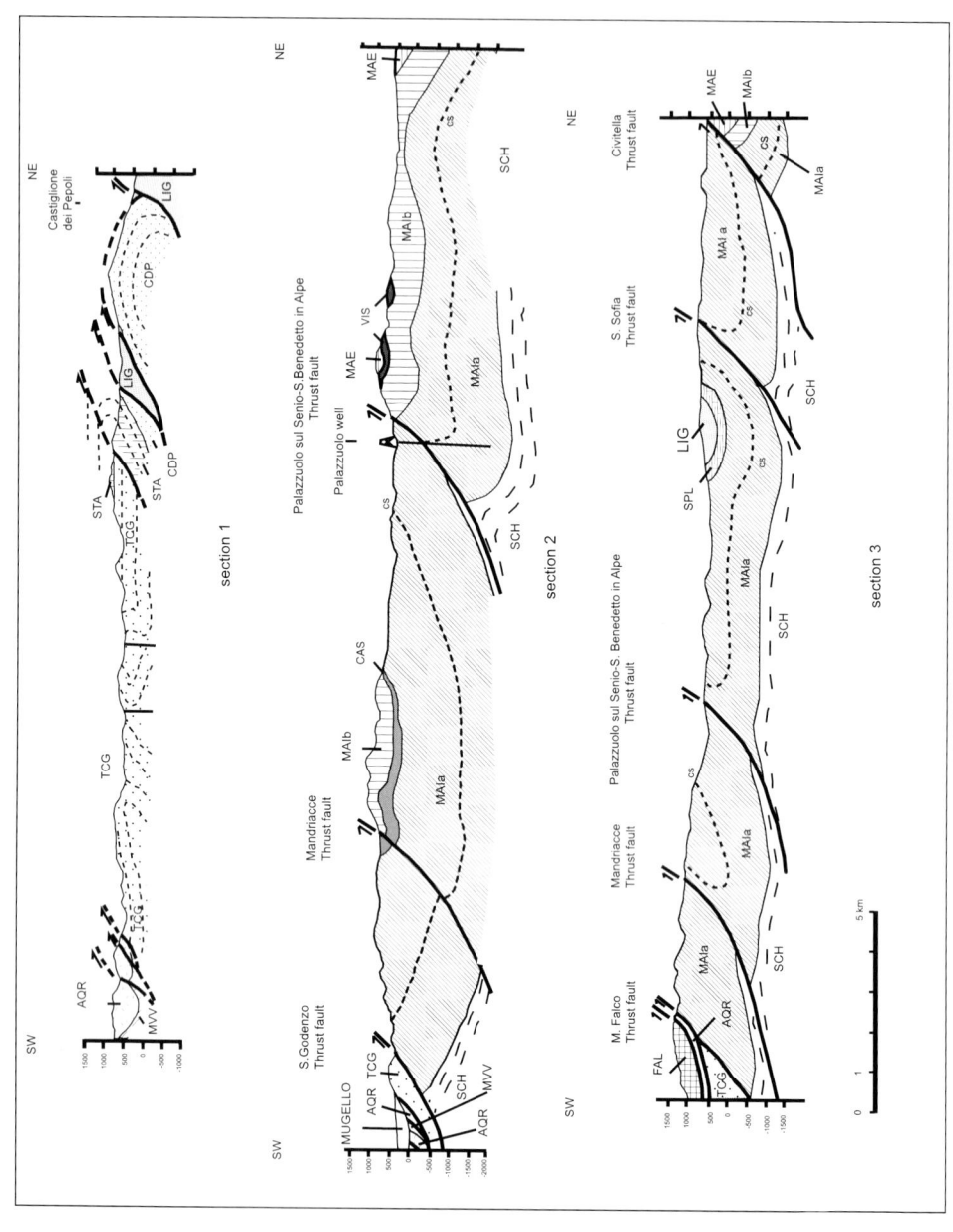

Fig. 6. Geological cross-section showing the overall tectonic arrangement of the study region. Symbols as in Figure 5.

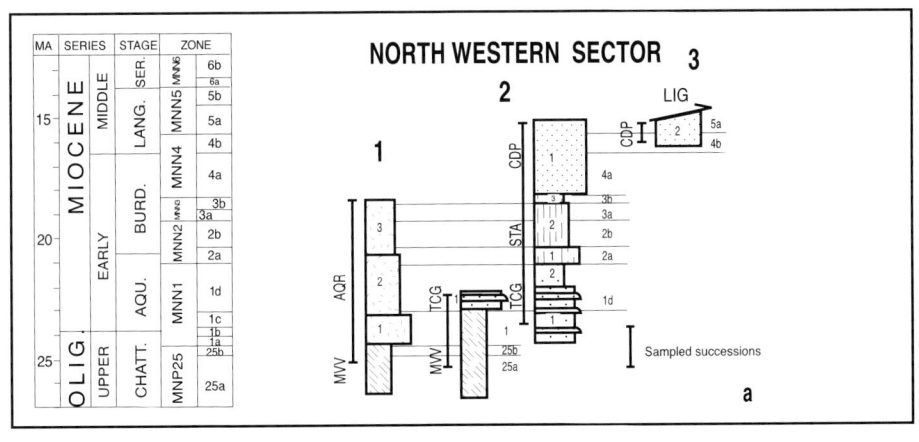

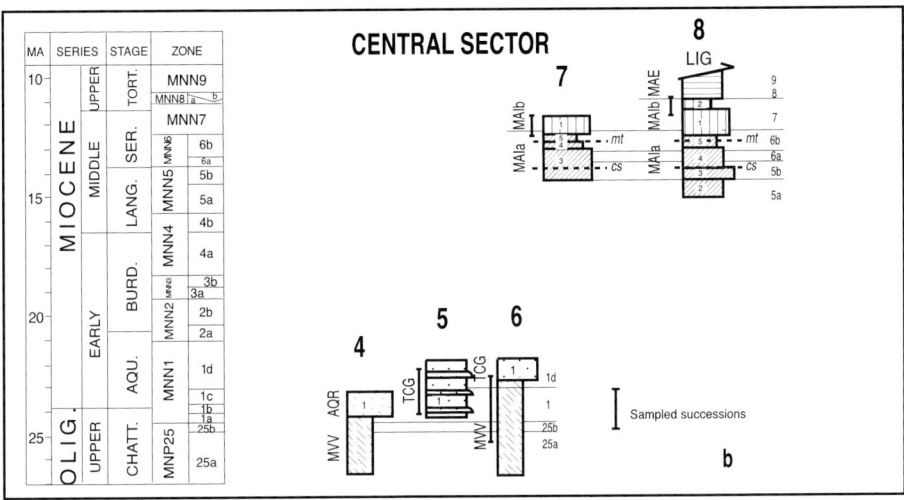

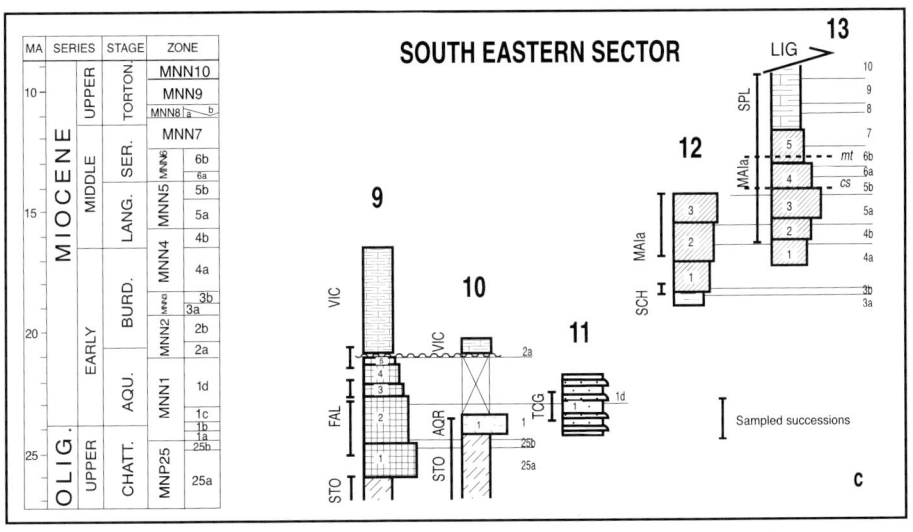

Fig. 7. Chronostratigraphical distribution of the studied turbidite systems and stages according to nannoplankton biostratigraphy.

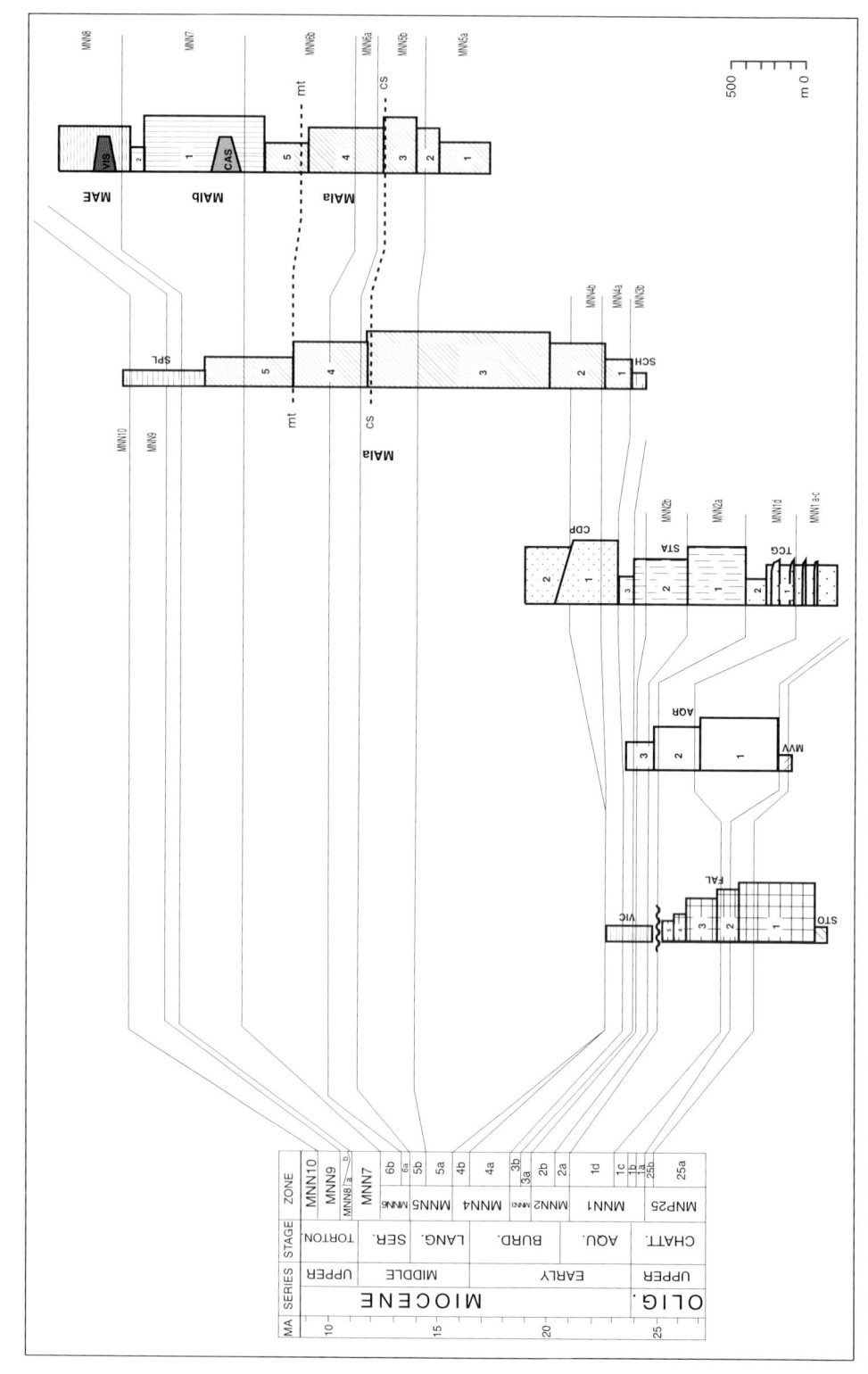

Fig. 8. Very simplified lithostratigraphy and correlation scheme of the studied turbidite systems and stages.

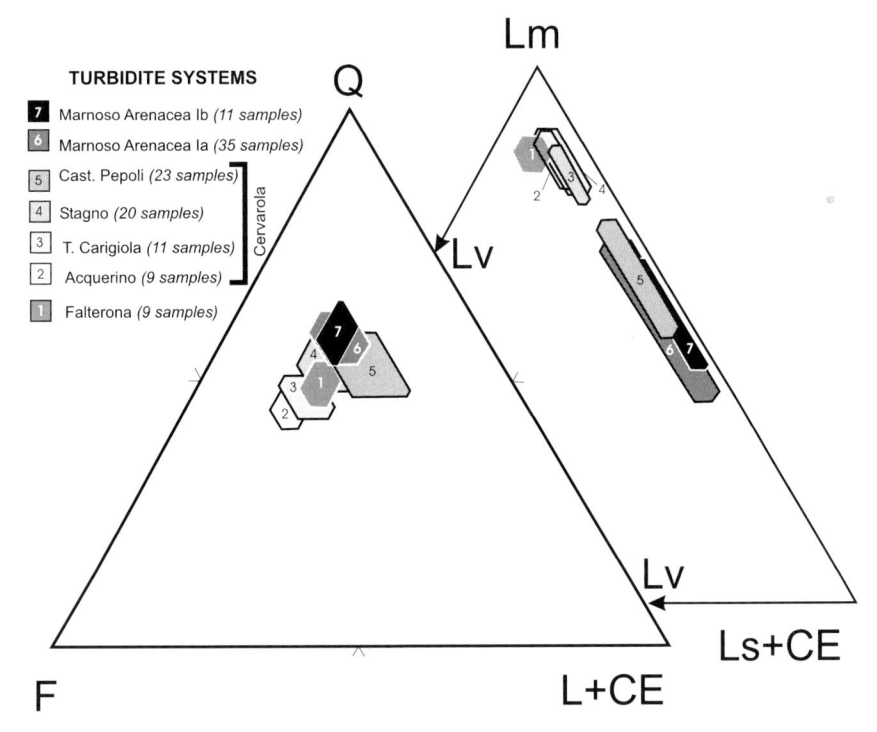

Fig. 9. Sandstone petrology of studied turbidite systems according to traditional QFL + C (quartz-feldspar-lithic + carbonates) and LmLvLs + C (fine-grained metamorphic-volcanic-sedimentary + carbonate lithics) plots. Note: considerably homogeneous QFL + C compositions are coupled with time-varying LmLvLs + C fine-grained lithic detrital modes.

zone), because of the presence of *Sphenolithus ciperoensis* and *Dictyococcites bisectus* and the lack of *Sphenolithus distentus* in the lower part of the succession, and the presence of *Sphenolithus disbelemnos* at the top (see also Bucefalo Palliani *et al.* 1997). Also, the zone and subzones between MNP25a and MNN1d are documented. The assemblage of MNP25b is characterized by the common and continuous occurrence of *D. bisectus* with *Cyclicargolithus abisectus*, *Cyclicargolithus floridanus*, *Coccolithus pelagicus*, *Sphenolithus moriformis*, *Discoaster deflandrei*, *Helicosphaera euphratis* and *Helicosphaera recta*, and the absence of *S. ciperoensis*. MNN1a Subzone is documented by the absence or rare and discontinuous presence of *D. bisectus*; MNN1b is characterized by *Sphenolithus delphix*, and MNN1c is documented by the absence of *S. delphix* and *S. disbelemnos*.

The sandstone petrology of Falterona TS samples point to a feldspathic–lithic composition (average Q49 F31 L + C 20), with a fine-grained rock fragment composition dominated by metamorphic rocks, with minor amounts of intermediate, strongly chloritized volcanic grains and even fewer fragments of sedimentary rocks (average Lm84 Lv9 Ls + C7; Fig. 9). Within the system, a certain change in fine-grained rock fragment composition emerged from samples of the lowermost stage (FAL1 stage; average composition Lm 85 Lv10 Ls + C5) to samples of younger stages (FAL 2 and 3; average composition Lm82 Lv6 Ls + C 12). Calcareous beds are present but their occurrence is very uncommon, and the sandstones are usually too fine for microscope point counting. Flute casts of calcareous beds suggest an E–SE provenance.

The Falterona TS is covered by an important mudstone horizon, the Vicchio Marl, Burdigalian–Serravallian in age (Delle Rose *et al.* 1994), having at its base a silica-rich horizon, which represents a very important regional scale stratigraphic marker (Amorosi *et al.* 1995). At the top, the Vicchio Marl is overthrust by the Ligurian Units, whose emplacement here is thus no older than Serravallian. In turn, the FAL TS tectonically overlies all the more external AQR, TCG and CDP TSs, also reaching the MAIa TS (Fig. 5)

Acquerino turbidite system (AQR)

The Acquerino system represents a single fully developed TS about 1400 m thick. It shows the complete FU-TU facies trend (Figs 7 & 8). It stratigraphically overlies deep-sea pelagic mudstones (Varicoloured Shales Formation).

Three TSGs have been identified in the AQR system. The lowermost (AQR1) is made up of thickly bedded, medium- to coarse-grained sandstone beds (sandstone/mudstone ratio $\gg 1$). Locally, thinly bedded volcaniclastic sandstones also occur. The middle interval (AQR2) is made up of sandstone/mudstone couplets ranging from sand-dominated to mud-dominated beds (average sandstone/mudstone ratio $\ll 1$), locally with centimetre-scale layers of chert. The upper interval (AQR3) is characterized by thinly bedded mud-dominated turbidites and hemipelagic mudstones. Palaeo-currents data suggest provenance predominantly from W–NW, rarely from E–SE.

The Acquerino TS has an age documented from latest Chattian to Burdigalian (MNN1–MNN3b zone). Zone MNN1, documented by an association with *C. pelagicus*, *Cy. abisectus*, *Cy. floridanus*, *Helicosphaera carteri* (rare) is present, with the subzones characterized by the absence of *D. bisectus* (MNN1a), the presence of *S. delphix* (MNN1b), the absence of MNN1b-MNN1d markers (MNN1c), and the presence of *S. disbelemnos* (MNN1d). Zone MNN2a is characterized by the common and continuous occurrence of *Helicosphaera carteri*, and the MNN2b zone is marked by the presence of *Helicosphaera ampliaperta*. The occurrence and extinction of *Sphenolithus belemnos* indicate zones MNN3a and MNN3b respectively.

The sandstone petrology of Acquerino TS highlights a feldspathic-lithic composition for framework grains (average Q46 F37 L+C17), coupled with a composition of fine-grained rock fragments still dominated by metamorphic grains, associated with minor fragments of sedimentary rocks and very few volcanic grains (average Lm83 Lv6 Ls+C11; Fig. 9). These results are consistent with biostratigraphical data, which indicate an equivalent age for the lower part of the Acquerino TS and FAL2-3 stages. Also in the Acquerino TS very fine-grained calcareous beds are present but very uncommon south from the Mugello area and are lacking westward. Flute casts of calcareous beds suggest an E–SE provenance. The AQR TS tectonically overlies more external TCG and MAIa Systems (Figs 5 & 6).

Torrente Carigiola turbidite system (TCG)

The Torrente Carigiola TS represents a single fully developed TS, about 1000 m thick, deposited above deep-sea pelagic mudstones (Scisti Varicolori Formation) and is composed of two stages (Fig. 8). The lower stage (TCG1) is characterized by the occurrence of megaturbidites up to 35 m thick, with massive, poorly sorted, chaotically textured, very coarse-grained sandstones and microconglomerates at the base and very thick mud intervals at the top (the sandstone/mudstone ratio of megabeds is nearly 1). These very peculiar beds are interbedded with arenaceous-pelitic thinly to thickly bedded medium- to coarse-grained turbidites alternating with pelitic-arenaceous beds (sandstone/mudstone ratio from $>$ to <1), which formed the background sedimentation in the basin. Slumps and centimetre-thick chert-rich horizons also occur locally. Palaeo-current data point out a W–NW provenance. The upper stage (TCG2) is made up of very thin, hardened marls and marly siltstones with minor slumpings.

The Torrente Carigiola TS dates back to the late Chattian (?)–Aquitanian (MNN1ind-MNN1d Zone). The MNN1 undifferentiated zone is characterized by an association with *Coccolithus miopelagicus*, *C. pelagicus*, *Cy. abisectus*, *Cy. floridanus*, *Dictyococcites hesslandii-scrippsae*, *H. euphratis*, *H. carteri* (rare), *Reticulofenestra* spp., *Sphenolithus dissimilis* (rare) and *D. deflandrei*. Only the MNN1d subzone is recognized for the presence of *S. disbelemnos*.

The sandstone composition of Torrente Carigiola TS is very close to that of the Acquerino TS (Fig. 9), with a feldspathic–lithic framework composition (average Q47 F34 L+C19) and a metamorphic-dominated fine-grained rock fragment population (average Lm80 Lv5 Ls+C15). In this respect, turbidite megabeds show exactly the same composition as 'background' turbidite beds, suggesting different triggering mechanisms but the same source. Also in this TS very fine-grained calcareous beds are present but very uncommon; they occur south from the Mugello area but are lacking westward, as in the AQR TS.

Stagno turbidite system (STA)

The Stagno system is a fully developed TS, about 1200 m thick, with a completely developed

FU-TU facies trend (Fig. 8); it is composed of three TSGs. The lower stage (STA1) is made up of arenaceous-pelitic turbidites with alternating fine-grained thin- to medium-bedded turbidite beds (sandstone/mudstone ratio <1) and medium- to coarse-grained thick to very thick beds with a sandstone/mudstone ratio >1. The middle stage (STA2) is made up of sandstone/ mudstone couplets (sandstone/mudstone ratio <1) forming thin to medium (subordinately thick) beds. The upper stage (STA3) is characterized by very thin beds of hardened marly siltstone and silty marlstone. In the basal part of STA3, thin to medium, fine- and very fine-grained sandstone beds also occur. Palaeo-currents data suggest W–NW provenance.

The Stagno TS dates from Aquitanian to Burdigalian times (MNN2a–MNN3b Zones). A normal assemblage in these zones contains *C. miopelagicus, C. pelagicus, Cy. floridanus, Cy. abisectus, Dictyococcites* spp., *H. carteri, H. ampliaperta, H. euphratis, Helicosphaera intermedia, Discoaster deflandrei, Reticulofenestra* spp. and *Sphenolithus* spp. The specimens that characterise the zones are: *H. carteri* (common and continuous occurrence, MNN2a), *Helicosphaera mediterranea* (presence, MNN2b), *S. belemnos* (presence, MNN3a) and *S. belemnos* (absence, MNN3b).

The composition of sandstones in the Stagno TS is still very close to that of the Acquerino and T. Carigiola systems (Fig. 9). The framework composition is feldspathic–lithic (average Q54 F26 L + C20) and the fine-grained rock fragment association is still metamorphic-dominated (average Lm83 Lv5 Ls + C12).

Castiglione dei Pepoli turbidite system (CDP)

The Castiglione dei Pepoli system is more than 1000 m thick, even though it is not fully developed, and it is truncated by a thick debris flow deposit directly overlain by a thrust surface. The composition of the debris flow deposit mirrors the overthrust unit, and thus suggests that the debris flow may represent the depositional precursor of the thrust emplacement. In addition the thrust geometry is roughly conformable with bedding planes for the few kilometres of the outcrop area. For these reasons it may be interpreted as a syndepositional thrust.

The lower stage of the Castiglione dei Pepoli System (CDP1) is made up of thick to very thickly bedded, medium-grained sandstone turbidite bed-sets with very thin shale interbeds, alternating with thin to medium mud-dominated bed-sets; the overall sandstone/mudstone ratio is ≫1. In the middle stage (CDP2), mud-dominated bed-sets prevail over sand-dominated sets (sandstone/mudstone ratio <1).

The age of the CDP system spans from Burdigalian(?) to Langhian (MNN4a?–MNN5a Zones). The age of the base is possibly related to the MNN4a zone judging by its stratigraphic relations with the top of the Stagno TS. The samples that indicate Langhian zones (MNN4b–MNN5a) contain this assemblage: *C. miopelagicus, C. pelagicus, Calcidiscus leptoporus, Cy. floridanus, Dictyococcites* spp., *H. carteri, Reticulofenestra* spp., *Sphenolithus heteromorphus*, and *Sphenolithus* spp.; the scattered and discontinuous occurrence and the common and continuous occurrence of *Sphenolithus heteromorphus* characterizes the MNN4b zone and the MNN5a subzone respectively.

Sandstones of the Castiglione dei Pepoli TS show a lithic–feldspathic framework composition (average Q53 F22 L + C25), and, most importantly, record an abrupt change in the composition of fine-grained rock fragments with respect to older TSs, due mostly to a strong increase in sedimentary rock fragments and a relative decrease in the amount of metamorphic grains (average Lm61 Lv4 Ls + C35). This is the most important compositional change recorded by the sandstones of the studied turbidite systems (Fig. 9).

The lower turbidite system of the pre-Tortonian Marnoso Arenacea turbidite complex (MAIa)

The pre-Tortonian part of the Marnoso Arenacea has been subdivided into two fully developed turbidite systems (MAI1a and MAI1b).

The lower system (MAI1a, corresponding to the 'inner stage' deposits of Ricci Lucchi 1975, 1981) is up to 3000 m thick and accumulated above slope marls (Schlier Formation). It is characterized by an unusual coarsening and thickening upward trend in its lowermost part, coupled with a sandstone/mudstone ratio that increases from $\frac{1}{5}-\frac{1}{3}$ to 2 from the lowermost (MAIa1) to the intermediate stage (MAIa3); upward, a typical FU-TU facies trend with a sandstone/ mudstone ratio decreasing to $\frac{1}{5}$ completes the system. This peculiar trend could be related to a gradual shift of the foredeep system in this case, different from the abrupt one recorded by the other TSs. The sandstones are normally medium- to fine-grained, rarely coarse-grained (Figs 7 & 8) with palaeo-current casts from the W–NW, and rarely from the E–SE.

The upper part is characterized by the frequent occurrence of mud-dominated calcareous beds with a SE provenance ('colombine' beds) and basin-wide hybrid turbidite beds (see, for instance, Ricci Lucchi 1981; Gandolfi et al. 1983). Among these well-known marker beds, the most recognizable is the Contessa layer (cs; Figs 2, 6 & 7). Frequent slump deposits occur at the top of the MAI1a system; they include Ligurian blocks and debris of older and penecontemporaneous foredeep sandstones in a muddy matrix.

The age of the MAI1a TS base is Late Burdigalian (MNN3b zone), for the presence of *H. ampliaperta* and for the lack of *S. belemnos* and *S. heteromorphus* at the base of the succession (see also Fornaciari 1996). Its top is Early Serravallian (MNN6b subzone) in the Santerno-Lamone area, for the common and continuous occurrence of *Reticulofenestra pseudoumbilicus* (>7 μm) and the lack of *Sphenolithus heteromorphus*, and Late Serravallian (MNN7 Zone) in the Savio Valley region, for the common and continuous occurrence of *Helicosphaera walbersdorfensis* and *Reticulofenestra pseudoumbilicus* (>7 μm) and the lack of *Calcidiscus premacintyrei*. In this area MNN4a, MNN4b, MNN5a and MNN5b are also documented, indicated respectively by: presence of *S. heteromorphus* (common and continuous, MNN4a); presence of *S. heteromorphus* (rare and discontinuous, MNN4b); presence of *S. heteromorphus* (common and continuous) in association with *Helicosphaera walbersdorfensis* (rare and discontinuous), and *Discoaster musicus* (MNN5a); the presence of *H. walbersdorfensis* (common and continuous) with *S. heteromorphus*, *Geminilithella rotula* (common), *R. pseudoumbilicus* (>7 μm) (rare and discontinuous, MNN5b).

The sandstone composition of samples from the oldest TS detected in the Marnoso Arenacea formation records a feldspathic–lithic composition (average Q58 F24 L+C18) and a rather variable fine-grained rock fragment association with average sedimentary rock fragments that almost equal the amount of metamorphic grains, with very few volcanic grains (average Lm53 Lv4 Ls+C 43). This composition records the continuation of the petrographic change abruptly begun with the Castiglione dei Pepoli TS (Fig. 9). Beds with palaeo-currents from E–SE generally have carbonate to hybrid composition (Gandolfi et al. 1983) and generally thin northwestward; in addition few siliciclastic beds with palaeo-currents from S and SE have been recognized, possibly as a result of re-sedimentation of penecontemporaneous or slightly older foredeep sediments.

The upper turbidite system of the pre-Tortonian Marnoso Arenacea turbidite complex (MAI1b)

The upper system detected in the pre-Tortonian Marnoso Arenacea turbidite complex (MAI1b) corresponds to the transition between the 'inner stage' and the 'outer stage' deposits of Ricci Lucchi (1975, 1981). It is a fully developed system, up to 1000 m thick, of arenaceous-pelitic beds whose thickness ranges from very thick to medium; sandstone intervals are generally medium- to fine-, rarely coarse-grained, and palaeo-currents are from the NW, seldom from the SE. The system is also characterized by the presence of slump deposits and a very thick debris flow and slide deposit at the base (Casaglia chaotic unit), draped by a mudstone horizon at the top, also frequently slumped.

The age of the MAI1b system is Late Serravallian (MNN7 zone). The zone is characterized by the common and continuous occurrence of *H. walbersdorfensis* and *R. pseudoumbilicus* (>7 μm), in association with *C. leptoporus*, *H. carteri*, *G. rotula*, *Helicosphaera orientalis*, *Helicosphaera intermedia*, *Discoaster variabilis*, and the presence of *Calcidiscus macintyrei* in its lower part, and the absence of *Coccolithus miopelagicus* in its upper part.

In the southeastern sector of the Romagna Apennines (Bidente and Savio Valleys) MAI1b is not present, and an important pelitic unit, the S. Paolo Marl, Late Serravallian–Late Tortonian in age (the upper part of the Verghereto Marls Auct.; see van Wamel and Zwart 1990), stratigraphically covers the MAI1a system.

Sandstone compositions of the MA1b turbidite beds are virtually the same as the MA1a TS (average Q59 F23 L+C18 and Lm54 Lv2 Ls+C 44), although a smaller variability of sample composition is recorded (Fig. 9). Calcareous beds are not present.

Discussion

Regional field-mapping shows that turbidite systems and their constituent turbidite stages represent the fundamental building blocks of Oligocene–Miocene Northern Apennines foredeep depositional sequences. Their definition in terms of types of bounding surfaces, lithology, thickness, depositional age range and detrital composition provides an integrated dataset that makes it possible to unravel the tectono/sedimentary evolution of the Northern Apennines foredeep since the latest Oligocene to Tortonian times. This allows us to discuss the relative influence on that evolution of the three major factors controlling foredeep turbidite deposition.

The starting point for the discussion is that all recognized turbidite systems, unless tectonically truncated (e.g. in the Castiglione de Pepoli TS), basically show the same overall stacking pattern, of a lower sand-rich, thickly bedded stage (*depocentre stage*) passing upward to mud-rich thinly bedded stages, eventually grading up to mostly pelitic units (*abandonment stage*). These pelitic units are often associated with slump facies that become increasingly frequent and thick from older (TCG) to younger (MAI) TSs; it is probably related to the increasing instability (more steep?) of the southwestern slope. This overall, rhythmically repeated pattern is interpreted as the result of the generally abrupt switching on and progressive switching off of coarse-grained input, coupled with an alternating increase/decrease of the depositional rate clearly recorded in the detected systems (Fig. 10).

The question that arises is: what are the factors controlling this pattern? As mentioned in the introduction, at first glance three different major causes can be invoked (Fig. 11): basin tectonics, source tectonics and relative sea-level fluctuations (both regional and global). These three factors are largely independent and can act separately or together during the life of the foredeep, guiding its complex stratigraphic evolution. An attempt to distinguish, step by step, the real cause for the switching on and off of each TS can be made by taking into account the fact that, besides the change of sediment type and accumulation rate within a TS, each of the mechanisms can be expected to have different detectable effects (Fig. 11).

In foredeep settings, basin tectonics is mostly expected to force the basin depocentre to migrate outwards and incorporate the older foredeep system at the base of the inner slope; in a relatively simple topographic configuration with dominant axial clastic input and turbiditic currents reaching the deepest depocentre of the basin, this has the effect of also shifting the point of maximum turbidite accumulation. Note that this kind of event does not significantly change the detrital input into the foredeep in a setting such as the Oligocene–Miocene Northern Apennines, where the clastic source was located in another independent belt. It should simply shift the depocentre location, switching on a

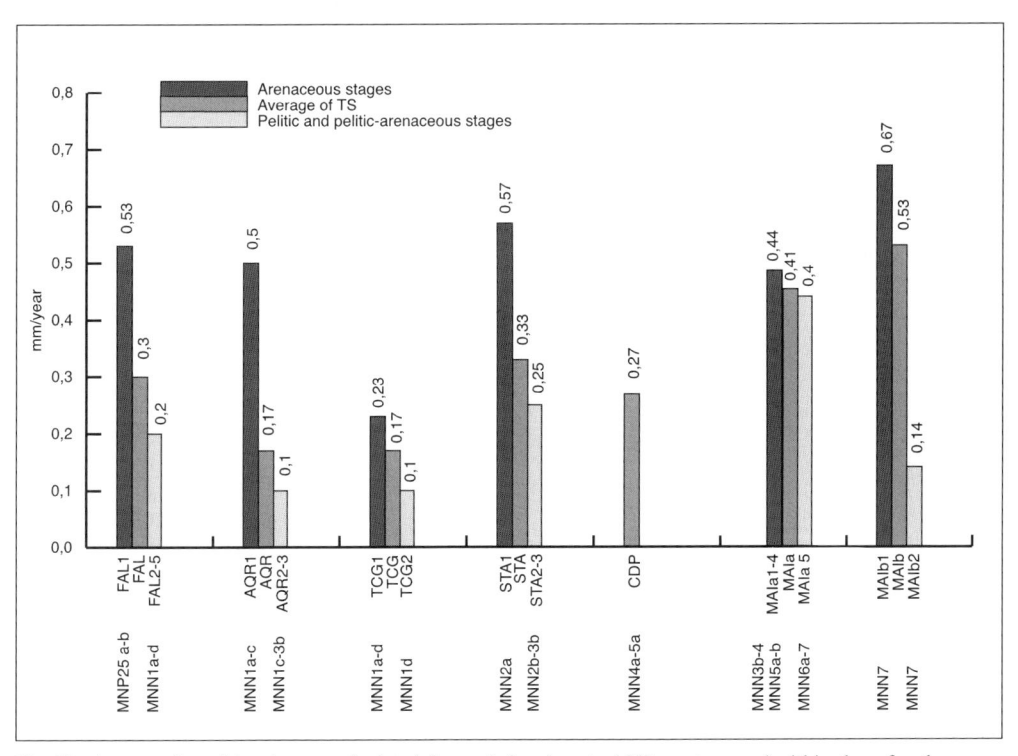

Fig. 10. Average depositional rates calculated for each foredeep turbidite system and within them for the depocentre stage and abandonment stage. Note that the calculation is based on actual, not decompacted, sediment thickness, so relative changes are meaningful more than absolute values.

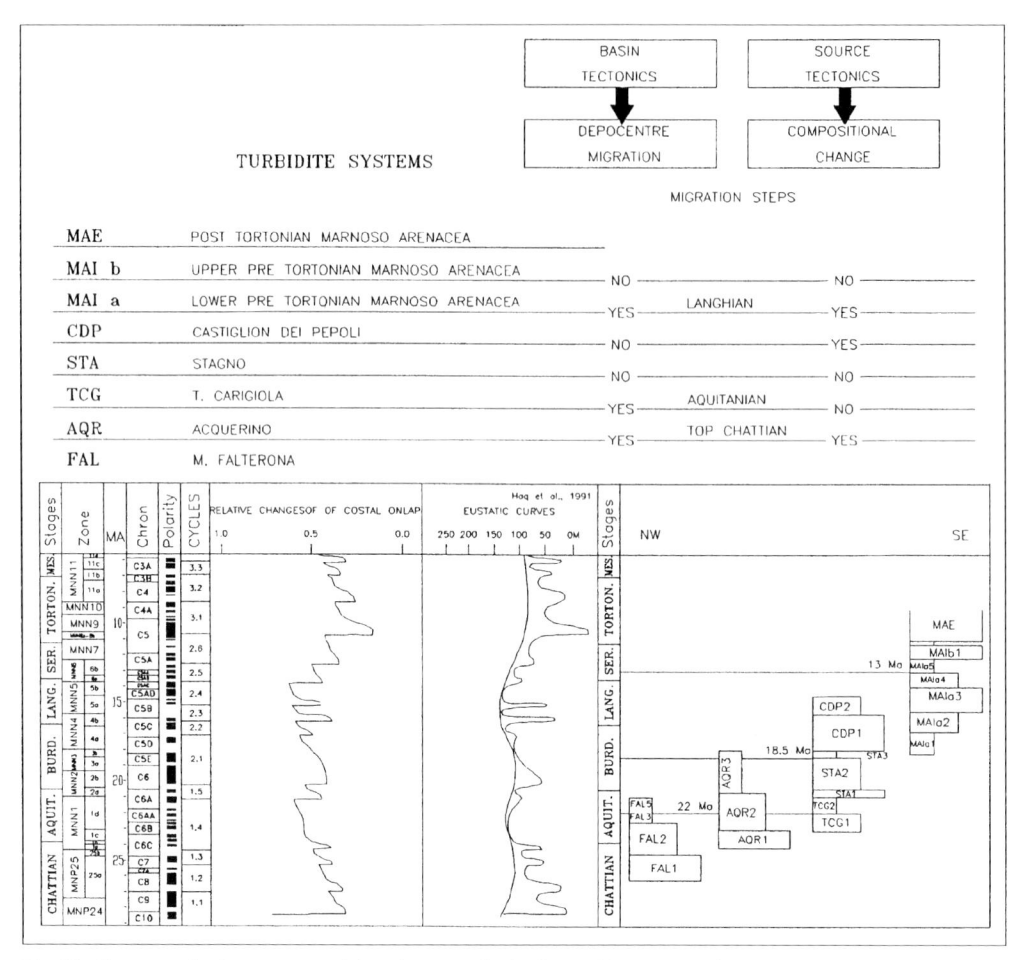

Fig. 11. Conceptual scheme summarizing the recognized effects of basin tectonics and source tectonics on the depositional evolution of the studied turbidite complexes. In addition, a comparison between the global sea-level curve and the timing of depositional variations in the studied foredeep turbidite complexes is reported.

new TS contemporaneously with the progressive or abrupt switching off of coarse input into the older and inner one. This can be recognized by integrated regional mapping and biostratigraphy.

In the area studied, for instance, this is the case for the relationship between the Falterona and Acquerino TSs, as well as between the Castiglione de Pepoli and MAIa TSs, which therefore record two major phases of depocentre migration dated to the end of the Chattian and the Langhian respectively.

An additional, peculiar situation is recorded by the T. Carigiola TS (Fig. 12): in it, the rapid lateral change of facies from sand-dominated to mud-dominated and the occurrence of three megabeds (up to 35 m thick), which thin north-northeastward and pinch out over an impressively short distance (approximately 2 km), consistently record a strong confinement effect due to deposition into a relatively small-scale depocentre, probably generated by a minor, possibly local, middle Aquitanian migration step of the foredeep. In this case the progressive retreat (S–SW shift) up section of pinch-out points of megabeds, as well as that of facies change, clearly reflect a syndepositional growth of the confining positive structure (syndepositional anticline) and its negative counterpart (syndepositional syncline; Fig. 12). The confinement effect of this local topography ceased to be efficient during the deposition of the Stagno lowermost stage, probably as a result of its smoothing by turbidite deposition.

Note also that, in the region studied, two of the recorded phases of depocentre shift (latest

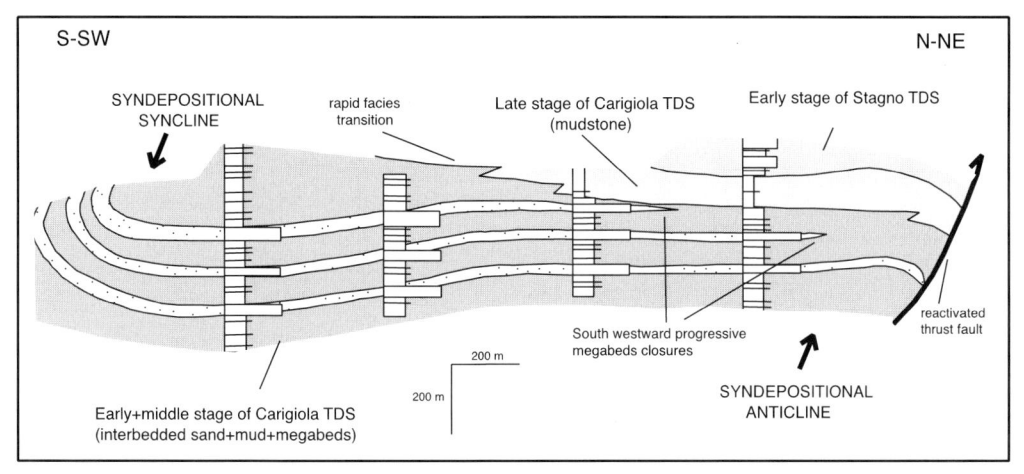

Fig. 12. Lithostratigraphical scheme for the Torrente Carigiola turbidite system. Note the occurrence of outward-thinning megabeds up to 35 m thick and lateral facies variations occurring at quite a small scale, which consistently indicate its deposition into a local, quite deep depocentre, under strongly topographically confined conditions.

Chattian and middle Aquitanian) are coupled with a gradual abandonment of the previous TS, through a stage of still-active turbidite deposition with reduced grain size and depositional rate and/or sediment supply (FAL2 and AQR2 stages). These stages probably record an 'open' piggyback depositional phase of the TS evolution, similar to that found by Ori et al. (1993) for Early-Middle Pliocene deposits of the Adriatic foredeep. This can be due to the nucleation of blind thrust surfaces below TS during sedimentation coupled with the growth on an anticline-syncline system in front of it (Jamison 1987; McNaught & Mitra 1993), before definitive switching off and transformation of depositional units into tectonic units (see also Boccaletti et al. 1990a,b). By contrast, this kind of evolution is not recorded during the Langhian step of depocentre migration, because of the early and abrupt tectonic switch-off of the Castiglion dei Pepoli TS.

On the other hand, source tectonics has the power to change contemporaneously both the volume and the composition of clastics funnelled into the basin, whereas basin tectonics only controls the routeing of clastics. This is especially true when that source is located in a rapidly evolving, geologically heterogeneous region, as is the case for the Alpine source of the Northern Apennines foredeep sediments, so that each tectonically induced change of the source location is likely to produce a change in the type of eroded rocks, or at least in their relative amounts.

Note that, in this case, foredeep deposition should record both volume and compositional changes of clastics without depocentre migration, unless coeval basin tectonic events occur. This would be recognized through integrated field mapping and sedimentary petrology.

In the studied region, this is the case for the Stagno to Castiglione dé Pepoli TS transition, and, to a lesser degree, for the Falteronal/ Acquerino and Castiglione de Pepoli/MAIa transitions, when coeval depocentre shift also occurred. It can hardly be a coincidence that the ages of these three petrographic changes fit well with the ages of major tectonically related unconformities recorded in the Tertiary successions of the central Southern Alps (Como-Bergamo area; Castellarin et al. 1992). At the same time, it should be stressed that the observed change of petrographic signature is the opposite with respect to the one expected for a simple deroofing of the source region, as it is due mostly to the increase in time of the amount of clastics fed by carbonate sedimentary rocks relative to the contribution from plutonic-metamorphic rocks. This is probably due to the lateral eastward shift of the source area from the Western Alps to the central Southern Alps during the Chattian-Early Miocene and Middle Miocene respectively.

Finally, sea-level changes (both regional or global) also have the potential to modulate the re-sedimentation of clastics to the deep sea via turbidity currents (lowering sea-level), or conversely causing their temporary storage on shallow-marine areas (rising sea-level) and the starvation in the deep sea, recorded by a drop both of sand input and depositional rate (Mutti

1985). In this case, presumably, neither depocentre shift nor petrographic change should occur unless regional sea-level change is due to or coeval with source or basin tectonics. As a result, the switch-on and off of petrographically monotonous turbidite systems should occur in a vertical stratigraphic section.

In the study region, this seems to be the case for the uppermost MNN1d biozone time span, for the MNN3b biozone time span, and possibly for the MNN6 biozone time span, corresponding to 21–22 Ma, 18.5 Ma and around 13 Ma respectively, according to the timescale of Cande & Kent (1992, 1995). In this respect, note that these ages fit surprisingly well with the three sea-level rises shown by the global sea-level curve (at 21.5–18.5 and 13.4 Ma; Haq 1991). At the same time note that other global sea-level rises, sometimes very marked according to the global curve (for instance, that at 15 and 16 Ma) are completely unrecorded in the Northern Apennines foredeep, as they coincide with periods of very high terrigenous input.

This makes it questionable whether the sea-level signal could be due to global events (see Miall & Miall 2001 for a discussion of that point), even if the lack of evidence for some global events could be due to their coincidence with phases of tectonics and rapid uplift in the sediment source area, overbalancing at a regional scale the effect of eustacy.

Conclusion

From a general point of view, this study documents how integrated field mapping, biostratigraphy and sedimentary petrography are able to shed some light on the complex interplay of the main factors controlling the large-scale architecture, geometry and evolution of the Northern Apennines foredeep turbidite complexes.

This kind of approach deals with that problem at the scale of the whole source–basin clastic routeing systems, and focuses mostly on large-scale external controls on deep-sea sedimentation, through the modulation in time and space of the volume and grain size of sediment load carried by turbidity currents. It provides a conceptual framework into which more detailed facies analyses and depositional models regarding turbidite dynamics at the scale of beds or bed-sets can be integrated (e.g. Mutti 1992).

Together, these different approaches provide the basis for producing predictive models of sandstone distribution into foredeep turbidite complexes, such as would be useful for petroleum exploration. For future research, the Northern Apennines case demonstrates that, for this task, knowledge of the source and its tectonic history, besides the basin tectonics and sea-level changes, is of great importance.

D. Hodgson (Liverpool) and B. Cronin (Aberdeen) are kindly acknowledged for their constructive reviews. Financial grants were received from CARG-Project (Regione Emilia Romagna) and FAR Funds (Pavia University).

References

AMOROSI, A., RICCI LUCCHI, F. & TATEO, F. 1995. The Lower Miocene siliceous zone: a marker in the palaeogeographic evolution of the northern Apennines (Italy). *Palaeogeography, Palaeoclimatology, Palaeoecology*, **118**, 131–149.

ANDREOZZI, M. & DI GIULIO, A. 1994. Stratigraphy and petrography of the M. Cervarola Sandstones in the type area, Modena Province. *Memorie Società Geologica Italiana*, **48**, 351–360.

BACKMAN, J. & SHACKLETON, N. J. 1983. Quantitative biochronology of Pliocene and Early Pleistocene calcareous nanoplankton from the Atlantic, Indian and Pacfic Oceans. *Marine Micropaleontology*, **8**, 141–170.

BARBERI, F., GASPARINI, P., INNOCENTI, F. & VILLARI, L. 1973. Volcanism of the southern Tyrrhenian Sea and its geodynamic implications. *Journal of Geophysical Research*, **78**, 5221–5232.

BERGREEN, W. A., KENT, D. V., SWISHER, C. C. III & AUBRY, M. P. 1995. A revised Cenozoic geochronology and chronostratigraphy. *In*: BERGREEN, W. A., KENT, D. V., SWISHER, C. C. III & HANDERBOL, J. (eds) *Geochronology, Time Scales and Global Stratigraphic Correlation*. Society of Economic Paleontology Mineralogy, Special Publications, **54**.

BIJU DUVAL, B., DERCOURT, J. & LE PICHON, X. 1977. From the Tethys Ocean to the Mediterranean Sea: plate tectonic model of the evolution of the western Alpine system. *In*: BIJU-DUVAL, B. & MONTADERT, L. (eds) *Structural History of the Mediterranean Basins*, 143–164.

BOCCALETTI, M. & GUAZZONE, G. 1972. Gli archi appenninici, il mar Ligure ed il Tirreno nel quadro della tettonica dei bacini marginali retro-arco. *Memorie Società Geologica Italiana*, **11**, 201–216.

BOCCALETTI, M. & GUAZZONE, G. 1974. Remnant arcs and marginal basins in the Cainozoic development of the Mediterranean. *Nature*, **252**, 18–21.

BOCCALETTI, M., CALAMITA, F., DEIANA, G., GELATI, R., MASSARI, F., MORATTI, G. & RICCI LUCCHI, F. 1990a. Migrating foredeep-thrust belt system in the Northern Apennines and Southern Alps. *Palaeogeography, Palaeoclimatology, Palaeoecology*, **77**, 3–14.

BOCCALETTI, M., CIARANFI, N. *et al*. 1990b. Palinspastic restoration and palaeogeographic reconstruction of the peri-thyrrenian area during the Neogene. *Palaeogeography, Palaeoclimatology, Palaeoecology*, **77**, 41–50.

BRUNI, P. & PANDELI, E. 1980. Torbiditi calcaree nel Macigno e nelle Arenarie del Cervarola nell'area del Pratomagno e del Falterona (Appennino Sett.). *Memorie della Società Geologica Italiana*, **21**, 217–230.

BUCEFALO PALLIANI, R., LUCHETTI, L., NINI, C., NOCCHI, M. & RETTORI, R. 1997. Age and palaeoecological inferences of the upper Monte Falterona Sandstone Formation (Lonnano Member, Early Miocene), Northern Apennines. *Giornale di Geologia*, ser. 3a, **59**, 143–168.

BURRUS, J. 1984. Contribution to a geodynamic synthesis of the Provençal basin (north-western Mediterranean). *Marine Geology*, **55**, 247–269.

BUTLER, R. W. H. 1987. Thrust sequences. *Journal of the Geological Society, London*, **144**, 619–634.

CAMPBELL, C. V. 1967. Lamina, laminaset, bed and bedset. *Sedimentology*, **8**, 7–26.

CANDE, S. & KENT, D. V. 1992. A new geomagnetic polarity time scale for the Late Cretaceous and Cenozoic. *Journal of Geophysical Research*, **13**, 917–971.

CANDE, S. & KENT, D. V. 1995. Revised calibration of the geomagnetic polarity time scale for the Late Cretaceous and Cenozoic. *Journal of Geophysical Reserch*, **100**, 6093–6095.

CASTELLARIN, A., CANTELLI, L. et al. 1992. Alpine compressional tectonics in the Southern Alps. Relationships with the N-Apennines. *Annales Tectonicae*, **6**, 62–94.

COWARD, M. & DIETRICH, D. 1989. Alpine tectonics: an overview. In: COWARD, M. P., DIETRICH, D. & PARK, G. (eds) *Alpine Tectonics*. Geological Society, London. Special Publications, **45**, 1–29.

DELLE ROSE, M., GUERRERA, F., RENZULLI, A., RAVASZ-BARANYAI, L. & SERRANO, F. 1994. Stratigrafia e petrografia della Marne di Vicchio (Unità tettonica Cervarola) dell'alta Val Tiberina (Appennino Tosco-Romagnolo). *Bollettino Società Geologica Italiana*, **113**, 675–708.

DEWEY, J. F., PITMAN, W. B., RYAN, F. & BONNIN, J. 1973. Plate tectonics and the evolution of Alpine system. *Geological Society of America Bulletin*, **84**, 3137–3180.

DICKINSON, W. R. 1970. Interpreting detrital modes of graywacke and arkose. *Journal of Sedimentary Petrology*, **40**, 695–707.

DI GIULIO, A. 1999. Mass transfer from the Alps to the Apennines: volumetric constraints in the provenance study of the Macigno-Modino source-basin system, Chattian-Aquitanian, north-western Italy. *Sedimentary Geology*, **124**, 69–80.

DI GIULIO, A. & VALLONI, R. 1992. Sabbie e areniti: analisi ottica e classificazione. *Acta Naturalia Ateneo Parmense*, **28**, 55–101.

FINETTI, I. & DEL BEN, A. 1986. Geophysical study of the Tyrrhenian opening. *Bollettino Geofisica Teorica e Applicata*, **28**(110), 75–156.

FORNACIARI, E. & RIO, D. 1996. Latest Oligocene to early middle Miocene quantitative calcareous nannofossil biostratigraphy in the Mediterranean region. *Micropaleontology*, **42**, 1–36.

FORNACIARI, E., DI STEFANO, A., RIO, D. & NEGRI, A. 1996. Middle Miocene quantitative calcareous nannofossil biostratigraphy in the Mediterranean region. *Micropaleontology*, **42**, 37–63.

GANDOLFI, G., PAGANELLI, L. & ZUFFA, G. G. 1983. Petrology and dispersal pattern in the Marnoso Arenacea Formation (Miocene, Northern Apennines). *Journal of Sedimentary Petrology*, **53**, 493–507.

GAZZI, P. 1966. Le arenarie del Flysch sopracretaceo dell'Appennino modenese; correlazioni con il Flysch di Monghidoro. *Mineralogica Petrografica Acta*, **16**, 69–97.

GHIBAUDO, G. 1980. Deep-sea fan deposits in the Macigno Formation (Middle-Upper Oligocene) of the Gordana Valley, Northern Apennines, Italy. *Journal of Sedimentary Petrology*, **50**, 723–742.

HAQ, B. U. 1991. Sequence stratigraphy, sea-level change, and significance for the deep sea. In: MACDONALD, D. I. M. (ed.) *Sedimentation, Tectonics and Eustasy*. IAS Special Publications, **12**, 3–39.

JAMISON, W. R. 1987. Geometric analysis of fold development in overthrust terraines. *Journal of Structural Geology*, **9**, 207–219.

MARTINI, E. 1971. Standard Tertiary and Quaternary nannoplankton zonation. In: FARINACCI, A. (ed.) *Proceedings II Planktonic Conference*, **2**, 85–739.

MCNAUGHT, M. A. & MITRA, G. 1993. A kinematic model for the origin of footwall synclines. *Journal of Structural Geology*, **15**, 805–808.

MIALL, A. D. & MIALL, C. E. 2001. Sequence stratigraphy as a scientific enterprise: the evolution and persistence of conflicting paradigms. *Earth-Science Reviews*, **54**, 321–348.

MUTTI, E. 1985. Turbidite systems and their relations to depositional sequences. In: ZUFFA, G. G. (ed.) *Provenance of Arenites*. NATO-ASI Series, 65–93.

MUTTI, E. 1992. *Turbidite Sandstones*. AGIP Spa, S. Donato Milanese.

MUTTI, E. & NORMARK, W. R. 1987. Comparing examples of modern and ancient turbidite systems: problems and concepts. In: LEGGETT, J. K. & ZUFFA, G. G. (eds) *Marine Clastic Sedimentology: Concepts and Case Studies*, 1–38.

OKADA, H. & BUKRY, D. 1980. Supplementary modification and introduction of code numbers to the low latitude coccolith biostratigraphic zonation. *Marine Micropaleontology*, **5**, 5–321.

ORI, G. G., SERAFINI, G., VISENTIN, C., RICCI LUCCHI, F., CASNEDI, R., COLALONGO, M. L. & MOSNA, S. 1993. Depositional history of the Pliocene-Pleistocene Adriatic foredeep (Central Italy) from surface and subsurface data. In: SPENCER, A. M. (ed.) *Generation, Accumulation and Production of Europe's Hydrocarbons III*. EAPG Special Publications, **3**, 233–258.

PRINCIPI, G. & TREVES, B. 1984. Il sistema corso-appenninico come prisma di accrezione. Riflessi sul problema generale del limite Alpi-Appennini. *Memorie Società Geologica Italiana*, **28**, 549–576.

RICCI LUCCHI, F. 1975. Miocene palaeogeography and basin analysis in the peri-Adriatic Apennines. In: SQUIRES, C. (ed.) *Geology of Italy*. PESL, Tripoli, **2**, 129–236.

Ricci Lucchi, F. 1981. The Miocene Marnoso-Arenacea turbidites, Romagna and Umbria Apennines. *Excursion No. 7. Excursion Guidebook*, 2nd European Regional Meeting IAS.

Ricci Lucchi, F. 1986. The Oligocene to recent foreland basins of the Northern Apennines. *In*: Allen, P. A. & Homewood, P. (eds) *Foreland Basins.* IAS Special Publications, **8**, 105–139.

Ricci Lucchi, F. 1990. Turbidites in foreland and on-thrust basins of the Northern Apennines. *Paleogeography, Paleoclimatology, Paleoecology*, **77**, 51–66.

Rio, D., Raffi, I. & Villa, G. 1990. *Pliocene-Pleistocene calcareous nannofossil distribution patterns in the western Mediterranean. In*: Kastens, K., Mascle, J. et al. (eds) *Proceedings of the ODP, Scientific Results.* ODP, College Station, TX, 513–533.

Valloni, R., Lazzari, D. & Calzolari, M. A. 1991. Selective alteration of arkose framework in Oligo-Miocene turbidites of Northern Apennines foreland: impact on sedimentary provenance analysis. *In*: Morton, A. C., Todd, S. P. & Haughton, P. D. W. (eds) *Developments in Sedimentary Provenance Studies.* Geological Society, London, Special Publications, **57**, 125–136.

Van Wamel, W. A. & Zwart, P. E. 1990. The structural geology and basin development of the Romagnan-Umbrian zone (Upper Savio and Upper Bidente valleys, N. Italy). *Geologie en Mijnbown*, **69**, 53–68.

Weimer, P. & Link, M. H. 1991. Global petroleum occurrence in submarine fans and turbidite systems. *In*: Weimer, P. & Link, M. H. (eds) *Seismic Facies and Sedimentary Processes of Submarine Fans and Turbidite Systems.* 9–67.

Impact of syndepositional faulting on gravity current behaviour and deep-water stratigraphy: Tabernas-Sorbas Basin, SE Spain

DAVID M. HODGSON[1,2] & PETER D. W. HAUGHTON[3]

[1] *Department of Geological Sciences, Gower Street, University College London, London WC1E 6BT, UK (e-mail: hodgson@liv.ac.uk)*
[2] *Present address: Stratigraphy Group, Department of Earth Sciences, Jane Herdman Laboratories, 4 Brownlow Street, Liverpool L69 3GP, UK*
[3] *Department of Geology, University College Dublin, Belfield, Dublin 4, Ireland*

Abstract: Seabed faulting can have a significant impact on the routeing and behaviour of gravity currents depositing sand on deepwater basin floors. The Neogene El Cautivo Fault in the Tabernas-Sorbas Basin, SE Spain, is a rare example of a fault that demonstrably propagated through to the seabed during turbidite deposition, allowing the interplay between deepwater sedimentation and tectonics to be explored. The fault is associated with a wide (up to 350 m) gouge zone that varies significantly in thickness along its length, reflecting upward expansion towards the original seabed and progressive burial as fault activity ceased. Kinematic and stratigraphic evidence indicate that the fault was a dextral oblique strike-slip fault that accommodated an area of deeper ponded bathymetry (a 'mini-basin') and accelerated subsidence on its southern flank. Active faulting controlled the routeing of turbidity currents (revealed by changing provenance across the structure), rates of seabed deformation (resulting in differential subsidence and 'growth' of the stratigraphy), and the behaviour of the ponded currents (producing distinctive bipartite beds when deposition was in localized ponded depressions). The seabed expression of the fault varied from a forced fold, which warped the surface causing local wedging and onlap in the vicinity of the structure, and an unstable scarp that locally collapsed. The fault gouge fabrics and vein arrays are consistent with faulting of soft, water-rich sediment close to the seabed.

Hydrocarbon exploration on morphologically complex slopes has emphasized the importance of bathymetry to understanding gravity-driven transport in deep water. Bathymetry can have an overriding control on processes of deposition and system style via gradient changes and flow obstruction. The Tabernas-Sorbas Basin, SE Spain, is a small Neogene fault-controlled basin (Fig. 1) where the impact of faulting on the mode of turbidite deposition has recently become apparent (Haughton 2000). Although previously thought to contain unconfined, basin-floor fan systems (Kleverlaan 1989b), recent work has shown that the basin floor was actively deforming during deposition, and this had a major impact on slope incision, the propensity of flows to bypass, and the development of ponded turbidite systems.

There are few outcrop examples where faults that have actively propagated through deepwater successions can be studied; yet seabed faulting can have a significant impact on the routeing and behaviour of turbidity currents. Although many workers have studied the faults bounding Upper Miocene turbidite basins in SE Spain, the significance of intrabasinal faults has been largely overlooked. This is partly because the overall marl-prone basin fill means the fine-grained gouge zones are typically poorly exposed. The lithological distinction between the rocks they separate is commonly more subtle than the range-bounding faults that demarcate the basin margins. The latter are typically demonstrably later faults or reactivated structures associated with basin inversion, and hence it is difficult to read the earlier syndepositional history and style of displacement during basin filling from them.

A new analysis of the turbidite systems in the Tabernas area of the basin has revealed an important intrabasinal fault zone, the El Cautivo fault zone. The fault zone cuts the early basin-fill, but is sealed by later deposits, indicating that it was active during deposition (Fig. 2). The significance of this fault and the character of the unusually thick gouge zone associated with it have become apparent following extensive excavations for a new motorway that cuts through the fault zone. This study focuses on the character of the fault zone itself, combining detailed structural observations on the geometry, the structure and the kinematics of the fault zone with sedimentological observations relating to the stacked turbidite systems either side of it. The associated seabed topography influenced the behaviour of turbidity currents entering the basin, with evidence for changing styles of flow containment during and after fault propagation.

From: Lomas, S. A. & Joseph, P. (eds) 2004. *Confined Turbidite Systems*. Geological Society, London, Special Publications, **222**, 135–158. 0305-8719/04/$15.00 © The Geological Society of London 2004.

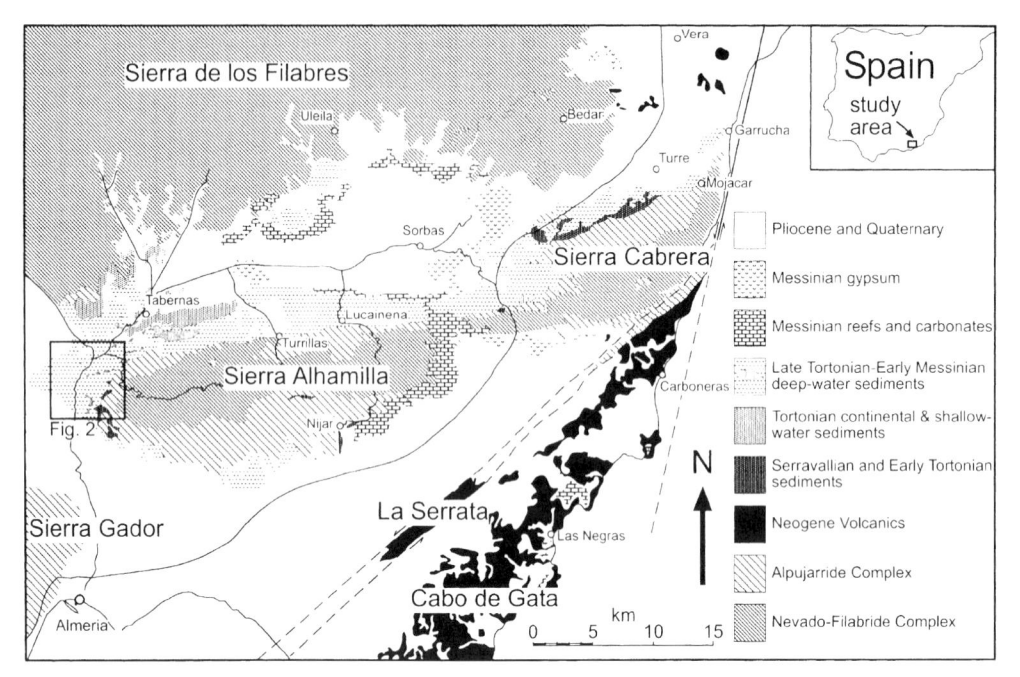

Fig. 1. Location map showing the simplified geology of the Tabernas-Sorbas Basin and the intervening basement ridges.

Geological setting

The Tabernas-Sorbas Basin is one of a series of Neogene intramontane basins resting on the Internal Zones of the Betic Cordillera, SE Spain (Fig. 1), the westernmost extension of the Tertiary Alpine Belt. During the Tortonian and Messinian, the basin was a narrow E–W trending elongate trough. Basement highs of variable metamorphic grade bound the basin and were source areas throughout the basin filling. The northern and northwestern margins (the Sierra de los Filabres) were the dominant sediment source with clasts derived from the high-grade Nevádo-Filabride Complex (Fig. 1). Sediment supplied from the southern and western margins (Sierra Alhamilla and Sierra Gador) had a very different provenance from lower-grade Alpujarride Complex lithologies (Fig. 1). Input from the southern margin waned through the Tortonian and Lower Messinian, although the submerged Alhamilla palaeo-high was critical to the stratigraphic evolution of the Tabernas-Sorbas Basin during this time.

The origin and development of the Neogene basins is contentious. Montenat et al. (1987) believed that the subsidence in these basins was due to a wide NE–SW trending left-lateral shear zone, linked to the N–S African-Iberian convergence, which produced pull-apart basins. Basin-forming, E–W trending dextral strike-slip faults were postulated by Sanz de Galdeano & Vera (1992) and Stapel et al. (1996). Vissers et al. (1995) suggested that, as the southern boundaries of the basins are typically faulted and the northern boundaries are in unconformable contact with basement metamorphic rocks, the basins probably originated as extensional half-graben. Poisson et al. (1999) interpreted the Neogene basins as lateral ramp basins oriented parallel to westward verging, deep-seated thrust faults (the Sierras being the structural culmination of these thrusts).

The western part of the Tabernas-Sorbas basin contains >1 km of predominately marine sedimentary rocks, in an overall transgressive-regressive succession, involving continental and shallow-water environments passing into deep-water and then fan-delta deposits (Figs 1 & 3). A regressive-transgressive pulse in the upper part of the basin infill is associated with the Mediterranean Messinian salinity crisis. The basin fill ranges in age from Serravallian to Pliocene (Kleverlaan 1987; Hodgson 2002). There is no high-resolution biostratigraphic framework for this basin despite previous work (e.g. Kleverlaan 1989a; Poisson et al. 1999), although the deepwater fill is known to straddle the Tortonian/Messinian boundary. The poor biostratigraphic control is probably due to the rapid subsidence and filling of the basin, and

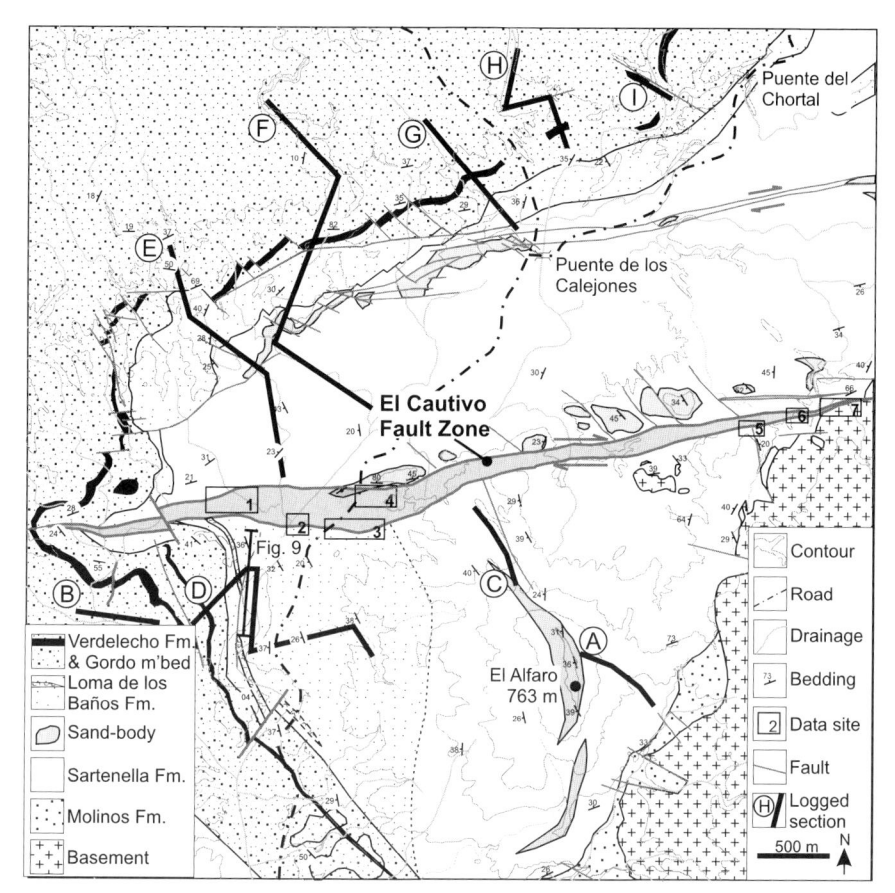

Fig. 2. Geological map of the study area in the western Tabernas-Sorbas Basin. The E–W trending El Cautivo Fault is a wide oblique-dextral fault that was active during the deposition of the Loma de los Baños Formation. Logging routes, A–I (Fig. 4), and kinematic data collection sites, 1–7 (Fig. 7), are shown.

the occasional restriction to open marine conditions. Generally, the deepwater successions are texturally and compositionally immature, consistent with a steep and tectonically active hinterland, which was separated from the marine basin by a steep cliff-lined coast, narrow shelves and fringing coral reefs (Kleverlaan 1989b). Sediment eroded from the uplifted basement source areas was rapidly transported and redeposited in the deepwater succession in a variety of environments (fan deltas, slope aprons, submarine channels and fans, and sheet turbidite systems).

Stratigraphic framework

Aspects of the deep-water sedimentology of the Tabernas area of the basin have been described by Kleverlaan (1987, 1989a,b), Cronin (1995), Haughton (2000), Pickering et al. (2001) and Hodgson (2002). Kleverlaan (1989a, 1994) recognized three essentially coeval Upper Tortonian turbidite systems within the Tabernas area: System I (Sandy System), System II (Mixed System), and System III (Solitary Channel), although this stratigraphic and depositional framework has been challenged recently (Haughton 2000; Hodgson 2002).

A new stratigraphic framework and nomenclature for the Tabernas area is proposed here, building on and extending the pilot scheme developed by Haughton (2000). The lower basin fill (i.e. the Tortonian and Lower Messinian) is split into four lithostratigraphic units. From base to top, these are the Molinos, Sartenella, Loma de los Baños and Verdelecho Formations (Fig. 3).

Molinos Formation

The Molinos Formation (Hodgson 2002) is composed of basal red alluvial fanglomerates that

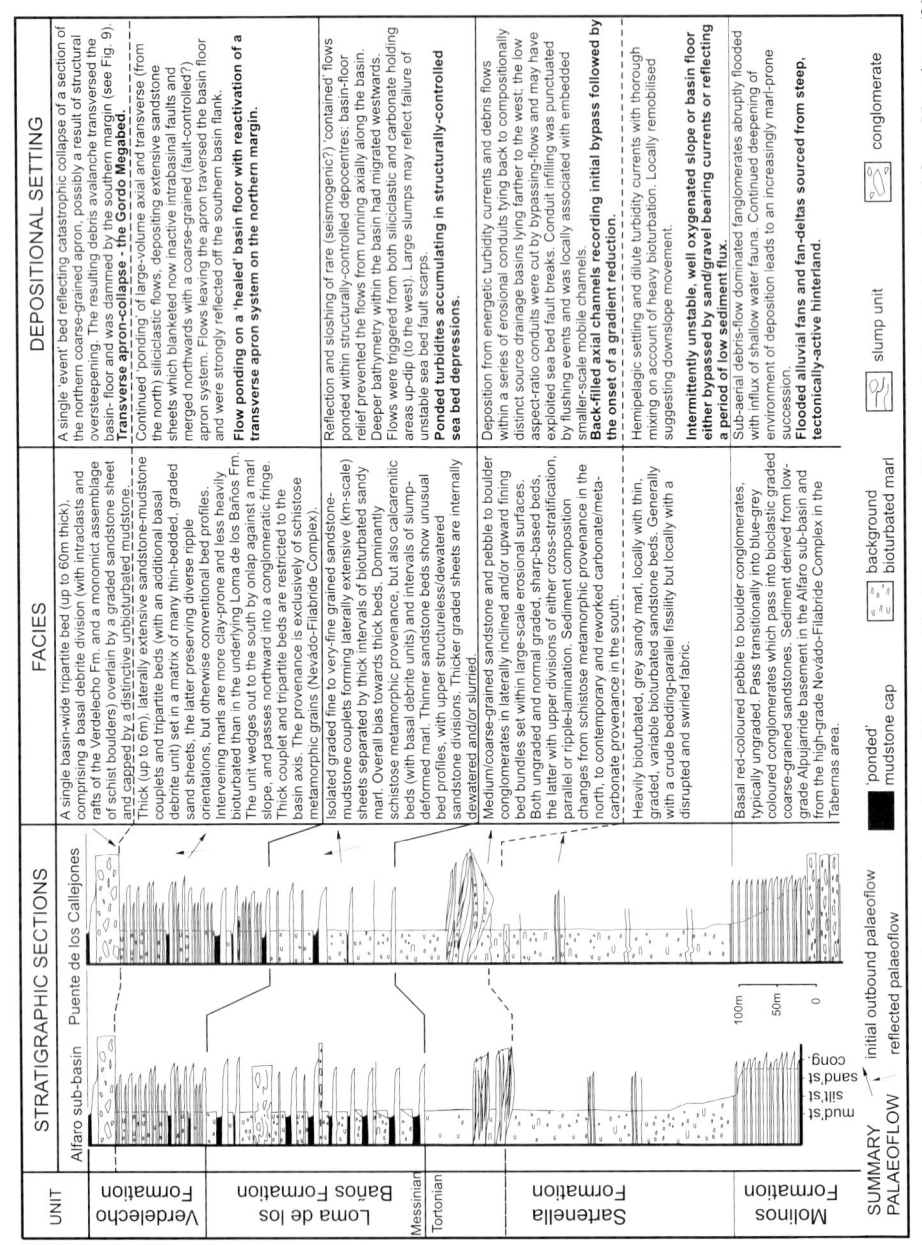

Fig. 3. Revised stratigraphic terminology for the Tortonian to Lower Messinian fill of the western Tabernas-Sorbas Basin, adapted from Haughton (2000). The first appearance datum of *Globorotalia conomiozea* is used to place the Tortonian/Messinian boundary at the base of the Loma de los Baños Formation.

transitionally pass into blue-grey, bioclastic, shallow-water fanglomerates, which in turn pass up into the marl-prone Sartenella Formation. The formation reflects the onset of tectonically controlled basin subsidence followed by abrupt flooding and deepening during the middle Tortonian.

Sartenella Formation

The Sartenella Formation forms the distinctive Tabernas 'badlands' and is composed of monotonous pale grey, micaceous and silty marls that typically are heavily bioturbated. Even in areas of intense bioturbation a crude stratification can still be identified. Local wedge-shaped packages (several metres thick) of thin-bedded (cm to dcm) turbidites preserve an eastward palaeo-flow direction. Other units, again several metres in thickness, have a distinctive pillowed fabric and contain rare pebble-grade mudstone rip-up clasts. Close to El Alfaro (Fig. 2), the Sartenella Formation includes common graded pebble-grade and bioclast-rich conglomerate beds and rare >5 m thick conformable bodies of contorted sandstone beds.

A series of linear, erosively contained, sandstone- and conglomerate-filled bodies up to 50 m thick are incised into the upper part of the marl-prone succession (Unit B of Haughton 2000, Figs 2 & 3). Five separate sand-bodies have been identified at this stratigraphic level (Hodgson 2002). These vary in their provenance, aspect ratio, palaeo-flow, and architecture. One of the sand-bodies is the western extension of the 'solitary system' previously described by Kleverlaan (1989a) and Cronin (1995). The solitary system was originally interpreted as a westward-flowing submarine channel (Kleverlaan 1989a; Cronin 1995). Recent work has reinterpreted this channel as an eastward-flowing slope channel that delivered sand to the basin floor, which lay to the east (Haughton 2000; Pickering et al. 2001). Palaeo-flow indicators measured from the four southernmost sand-bodies all preserve a dominant eastward flow direction. Only the most northerly sand-body does not follow this trend and is dominated by a southwesterly-directed palaeo-flow (Hodgson 2002). The Sartenella Formation is 330 m thick close to El Alfaro, and at least 300 m thick in the centre of the basin around Puente de Los Callejones (Figs 2 & 4).

Loma de los Baños Formation

The underlying channelized sand-bodies are covered by a markedly different depositional system (ranging from <10 to 300 m thick) composed of sheet turbidites and breccia beds, each isolated in a mud-prone succession (Unit C of Haughton 2000, Fig. 3). The character of the sandstone beds implies flow containment, not by erosional topography, as locally in the Sartenella Formation, but by the full width of the basin floor. Large-volume turbidity currents entered the basin from the west, with either a siliciclastic (Nevádo-Filabride) or carbonate (Alpujarride/contemporary and reworked Miocene) provenance. Significant lateral changes occur in this unit and are the key to identifying syndepositional tectonism on the basin floor. The character of this formation will thus be discussed in more detail in a subsequent section.

A critical taxon present in the Loma de los Baños Formation, identified from micropalaeontological studies of the Tabernas-Sorbas Basin (Hodgson 2002), is the marker species *Globorotalia conomiozea*. Berggren et al. (1995) adopt the first appearance datum (FAD) of *Globorotalia conomiozea* to identify the Tortonian/Messinian boundary in the Mediterranean, dated as 7.12 Ma (Krijgsman et al. 1994). Kleverlaan (1989a) used *Globorotalia conomiozea* to identify the Tortonian/Messinian boundary in the Tabernas area of the Tabernas-Sorbas Basin, as ~230 m above the top of Sartenella Formation and ~100 m above the Gordo megabed. Hodgson (2002), however, places the boundary close to the transition from the Sartenella Formation to the Loma de los Baños Formation, although it could be even lower in the stratigraphy.

Verdelecho Formation

The conformable base of the Verdelecho Formation, which comprises Units D and E of Haughton (2000), marks another abrupt change in the style of deposition (Figs 2 & 3). The Verdelecho Formation is sandier than the underlying Loma de los Baños Formation, with common thin-bedded turbidites, which exhibit a different internal structure from the turbidites in the Loma de los Baños Formation (see discussion below). Carbonate- and Alpujarride-rich beds are completely absent. The Verdelecho Formation also contains a number of very thick beds, including the Gordo Megabed of Kleverlaan (1987), a 60 m thick marker bed that has the same tripartite structure (basal breccia unit, graded sandstone and thick mudstone cap) reported from megabeds elsewhere (Figs 2 & 4). Palaeo-flow measurements indicate a dominant eastward trend at bed bases with two ripple lamination

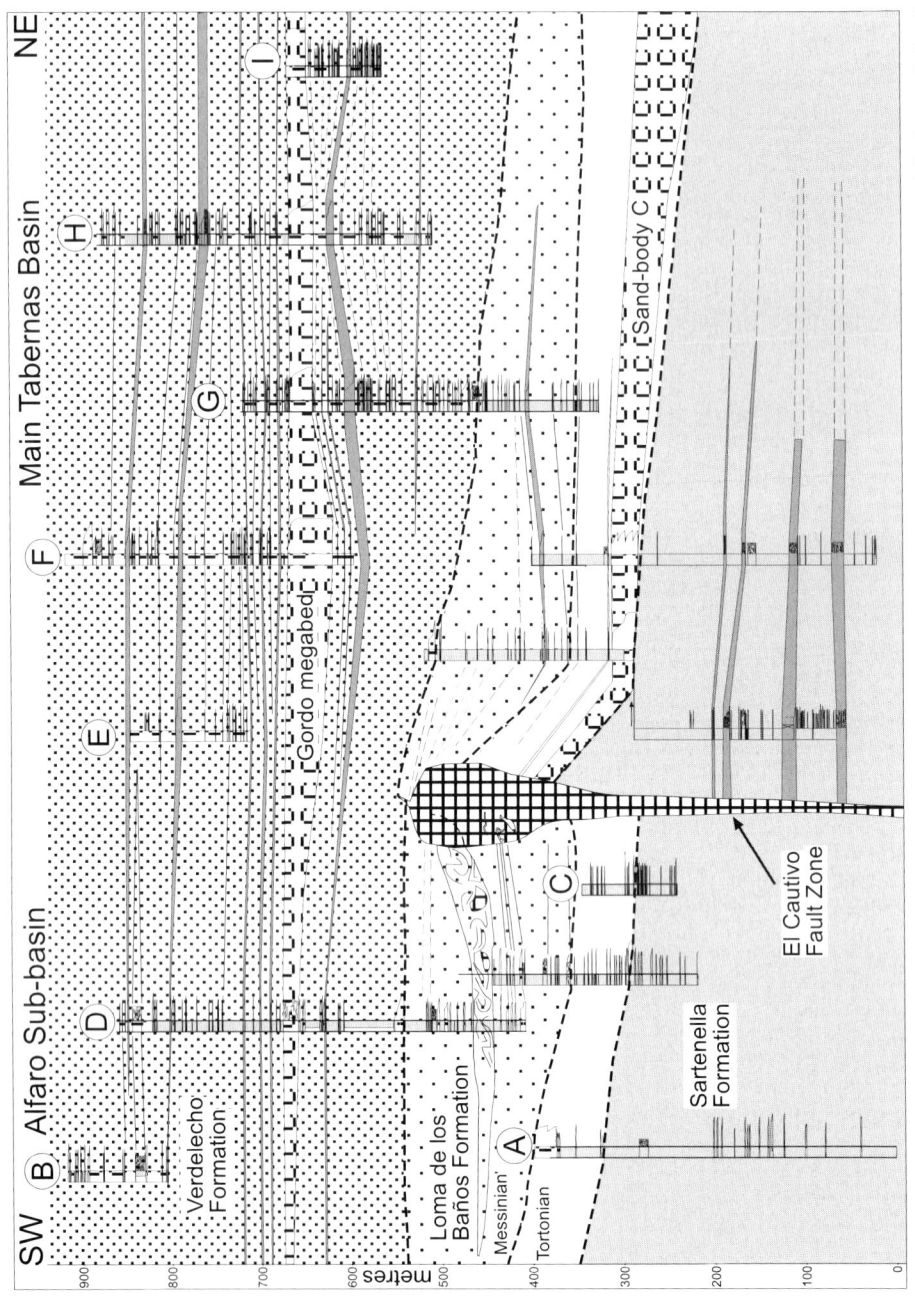

Fig. 4. A panel of logged section from the western Tabernas-Sorbas Basin (see Fig. 2 for log locations). Note the change in thickness of the Loma de los Baños Formation and the lack of displacement in the Verdelecho Formation across the El Cautivo Fault. The El Cautivo Fault also separates sequences of different provenance and sedimentology. The beds in the Verdelecho Formation are correlated over greater distances than in the Loma de los Baños Formation.

trends, a southeasterly and a less well-defined mode directed to the north.

There are no abrupt lateral or vertical changes in the style of deposition within the Verdelecho Formation, even above the Gordo Megabed, although the unit becomes more coarse-grained and proximal in nature towards the northern basin margin (Haughton 2000). The top of the Verdelecho Formation is marked by an angular unconformity that separates coarse-grained clastics and overlying gypsum and laminates. The gypsum was deposited during the Messinian Salinity Crisis. Because of the restricted nature of the Tabernas area, however, evaporate deposition might not have been synchronous with the Sorbas area to the east, where it is dated as ~5.9–5.5 Ma (Riding et al. 1998).

Summary of depositional evolution

Hemipelagic and pelagic settling of silty marls with occasional low-volume turbidity currents dominate the lower Sartenella Formation. An oxygenated and nutrient-rich environment is inferred from the intense bioturbation of the lower Sartenella Formation. The conformable pillowed horizons and contorted sandstone bed units are interpreted as the result of gravity-induced mobilization and/or slumped units indicating a slope setting. Palaeo-current readings (toolmarks) indicate that the basin slope was eastward facing. Sands either bypassed the slope or there was a low sediment flux. The wedge-shaped sandstone packages are thought to represent the infilling of slump scars or topographic depressions on the slope that trapped gravity flows that would otherwise have bypassed.

Incision of the sand-starved and bypassed silty marls of the lower Sartenella Formation by bypassing gravity flows led to the development of a series of linear slope channels that were backfilled with sandstone- and conglomerate-grade turbidites and debrites (Fig. 3). The main eastward-facing slope was maintained, although a local southwest-facing slope developed to the north of Tabernas. The erosional channels contain multiple phases of erosion and aggradation (Haughton 2000; Pickering et al. 2001), and are sandwiched within marl-prone sediments: they are therefore interpreted as slope canyons or base-of-slope submarine channels that were sufficiently deep to prevent flow overspill. The clast provenance in each channel fill reflects a south-to-north progression from low to high metamorphic grade in the hinterland. Westward onlap of the internal stratification, and increasing shale content towards the west, suggest the conduits were backfilled. Backfilling bedforms have been identified in two of the conduits (Pickering et al. 2001), although the backfill phase was punctuated by erosive flushing events. The change from bypass to backfill is thought to relate to the onset of the basin-floor deformation (Haughton 2000), recording a reduction in basin floor gradients that preceded the switch to the overlying sheet turbidite systems. At this time (the latest Tortonian), an angular unconformity was forming in the 'downslope' Sorbas area. A structurally related unconformity has also been identified farther to the east in the Tabernas area (Hodgson 2002).

The progressive change from active slope bypass to backfill and basin aggradation continues with the Loma de los Baños Formation. The thick sheet-like beds are interpreted as large-volume contained turbidites that were ponded within structurally controlled depocentres. The appearance of basin-floor, ponded turbidite beds indicates that the deep basin has migrated westwards to replace what was previously a slope setting.

Although the beds of the Verdelecho Formation still display contained flow characteristics they are laterally more extensive than beds of the Loma de los Baños Formation. Some thick beds, for example the Gordo megabed, are basin-wide and are interpreted as seismites (Kleverlaan 1987). The Verdelecho Formation merges with basin margin apron systems to the north (Haughton 2000) and pinches out onto a submerged palaeoslope to the south.

Tectonic activity was the primary control on the stratigraphic evolution of the Tabernas-Sorbas Basin during the Tortonian and Lower Messinian, driving the overall evolution from slope to basin-floor (Fig. 5). This paper will focus on the detailed stratigraphic evolution of the Loma de los Baños Formation in the zone flanking an important intrabasinal fault zone, the El Cautivo Fault Zone (Fig. 2). Basin-floor faulting accompanied the changing environments, and facilitated the collapse of the early eastward-facing slope (Fig. 5). The characteristics of this fault zone are first described, and then the geometry of the flanking stratigraphy is assessed.

El Cautivo Fault Zone

An important intrabasinal fault zone has been identified in the course of reconstructing the basin fill in the area immediately north and west of El Alfaro, in the southwestern part of the Tabernas-Sorbas Basin (Fig. 2). It comprises

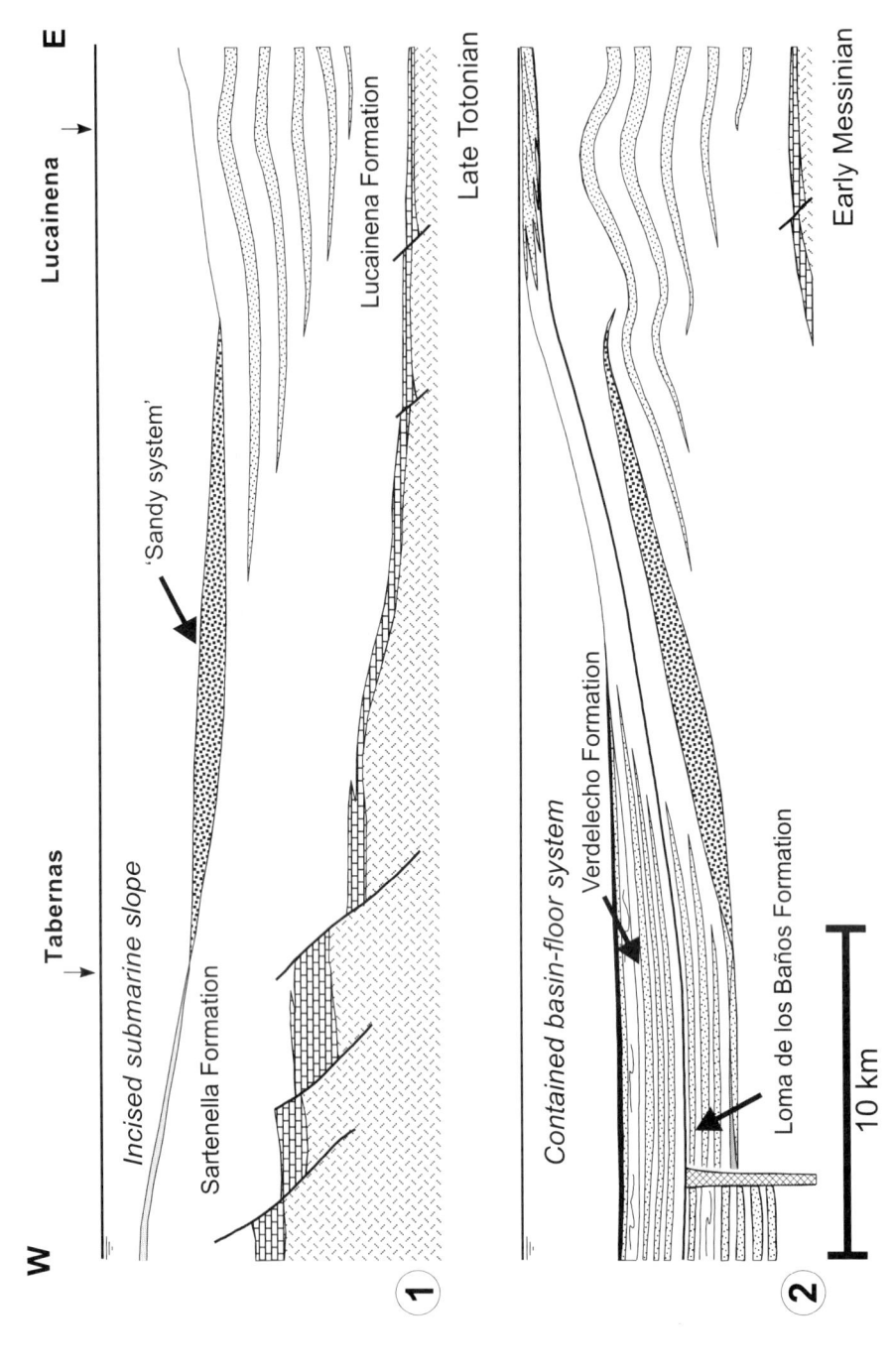

Fig. 5. Schematic cross-section to illustrate the evolution of the western Tabernas-Sorbas Basin during the Tortonian and Lower Messinian from (1) a submarine slope system (Sartenella Formation) to (2) a contained basin floor system (Loma de los Baños Formation and the Verdelecho Formation). This unusual stratigraphic evolution was driven primarily by westward-stepping oblique-compressional tectonics. The Lucainena Formation was folded and uplifted in the late Tortonian and formed an angular unconformity in Sorbas area, whereas the Tabernas area continued to subside and become increasingly contained.

an E–W trending zone of highly deformed sediments with a distinctive scaly fabric (Fig. 6). The gouge zone associated with the El Cautivo Fault Zone is up to 350 m in thickness, although in map view it is strongly lenticular, thinning to the east and west (Fig. 2). In the east, the fault offsets low-grade Alpujarride basement in the Sierra Alhamilla by >100 m with a right-lateral sense of displacement. In the westernmost exposures the fault zone thins abruptly, and the basal parts of the Verdelecho Formation are displaced vertically (down to the south) by only 5 m. The fault zone has a subvertical orientation, and therefore trends at a steep angle to bedding. It runs approximately along the hinge-line of a large antiform that formed during the later uplift of the Sierra Alhamilla (Fig. 2).

Typically, sandstone sheets of Loma de los Baños Formation truncate against the margins of the El Cautivo Fault. Some beds towards the top of the Loma de los Baños Formation, however, can be traced on either side of the structure. Verdelecho Formation beds continue across the El Cautivo Fault with only minor brittle displacement (~5 m, Figs 2 & 4). Upper Loma de los Baños Formation beds can be seen to onlap northwards against a marl-prone slope just south of the trace of the fault.

Kinematic indicators

The El Cautivo Fault is associated with a zone of intensely deformed argillaceous rocks that have a distinctive scaly and swirled fabric (Fig. 6). The intensity, longevity and/or temporal variations in the sense of displacement during faulting have overprinted and destroyed clear and consistent sets of kinematic indicators towards the centre of the fault zone where there is no systematic pattern of shear planes or gouge fabric preserved.

Kinematic indicators have a far higher preservation rate towards the margins of the El Cautivo Fault zone, where gouge fabrics and Riedel shear planes are commonly identified, and sandstone beds are highly contorted and dissected by vertical shear planes (Fig. 6). Bedding is sub-vertical towards the margins of the fault zone where fault drag increased bed dips. Large calcarenite blocks (tens of metres across) entrained within the gouge zone are highly deformed, with abrupt bed thickness changes and tight folds implying deformation while the sediment was still unlithified. Away from the fault these calcarenite bodies form essentially sheet-like beds. The most intensely deformed areas are also the focus for concentrations of fibrous calcite veins. The calcite fibres typically trend sub-horizontally, although several phases of activity are indicated by complex cross-cutting relationships. The calcite veins, however, typically cut shear planes and the fault gouge fabric and appear to have developed late in the evolution of the El Cautivo Fault, although they themselves are locally displaced on late shear planes.

Kinematic data groups

Measurements were taken at a series of sites along the length of the El Cautivo Fault in order to define the tectonic regime(s) active during the evolution of the fault (Figs 2 & 7). The dataset acquired from the measurement of kinematic indicators were grouped depending on their orientation, type, and sense of displacement and cross-cutting relationships with other structures. Most of the measurements were taken from the margins of the fault zone.

(1) Measurements of the well-developed fault gouge fabric, best preserved towards the fault zone margins, gave an average orientation of 058° ($n = 83$, Fig. 7). Sandstone and siltstone blocks entrained within the El Cautivo Fault are deformed and stretched into ellipsoids and lozenges, and are oriented about the same direction (Fig. 6).
(2) Sub-vertical fault planes cut the gouge fabric (Fig. 6). The fault planes preserve rare sub-horizontal lineations. A right-lateral sense of displacement is present where sandstone and siltstone ellipsoids and lozenges are cut. The average orientation of the (right-lateral) fault planes is 110° ($n = 67$, Fig. 7).
(3) Rare, poorly developed sub-vertical fault planes trending at 027° ($n = 7$), preserve a left-lateral sense of movement where they displace sandstone and siltstone ellipsoids (Fig. 7).
(4) Calcite veins are typically sub-vertical, with sub-horizontal crystal fibres. The dominant orientation of the veins is ENE–WSW, although there is a significant amount of scatter associated with the calcite vein measurements (Fig. 7).
(5) Several faults trending at ~120° with an oblique right-lateral sense of displacement cut the sand-body to the north of the El Cautivo Fault (Fig. 2).

Interpretation of kinematic data

The consistent E–W orientation of the El Cautivo Fault simplifies the kinematic analysis

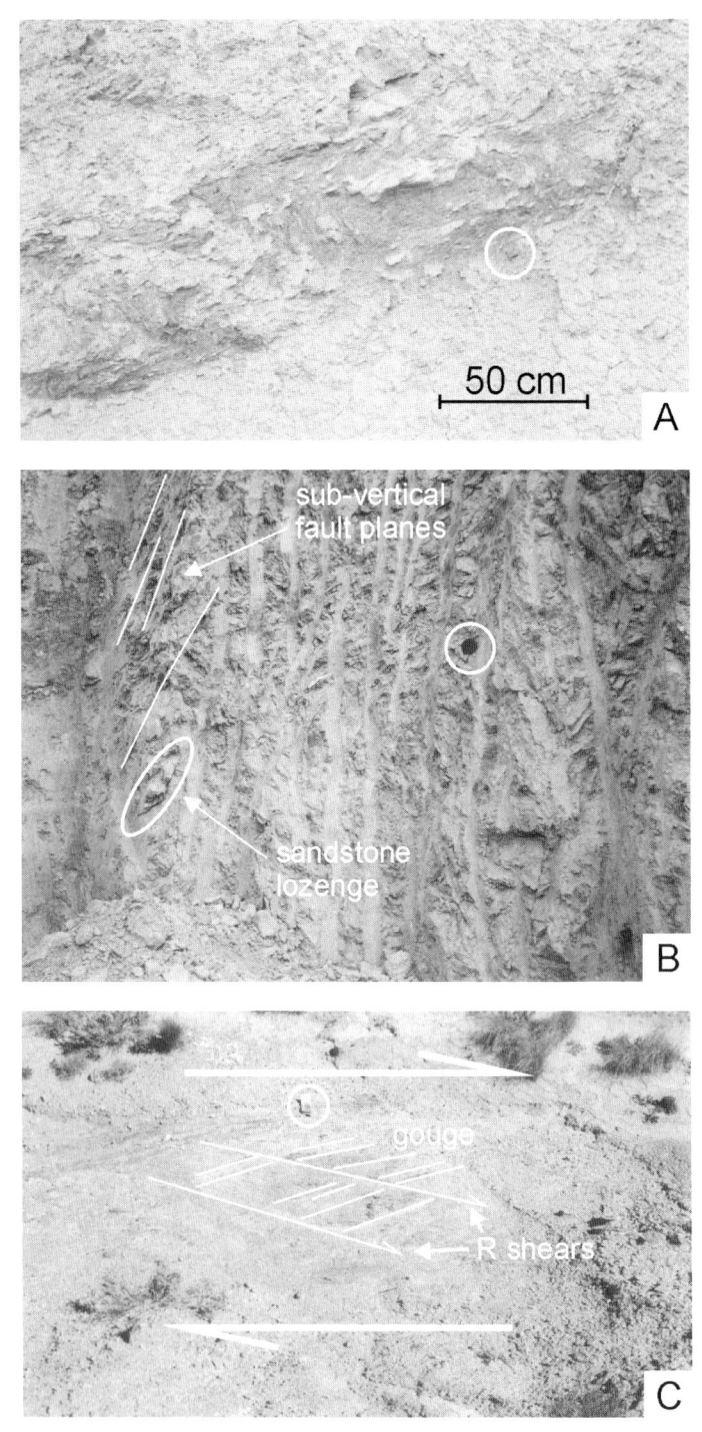

Fig. 6. Representative outcrops from the El Cautivo Fault. (A) Intense deformation of marls towards the middle of the El Cautivo Fault (compass circled). (B) A fresh cut through the fault zone. Note the sub-vertical shear planes and the sandstone lenses (camera lens circled). (C) Southern margin of the El Cautivo Fault, exposed near to the basement to the east, where the gouge fabric and hear with right-lateral displacement can clearly be identified (compass circled).

Fig. 7. Lower-hemisphere equal-area stereographic projections constructed from kinematic data collected from seven sites along the El Cautivo Fault zone (Fig. 2). A gouge fabric (solid) trends at 58° ($n = 83$), sub-vertical shear planes with right-lateral sense of displacement (dashed) trend at 110° ($n = 67$) and rare shear planes with a left-lateral sense of displacement (dashed and dotted) trend at 027° ($n = 7$). The right-lateral fault planes that cut the gouge fabric are interpreted as R_1 Riedel shear planes, and the rare left-lateral fault planes are interpreted as X shears. The gouge fabric, therefore, is interpreted as forming under dominantly right-lateral conditions. The stratigraphic evidence suggests that there was a dip-slip component to the fault motion; although the amount of strike-slip displacement cannot be ascertained, the structure is interpreted as an oblique-dextral strike-slip fault. Calcite vein measurements are disparate (box 8), although generally lineations are generally sub-horizontal.

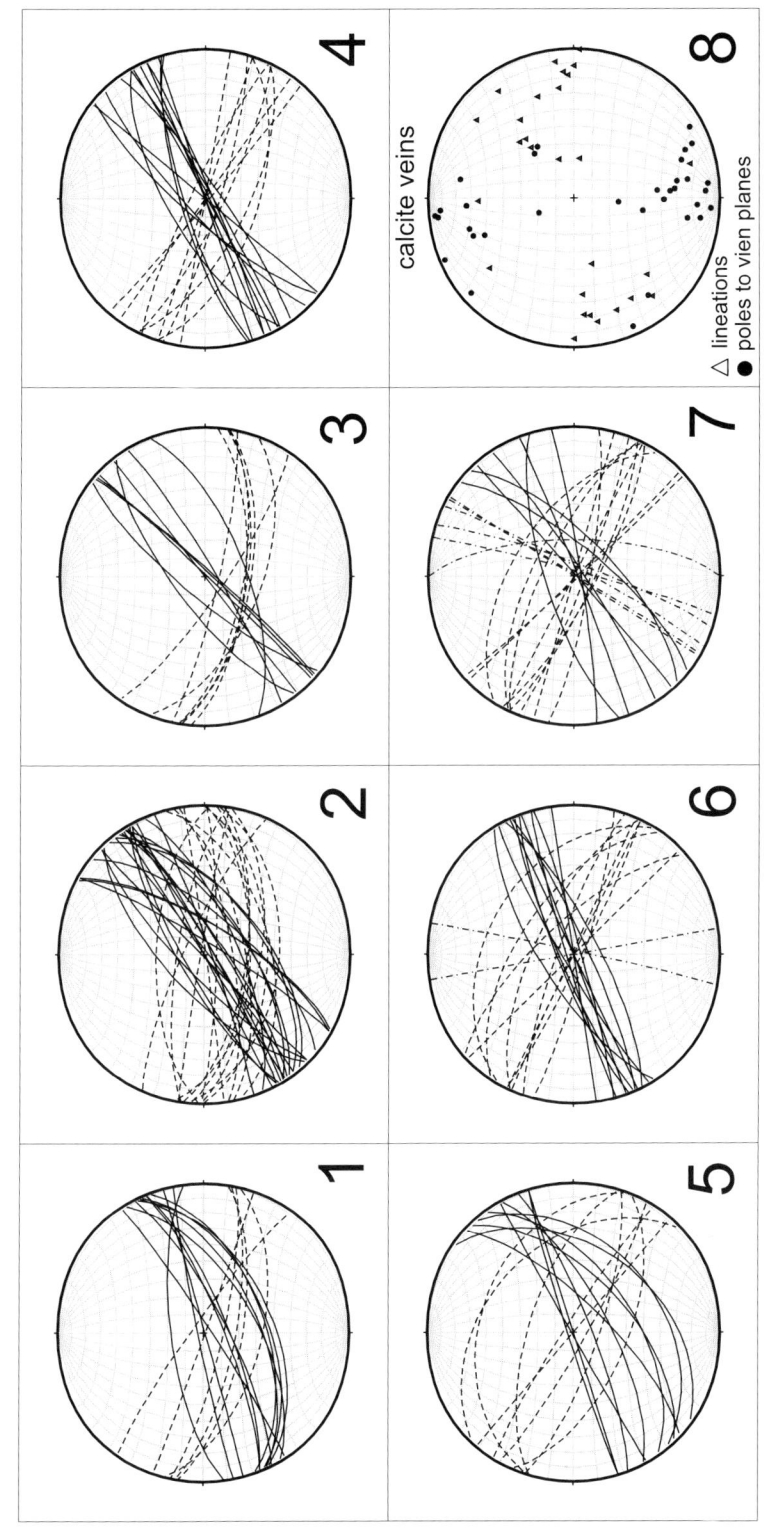

of the fault. The gouge fabric that trends at 058° is interpreted as forming under dominantly right-lateral conditions. The right-lateral fault planes that cut the gouge fabric are, therefore, interpreted as R_1 Riedel shear planes, and the rare left-lateral fault planes are interpreted as X shears. The stratigraphic evidence (see below) suggests that there was a dip-slip component to the fault motion, although the amount of strike-slip displacement cannot be ascertained. Kinematic indicators with sub-horizontal right-lateral motion dominate the margins of the El Cautivo Fault Zone: therefore the structure is interpreted as an oblique-dextral strike-slip fault.

The El Cautivo Fault may have originated from the reactivation of an E–W basement lineament or as the intrabasinal propagation of the dominantly dextral strike-slip fault found at the northern margin of the Sierra Alhamilla (Sanz de Galdeano 1989). It is very difficult to trace the fault across the basement in the east. The position and trend of the El Cautivo Fault, however, are intriguing. It closely follows the hinge-line of a regional-scale antiform that plunges at ~30° towards the WSW. The latter formed after motion on the fault had ceased through the uplift of the Sierra Alhamilla (the Verdelecho Formation is folded but seals the El Cautivo Fault, Figs 2 & 4). This means that the El Cautivo Fault has been tilted to the WSW by ~30° with the stratigraphy, so that an oblique section through the original fault zone is exposed at the surface today (Fig. 2). Although one of the Sartenella Formation sand-bodies trends parallel to the El Cautivo Fault, it does not preserve evidence of synsedimentary deformation by the fault in the form of thickness differences or the ductile deformation of sandstone beds. Therefore it is not clear whether the precise location of the sand-body is a coincidence, or whether there is a causal relationship between its position and that of the developing fault. The oblique right lateral faults trending at 120° that cut the sand-body are interpreted as synthetic splays to the main fault (Fig. 2).

Turbidite deposition adjacent to an active fault: the Loma de los Baños Formation

The Loma de los Baños Formation is dominated by sheet turbidites, typically with a complex internal structure (Haughton 2000; see below). Lateral changes in the thickness of the formation, the provenance of the sand-grade sediment, and the types of gravity flow across the El Cautivo Fault zone imply fault activity and basin-floor deformation during sedimentation. The Loma de los Baños Formation is described initially from where the unit is thickest (>300 m) to the south of the fault in an area referred to as the Alfaro sub-basin (Haughton 2000; Fig. 4). The lateral equivalent succession north of the fault zone is then considered.

Alfaro sub-basin

A characteristic feature of the Loma de los Baños Formation is the large number of very thick sandstone-mudstone couplets that range from 2 to 11 m in thickness. Sandstone units grade up into very thick, unbioturbated and continuously graded silty mudstone caps. The caps are as thick as, or sometimes even thicker than, the sandstone part of the couplet. There are also an anomalous number of thick beds, which leads to a non-power law distribution of bed thicknesses. The thick-bedded sandstones are interbedded with thin-bedded sandstones, and the background sediments are bioturbated silty marls. Haughton (2000) pointed out that the detailed vertical pattern of sedimentary structures preserved in both the thick and the thin beds is unlike those proposed in the model of Lowe (1982) for 'low' and 'high' density turbidites. Most of the thicker (>1 m) beds grade abruptly from a sharp and coarse-grained bed base to fine- and very-fine-grained sandstones. The normal grading is commonly disrupted by grain-size breaks, and either a contorted internal structure or parallel laminations, which are increasingly disrupted by water-escape structures (pipes and dewatered sheets, Haughton 2000). Charcoal fragments commonly occur in the thicker beds, and because of their higher relative buoyancy are present towards the top of graded units.

The thinner, decimetre-scale sandstone sheets are commonly bipartite, with a lower graded and parallel and/or ripple laminated division that is sharply overlain by a structureless, dewatered and ungraded division (Fig. 8). The thinner beds also have either a thin, or lack a mudstone cap, which may have been completely bioturbated. These thin beds can be traced laterally on a kilometre-scale with little variation in thickness or structure (Haughton 2000), and have a more simple vertical structure than turbidite beds with a similar thickness range described from the Lucainena area in the east (Haughton 1994, 2001). A thick conformable body composed of contorted blocks and scaly material is intercalated with the Loma de los Baños Formation to the south of the El Cautivo Fault (Fig. 9).

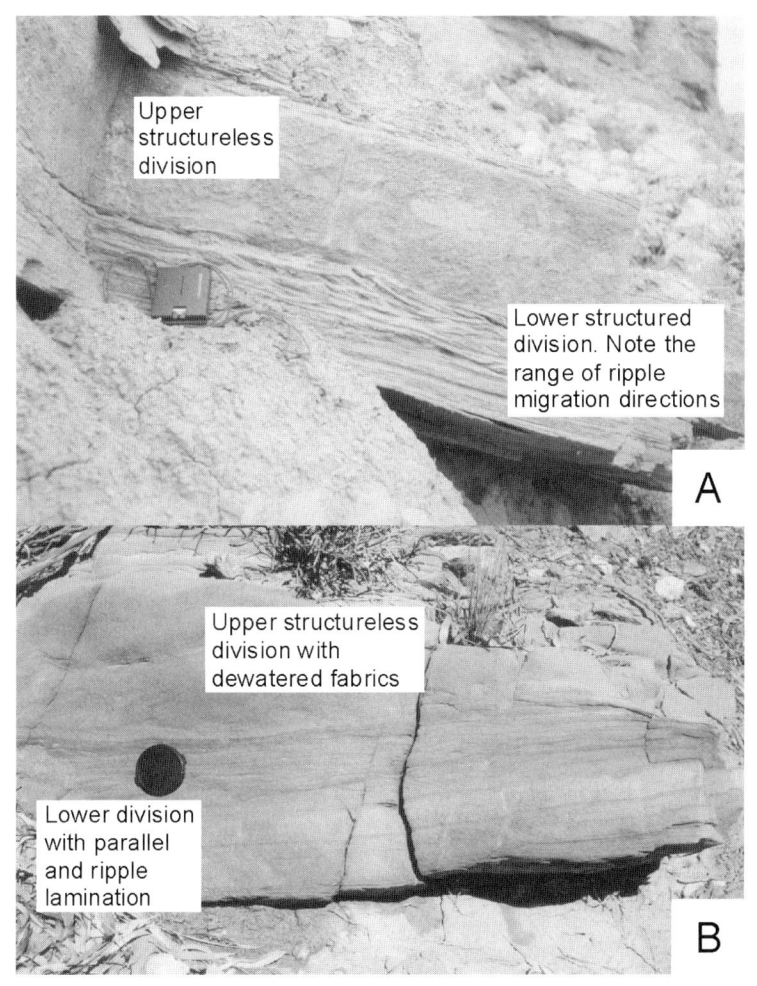

Fig. 8. Selected representative photographs from the Loma de los Baños Formation. (**A**) and (**B**) both illustrate distinctive bipartite turbidite beds, found in the Loma de los Baños Formation, but not the Verdelecho Formation. The lower division is planar and/or ripple laminated, typically with more than one ripple-migration direction, and the upper division is structureless or preserves dewatering fabrics.

Lateral variations

The thickness of the Loma de los Baños Formation in the Alfaro sub-basin is significantly greater (~300 m, Haughton 2000) than in the Rambla de Carrizalejo (~60 m), the Puente de los Callejones section (~110 m), and at the Puente del Chortal (~70 m) (Figs 2 & 4). The El Cautivo Fault separates the thinner northerly successions from the thick Alfaro sub-basin succession (Fig. 4). Although the sandstone sheets are laterally continuous up to distances of 2 km (Haughton 2000), they locally thin and pinch out against marl-cored palaeo-highs in the vicinity of the fault. The Loma de los Baños Formation is not present at the southern margin of the Alfaro sub-basin, where thinned Verdelecho Formation sediments unconformably overlie Sartenella Formation sediments (Fig. 10).

As well as differences in thickness north and south of the El Cautivo Fault, there are also critical sedimentological and provenance contrasts. To the north of the El Cautivo Fault there are none of the distinctive thin bipartite beds commonly seen to the south. The rare thin beds to the north are sheet-like, graded and/or laminated throughout, and do not have a mudstone cap. Slump and slide bodies and the carbonate- and Alpujarride-rich megabeds are also absent to the north of the El Cautivo Fault. Indeed, the only carbonate- and Alpujarride-rich beds

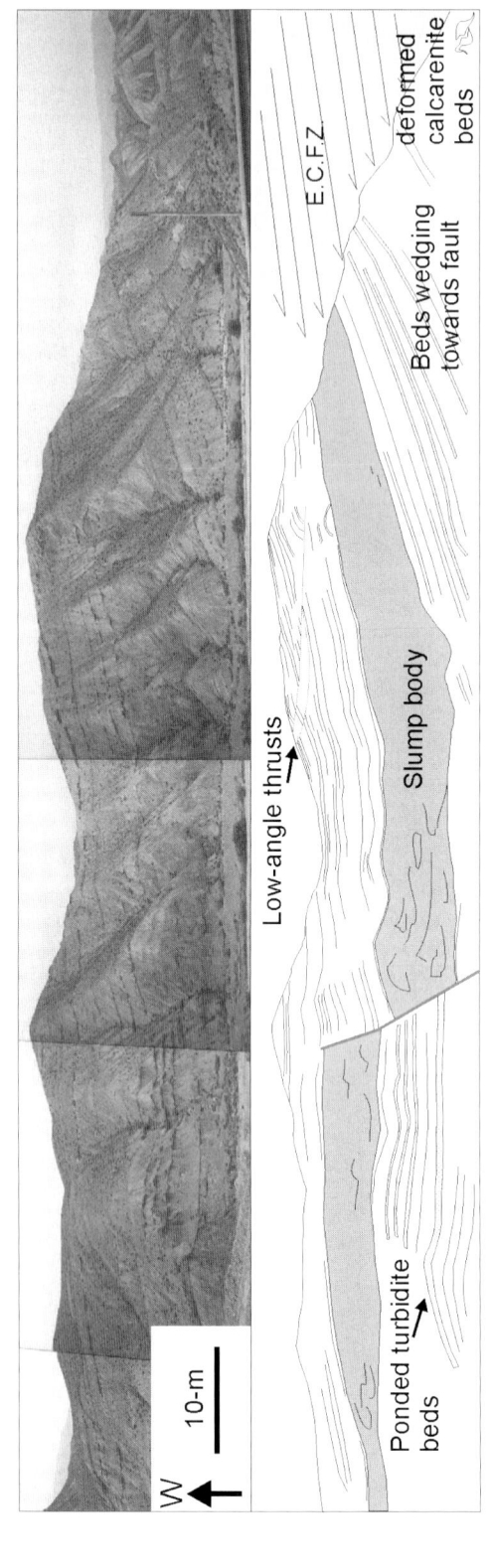

Fig. 9. A cliff section through the Loma de los Baños Formation close to the El Cautivo Fault Zone along the western bank of the Rambla Tabernas. Note the thinning and onlapping of beds towards the fault, the large slump package that contains scaly material, and the low-angle thrusting towards the top of the outcrop.

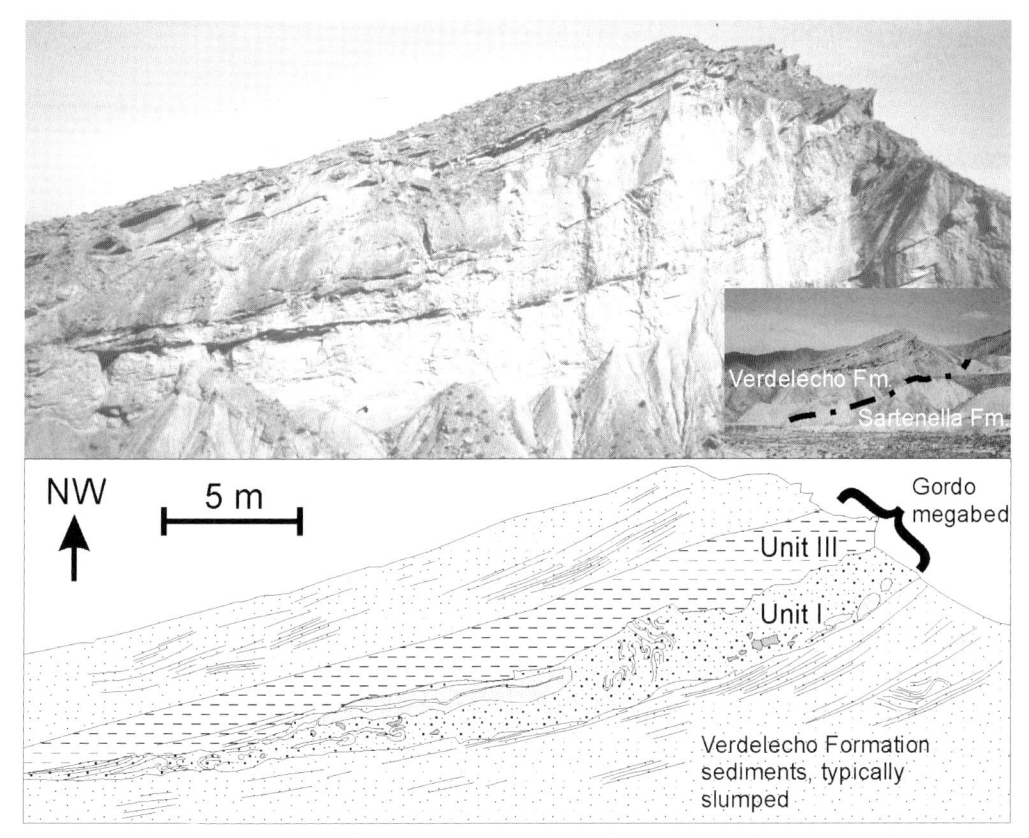

Fig. 10. The onlap and pinchout of the Gordo megabed onto marl-prone slope sediments exposed to the south of El Alfaro. The inset shows the unconformable relationship between the Sartenella Formation and the Verdelecho Formation at the southern margin.

found to the north of the fault are rare thin (10–20 cm thick) normally graded horizons. The dominant bed types to the north of the El Cautivo Fault are thick (2–8 m thick) and structureless or highly convoluted medium- and coarse-grained sandstone beds passing gradationally up into thick unbioturbated mudstone caps.

Provenance

Two distinct source areas supplied sediment to the Loma de los Baños Formation. Siliciclastic beds, rich in Nevádo-Filabride Complex schist fragments sourced from the northern and northwestern basin margin, dominate the succession. The subordinate carbonate-rich beds, with a clastic component of Alpujarride detritus, were sourced from the southern and southwestern margins of the basin. This is the first time that the sedimentary system in the Alfaro sub-basin was supplied from a Nevádo-Filabride source area, as beds rich in carbonate grains and Alpujarride detritus dominate the underlying Sartenella Formation here. This change reflects the increasing influence of the more deeply eroded northern margin and/or the decreasing influence of the southern and western margins as a source area.

The carbonate-rich beds occur either as distinctive tripartite beds (ungraded conglomerate—graded sandstone sheet—mudstone cap; Haughton 2000) or as thin graded sandstones with planar laminations and bioturbated tops. Note that the distinctive bipartite structure referred to above is present only in beds with a siliciclastic provenance. The beds of different provenance appear randomly intercalated with no systematic vertical pattern (Haughton 2000).

Palaeoflow

Toolmarks and flutes commonly indicate a well-constrained axial palaeoflow towards the east

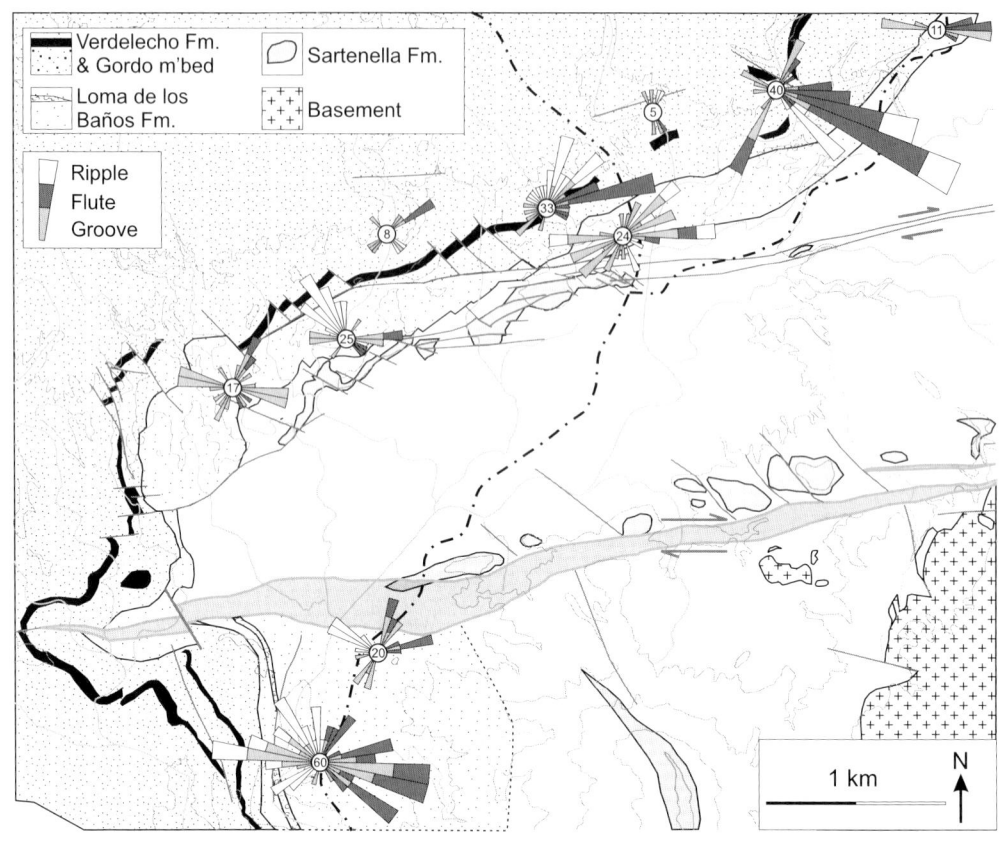

Fig. 11. Palaeo-current map constructed from measurements collected in the Loma de los Baños Formation and the Verdelecho Formation. The dominant trend for toolmarks is to the ENE and SE, although a more northerly component is preserved to the south of the El Cautivo Fault. Although ripple migration measurements are more diverse there is a strong trend to the NW. The higher spread of measurements in the Loma de los Baños Formation supports the inference of higher amplitude of basin bathymetry at this time.

and southeast (Fig. 11). The exception is where Loma de los Baños Formation deposits overlie a Sartenella Formation sand-body to the north of Tabernas, where they preserve a very different palaeo-flow towards the southwest. Ripple lamination azimuths are more disparate, although typically the ripple migration direction is dominated by flow to the N and NW, with a subordinate flow component to the SE (Figs 8 & 11). The variation in palaeo-flow within a single turbidite bed is typical of flow containment (*sensu* Pickering & Hiscott 1985), and indicates the presence of basin-floor topography that formed obstacles to the gravity flows. The palaeo-flow data to the south of the El Cautivo Fault are generally more dispersed. For example, adjacent to the El Cautivo Fault there is a set of north-directed toolmarks preserved in the lower division of bipartite beds (Fig. 11). Ripple migration directions also have a greater spread to the south of the El Cautivo Fault (Fig. 11).

Transition to the overlying Verdelecho Formation

There is a transitional, although abrupt, change in sedimentation style in the overlying Verdelecho Formation, which is only slightly deformed by the El Cautivo Fault (Fig. 4), so lateral trends in sedimentology across the fault are not present at this level. There is a marked increase in sandstone sedimentation with common thin- and medium-bedded, medium-grained sandstones interpreted as low-density turbidites (Haughton 2000). Importantly there are no distinctive thin bipartite beds. Haughton (2000) identified two directions of ripple migration from the well-structured thin-bedded sandstones: a southeasterly mode and a smeared northerly mode. The palaeo-flow readings are generally less dispersed in the Verdelecho Formation than in the Loma de los Baños Formation (Fig. 11).

Thick graded sandstone beds, with some examples preserving a convoluted basal horizon and overlain by thick graded mudstone caps, are still common (Fig. 12). The most dramatic bed of this type is the black graphitic schist-rich Gordo megabed (Kleverlaan 1987; Fig. 12). Alpujarride- and carbonate-rich beds are absent in the Verdelecho Formation. The Verdelecho Formation, and the Gordo megabed, progressively thin and pinch out onto older basin-fill (Sartenella Formation) towards the southern basin margin (Fig. 10). To the north the Verdelecho Formation attains a more proximal character, with the loss of the thick beds.

Interpretation

The anomalously thick sandstone beds of the Loma de los Baños Formation with their unusual vertical organization, capped by thick graded mudstones, are interpreted as turbidity currents that were contained by (intra) basinal topography (Haughton 2000; *sensu* Pickering & Hiscott 1985; Marjanac 1987). The mudstone caps indicate that the basin was restricted downslope to the east, as the muddy suspension clouds generated by the gravity flows were not allowed to escape from the immediate area. A contained basin-floor geometry is supported by the change in palaeo-current directions within the thinner bipartite beds where the basal part of the flows ran axially to the east, but the rest of the flow was obliquely reflected upon encountering approximately north-facing slopes. The reflected part of the turbidity currents preserves flow dominantly to the N and NW, suggesting that the orientation of the slope they encountered was ENE–WSW. The lack of mudstone caps on the thin carbonate- and Alpujarride-rich turbidite beds sourced from the S and SW margins suggests that this sediment was derived from clean, well-washed littoral sands, thereby lacking the mud and mica of the other beds sourced from the north prior to delivery into the deep-basin. The subordinate amount of carbonate- and Alpujarride-rich beds to the north of the El Cautivo Fault suggests that the structure limited south-to-north sediment exchange.

The El Cautivo Fault is interpreted as an oblique-dextral strike-slip fault based on kinematic measurements and stratigraphic evidence. Several strands of evidence indicate that this structure was active during the deposition of the Loma de los Baños Formation, but not during the deposition of the overlying Verdelecho Formation. An alternative interpretation would be that the El Cautivo Fault was active only during the later stages and/or immediately after the deposition of the Loma de los Baños Formation. This would mean that the observed variations across the oblique-dextral fault are due simply to the juxtaposition of different parts of the succession that were originally laterally disparate: that is, more upslope succession to the north against more downslope succession to the south. The stratigraphic succession on either side of the El Cautivo Fault, however, shows the same stratigraphic evolution from muddy slope to bypass/erosionally confined sedimentation to isolated contained sheets, with dominantly west-to-east transport (Fig. 4). This suggests that the fault has not brought together two completely unrelated parts of the basin fill. Sedimentological variations across the fault support the preferred interpretation that the El Cautivo Fault was active during the deposition of the Loma de los Baños Formation succession as an oblique-dextral fault with a positive surface expression.

Immediately to the north of the El Cautivo Fault the thickness of the Loma de los Baños Formation is greatly reduced (~30 m, Haughton 2000) compared with the Alfaro sub-basin, where it is >300 m thick (Figs 2 & 4). The Loma de los Baños Formation thickens northwards away from the El Cautivo Fault and then thins again. Differences in provenance (calcarenite turbidites are rare to the north) and sedimentology (ponded caps and very thick beds are absent immediately to the north) indicate a restricted connection between the Alfaro sub-basin and the rest of the Tabernas Basin fill at this time. Thick ponded sandstone sheets occur farther to the north in the Puente de los Callejones and the Puente del Chortal areas (Fig. 2). Sandstone sheets can be traced either side of the El Cautivo Fault, but only near the top of the Loma de los Baños Formation succession. South of the fault, thinning of sandstone sheets and the intervening marl succession as the fault is approached support a bathymetric expression for the fault zone during deposition (Fig. 9).

Large rafts of highly contorted calcarenite entrained in the gouge zone indicate deformation prior to lithification. The rafts are tightly folded with abrupt changes in thickness along their length. A thick slump body crops out on the western margin of the Rambla Tabernas immediately to the south of the fault zone where the

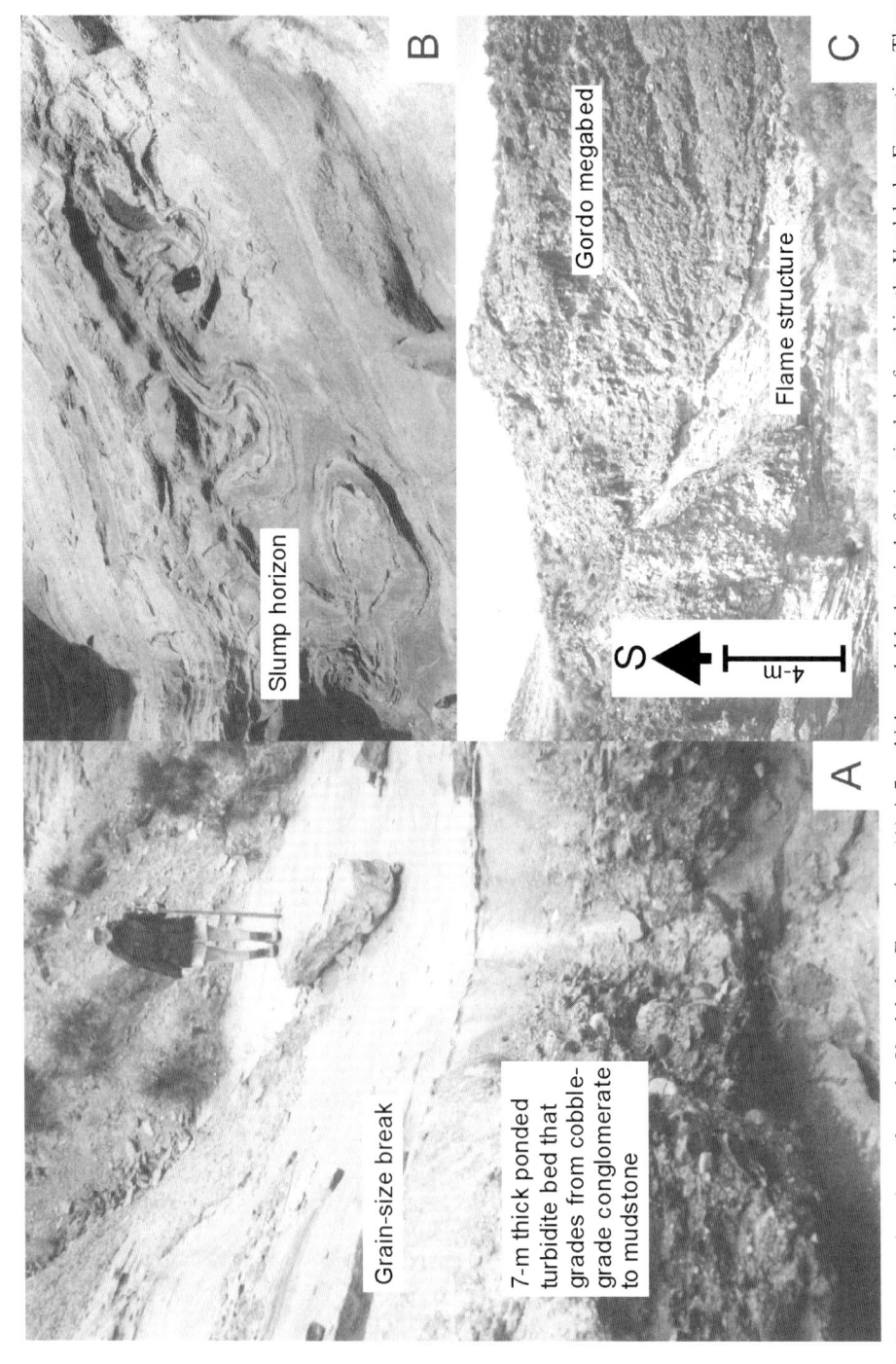

Fig. 12. Representative photos from the Verdelecho Formation. (**A**) A 7 m thick graded bed typical of tripartite beds found in the Verdelecho Formation. These beds are composed of a muddy slump or mudclast conglomerate—sandstone sheet—mudstone cap. (**B**) The slumped basal division of a tripartite bed. (**C**) A 5 m high flame structure formed at the base of the Gordo Megabed.

gouge zone is at its thickest. This is intercalated with the thicker Loma de los Baños Formation succession in the Alfaro sub-basin (Fig. 9). Haughton (2000) suggests that this slump body may have originated from the collapse of a poorly lithified fault scarp, or the venting of mud volcanoes sitting above active faults (cf. Cronin *et al.* 1997). The slump body is composed of significant amounts of material with a scaly fabric similar to that observed in the El Cautivo Fault, and therefore the collapse of the fault scarp is preferred. The Verdelecho Formation seals the El Cautivo Fault, with only a slight displacement of 5 m by a small brittle fault. There is no thickness variation in the Verdelecho Formation across the structure (Fig. 4).

The presence of thick structureless or convoluted beds to the north of the El Cautivo Fault indicates that only the largest volume flows were able to pass over or find their way around local bathymetry associated with the seabed faulting, but were then deposited abruptly. The contorted sandstone sheets reflect shearing and liquefaction induced associated with the collapse of gravity flows that sloshed to and fro within the high-amplitude bathymetry (Haughton 2000, Fig. 13). The north-directed palaeo-flow readings preserved in lower divisions of bipartite beds to the south of the El Cautivo Fault are interpreted as flows that rotated towards deeper hanging-wall bathymetry (Fig. 11). The smaller-volume flows were not able to overtop the structure, and this led to their rapid collapse, forming the distinctive bipartite beds. The structureless upper parts of the bipartite beds are inferred to have formed when the returning reflected flows surged back off steep fault-controlled submarine topography (Fig. 13). The returning flow, reflected from the fault-high, appears not to have been dense enough to underun the collapsing tail of the flow (*sensu* Edwards *et al.* 1994). The resultant interflow would mean deposition by rapid suspension fallout, leading to the characteristic dewatered sheets and structureless fabric of the upper bed divisions on top of the structured lower division.

Not only do the observations above support the interpretation that the El Cautivo Fault was active during sedimentation, but they also imply that the structure had a vertical expression and formed a south-facing submarine fault-scarp and/or blind fault monocline. The presence of a surface expression to the fault explains the thickness variations (thin footwall deposits to the north and thick hanging-wall deposits to the south), and the near absence of sediments with SW- and S-margin provenance to the north of the El Cautivo Fault.

Discussion

The recognition of a fault that was demonstrably active during deepwater sedimentation has importance both for the Neogene basin evolution in the area and for the interplay between faulting and turbidite sedimentation in general. There are also implications for the nature and geometry of fault rocks associated with the propagation of faults through wet and soft sediment close to the seabed.

Basin subsidence

A number of basin models have been proposed to explain the evolution of the Miocene Basins in the Eastern Betic Cordillera including pull-apart (Montenat *et al.* 1987), dextral strike-slip bounding faults (Sanz de Galdeano & Vera 1992; Stapel *et al.* 1996), half-graben (Vissers *et al.* 1995) and lateral ramp basins (Poisson *et al.* 1999). Such wide variations in the interpreted evolution of these basins are, in part, due to their tectono-stratigraphic complexity and the overemphasis on the analysis of basin-bounding faults, rather than intrabasinal structures that were active during sedimentation. The tectonic regime responsible for basin subsidence varied spatially as well as temporally, thereby complicating the classification of the Tabernas-Sorbas Basin.

The identification of an oblique-dextral fault active in a deepwater setting during latest Tortonian and earliest Messinian time aids our understanding of the mechanisms behind the subsidence of the Tabernas-Sorbas Basin. Oblique-extension continued in the western Tabernas area while coeval oblique-compressional tectonics deformed the basin in the Lucainena (Haughton 2000) and south Tabernas (Hodgson 2002) areas. This indicates a complex basin subsidence history that drove the stratigraphic change from slope bypass to conduit backfill to contained sheets. To the east, oblique-compressional structures formed SW-NE trending antiforms and synforms (e.g. Stapel *et al.* 1996) that grew during deepwater sedimentation (Hodgson 2002). The extensional structures in an E-W oblique-extensional dextral setting might be expected to produce NW-SE trending faults. The E-W orientation of the El Cautivo Fault Zone, however, does not fit this scheme. The fault zone is coincident with the trend of the basement highs, and might be an intrabasinal segment of the dextral strike-slip system preserved at the northern margin of the Sierra Alhamilla (Sanz de Galdeano 1987). Lower Messinian

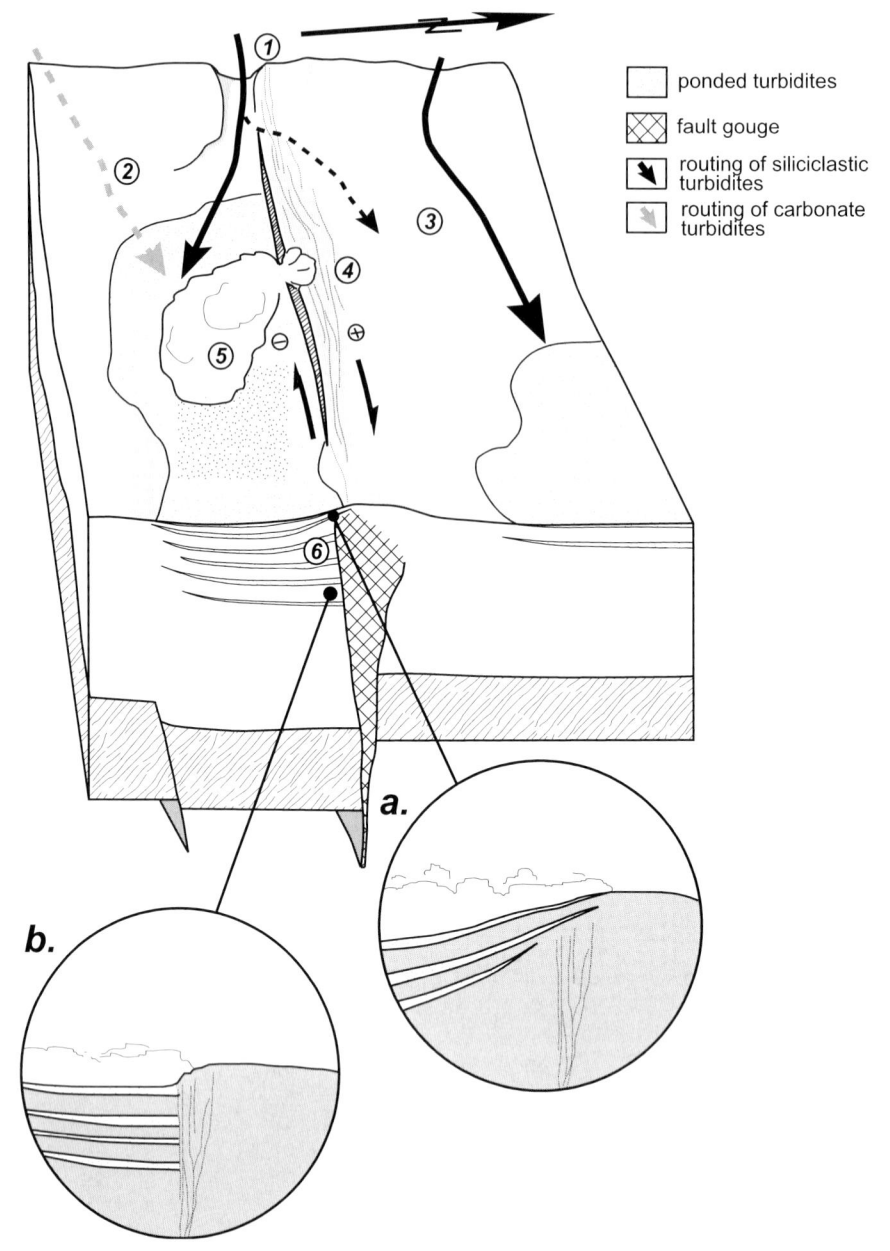

Fig. 13. Cartoon illustrating the impact of oblique slip, syndepositional faulting on gravity current deposition incorporating 'lessons' from the El Cautivo fault zone, SE Spain. The propagation of a basement fault through the overlying cover produced a wide zone (cross-hatched) of deforming wet sediment with either a draped fold or unstable scarp expression at the seabed. Where faulting extended onto the delivery slope (1), this may have controlled the location of the feeder slope incisions. Some carbonate-bearing turbidity currents (2) were unable to breach/overtop the slope topography, and their deposits are preferentially found only on one side of the fault. Flows were preferentially trapped in deeper topography flanking fault to the south, although large-volume flows were either able to escape locally over the fault topography or were delivered to a northern depocentre by alternative routeways (3). The developing scarp slope (4) was cored by extensively sheared and scaly gouge that collapsed locally to produce large slumps (5). Ponded turbidites (6) accumulating in an area of deeper bathymetry associated with displacement of the seabed and either thinned and onlapped onto a forced fold (inset a) or thickened into and abutted against a steeper fault scarp (inset b).

carbonates cap this fault in the Lucainena area, which suggests that the fault propagated to the west.

The late Tortonian to early Messinian period in the evolution of the Tabernas-Sorbas Basin is dominated by both oblique-extensional and oblique-compressional dextral strike-slip tectonics. This developed within the context of differential compression and basement block rotation from the south (Hodgson 2002). The E–W dextral oblique extensional deformation of the Tabernas-Sorbas Basin does not fit mechanically or temporally with the major (N)NE–(S)SW trending sinistral strike-slip faults (e.g. the Carboneras Fault and the Palomares Fault) as suggested by Montenat et al. (1987).

The initiation of subsidence under oblique-extensional conditions in the Tabernas-Sorbas Basin correlates closely with a change in the relative plate motion between the African and Iberian plates. A major change in boundary conditions occurred at Anomaly 5 (8.9 Ma) when Africa's motion became directed to the NW (Dewey et al. 1989). This resulted in enhanced extension in the Tyrrhenian, Linosa and Pantellaria basins, strike-slip faulting throughout Calabria and Sicily, and the reactivation of older thrusts along the North African margin (Dewey et al. 1989). It is possible that this change in relative plate motion led to the initiation of widespread oblique-slip tectonics in the mid-Tortonian. This is consistent with continued westward translation of the Alboran domain. The dextral oblique-slip is therefore a consequence of the position of the basin near to the northern boundary of the Alboran domain.

Impact of seabed faulting on flow containment

Inferred displacements of the seabed by faulting drove switches in the behaviour of the turbulent currents entering the deep basin. The inferred activity on the fault overlaps with the major switch between erosion and incision to backfilling (upper Sartenella Formation) and the establishment of contained turbidite deposition (Loma de los Baños Formation). When the intrabasinal faulting was active, the basin floor would have been uneven and the areas of deep bathymetry localized. The steep slopes and narrow depressions meant that the smaller-volume, contained flows decelerated rapidly, producing bipartite beds. However, these distinctive beds are absent from the overlying Verdelecho Formation, where the intrabasinal fault was blanketed and inactive, although the flows were still contained. Palaeo-flow readings in the Verdelecho Formation are less dispersed than in the Loma de los Baños Formation (Fig. 11), which supports the development of a less morphologically complex seabed at this time. The sub-metre scale turbidites at this level are laminated throughout, showing upward transitions from parallel to ripple lamination, the latter preserving diverse flow orientations (Haughton 2000). This suggests that the style of 'contained' turbidite deposited might reflect the geometry of the basin floor, with the enlarged basin allowing flows to collapse progressively following reflection.

Northward thinning of Loma de los Baños sheet turbidite bundles as they abut against the El Cautivo Fault from the south, and evidence of local onlap of sandstone sheets against a marl high just north of the fault zone, indicate that the fault may at times have been the site of a forced fold that warped the seabed. The evidence for thick slump sheets adjacent to the fault indicates that at other times the fault may have broken the surface and formed an unstable scarp.

Syn-tectonic deepwater sedimentation

Although there is a range of models accounting for the interrelationship between faulting and deposition in terrestrial and shallow-water settings (e.g. Gawthorpe et al. 1997; Gawthorpe & Leeder 2000), those for deepwater successions are less sophisticated and focus mainly on the role of marginal fault systems in undersupplied basins (Ravnås & Steel 1998). Faults commonly control sedimentation by governing the sites and sizes of hinterland drainage basins or by inducing tilting of floodplains and asymmetric channel migration in the case of terrestrial successions. The El Cautivo Fault provides evidence of the response seen in deepwater successions, where the tectonic influence is felt largely via seabed deformation that governs flow behaviour and deposition. The main characteristics of syn-tectonic sedimentation in this sort of setting are adduced to be:

(1) potentially wide and unstable fault zones at the surface, reflecting fault propagation through water saturated and clay-prone deep water sediment;
(2) evidence for induced gradient changes when the faults were active in the form of switching between bypassing (steeper), backfilling (reduced gradient) and flow containment ('mini-basin' development);

(3) aggradational infilling of hanging wall depocentres by contained turbidites where enclosed bathymetry develops, with onlap onto surrounding slopes and the developing scarp (or a monocline where the scarp is draped);
(4) local slumps and debris flows shed from the emergent scarps or monoclines;
(5) unusual bed thickness distributions reflecting flow filtering by topography and the collection of steadier currents to form trapped and perhaps even sustained suspension clouds;
(6) changing turbidite provenance across active fault zones, reflecting partitioning of flow pathways by fault-controlled highs; and
(7) evidence for rapid increases in the rate of suspension fallout in the deposits as the flows rebound off steep fault-induced topography.

Deformation of wet argillaceous sediments

Stratigraphic and sedimentological evidence suggests that the El Cautivo Fault was active during sedimentation. The fault zone is associated with an anomalously wide gouge zone that varies in thickness along its length. The observed fault length:fault zone thickness ratio is 15:1, much higher than typical ratios for oblique-slip fault zones (c. 2000:1 Walsh, J. pers. comm. 2001). This partly reflects the westward draping of the fault by younger stratigraphy that postdates fault activity, and it is likely that the fault continues westwards at depth to link with dextral strike-slip faults in the Alpujarride corridor. The east-to-west thickening and then thinning of the fault zone may reflect the oblique slice through the fault afforded by later uplift of the Alhamilla block. The westward thickening of the fault zone north of Alfaro might be related to the near-surface propagation of the fault with the pervasive deformation of water-saturated argillaceous sediments and increasingly distributed shear deformation as the softer, more water-rich sediment close to the contemporary seabed became involved in the deformation. Farther to the west, the fault zone narrows as activity waned and it was progressively buried by younger sediment.

Experimental studies of wet argillaceous sediments have shown that the water content governs whether the deformation is brittle or ductile. Maltman (1987) showed that although argillaceous sediments with water content of 60% are extremely weak, and despite the appearance of pervasive homogenous flow, narrow zones of concentrated displacement form. These 'shear zones' were found to be the fundamental deformation mechanism over a range of clay minerals and experimental configurations. In experimental studies using mica-rich synthetic muds it was found that to effectively enhance permeability and to channel fluids along shear zones in wet sediments, mean effective stress needs to be less than several tens of MPa (Zhang & Cox 2000). Coring of sediments in active accretionary prisms by the Ocean Drilling Program (ODP) has revealed a wide variety of deformation features (Maltman 1998), including similar features to those identified in the El Cautivo Fault. Narrow zones of faulting are common, and grade into planar bands of breccia. A distinctive scaly fabric results from microscopic zones of flattening, which become supplanted by shear-induced rotation and intensification into CS bands. The surfaces associated with shear in the above structures commonly bear lineations, which can be used for stress tensor analysis even in these incompletely lithified materials (Maltman 1998).

If the silty marls of the Loma de los Baños Formation had a high water content during faulting then ductile processes would dominate deformation, yet fault planes with lineations would still form. High parallel fault zone permeability would contribute to the pervasive deformation of the central areas of the El Cautivo Fault and the formation of a scaly and swirled fabric. The thickness of the overburden and material properties of the deforming sediments are other important factors to be considered in the development of water-saturated deepwater structures.

Conclusions

Combined structural, sedimentological and stratigraphical evidence point to oblique slip dextral faulting of the seabed during Late Tortonian-Early Messinian turbidite sedimentation in the Tabernas Basin. The faulting overlapped and probably drove a major change in basin physiography (slope to basin floor, Fig. 5) and resulted in localized ponded accommodation in which distinctive sheet turbidites aggraded (Fig. 13). A seabed expression for the faulting (as a monoclinal flexure or fault scarp, depending on the rates of vertical displacement vs. sedimentation) is evident from provenance changes across the fault, the structure of the ponded turbidites, wedging and onlap of sandstone sheets as the fault is approached, and thickness variations. The fault zone itself is characterized by a complex and locally very thick gouge zone, which is attributed to propagation of a basement fault through wet and soft sediment close to the seabed.

References

BERGGREN, W. A., KENT, D. V., SWISHER, C. C. III & AUBRY, M.-P. 1995. A revised Cenozoic geochronological and chronostratigraphy. *In*: BERGGREN, W. A., KENT, D. V., AUBRY, M.-P. & HARDENBOL, J. (eds) *Geochronology, Time Scales and Global Stratigraphic Correlation.* SEPM Special Publications, Tulsa, OK, **54**, 129–212.

CRONIN, B. T. 1995. Structurally-controlled deep sea channel courses: examples from the Miocene of southeast Spain and the Alboran Sea, southwest Mediterranean. *In*: HARTLEY, A. J. & PROSSER, D. J. (eds) *Characterization of Deep Marine Clastic Systems.* Geological Society, London, Special Publications, **94**, 115–135.

CRONIN, B. T., IVANOV, M. K. *et al.* 1997. New discoveries of mud volcanoes on the eastern Mediterranean Ridge. *Journal of the Geological Society, London*, **154**, 173–182.

DEWEY, J. F., HELMAN, M. L., TURCO, E., HUTTON, D. H. W. & KNOTT, S. D. 1989. Kinematics of the western Mediterranean. *In*: COWARD, M. P., DIETRICH, D. & PARK, R. G. (eds) *Alpine Tectonics.* Geological Society, London, Special Publications, **45**, 265–283.

EDWARDS, D. 1994. On experimental reflected density currents and the interpretation of certain turbidites. *Sedimentology*, **41**, 437–461.

GAWTHORPE, R. L. & LEEDER, M. R. 2000. Tectono-sedimentary evolution of active extensional basins. *Basin Research*, **12**, 195–218.

GAWTHORPE, R. L., SHARP, I., UNDERHILL, J. R. & GUPTA, S. 1997. Linked sequence stratigraphic and structural evolution of propagating normal faults. *Geology*, **25**, 795–798.

HAUGHTON, P. D. W. 1994. Deposits of deflected and ponded turbidity currents, Sorbas Basin, Southeast Spain. *Journal of Sedimentary Research*, **A64**, 233–246.

HAUGHTON, P. D. W. 2000. Evolving turbidite systems on a deforming basin floor, Tabernas, SE Spain. *Sedimentology*, **47**, 497–518.

HAUGHTON, P. D. W. 2001. Contained turbidites used to track sea bed deformation and basin migration, Sorbas Basin, south-east Spain. *Basin Research*, **13**, 117–139.

HODGSON, D. M. 2002. *Tectono-stratigraphic evolution of a Neogene oblique extensional orogenic basin, southeast Spain.* Ph.D. thesis, University of London.

KLEVERLAAN, K. 1987. Gordo megabed: a possible seismite in a Tortonian submarine fan, Tabernas Basin, province Almeria, southeast Spain. *Sedimentary Geology*, **51**, 181–213.

KLEVERLAAN, K. 1989a. Neogene history of the Tabernas basin (SE Spain) and its Tortonian submarine fan development. *Geologie en Mijnbouw*, **68**, 421–432.

KLEVERLAAN, K. 1989b. Three distinctive feeder-lobe systems within one time slice of the Tortonian Tabernas fan, SE Spain. *Sedimentology*, **36**, 25–45.

KLEVERLAAN, K. 1994. Architecture of a sand-rich fan from the Tabernas submarine fan complex, southeast Spain. *GCSSEPM Foundation 15th Annual Research Conference, Submarine Fans and Turbidite Systems*, 4–7 December, 209–215.

KRIJGSMAN, W., HILGEN, F., LANGEREIS, C. G. & ZACHARIASSE, W. J. 1994. The age of the Tortonian/Messinian boundary. *Earth and Planetary Science Letters*, **121**, 533–547.

LOWE, D. R. 1982. Sedimentary gravity flows II: Depositional models with special reference to the deposits of high-density turbidity currents. *Journal of Sedimentary Petrology*, **52**, 279–297.

MALTMAN, A. 1987. Shear zones in argillaceous sediments: an experimental study. *In*: JONES, M. E. & PRESTON, R. M. F. *Deformation of Sediments and Sedimentary Rocks*, Geological Society, London, Special Publications, **29**, 77–87.

MALTMAN, A. 1998. Deformation structures from the toes of active accretionary prisms. *Journal of the Geological Society, London*, **155**, 639–650.

MARJANAC, T. 1987. Ponded megabeds and some characteristics of the Eocene Adriatic basin (Middle Dalmatia, Yugoslavia). *Memorie della Societa Geologica Italiana*, **40**, 241–249.

MONTENAT, C., OTT D'ESTEVOU, P. & MASSE, P. 1987. Tectono-sedimentary characters of the Betic Neogene Basins evolving in a crustal transcurrent shear zone (SE Spain). *Bulletin du Centre de Recherches Exploration Production Elf-Aquitaine*, **11**, 1–22.

PICKERING, K. T. & HISCOTT, R. N. 1985. Contained (reflected) turbidity currents from the Middle Ordovician Cloridorme Formation, Quebec, Canada: an alternative to the antidune hypothesis. *Sedimentology*, **32**, 373–394.

PICKERING, K. T., HODGSON, D. M., PLATZMAN, E., CLARK, J. D. & STEPHENS, C. 2001. A new type of bedform produced by backfilling processes in a submarine channel, late Miocene, Tabernas-Sorbas Basin, SE Spain. *Journal of Sedimentary Research*, **71**, 692–704.

POISSON, A. M., MOREL, J. L., ANDRIEUX, J., COULON, M., WERNLI, R. & GUERNET, C. 1999. The origin and development of Neogene basins in the SE Betic Cordillera (SE Spain): a case study of the Tabernas-Sorbas and Huercal Overa basins. *Journal of Petroleum Geology*, **22**, 97–114.

RAVNÅS, R. & STEEL, R. J. 1998. Architecture of marine rift-basin successions. *American Association of Petroleum Geologists Bulletin*, **82**, 110–146.

RIDING, R., BRAGA, J. C., MARTÍN, J. M. & SÁNCHEZ-ALMAZO, I. M. 1998. Mediterranean Messinian Salinity Crisis: constraints from a coeval marginal basin, Sorbas, southeastern Spain. *Marine Geology*, **146**, 1–20.

SANZ DE GALDEANO, C. 1989. Las fallas de desgarre del borde Sur de la cuenca de Sorbas-Tabernas (Norte de Sierra Alhamilla, Almeriá, Cordilleras Beticás). *Boletín Geológico y Minero*, **101**, 73–85.

SANZ DE GALDEANO, C. & VERA, J. A. 1992. Stratigraphic record and palaeogeographical context of the Neogene basins in the Betic Cordillera, Spain. *Basin Research*, **4**, 21–36.

STAPEL, G., MOEYS, R. & BIERMANN, C. 1996. Neogene evolution of the Sorbas basin (SE Spain) determined by paleostress analysis. *Tectonophysics*, **255**, 291–305.

VISSERS, R. L. M., PLATT, J. P. & VAN DER WAL, D. 1995. Late orogenic extension of the Betic Cordillera and the Alboran Domain: a lithospheric view. *Tectonics*, **14**, 786–803.

ZHANG, S. Q. & COX, S. F. 2000. Enhancement of fluid permeability during shear deformation of a synthetic mud. *Journal of Structural Geology*, **22**, 1385–1393.

Comparing the depositional architecture of basin floor fans and slope fans in the Pab Sandstone, Maastrichtian, Pakistan

R. ESCHARD[1], E. ALBOUY[1], F. GAUMET[1] & A. AYUB[2]

[1]*Institut Français du Pétrole, 1, 4 Avenue de Bois Préau, 92852 Rueil-Malmaison Cedex, France*
[2]*Premier Pakistan, 4th floor, Jang building, Fazal-e-Haq road, Blue area, Islamabad, Pakistan*

Abstract: Basin-floor fans and slope fans present major differences in their internal architecture related to changes in: (1) margin morphology, (2) relative sea-level change, and (3) sediment supply. These variations are illustrated in the outcrops of the Pab Sandstone in Pakistan. The Pab Sandstone third-order sequence was deposited on the Indo-Pakistani margin during the Upper Maastrichtian. Uplift of the margin induced erosion on the shelf, incision of submarine canyons on the slope and the development of a sand-rich, high-efficiency basin-floor fan extending over hundreds of kilometres on the basin floor. During transgression, sediment accumulated in backstepping shoreface deposits on the shelf, and a minor mud-rich slope fan was deposited in the basin. Finally, a sand-rich braided delta prograded across the shelf, feeding a sand-rich slope fan where it reached the shelf margin. This slope fan was of more limited lateral extent. The Lower Pab basin-floor fan shows the effects of flow funnelling and confinement due to a canyon incised into the slope. It consists mainly of channel complexes deposited by superconcentrated density flows to low-density turbidity currents. In contrast, the Upper Pab slope fan shows little confinement and low transport efficiency. It consists of tabular lobes, aggrading mid-fan channels and conglomeratic channels in the upper fan. The low transport efficiency of the gravity flows probably explains the low degree of organization of the slope fan.

The exploration of passive margins has recently brought to light the great diversity of turbidite fan systems, reflecting the variable geological contexts of these margins. Sediment supply variations through time and slope and margin morphology are two of the main controlling factors. The sediment supply variations result from the combination of both basin physiography and climatic factors (Eschard 2001). The amount of sediment arriving in the deep basin depends on the presence or absence of large deltas on the margin, the size and nature of the fluvial network upstream of the delta, and the nature of the clastic material transported by rivers (sand or mud-dominated) (Steel *et al.* 2000; Mutti *et al.* in press). Shelf width is, in itself, a critical factor and can influence the depositional processes of the turbidite fan. In the case of a narrow shelf, turbidites can be induced by fluvial floods (Mutti *et al.* 1996) and hyperpycnal flow (Mulder *et al.* in press). Shelf-width changes through time with the variations in relative sea-level; sediment accumulates on the shelf during periods of relative rise or high-stand and transits through incised valleys during periods of low relative sea-level (Vail *et al.* 1977). Differences in slope morphology, such as between a margin with a well-defined shelf break or a ramp-type margin (Van Wagoner *et al.* 1988), and the occurrence of single or multiple source points for clastic input to the basin, also influence the fan morphology (Reading & Richards 1994).

Deformation of the slope and deep basin are also critical in determining the architecture of the fans, gravity flows being preferentially trapped in the topographic lows. Intraslope basins, often due to salt tectonics, form sediment traps along many passive margins (Prather *et al.* 1998). In foreland basins, the fan architecture is directly controlled by the foredeep basin morphology and the location of the sediment source (longitudinal or lateral) (Mutti *et al.* 1999).

Basin-floor fans and slope fans constitute two major types of deep-sea fan system that present contrasting architecture. Basin floor fans (Vail *et al.* 1977) are deposited in the basin plain setting along passive margins. They correspond to high transport efficiency systems (Mutti *et al.* 1992), and sediment is transported over hundreds of kilometres from the canyon mouth to the lobe pinch-out (over 500 km in the case of the Congo fan, according to Vittori *et al.* 2000). The different elements of the fans (canyon, channel complexes, levees, terminal lobes) are well differentiated although their deposition may be diachronous (Mutti *et al.* 1992). Large channel complexes which often present a sinuous shape and complex internal architecture (Kolla

From: LOMAS, S. A. & JOSEPH, P. (eds) 2004. *Confined Turbidite Systems.* Geological Society, London, Special Publications, **222**, 159–185. 0305-8719/04/$15.00 © The Geological Society of London 2004.

et al. 2001; Abreu *et al.* in press), pass distally to depositional lobes.

In contrast, a slope fan is deposited in a slope apron setting and is of more limited spatial extent (Posamentier & Vail 1988; Mutti *et al.* 1992). Slope fans are often fed by multiple source points. The component elements of slope fans are not as clearly differentiated as in the basin-floor fan: transport efficiency is lower, so the upstream–downstream distance is shorter between the proximal fan, made up principally of feeder channels, and the distal fan, consisting of tabular lobes. A significant proportion of the slope fan results from debris flows and associated turbidite surges caused by margin collapse (Reading & Richards 1994).

The aim of this paper is to discuss the geological factors controlling the architecture of

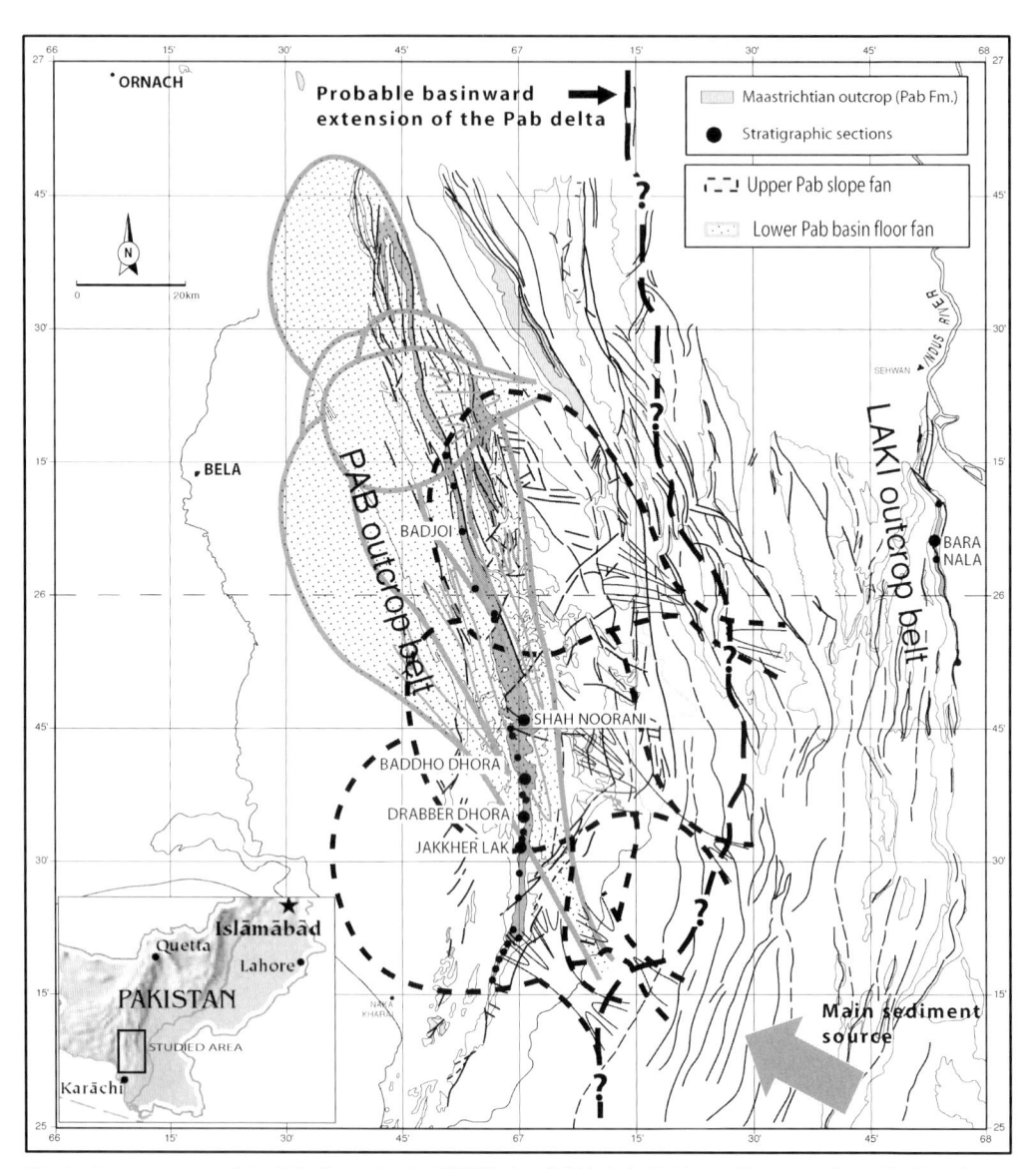

Fig. 1. Location map of the Pab Range in the SW Kirthar fold belt in Pakistan. Base map from the Huntings Geological Survey (1958). The map shows the location of stratigraphic sections and the extent of the Late Maastrichtian delta and turbidite systems. The Lower Pab basin-floor fan extends sub-parallel to the margin over more than 150 km. It is overlain by the Upper Pab slope fan, trapped at the base of the slope. The hypothetical extent of the coeval Pab delta is shown as a dotted line.

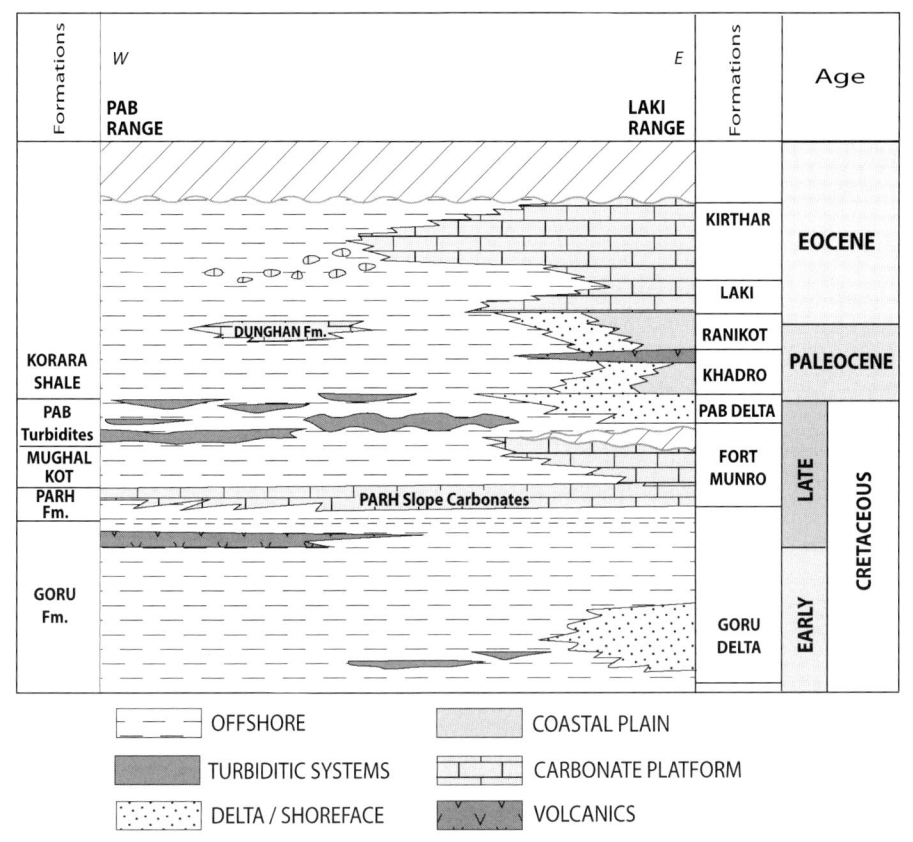

Fig. 2. Chronostratigraphy of the Cretaceous to Eocene in the southern Kirthar fold belt.

basin-floor fans and slope fans. The discussion will be based on the Pab Sandstone depositional system, which crops out in southern Pakistan (Fig. 1). The Pab depositional system was deposited in a passive margin setting along the Indo-Pakistani plate during the Late Maastrichtian (Fig. 2). In this depositional system, a basin-floor fan is overlain by a mud-rich slope fan (Lower Pab), then by a sand-rich slope fan (the Upper Pab; Fig. 3). The exceptional outcrop continuity allows reconstruction of the geometry of the margin, and the internal architecture of the three different turbidite systems can be compared. The shelf equivalent of the Pab turbidite systems is also well known at a regional scale (Khan et al. 2002), and is a petroleum reservoir in the Kirthar fold belt.

Regional geological context

The studied outcrops are located in southern Pakistan in the southern part of the Kirthar fold belt (Fig. 1). In this area, the Pab Formation crops out in two main mountain belts: the Laki Range in the east, and the Pab Range in the west. In the Laki Range the Pab Formation is represented by delta facies (Fig. 4), whereas in the Pab Range it consists of turbidite and gravity-flow deposits in a deep slope to basin setting (Fig. 5). The transition between the delta and the turbidite systems does not crop out, and is known only in subsurface data in the Kirthar fold belt.

In the Pab Range the turbidite system is exposed in an outcrop belt over 200 km long. Exposure is almost continuous all along the Pab mountain belt, and the outcrop quality is excellent because of the semi-desertic conditions (Fig. 6). The Cenomanian to Eocene sedimentary series crop out in a N–S trending monocline. The Pab mountain belt is incised by dry river gullies, which provide good three-dimensional exposures of the architecture of the Pab Formation at a kilometric scale. These outcrop conditions provide a unique opportunity to study the sequential

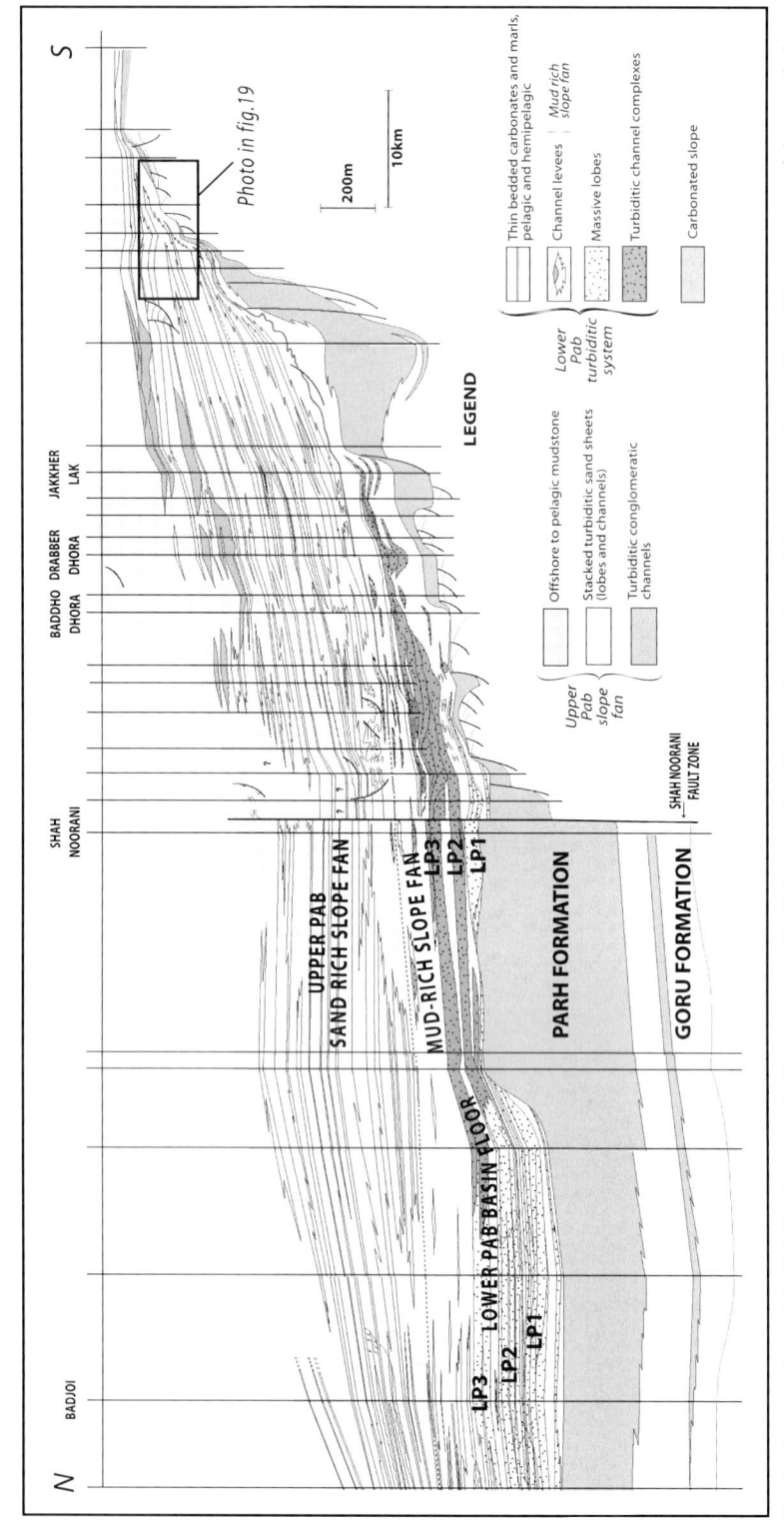

Fig. 3. Cross-section of the Pab Sandstone turbidite systems in the Pab range. See Figure 1 for location of the stratigraphic sections. The geometry of the margin is preserved from the slope to the deep basin, and the turbidite systems onlap it and pinch out southwards. The photograph shows the detail of the onlap of the Upper Pab lobes onto the slope. Modified from Eschard *et al.* (in press).

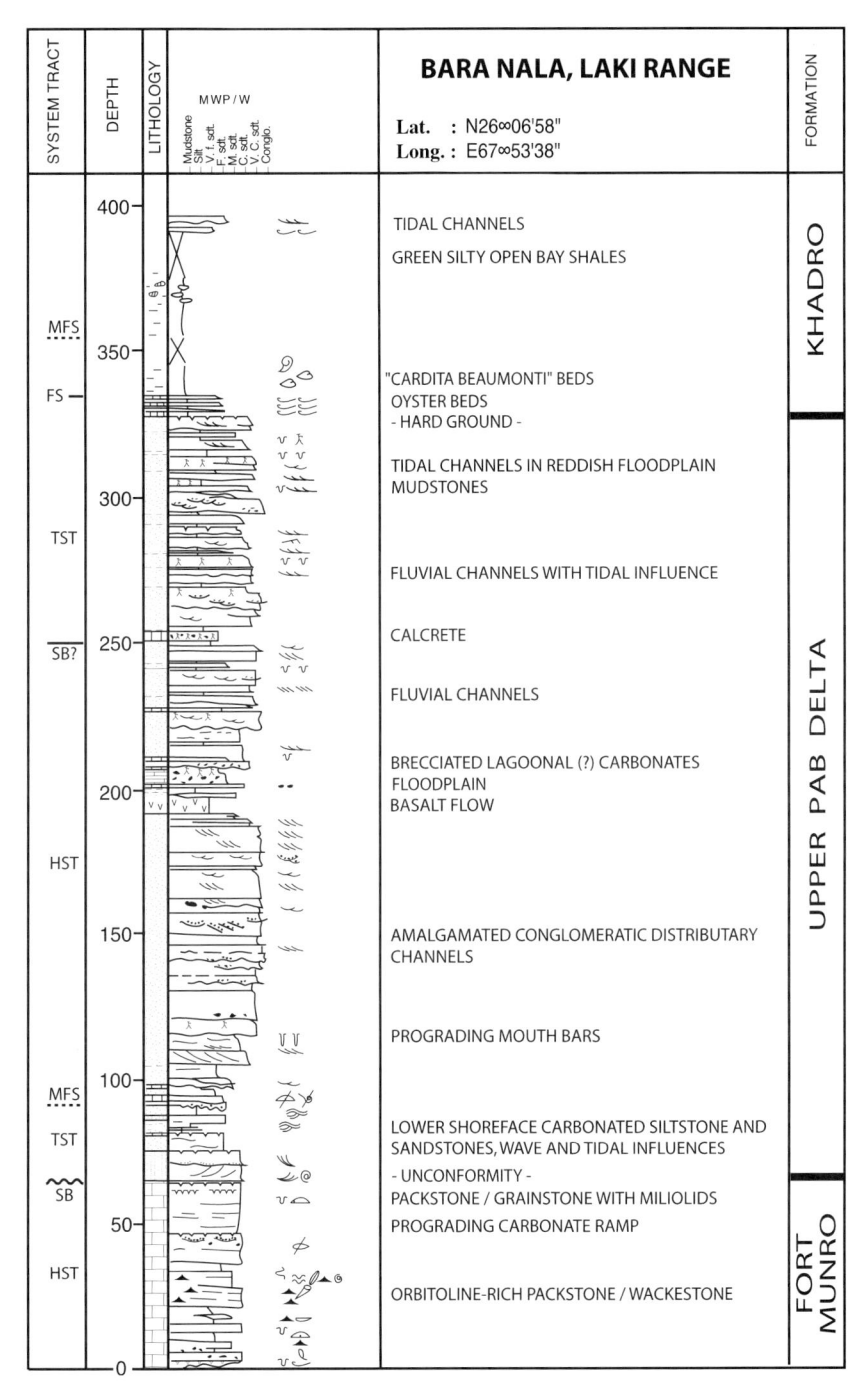

Fig. 4. Stratigraphic section of the Fort Munro, Pab and Khadro Formation in the Laki range (Bara Nala; see Fig. 1 for location). Above a prograding carbonated ramp, a sharp surface probably marks an unconformity, even though no evidence of incision has been observed. Above this surface, clastic supply abruptly increased, first deposited in backstepping shoreface deposits and tidal channels. A sand-rich braided delta then prograded basinwards, and coastal plain deposits developed behind the delta front. The backstep of the delta is shown by the increasing tidal influence in the upper part of the Pab sandstone, before a regional flooding marked by open bay shales.

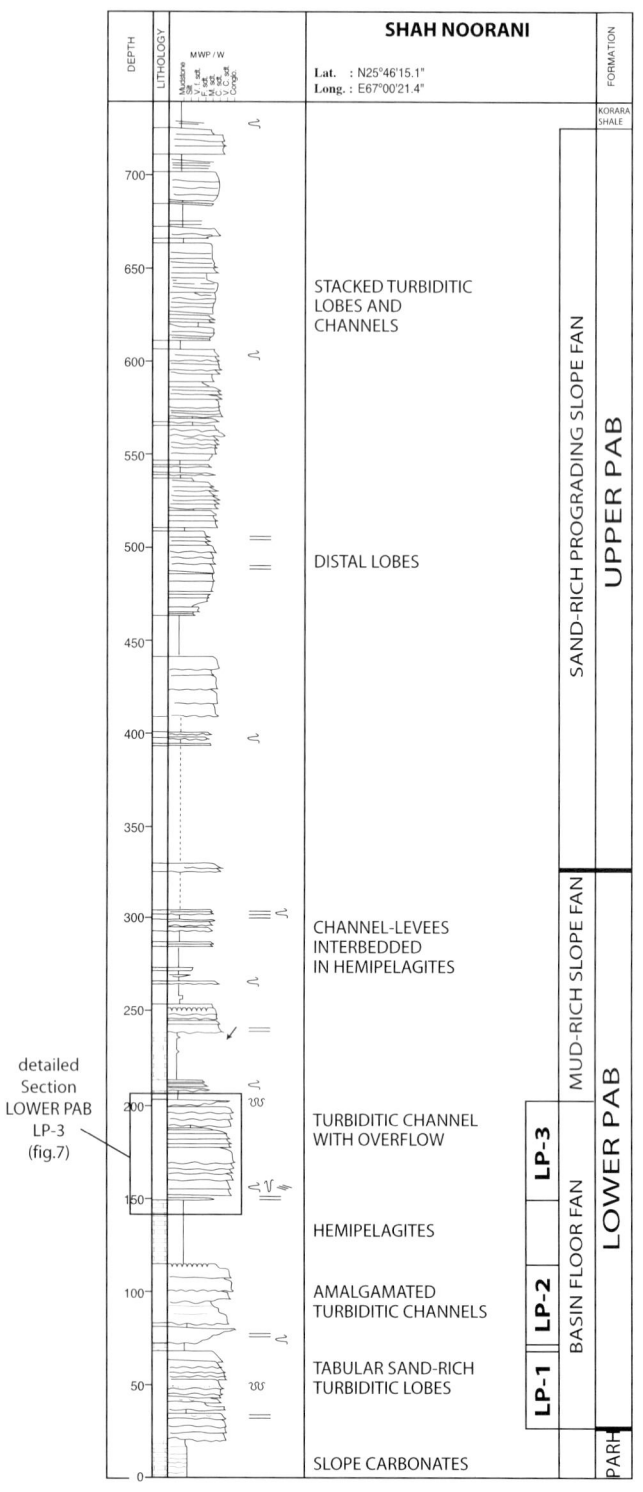

Fig. 5. Stratigraphic section of the three Pab Sandstone turbidite systems in Shah Noorani (see Fig. 1 for location). LP1, LP2 and LP3 refer to the three units of the Lower Pab basin-floor fan.

Fig. 6. General outcrop view south of Shah Noorani showing the Lower Pab basin-floor fan, the mud-rich slope fan and the Upper Pab sand-rich slope fan. The Parh Formation pinches out towards the northeast.

evolution and the reservoir architecture of both the turbidite systems and their deltaic equivalent on the shelf, at the scale of the margin and at a more detailed reservoir scale. Furthermore, the longitudinal evolution of the fan can be studied in these outcrops, as current directions, directed towards the N to NW, were oblique to the main outcrop belt (Fig. 1). In the Pab range the Lower Pab sand-rich turbidite system is preserved from the canyon setting to the lobes of the basin-floor fan. The Upper Pab system shows a full progradational trend from distal lobes to conglomeratic feeder channels, and shales out northwards along the Pab range outcrops.

History of the Indo-Pakistani margin in the Kirthar belt

The history of the Pakistani margin and its lithostratigraphy have been described by Shah (1977), Kadri (1995) and more recently by Hedley et al. (2001). Regional stratigraphy was established by Butt (1992) and Nomura & Brohi (1995) and revised with new nannofossil dating in the course of this work. The lithostratigraphy and chronostratigraphy of the area are summarized in Figure 2.

The palaeomargin exposed in the Kirthar fold belt is the consequence of the break-up between the Indo-Pakistan-Madagascar plate and the African plate. Rifting started in the Early Jurassic, and break-up was complete by the Middle Jurassic. From this time onward, a passive margin striking roughly N–S existed to the northwest of the Indo-Pakistani plate until the start of collision with the Eurasian continent (Waheed & Wells 1990). During the Early Cretaceous, the basinal setting was the site of pelagic sedimentation (Sembar and Goru Formations), while delta systems accumulated on the shelf (Hedley et al. 2001). In the Coniacian-Santonian, break-up occurred between India and Madagascar. This marked the beginning of rapid northward drift of the Indo-Pakistani plate (Scotese et al. 1988; Gnos et al. 1997). The margin showed a complex geometry as a result of its segmentation by wrench faults. Progradation of the margin began during the Santonian and Campanian with the deposition of slope carbonates of the Parh Formation, affected by active normal faults combined with more minor growth faults due to gravity sliding.

During its northward journey, the Indian continent approached, and then passed above the 'Réunion' mantle plume. Volcanic activity began in the Upper Maastrichtian, probably during early India–Seychelles rifting, and reached a paroxysm with the eruption of the Deccan trap basalt flows at the Cretaceous-Tertiary boundary (Courtillot et al. 2000). It is probable that the rapid increase of the sediment supply during deposition of the Pab Formation in the Upper Maastrichtian is related to regional doming due to the mantle plume. An unconformity was formed on the shelf, and most of the clastic material bypassed the shelf through incised valleys (Fig. 7a). Major submarine canyons were also incised in the slope, and supplied coarse-grained clastic sediments to a sand-rich basin-floor fan in the pelagic domain. A regional transgression then led to drowning of the shelf. The rate of sediment supply to the turbidite basin dramatically decreased. On the shelf, shoreface deposits backstepped during the transgression. In the basin, a mud-rich slope fan with channel levees also backstepped and onlapped the slope. Channel-levee complexes of the mud-rich slope fan were interbedded with background calcareous pelagic sediments. The basin-floor fan and the overlying mud-rich slope fan will be referred to as the Lower Pab.

As plate uplift continued, a sand-rich braided delta prograded rapidly across the shelf, and soon reached the shelf-break. Sandy turbidite slope fans were directly supplied by the delta in the areas where the shelf was narrow. These fans consisted of sand-rich lobe/channel complexes. They were organized in a thickening and coarsening-upward succession of turbidite lobes overlain by channels. This evolution suggests a major progradation of the turbidite slope fans to the NW, the equivalent in time of the delta progradation on the shelf. The sandy slope fans and the coeval delta systems will be referred to as the Upper Pab. Volcanic activity also increased in relation with the thermal doming, and both fine- and coarse-grained material reworked in the slope fan is rich in volcanic clasts and ashes, with some samples of sandstone from the Upper Pab containing up to 30% volcanic material.

During the Palaeocene, a major transgression was recorded in the southern Kirthar belt. Sediment supply was reduced and most of the sediment accumulated in aggrading coastal sediments on the shelf (Khadro and Ranikot formations). The Upper Pab slope fan rapidly backstepped, and was finally abandoned. Pelagic ramp sedimentation (the Korara Shale Formation) then established itself in the area now exposed in the southern Pab Range. During Paleocene times, obduction of the Bela ophiolites occurred to the north-west of the Indo-Pakistani

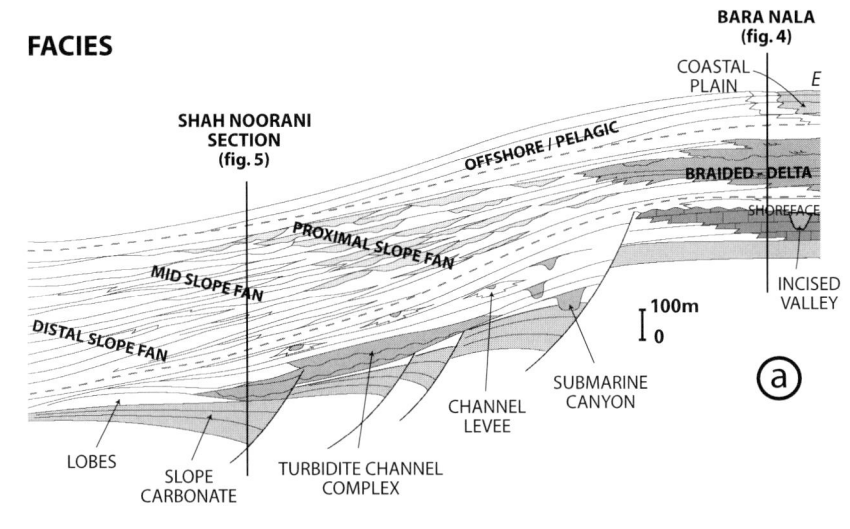

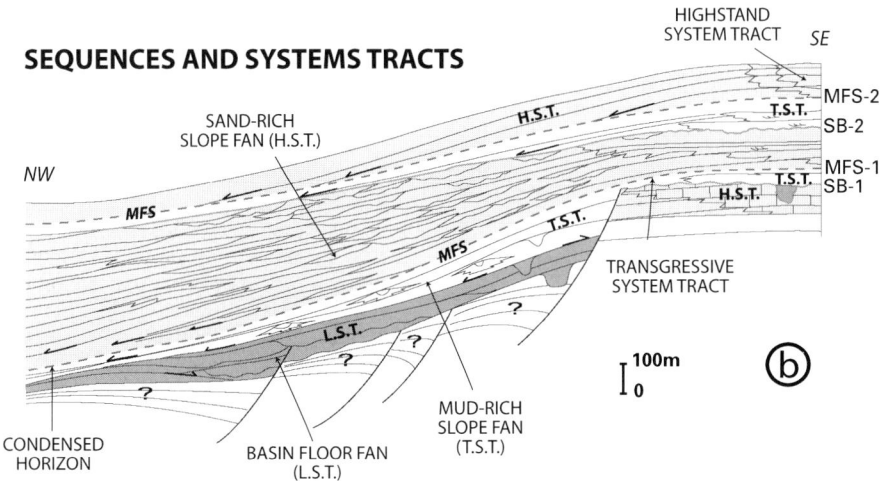

Fig. 7. Reconstruction of the sequence architecture of the Pab Formation in (a) facies and (b) systems tracts. The main transport direction of the Pab basin-floor fan is to the northwest. The Lower Pab basin-floor fan is the time equivalent of an unconformity on the shelf, the mud-rich slope fan is the time equivalent of backstepping shoreface deposits during regional transgression, and the Upper Pab slope fan is coeval with a prograding braided delta.

plate (Allemann 1979; Tapponier et al. 1981), and the margin was affected by compressive deformation (Bannert & Raza 1992). After a return to a passive margin setting during Early Eocene times, the collision of the Indo-Pakistani plate with the Eurasian plate started, with the initiation of the proto-Himalayan thrust belt and the closure of the last remnant of the Tethys ocean during the Late Eocene (Searle et al. 1987). In the southern Kirthar Fold Belt, the first signs of compression occurred during Oligocene times, and the rate of compressional deformation progressively increased from that time onwards. The age of formation of most of the thrust sheets forming the Southern Kirthar Fold Belt is fairly recent (Pliocene to Pleistocene; Schelling 1999).

Methodology

The regional east-to-west reconstruction of the Pab Sandstone systems is based on the acquisition and the correlation of numerous regional stratigraphic sections in the Pab Range (23 sections) and in the Laki Range (3 sections)

(Fig. 1). Subsurface data in the Kirthar belt (well and seismic data) were used in correlating between the outcrop belts.

The reconstruction of the slope to margin profile and of the Lower and Upper Pab turbidite systems is based on the correlation along the Pab range of the stratigraphic sections mentioned above, along with dozens of more detailed sections in the Lower Pab, and on a continuous photomosaic of the outcrop and high-resolution satellite images in the Pab range. The correlation is also controlled by nannofossil dating.

Description of the Lower Pab sand-rich turbidite system

A sand-rich turbidite system lying just above the Parh slope limestones crops out continuously for over 100 km along the Pab range. Given the orientation of the paleocurrents (N to NW) compared with the reconstructed strike of the paleo-margin (N–S), the outcrop transect corresponds to an oblique upstream-to-downstream image of a turbidite system lying sub-parallel to the paleomargin.

Three main sandy units (named LP1–3 in stratigraphic order) have been identified (Figs 3 & 5), separated by regionally extensive hemipelagite drapes. The two lower units (LP1 and LP2) lap onto the Parh limestones in the north of the studied zone (Fig. 3), and it is the upper unit (LP3) that has been described in the most detail, along an outcrop transect 25 km long. The detailed upstream to downstream evolution of this unit and the facies of the turbidite sandstones have been described in detail by Eschard *et al.* (in press).

(1) *LP1: sand-rich lobes and bypass surface.* The unit consists of coarse-grained sandstone organized in stacked tabular beds with normal grading and occasional plane parallel laminations. Bedsets do not show any basal incision. LP1 is interpreted as sand-rich lobes deposited on the basin floor. The stacked lobes are 40 m thick in their proximal part and show a massive vertical profile in stratigraphic logs (Fig. 5). The LP1 unit shales out northwards, the amount of heterolithics interbedded with the sandstones progressively increasing in the basinward direction. The proximal equivalent of these lobes has not been identified. They lap directly onto the Parh slope carbonates and pinch out southwards. It is possible that channels that are the time equivalent of the lobes are present in the subsurface, east of the Pab Range. An alternative explanation would be to consider that the deposition of the LP1 lobes was the time equivalent of a major period of bypass in the slope setting, where a canyon incised into the Parh limestone has been observed. The basal canyon fill is made up of fine-grained sediments, suggesting that, initially, coarse-grained material transited through the canyon to accumulate in the lobes. The canyon erosional surface is interpreted as being the proximal equivalent of these lobes.

(2) *LP2: sandy channel complexes passing distally to sand-rich lobes.* The proximal part of this unit consists of coarse-grained to pebbly sandstones filling deeply incised channels. Overflow deposits are not very well defined. The downstream equivalent of the channels corresponds to stacked, tabular beds of medium to coarse sandstone, interpreted as well-developed lobes that pinch out distally (northwards). LP2 is separated from LP1 by a laterally continuous layer of hemipelagic sediments 5 m thick. The LP2 channel complex is around 30 m thick (Fig. 5). Part of the sediment must have bypassed the proximal part of the fan to accumulate in the sand-rich lobes basinwards. The deep incision at the channel base reflects these bypass processes. A large amount of sediment then accumulated in the channels.

(3) *LP3: aggrading channel complex with poorly developed lobes.* This unit consists of coarse to very coarse sandstones filling erosional surfaces, with well-defined associated tabular medium to coarse sandstone bodies spilling out beyond the erosional surfaces and interpreted as terminal lobe deposits. Laterally to the channels, tabular fine to medium sandstones intercalated with siltstones are interpreted as crevasse deposits and levees (Fig. 8). Sole marks are frequent at the bases of the turbidites. This channel complex passes distally (northward) poorly developed lobes. This channel complex can be over 60 m thick (Figs 5 & 8) and pinches out southwards, lapping onto the Parh slope carbonates. It consists of dozens of amalgamated turbidite channels that have been correlated and mapped along the cliff face. The incision at the base of the individual channels is generally shallower than in the LP2 complex, and overflow deposits are better developed. This architecture shows that the channels were more aggradational and constructive than in the LP2 channel complex. Bypass processes were also more

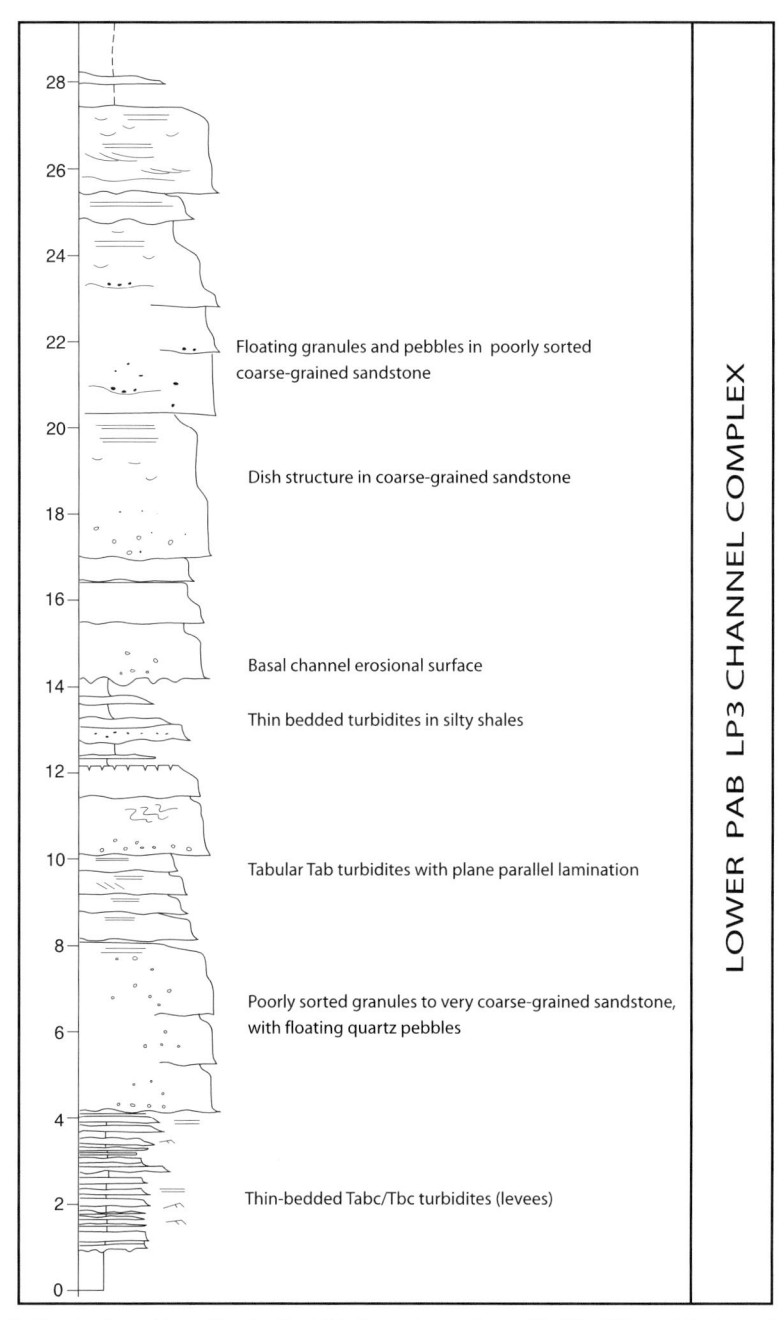

Fig. 8. Detail of a stratigraphic section in the LP3 channel complex at Baddho Dhora. The levee deposits are cut by a sharp erosional surface at the base of the channel complex. The channel fill consists mostly of very coarse to coarse grained sandstones deposited by high-density gravity flows.

limited. In consequence, lobes are poorly developed in the distal part of the fan, most of the sediment being trapped in the channel complexes of the proximal to mid fan setting.

Depositional system architecture

The turbidite fan forming the Lower Pab depositional system corresponds to a deep-sea fan of

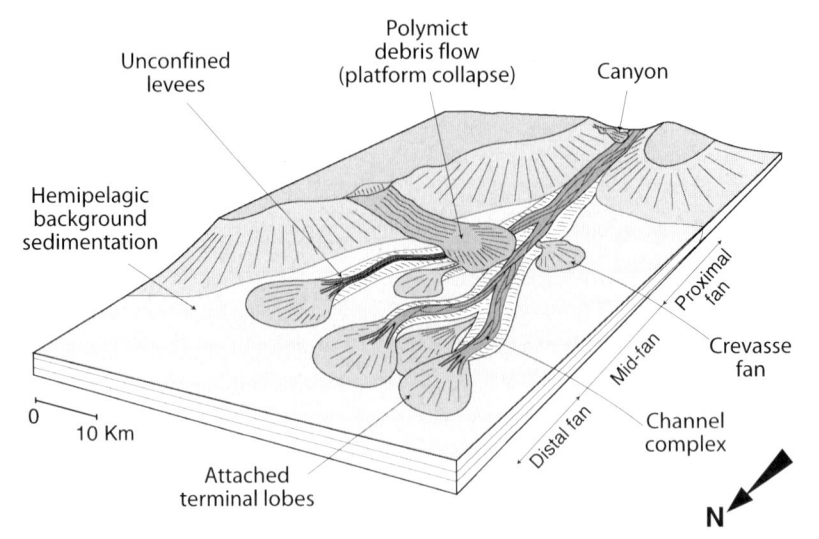

Fig. 9. Reconstruction of the Lower Pab basin-floor fan depositional system, modified from Eschard *et al.* in press.

high transport efficiency, deposited on the basin floor (Fig. 9). The background sedimentation of the fan consisted of thin-bedded hemipelagic wackestones alternating with marls. These facies were very rich in pelagic foraminifera, which indicate deep to very deepwater depositional conditions. The *Zoophycos* and *Nereites* ichnofacies that are found in the Lower Pab also support this interpretation. More than 150 km separated the canyon mouth from the distal lobe pinch-out. Sediment arrived in the basin through a submarine canyon at least 300 m deep, which was incised in the Parh slope carbonates (Eschard *et al.* in press). The extension of the basin-floor fan was controlled by the segmentation of the margin. Palaeocurrents flowed northwards in the canyon, and then turned to the northwest downstream of the canyon mouth in the fan.

The fan itself consisted of channel complexes passing northwards to sand-rich attached lobes. These channel complexes were made of tens of amalgamated low-sinuosity channels (see Figs 8 & 10). They were flanked by levees whose thickness and extent increased downstream. Facies and detailed architecture of these channels have been described by Eschard *et al.* (in press).

In the proximal fan the channel complexes are 50–70 m thick and 1–2 km wide (Fig. 11). The channel fill consisted of coarse- to very

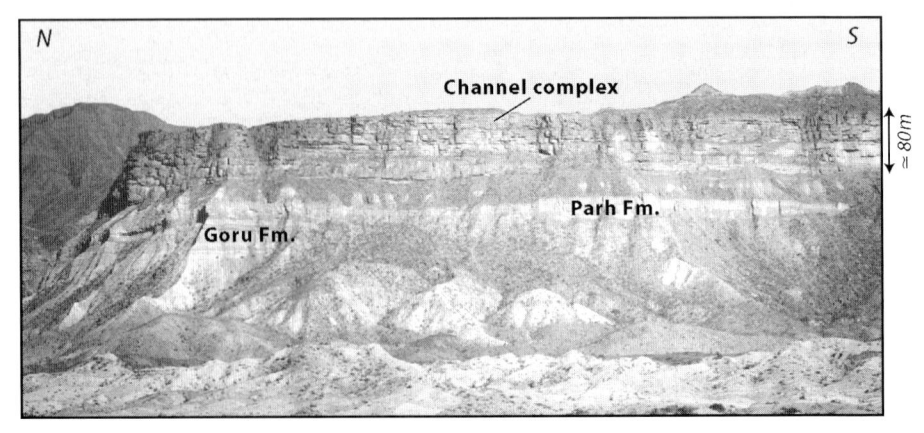

Fig. 10. The LP3 channel complex of the Lower Pab in the mid-fan setting. The channel complex results from the amalgamation of several individual channels. Photograph located north of the Baddho Dhora section.

Fig. 11. The Lower Pab LP3 channel complex close to the canyon mouth at Drabber Dhora. The channel complex is 60 m thick in the channel axis (to the right in the picture) and fringes out to the left. Modified from Eschard *et al.* (in press).

coarse-grained sandstones deposited by high-density gravity flows. Evidence of bypass is frequent, especially at the canyon mouth, in the form of thick mudclast lags, gravel lags and tractive megadunes at the bases of the channels. The individual channels making up the channel complexes were deep and narrow. The degree of amalgamation of the component channels was high, and the resulting channel complexes preserve a high sand/shale ratio. Within the channel complexes overflow deposits were rarely preserved between the individual channels, and the levees flanking the channel complexes were poorly developed.

In the mid-fan setting the density flows are much less confined than in the proximal fan setting, and as a result the channel complexes widened out on the basin plain. The channel complexes were 40 m thick, 2–4 km wide and made up of tens of amalgamated channels (Fig. 10). Individual channels showed a well-developed aggradation phase with construction of spillover deposits in the upper half of the channel sequence. Sandy overflow deposits are preserved between the individual channel conduits within the general envelope of the channel complex. Evidence of lateral migration of the channels can also be frequently observed, the channel axis laterally shifting while aggrading, resulting in multiple inclined erosional surfaces draped with mudclasts ('shingle' geometry; Eschard *et al.* in press). Incision was moderate at the base of the channel complexes and at the base of each single channel. Sand-rich levees extended laterally to the channel complex margin, and their development progressively increased downstream.

In the distal fan, proximal lobe facies presented a massive character and were made up of stacked, tabular, coarse-grained sandstone beds (Fig. 12). Lobes present a coarsening and thickening-upward trend. The number and thickness of heterolithic interbeds progressively increased downstream, and the lobes progressively shale out distally. These lobes were attached to the channel complexes, without a significant bypass area between lobes and channels. Distal lobe pinch-out has not been observed but should be visible in outcrop in the northern part of the Pab Range. The proximal sand-rich lobe unit at the base of the Lower Pab presents a cumulative thickness of 40 m.

Backstepping sequential evolution of the basin-floor fan

At a regional scale, the Lower Pab turbidite system consists of three sandstone units (LP1–3) with a high sand/shale ratio. The top of each sandstone unit is a major abandonment surface with a well-developed hardground, overlain by laterally extensive intervals of hemipelagic wackestones and marls, which drape the sandy turbidite deposits (Fig. 13). The hemipelagic sedimentation records long periods of time during which the fan was abandoned. These periods can either be related to major lateral avulsion of the main channel feeders or to fourth-order cyclicity. Allocyclic parameters

Fig. 12. Lower Pab depositional lobes at Badjoi. Note the tabular organization of the lobes, which are organized in sand-bodies around 30 m thick.

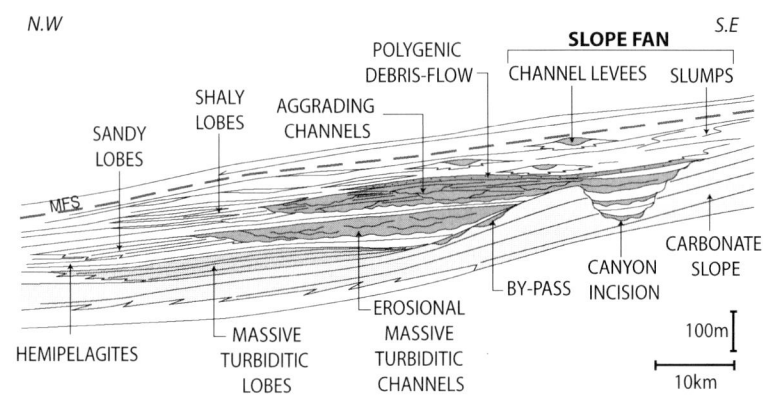

Fig. 13. Sequential organization of the Lower Pab basin-floor fan. The fan shows an overall backstep and finally complete abandonment as the relative sea-level rises; Note the evolution in facies partition between lobes and channels from LP1 to LP3. Sediment first accumulated in lobes then in the channel complexes.

may have controlled a high-frequency variation of the sediment supply to the fan.

The evolution of the turbidite systems from LP1 to LP3 is characteristic of an overall backstep of the basin-floor fan, which we interpret as being due to a decrease in sediment supply through time. Bypass phenomena are dominant in the proximal part of the margin during the deposition of the massive lobes in more distal settings (LP1 complex). Then, the importance of bypass decreases in LP2 and in LP3, where most of the sediment accumulated in the channel complexes. The total amount of sediment arriving in the distal fan setting therefore decreased through time. The size of the terminal lobes accordingly decreased from LP1 to LP3.

Description of the Lower Pab mud-rich turbidite system

A unit consisting principally of hemipelagic deposits is developed above the Lower Pab channel complexes (Figs 3, 5 & 13). The unit is made up mostly of thin-bedded carbonates and marls, in which turbidite channels and lobes are embedded. The hemipelagic deposits consist of thin-bedded wackestones very rich in pelagic foraminifera, interbedded with mudstones. The wackestones present a tabular geometry occasionally deformed by slumping.

Isolated channel–levee and lobe sand-bodies are interbedded with the hemipelagic deposits. They consist of medium- to fine-grained

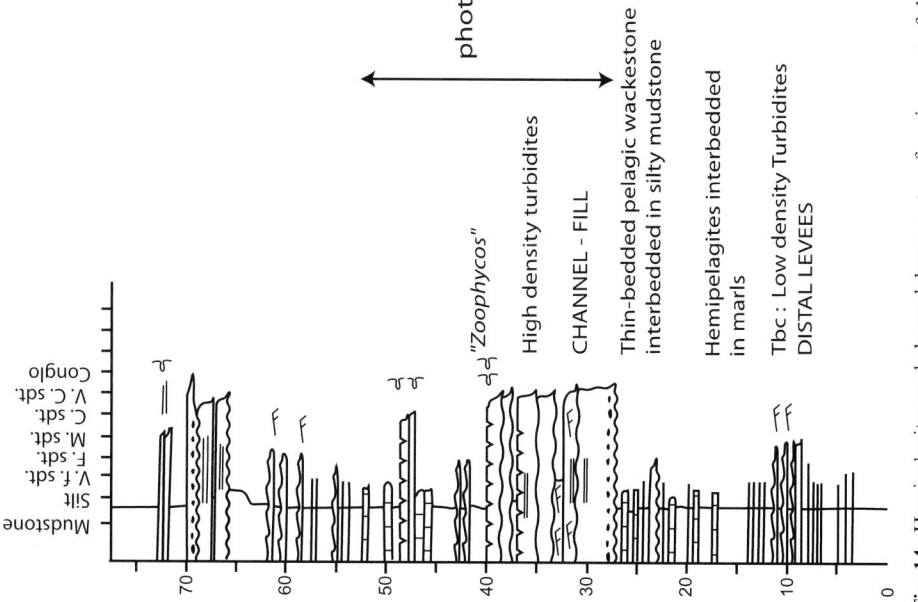

Fig. 14. Hemipelagites and channel–levee system forming part of the mud-rich slope fan at Shah Noorani. The sandstone body is 12 m thick and has an erosive base and a convex upper surface. The detailed stratigraphic section shows the channel fill interbedded with hemipelagites and thin-bedded levees.

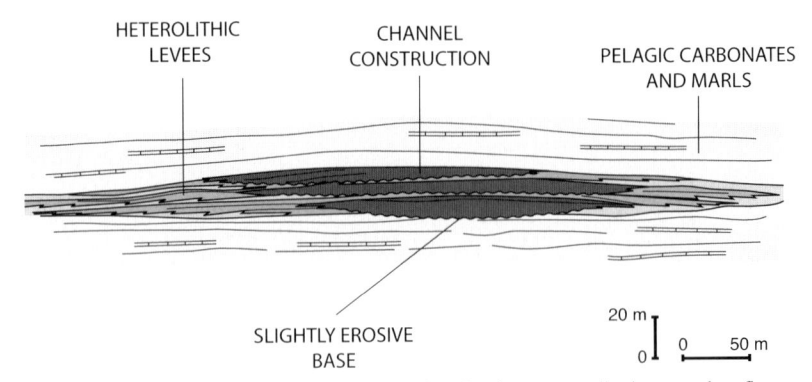

Fig. 15. Sketch of a channel–levee in the mud-rich slope fan. Section perpendicular to palaeoflow.

sandstones with limited incision at the base of the beds; it is often difficult to determine whether these sedimentary bodies are lobes or channels. Channel thickness varies between 2 and 10 m, and channel width is less than 200 m. The top of the channels often shows positive relief attributed to spillover lobes ending the channel fill, probably reinforced by differential compaction (Figs 14 & 15). Thin-bedded fine-grained rippled sandstones (Tbc turbidites) interbedded with siltstones extend laterally to the channel axes. These are interpreted as levees (Fig. 14). The upper surfaces of the beds are highly bioturbated with trace fossils of the *Zoophycos* ichnofacies. Mapping and reconstruction of some of these channels shows that they are of moderate sinuosity.

Polymict debris-flow deposits have been found interbedded with hemipelagic deposits in this slope fan. These debris flows are generally matrix-supported and rework benthic macrofauna from the shelf to the basin.

This unit can be correlated regionally along the central part of the Pab Range. It onlaps the slope southwards and distally passes to hemipelagites northwards. A reddish level in the hemipelagites is found in the upper part of the unit. At this level the hemipelagite is made up exclusively of pelagic foraminifera, indicating that this is a major interval of condensation.

Supply shut-off during channel–levee deposition

The development of this unit at the base of the slope supports its interpretation as a slope fan (Fig. 16). The association between sinuous channels, well-developed levees and hemipelagic background sedimentation suggests that this is a low-energy channel-levee system, feeding terminal lobes of limited lateral extension (Mutti *et al.* 1992).

The mud-rich slope fan marks the reduction in clastic supply as the shelf was drowned during regional transgression. As a result, the basin-floor fan was abandoned, and sediment accumulated principally on the shelf, with small, narrow

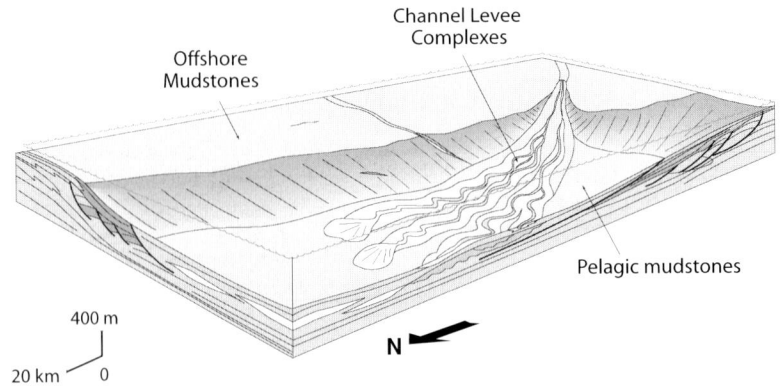

Fig. 16. Reconstruction of the Lower Pab mud-rich slope fan. Small, sinuous channels feed lobes of limited extent.

channel–levee systems at the base of the slope (Figs 7a & 7b). The condensed red level is interpreted as the maximum flooding surface of the third-order Pab sandstone sequence.

Description of the Upper Pab sand-rich turbidite system

The Upper Pab sedimentation shows a significant change in the style of turbidite system. In outcrop, the sheet-sand geometry of the Upper Pab system contrasts with the channelled architecture of the Lower Pab system. In outcrop, the Upper Pab sandstone shows a thickening and coarsening-upward evolution, with distal fan facies passing progressively upwards to proximal fan facies (Figs 17 & 18). The ichnofacies evolves upwards from *Zoophycos/Nereites* to *Skolithos*, suggesting a concomitant shallowing-upward evolution. At a regional scale, the Upper Pab shales out northwards along the Pab range outcrops, interfingering with silty mudstones. The Upper Pab onlaps the slope and pinches out southwards (Figs 3 & 19). In contrast with the Lower Pab, the Upper Pab is greenish in colour and frequently contains reworked basaltic clasts, indicating a significant increase in volcanic material. The rare sole marks found in the Upper Pab turbidites indicate a NW trending flow direction.

Distal slope fan

The lower third of the Upper Pab consists of thin-bedded, normally graded tabular turbidites consisting of medium- to fine-grained sandstone. Plane parallel laminations are frequent, and ripple structures are not very common. The turbidites are interbedded with silty mudstones and occasionally with hemipelagic marls. They are organized in thickening and coarsening-upward bundles around 10–20 m thick. Slumps and oligomict debris flows are frequent. They form poorly organized beds consisting of shaly, unconsolidated, fine- to medium-grained sandstones. Mud clasts are often extremely abundant. The sediments reworked in these deposits are of local origin, and no exotic material transported from the shelf was found in these debris flows. The thin-bedded turbidites are interpreted as distal lobes deposited in a distal slope fan.

Mid-slope fan

The middle part of the Upper Pab consists of both tabular and channelled sand-bodies. The tabular deposits consisted of stacked beds of coarse-grained sandstone intercalated with siltstones. Inverse grading is very frequent at the base of the beds, passing upwards to normal grading. Plane parallel laminations and large-scale

Fig. 17. Lobe stacking in the mid-fan setting of the Upper Pab slope fan at Jakkher Lak. People for scale, lower right-hand corner. See Fig. 18 for stratigraphic location.

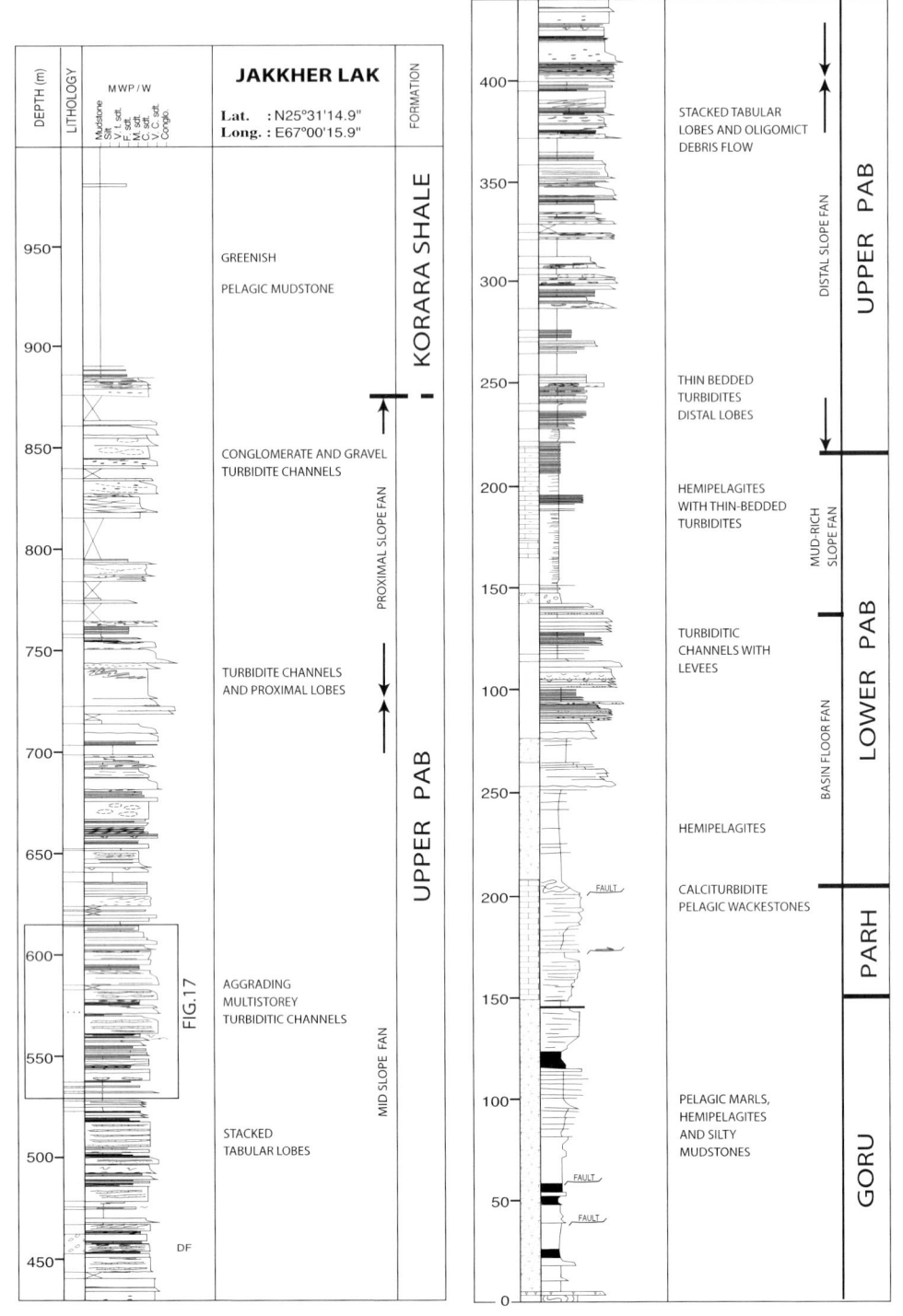

Fig. 18. Stratigraphic section of the Pab Sandstone at Jakkher Lak from the Goru Formation to the Korara shales. See Figure 1 for location. Above the Parh limestones, the Lower Pab has already practically pinched out. The prograding organization of the Upper Pab is clearly expressed by its thickening and coarsening-upward trend, before the complete abandonment of the system, covered by the Korara shales.

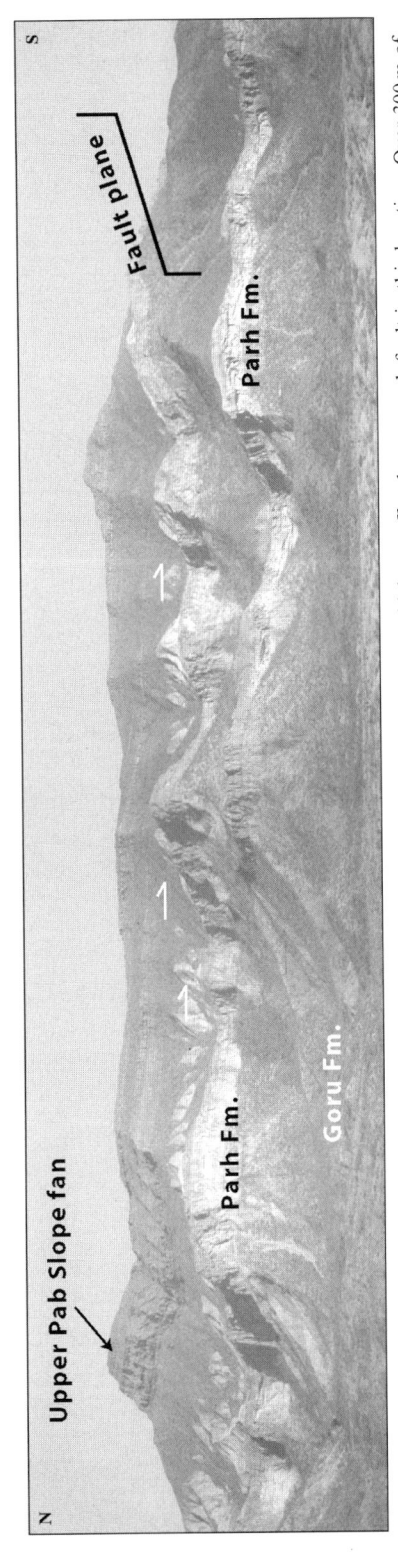

Fig. 19. Onlap of the Upper Pab sand-rich slope fan onto slope carbonates of the Parh Formation, which are offset by a wrench fault in this location. Over 300 m of turbidites pinch out in less than 7 km. Arrows indicate the lateral onlap of the slope fan lobes onto the depositional slope.

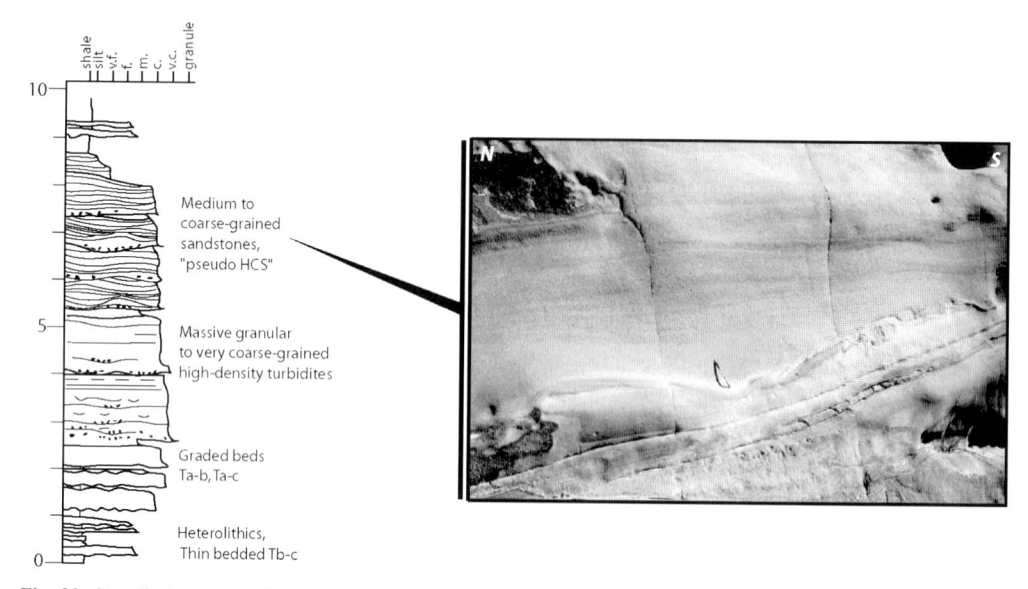

Fig. 20. Detailed stratigraphic section of mid-fan lobes in the Upper Pab slope fan, at Jakkher Lak. The photograph shows long-wavelength undulation of lamina sets in very coarse-grained sandstone interpreted as pseudo hummocky cross-stratification ('pseudo-HCS') in a high-density turbidite deposit. Upper Pab mid-fan lobes.

undulation of parallel laminae are often present in the massive beds. The undulating lamina sets have a wavelength of several metres. These structures resemble hummocky cross-stratification (HCS) and will be referred to as *pseudo-hummocky cross-stratification* (Fig. 20). The main difference from true HCS lies in the coarse grain-size of these structures, in the poor sorting of the sandstone material, and in the absence of associated oscillation ripples. These tabular sand-bodies are interpreted as proximal lobes.

Channelled bodies occasionally cut the uppermost surface of the tabular deposits. They consist of normally graded medium to very coarse-grained structureless sandstones interpreted as being the product of high-density turbidites. Individual beds have erosional bases in the channel axis and pinch out laterally, interfingering with silty shales. Dozens of individual beds 1–3 m thick can be vertically stacked in the channel fill. The channels are narrow (200 m wide) and typically show little basal erosion, with a strongly aggradational stacking pattern.

The stacked lobes and aggrading channels are organized in prograding sand-rich packages 40 m thick separated by distal lobe deposits and silty shales (Fig. 18). The lobes extended over several kilometres laterally, and compensation cycles were frequently observed.

This association between lobes and aggrading channels suggests a mid-fan setting.

Proximal slope fan

Channels are frequent in the upper third of the Upper Pab. The channel fills generally consist of very coarse-grained to conglomeratic polymict material. The channels are around 20 m thick and 300–500 m wide, and are made up of two or three amalgamated single channels, with relatively low incision at their base. The channels do not show any levee development. The channels erode structureless tabular sand-bodies interpreted as proximal lobes. Slump scars and erosional surfaces are often found in the proximal and mid-fan settings. These erosional surfaces are draped or onlapped by thin-bedded turbidites and shales.

The conglomeratic channels are interpreted as slope feeder channels in a proximal fan setting. The close association between the channels and the proximal lobes suggests that the lobes were attached to the channels without an intermediate bypass zone.

General architecture of the slope fan depositional system

The slope fan shows a low degree of organization compared with the basin-floor fan. Channel complexes such as the ones observed in the Lower Pab are absent even in the proximal part of the

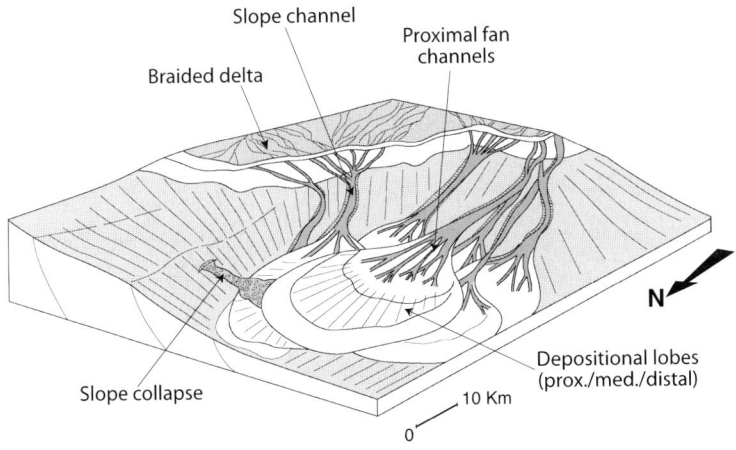

Fig. 21. Reconstruction of the Upper Pab slope fan depositional system. The Pab braided delta directly feeds the slope fan channels.

system. In the Upper Pab, lobes and channels are closely associated in space. Incised conglomeratic slope channels in the proximal part of the system branch to narrower, aggrading channels in the mid-fan setting, directly feeding the lobes, which interfinger distally with shales (Fig. 21). The lack of sole marks and of typical turbidite facies (Bouma 1962) associated with truly turbulent density flows suggests a different mechanism of transport. The pseudo-HCS facies is similar to the hummocky cross-stratification in flood-generated shelfal lobes described by Mutti *et al.* (1996, 1999, in press). These authors consider that hyperpycnal flows generated by rivers in flood can directly transform into long-lived turbidity currents through a phase of extensive bed erosion and bulking. This suggests that the Upper Pab turbidite system was probably connected to the Pab delta, in a ramp setting.

A prograding sand-rich slope fan

The thickening, coarsening and shallowing-upward evolution of the Upper Pab turbidite system is linked to the progradation of the slope fan from the distal fan lobes to proximal fan conglomeratic channels (Fig. 22). Higher-frequency cycles can be observed in the long-term progradational trend of the Upper Pab. In the mid-fan setting, the Upper Pab turbidites are organized in thickening and coarsening-upward units 30–40 m thick. These are interpreted as the progradation of lobe complexes, and often end with the development of aggrading channels. They are separated by shale-rich units with thin-bedded turbidites, and occasionally hemipelagic carbonate beds. These shale-rich units are 10–20 m thick, and can be correlated laterally over tens of kilometres. The shales and hemipelagic deposits correspond to periods during which the sediment supply decreased and the slope fan was temporally abandoned.

This sequential organization could be due either to major avulsion of the slope fan feeder channels, or to fourth-order sequences related to regional sediment supply variations. On the shelf, high-frequency transgressive events have been observed during the overall delta progradation (Fig. 4). These transgressive events were marked by the deposition of lagoonal, carbonate facies interbedded within the clastic delta sediments. This reduction of the sediment supply during transgression could have induced the abandonment of the slope fan during limited periods of time.

Sequence stratigraphic analysis

An interpretation in terms of sequences, depositional environments and systems tracts is proposed in Figures 7a and 7b. The Pab Formation is considered to have been deposited as a single third-order depositional sequence *sensu* Vail *et al.* (1977). According to the Haq *et al.* cycle chart (1988), the Late Maastrichtian lasted between two and three million years, a typical period for third-order depositional sequences.

The lower sequence boundary was due to a relative sea-level fall, which induced the emersion of the Fort Munro carbonate platform and the

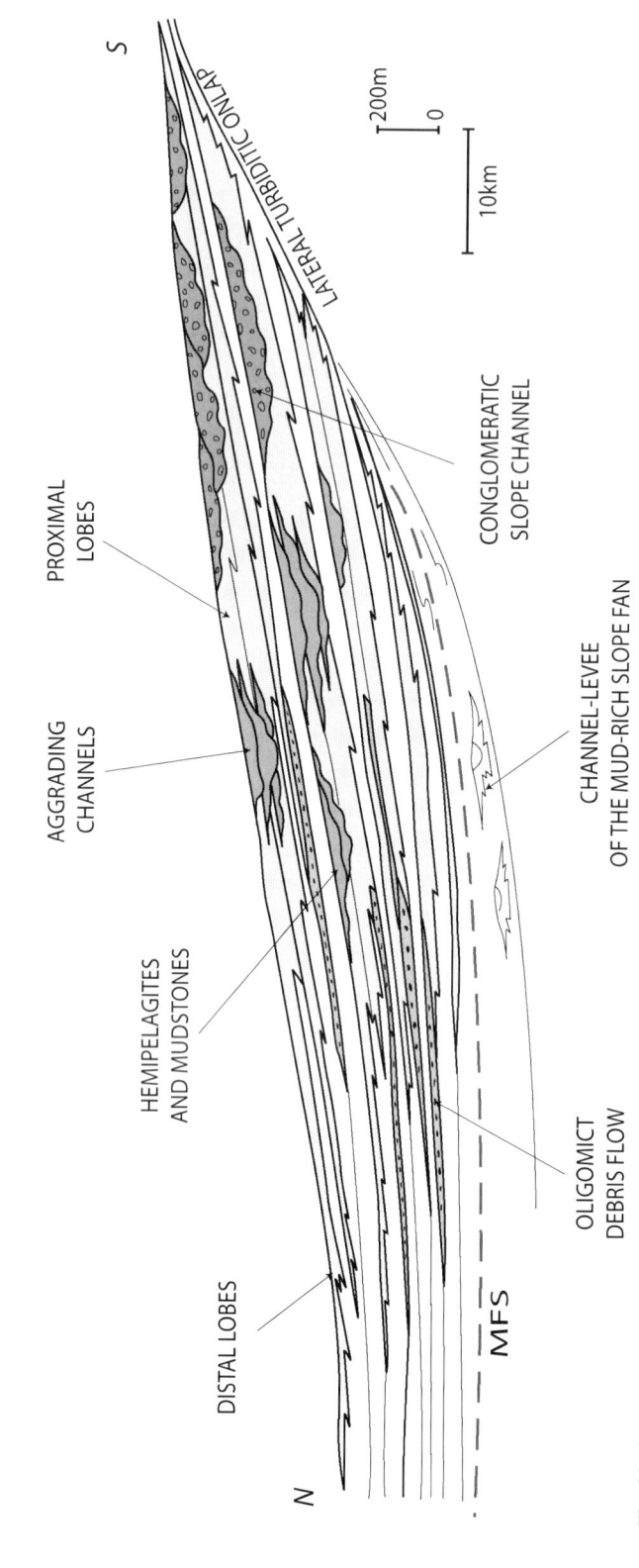

Fig. 22. Sequential organization and facies distribution along the Pab range of the Upper Pab slope fan. The flow direction is northwest. See text for comments.

development of an erosional unconformity (Fig. 4). Clastic material transiting through incised valleys abruptly arrived in the deep basin, feeding the Lower Pab sand-rich basin-floor fan and forming a lowstand systems tract (LST). No evidence of an incised valley has been observed in the Laki range outcrops, but Khan et al. (2002, Fig. 3a) p. 388, also describe a major erosional surface cutting into the Parh limestones in the Khuzdar area, showing that the erosional unconformity underneath the Pab sandstones is of regional extension.

The transgressive system tract (TST) corresponds to deposition of the mud-rich slope fan above the basin-floor fan, while a transgression drowned the shelf. On the shelf, backstepping carbonate shoreface deposits were deposited (Fig. 4). In the basin, thick debris flow deposits reworking benthic fauna are also found at the base of the TST. These probably resulted from the destabilization of the slope and shelf during the transgression because of the increasing load of the water column. The maximum flooding surface (MFS) corresponds to offshore deposits on the shelf and to a reddish pelagic condensed horizon in the basin.

The highstand system tract (HST) corresponds to the progradation of the Pab delta on the shelf and to the Upper Pab prograding sand-rich slope fan in the basin. The Pab delta probably prograded out across the shelf break in the narrowest parts of the shelf, directly supplying the deep basin with sediments. This marks a significant departure from the classical sequence stratigraphic models according to which only thin-bedded low-density turbidites and pelagic mudstones are deposited in the basin during the HST (Vail et al. 1977).

The turning point between the progradation of the slope fan and the subsequent backstep during Paleocene times corresponds to the third-order sequence boundary of the Pab system. The conglomeratic turbidite channels found at the very top of the Upper Pab slope fan could have been deposited at the time of maximum progradation of the delta. Alternatively, the basal incision of these conglomeratic channels may correspond to a regional erosional unconformity linked to relative sea-level fall at the third order sequence boundary. However, in the shelf setting, it has not proved possible to identify any major erosional unconformity at this level, except in the form of a mature calcrete level (Fig. 4).

The next third-order sequence started with the backstep of the Pab delta while thin-bedded turbidites were deposited in the basin, interbedded in the pelagic deposits of the Korara Shale Formation. On the shelf, the MFS was marked by a condensed oyster-rich hardground associated with open bay sediments as the shelf was drowned. The prograding delta of the Ranikot Formation constituted the HST of this third-order sequence.

Discussion

The basin floor and slope fans of the Pab turbidite sandstones present contrasting architectural styles, which are linked to the dynamic evolution of the Indo-Pakistani margin during the Late Maastrichtian and Early Paleocene.

Well-defined shelf break to ramp evolution

The evolution from the basin-floor fan to the slope fan was accompanied by a major change in the slope morphology. During deposition of the Lower Pab basin-floor fan, a well-marked shelf break separated an emerged shelf from a pelagic basin. The shelf break was probably controlled by normal faults, and regional correlation suggests that a topographic relief of at least 1000 m existed at that time between the shelf and the basin. The steep slope increased the velocity and hence the transport distance of the gravity flows. The change in angle between the slope and the basin floor induced a hydraulic jump, favouring flow transformation from a high-density, more laminar gravity flow to lower-density, turbulent flow. Downstream of this point, the extent and volume of levee deposits increased dramatically.

Draping of the margin started during the deposition of the Lower Pab mud-rich slope fan. The slope angle was smoothed, and a ramp morphology was initiated. The ramp developed fully during the Upper Pab slope fan, as the whole margin prograded because of the increase in the rate of sediment supply in relation with the uplift of the margin. The gravity flows travelled through shorter distances on this low-angle ramp, accumulating at the base of the slope. The increase in sediment supply and high pore pressures in the rapidly deposited sedimentary pile favoured slumping on the slope, the slumps transforming into oligomict debris flows.

Single canyon to multiple slope channel feeders

In the Lower Pab basin-floor fan the relative sea-level fall was accompanied by the creation of incised valleys and the incision of a canyon in the slope setting (Eschard et al. in press). The basin-floor fan extends for over 100 km from the mouth of the canyon. The confinement of the gravity flows within the canyon explains the high velocity of the flows and the deep erosional surfaces at the base of the channel complexes. As the density flows erode the substrate, they incorporate fine-grained material that increases flow efficiency and hence transport distance (Mutti et al. 1992; Gladstone et al. 1998).

The Upper Pab slope system shows a lack of confinement. Multiple-slope channel feeders were found in the proximal part of the system instead of erosional canyons. These slope channels directly fed the lobes in the slope apron without developing major intermediate channel complexes.

Shelf to basin sediment transfer

The mass balance of sediment transport between the shelf domain and the deep basin is probably one of the main factors controlling fan architecture. The Lower Pab basin-floor fan was deposited during a period of margin uplift, during which sediments bypassed the shelf domain to accumulate in the basin. A channel–levee slope fan was then deposited during the subsequent transgression, and only small amounts of clastic sediment arrived in the basin as most was trapped on the shelf, in the incised valleys (Posamentier & Vail 1988) and in the shoreface sands. The Upper Pab slope fan was thick and sand-rich because the sediments were fed directly by the delta progradation across the shelf margin to the slope apron. The width of the shelf was also a critical parameter for the volume of sand supplied to the slope fan. In the north of the studied area, where the shelf was wider, deltas did not reach the shelf margin and only muddy ramp deposits accumulated in the basin setting, with no slope fan development.

Transport capacity of the gravity flows

According to Mutti (1992), the efficiency of a density flow depends on both its mud content and the volume of the flow. The steady or catastrophic nature of the flow, the slope angle, the confinement of the flow and the amount of fine sediment eroded by the flow and incorporated into it during transport will also influence the transport capacity (Kneller & Buckee 2000; Mutti et al. in press).

In the Lower Pab basin-floor fan, we observe a full spectrum of deposits from gravelly sandstones deposited by superconcentrated flows to typical low-density Bouma-type turbidites (Bouma 1962). The confinement induced by the canyon funnels the flows, increasing their erosional capacity, and favouring the incorporation of fine-grained material. As said above, the hydraulic jump at the canyon mouth favours an increase in flow turbulence. As a consequence, levees and overflow deposits develop laterally to the channel axes. Given the thickness of individual high-density turbidite deposits in the Lower Pab, the size of an individual flow must have been greater than in the Upper Pab, characterized by thinner-bedded deposits.

The mud-rich slope fan is characterized by much smaller channels. This can be attributed to the trapping of sediment on the shelf during regional transgression, resulting in fewer and smaller density flows.

The Upper Pab slope fan shows a very different type of density flow deposit. The upstream-to-downstream sediment sorting is less efficient. The flows evolved on a gentler slope, and flow confinement was very reduced. Erosion is minimal. A direct hydraulic connection probably existed between the delta and the slope fan, with an increase of hyperpycnal turbidity currents in the slope fan (Mulder & Alexander 2000; Mulder et al. in press).

Sequential evolution of basin-floor and slope fans

The Lower and Upper Pab turbidite systems also show a very different sequential evolution. The Lower Pab basin-floor fan typically shows a backstepping evolution ending in the complete abandonment of the system. The vertical evolution of facies partitioning between lobes and channels is due to a variation in the ratio between bypass and accumulation processes. In the lowest sequence of the Lower Pab (LP1), lobes accumulated in the distal setting several tens of kilometres from the canyon mouth, and their proximal equivalent was probably a bypass surface. Channel fill starts to be preserved in the LP2 unit, but bypass processes were still dominant and lobe construction was possible. Finally, in the LP3 unit, most of the sediment accumulated in channel complexes and their overflow deposits, and only a small amount of sediment reached the terminal lobes, which were poorly developed. In contrast, the Upper Pab slope fan shows an overall progradational trend from the distal to the proximal fan. The progradation was interrupted by fourth-order cycles or major avulsion periods, which temporarily induced the abandonment of the fan.

Conclusions and perspectives

The Lower Pab basin-floor fan presents many analogies with modern and ancient fan systems deposited in passive margin settings associated with a wide oceanic domain (Mayall & Stewart 2000; Vittori et al. 2000; Kolla et al. 2001), even though the Lower Pab Sandstone is more sand-rich than the systems described by the aforementioned authors. The principal architectural element is the channel complex and its associated levees, which is itself composed of dozens of amalgamated individual channels (Eschard et al. in press). The connection with the lobes remains relatively unknown, although there is no evidence of channel–lobe detachment. The lobes are sand-rich and are certainly of considerable regional extension, although their distal pinch-out has not been observed.

The mud-rich slope fan is characterized by channel–levee complexes and their associated lobes deposited during a time of very low sediment supply to the basin. The abundance of hemipelagic deposits and bioturbation (*Zoophycos* ichnofacies) in this interval suggests considerable condensation. The fan is overlain by a maximum flooding surface marked by a reddish foram-rich condensed level.

The Upper Pab slope fan accumulated at the base of the slope and is of more limited extent than the Lower Pab basin-floor fan. Regional correlation suggests that this turbidite system is the time equivalent of a prograding delta on the shelf. However, the direct connection between the delta and the slope fan does not crop out, and the possibility of a bypass zone between the two cannot be excluded. The hydraulic connection between the delta and the turbidite system conditions the dynamics of the gravity flows, and hyperpycnal-type deposits dominate this system. The development of a sand-rich slope fan in the highstand systems tract constitutes a departure from the classical third-order depositional sequence model, according to which only thin-bedded low-density turbidites can be deposited during delta progradation.

The Upper Pab slope fan is very comparable to many of the turbidite systems found in confined basins (Joseph et al. 2000; Mutti et al. in press). The high fluvial discharge and narrow shelf are the two factors determining the dynamics of this slope fan.

The results presented here are part of the Pab consortium sponsored by BP, Eni-Lasmo, Petrobras, Premier Oil Pakistan, Shell and TotalFinaElf, and we thank all the company representatives involved for the very useful discussions we had with them during workshops and field excursions. Logistics in the field were provided by Premier-Pakistan and Eni-Lasmo. We also thank M. C. Cacas, J. M. Daniel, R. Deschamps, T. Euzen, D. Granjeon, C. Ravenne and J. L. Rudkiewicz of the IFP for helping us during a number of field sessions. Nannofossil datings were performed by C. Müller. We benefited from H. White's 1981 Masters dissertation on the petrography of the Pab system. Professor Haneef of Peshawar University also provided invaluable assistance in the field. The authors benefited from reviews by P. Haughton and J. Janocko, whose suggestions greatly influenced the final form of this paper.

References

ABREU, V., SULLIVAN, M. ET AL. in press. Architectural analysis of sinuous channels: the underappreciated deepwater channel type. *Marine and Petroleum Geology*.

ALLEMANN, F. 1979. Time of emplacement of the Zhob valley ophiolites and Bela ophiolites. *In*: FARAH, A. AND DEJONG, K. A. (eds) *Geodynamics of Pakistan*. Geological Survey of Pakistan, Quetta, 243–250.

BANNERT, D. & RAZA, H. A. 1992. The segmentation of the Indo-Pakistan Plate. *Pakistan Journal of Hydrocarbon Research*, **4**(2), 5–18.

BOUMA, A. H. 1962. *Sedimentology of some Flysch Deposits: A Graphic Approach to Facies Interpretation*. Elsevier, Amsterdam.

BUTT, A. 1992. The Upper Cretaceous biostratigraphy of Pakistan: a synthesis. *Géologie Méditerranéenne*, **19**(4), 265–272.

COURTILLOT, V., GALLET, Y. ET AL. 2000. Cosmic markers, (super 40) Ar/ (super 39) Ar dating and paleomagnetism of the KT sections in the Anjar area of the Deccan large igneous province. *Earth and Planetary Science Letters*, **182**(2), 137–156.

ESCHARD, R. 2001. Geological factors controlling sediment transport from platform to deep basin: a review. *Marine and Petroleum Geology*, **18**(4), 487–490.

ESCHARD, R., ALBOUY, E., DESCHAMPS, R., EUZEN, T. & AYUB, A. in press. Downstream evolution of turbiditic channel complexes in the Pab Range outcrops (Maastrichtian, Pakistan). *Marine and Petroleum Geology*.

GLADSTONE, C., PHILLIPS, J. C. & SPARKS, R. S. J. 1998 Experiments on bidisperse, constant-volume gravity currents: propagation and sediment deposition. *Sedimentology*, **45**, 833–843.

GNOS, E., IMMENHAUSER, A. & PETERS, T. J. 1997. Late Cretaceous/Early Tertiary convergence between the Indian and Arabian plates recorded in Ophiolites and related sediments. *Tectonophysics*, **271**, 1–19.

HAQ, B., HARDENBOL, J. & VAIL, P. R. 1988. Mesozoic and Cenozoic chronostratigraphy and cycles of sea-level change. *In*: WILGUS, C. K. ET AL. (eds) *Sea Level Changes: An Integrated Approach*, SEPM Special Publications, **42**, 71–108.

HEDLEY, R., WARBURTON, J. & SMEWING, J. 2001. Sequence stratigraphy and tectonics in the Kirthar Foldbelt, Pakistan. In: Proceedings of the SPE-PAPG Annual Technical Conference, Islamabad 2001, 61–72.

HUNTING PHOTOGRAPHIC SURVEY CORPORATION LIMITED. 1958. Reconnaissance Geology of Part of West Pakistan. Toronto, Government of Canada, Geological maps no. 6 & no. 11.

JOSEPH, P., BABONNEAU, N. ET AL. 2000. The Annot Sandstone outcrops (French Alps): architecture description as input for quantification and 3D reservoir modeling, In: WEIMER, P., SLATT, R. M. ET AL. (eds) Deep-Water Reservoirs of the World. Gulf Coast Section Society of Economic Palaeontologists and Mineralogists Foundation 20th Annual Research Conference, Houston, TX, 422–449.

KADRI, I. B. 1995. Petroleum Geology of Pakistan, Pakistan Petroleum Limited, Ferozesons (Pvt) Ltd., Lahore.

KHAN, A. S., KELLING, G., UMAR, M. & KASSI, A. M. 2002. Depositional environments and reservoir assessment of Late Cretaceous sandstones in the South Central Kirthar Foldbelt, Pakistan. Journal of Petroleum Geology, 25(4), 373–406.

KNELLER, B. & BUCKEE, C. 2000. The structure and fluid mechanics of turbidity currents: a review of some recent studies and their geological implications. In: BEST, J. L. ET AL. (eds) Millenium Reviews. Sedimentology, 47, Suppl. 1, 62–94.

KOLLA, V., BOURGES, P., URRUTY, J. M. & SAFA, P. 2001. Evolution of deep-water Tertiary sinuous channels offshore Angola (West Africa) and implications for reservoir architecture. American Association of Petroleum Geologists Bulletin, 85(8), 1373–1405.

MAYALL, M. & STEWART, I. 2000. The architecture of turbidite slope channels. In: WEIMER, P., SLATT, R. M. ET AL. (eds) Deep-Water Reservoirs of the World. Gulf Coast Section Society of Economic Palaeontologists and Mineralogists Foundation 20th Annual Research Conference, Houston, TX, 578–586.

MULDER, T. & ALEXANDER, J. 2001. The physical character of subaqueous sedimentary density currents and their deposits. Sedimentology, 48, 269–299.

MULDER, T., SYVITSKI, J. P. M., MIGEON, S., ALEXANDER, J., FAUGÈRES, J. C. & SAVOYE, B. in press. Marine hyperpycnal flows: initiation, behavior and related deposits. A review. Marine and Petroleum Geology.

MUTTI, E., DAVOLI, G. ET AL. 1992. Turbidite Sandstones. AGIP SpA, S. Donato Milanese, Istituto di Geology Università di Parma.

MUTTI, E., DAVOLI, G., TINTERRI, R. & ZAVALA, R. 1996. The importance of ancient fluvio-deltaic systems dominated by catastrophic flooding in tectonically active basins. Memorie di Scienze Geologiche, 48, 233–291.

MUTTI, E., TINTERRI, R., REMACHA, E., MAVILLA, N., ANGELLA, S. & FAVA, L. 1999. An Introduction to the Analysis of Ancient Turbidite Basins from an Outcrop Perspective. AAPG Continuing Education Course Note Series, 39.

MUTTI, E., TINTERRI, R., BENEVELLI, G., ANGELLA, S., DI BIASE, D., CAVANNA, G. & COTTI, A. in press. Deltaic, mixed and turbidite sedimentation of ancient foreland basins. Marine and Petroleum Geology.

NOMURA, R. & BROHI, I. A. 1995. Benthic foraminiferal fauna during the time of the Indian-Asian contact, in southern Balochistan, Pakistan. Marine Micropalaeontology, 24, 215–238.

POSAMENTIER, H. W. & VAIL, P. R. 1988. Eustatic controls on clastic deposition. II, Sequence and systems tract models. In: WILGUS, C. K. ET AL. (eds) Sea Level Changes. An Integrated Approach. Society of Economic Paleontologists and Mineralogists, Special Publications, 42, 125–154.

PRATHER, B. E., BOOTH, J. R., STEFFENS, G. S. & CRAIG, P. A. 1998. Classification, lithologic calibration, and stratigraphic succession of seismic facies of intraslope basins, deep-water Gulf of Mexico. American Association of Petroleum Geologists Bulletin, 82(5a), 701–728.

READING, H. G. & RICHARDS, M. 1994. Turbidite systems in deep-water basin margins classified by grain size and feeder system. American Association of Petroleum Geologists Bulletin, 78(5), 792–822.

SCHELLING, D. 1999. Frontal structural geometries and detachment tectonics of the northeastern Karachi arc, southern Kirthar range, Pakistan. Geological Society of America Special Papers, 328, 287–302.

SCOTESE, C. R., CAHAGAN, L. M. & LARSON, R. L. 1988. Plate tectonic reconstructions of the Cretaceous and Cenozoic basins. Tectonophysics, 155, 27–48.

SEARLE, M. P., WINDLEY, B. F. ET AL. 1987. The closing of Tethys and the tectonics of the Himalayas. Geological Society of America Bulletin, 98, 678–701.

SHAH, I. S. M. 1977. Stratigraphy of Pakistan. Memoirs of the Geological Survey of Pakistan, 12.

STEEL, R. J., CRABAUGH, J., SCHELLPEPER, M., MELLERE, D., PLINK-BJORKLUND, P., DEIBERT, J. & LOESETH, T. 2000. Deltas versus rivers on the shelf edge: their relative contributions to the growth of shelf-margins and basin-floor fans (Barremian and Eocene, Spitsbergen). In: WEIMER, P., SLATT, R. M. ET AL. (eds) Deep-Water Reservoirs of the World. Gulf Coast Section Society of Economic Palaeontologists and Mineralogists Foundation 20th Annual Research Conference, Houston, TX, 981–1009.

TAPPONNIER, P., MATTAUER, M., PROUST, F. & CASSAIGNEAU, C. 1981. Mesozoic ophiolites, sutures, and large-scale tectonic movements in Afghanistan. Earth and Planetary Science Letters, 52, 355–371.

VAIL, P. R., MITCHUM, R. M. & THOMPSON, S. 1977. Seismic stratigraphy and global changes of sea level. Part 3: Relative changes of sea level from coastal onlap. In: PAYTON, C. E. (ed.) Seismic Stratigraphy: Applications to Hydrocarbon Exploration. AAPG Memoirs, 26, 63–81.

VAN WAGONER, J. C., POSAMENTIER, H. W., MITCHUM, R. M., VAIL, P. R., SARG, J. F., LOUTIT, T. S. & HARDENBOL, J. 1988. An overview of the fundamentals of sequence stratigraphy and key definitions. In: WILGUS, C. K. ET AL. (eds) Sea-Level Changes: An Integrated Approach. SEPM Special Publications, 42, 39–45.

VITTORI, J., MORASH, A., SAVOYE, B., MARSSET, T., LOPEZ, M., DROZ, L. & CREMER, M. 2000. The Quaternary Congo deep-sea fan: preliminary results on reservoir complexity in turbiditic systems using 2D high resolution seismic and multibeam data. *In*: WEIMER, P., SLATT, R. M. *ET AL.* (eds) *Deep-Water Reservoirs of the World. Gulf Coast Section Society of Economic Palaeontologists and Mineralogists Foundation 20th Annual Research Conference, Houston, TX*, 1045–1058.

WAHEED, A. & WELLS, N. A. 1990. Changes in paleocurrents during the development of an obliquely convergent plate boundary (Sulaiman fold-belt, southwestern Himalayas, west-central Pakistan). *Sedimentary Geology*, **67**, 237–261.

Basin-floor fans of the Central Tertiary Basin, Spitsbergen: relationship of basin-floor sand-bodies to prograding clinoforms in a structurally active basin

JEFF P. CRABAUGH & RONALD J. STEEL

Dept. of Geology and Geophysics, University of Wyoming, Laramie, Wyoming 82071-3306, USA (e-mail: jpatrick@uwyo.edu)

Abstract: Lower Eocene shelf-slope clinoforms are exposed in 1×10 km mountainside outcrops in the Central Tertiary Basin, Spitsbergen. Where clinoforms are sand-prone they include a deepwater sand complex. Submarine fans represent an early, basin-floor aggradational phase of clinoform growth, whereas later growth of the same clinoform involves a phase of shelf-margin accretion. Individual fans, within stacked series, can be distinguished when traced towards the slope, where a thickening wedge of mudstones separates successive fan bodies. The sand-prone parts of basin-floor fans are some 15–60 m thick and extend into the basin by up to 10–12 km. The lower levels of any fan consist of ripple- to parallel-laminated thin-bedded turbidites interbedded with some thick-bedded turbidites. This association changes irregularly upwards to a succession dominated by thick beds that are structureless and parallel-laminated. The thin-bedded facies are interpreted as turbidite sheets that formed as channel-mouth sandy lobes, sandy levees and crevasse splays. The erosively based, thick-bedded facies are interpreted as constructional channel-fill sandstones. The shallow channels fed sheet-complexes both laterally and distally. The apparent short basinward extent and longitudinal palaeocurrents for the youngest fans suggest that downslope sediment transport became longitudinally deflected by anticlinal topography once sediment reached the basin floor.

Outcrops located along the northern side of Van Keulenfjorden, in Western Spitsbergen, reveal a Lower Eocene accreting shelf margin, the progradation of which appears to be time-equivalent with overthrusting in the West Spitsbergen Orogenic Belt, some 30 km to the west (Fig. 1). The component elements of this accreting shelf margin are a series of spectacular, large-scale clinoforms (Fig. 2) that can be walked out almost continuously along mountainsides (Palfjellet, Brogniartfjellet, Storvola, Hyrnestabben) on the northern shores of Van Keulenfjorden (Kellog 1975; Steel *et al.* 1985; Helland-Hansen 1992). Along the length of individual clinoforms (terminology modified from Rich 1951), a clear distinction can be made between a sub-horizontal sandy shelf segment, an eastward-dipping, muddy slope segment, and a sub-horizontal basin-floor segment (Fig. 2B). Some basin-floor segments have thick, sandy turbidite fans.

This study aims to:

(1) show how basin-floor fans relate upslope to their host clinoforms;
(2) describe and interpret the sand-prone parts of several basin-floor fans exposed on the mountainsides of Storvola and Hyrnestabben (Figs 2A & 2B); and
(3) show how the configuration of the youngest fans suggests some basin-floor confinement and development of longitudinal fan growth, probably caused by synsedimentary thrusting and the creation of basin-floor topography in front of the accreting shelf margin.

Tectonic setting of the basin

The Eocene study succession developed within a large north-to-south-trending foreland basin referred to as the Central Tertiary Basin of Spitsbergen (Kellogg 1975; Steel *et al.* 1981, 1985; Steel & Worsley 1984; Helland-Hansen 1990; Fig. 1). This basin formed during the late Paleocene and early Eocene, contemporaneous with the development of the Western Spitsbergen Orogenic Belt, along the western coast of Spitsbergen (Harland 1969; Lowell 1972; Craddock *et al.* 1985). The larger-scale setting was one of transpressional plate movement as the Eurasian and Greenland plates slid past each other, and as new oceanic crust was created in the Norwegian-Greenland Sea (Myhre *et al.* 1982; Eldholm *et al.* 1984; Spencer *et al.* 1984).

During the early Eocene, the Central Basin filled west to east with marine and non-marine clastic sediment. Basin infilling appears to have been asymmetric, enacted by a series of more than 20, eastward-sloping clinoforms with

From: LOMAS, S. A. & JOSEPH, P. (eds) 2004. *Confined Turbidite Systems.* Geological Society, London, Special Publications, **222**, 187–208. 0305-8719/04/$15.00 © The Geological Society of London 2004.

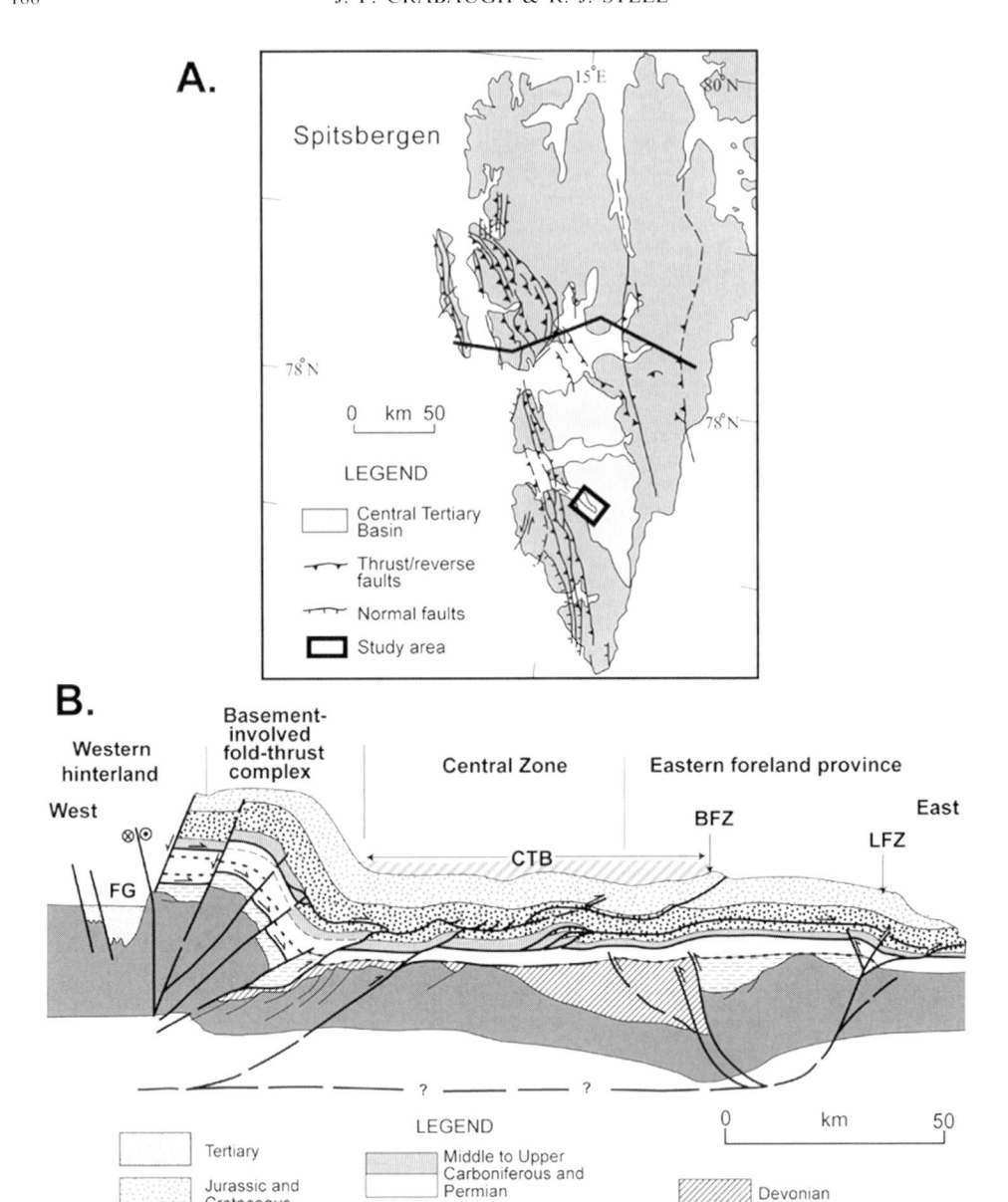

Fig. 1. (**A**) Central Tertiary Basin and Spitsbergen Orogenic Belt. Box marks the study area in Van Keulenfjorden. Bold black line marks position of cross-section shown in Fig. 1B. Modified from Blythe and Kleinspehn (1998). (**B**) Cross-section showing the different zones of the Spitsbergen Orogenic belt. The projected position of the Central Tertiary Basin (CTB) within the cross-section line is shown by a cross-hatched pattern. Modified from Braathen *et al.* (1999).

amplitudes of up to 300 m that record the eastward migration of the depocentre (Helland-Hansen 1992; Nyberg *et al.* 1995; Steel and Olsen 2002). The clinoforms, reflecting a basinward accretion of the Eocene shelf margin, highlight timelines through the coastal-plain, marine-shelf, slope, and basin-floor stratigraphy. Of particular interest, and the focus of the present study, are the occasional clinoforms that delivered significant volumes of sand beyond the

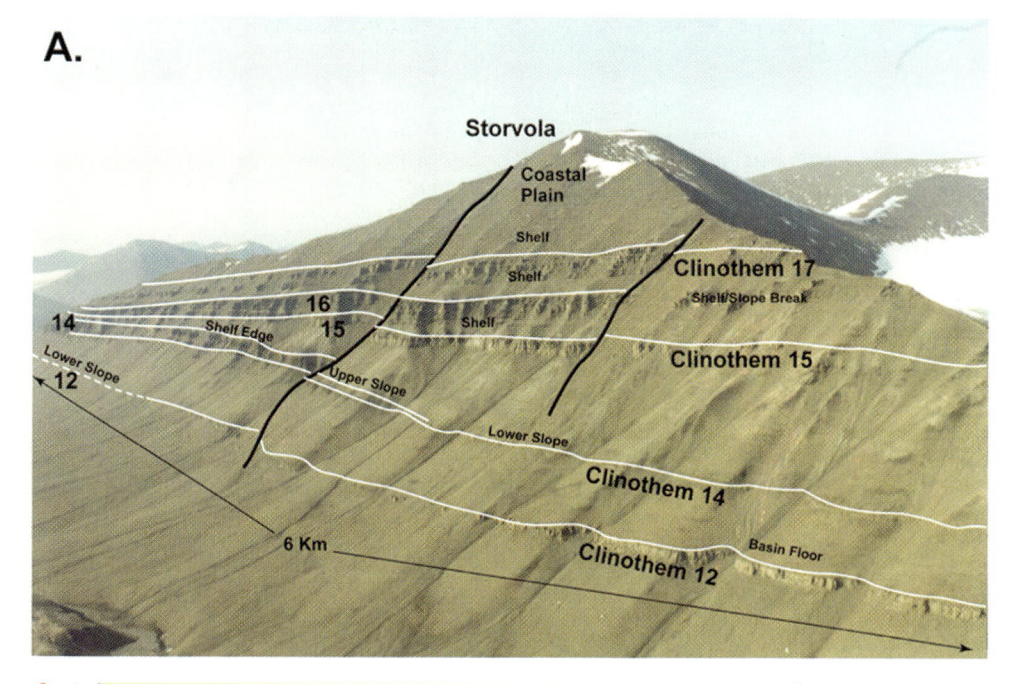

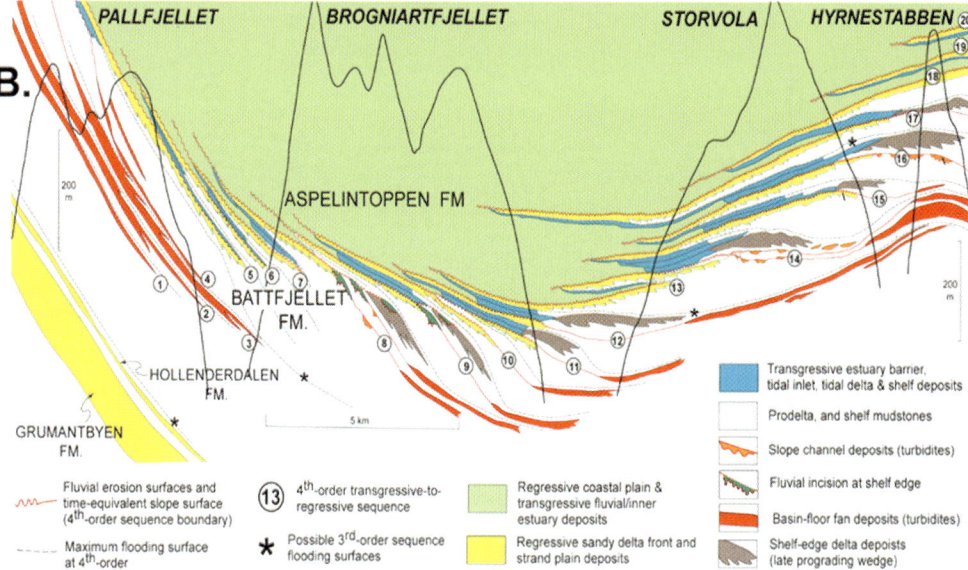

Fig. 2. (A) Southwest face of the mountain Storvola (*ca.* 1 km high), showing parts of Clinoforms 12, 14 and 15. Clinoform 14 shows a flat-lying, shelf segment (left end of mountain), and a slope segment (sloping to right at 3–4°). Clinoform 12 displays a flat-lying basin-floor segment on Storvola that rises slopewards off to the left of photo. Clinoform 15 shows mainly a flat-lying shelf segment on Storvola, but begins to roll over into a slope at right-hand end. Context of clinoforms is shown in Figure 2B. (B) Transect along Van Keulenfjorden showing the sub-regional development of the entire series of Lower Eocene clinoforms. Note the overall stratigraphic 'rise' of the clinoforms through time. Clinoforms 12 and 14 are dealt with in this work. The profiles of the mountains Palfjellet, Brogniartfjellet, Storvola and Hyrnestabben are outlined in thin black line. The stratigraphy between these peaks is projected based on exposed strata on these four mountains (from Steel & Olsen 2002).

shelf edge, i.e. down onto the deepwater slope and basin-floor areas (Steel et al. 2000; Plink-Bjorklund et al. 2001; Mellere et al. 2002).

Stratigraphic setting and age of study interval

The stratigraphy of the Central Tertiary foreland basin on Spitsbergen is outlined in Figure 3, though only the three uppermost stratigraphic units, the Aspelintoppen and Battfjellet Formations, and the Gilsonryggen Member, are dealt with here. These Lower Eocene stratigraphic units can be seen clearly in Figures 2A and 2B, where the lower third of the mountainside is shale prone (deepwater deposits of Gilsonryggen Member), the middle reaches are sand prone (shore zone and shelf deposits of Battfjellet Formation), and the upper third is mixed sandstones and shales (coastal-plain and estuarine deposits of Aspelintoppen Formation).

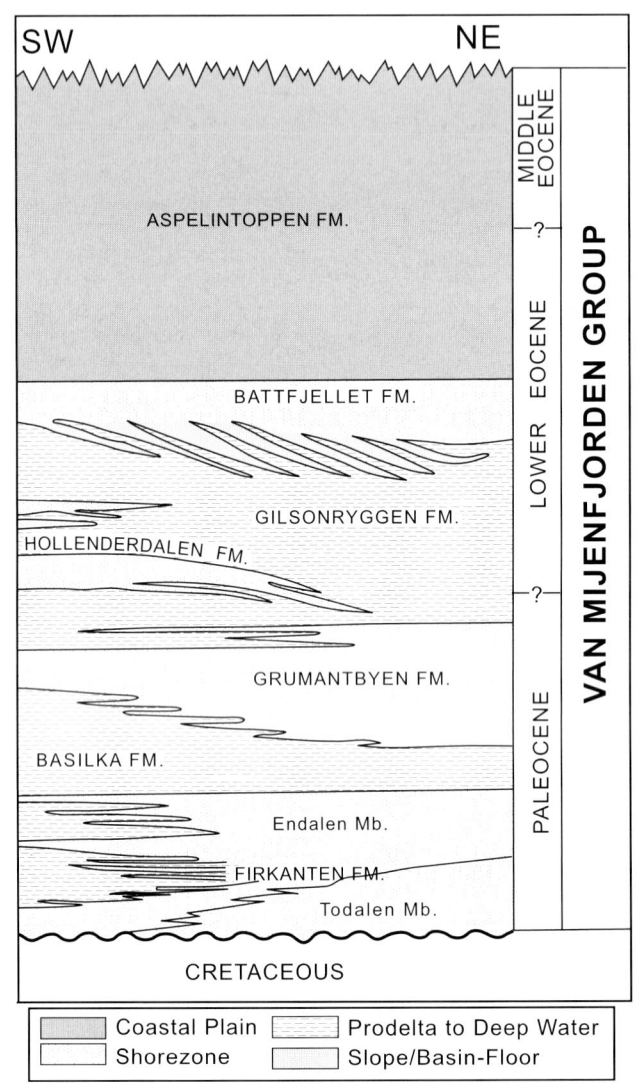

Fig. 3. Stratigraphy of the Central Tertiary Basin in Spitsbergen, with a schematic representation of the geometry of the major sandstone units. This study deals only with the Lower Eocene part of the succession, and focuses on some of the sandy clinoforms that penetrate from the shallow-water Battfjellet Formation down into the deepwater Gilsonryggen Member (modified from Steel et al. 1985).

The focus of the present study is the occasional sandy tongue that penetrates from the Battfjellet Formation down into the shales of the Gilsonryggen Member (Fig. 3). These clinoforms can be sandy in their slope reaches only (e.g. Plink-Bjorklund *et al.* 2001; Mellere *et al.* 2002), or can have basin-floor sandstones that are apparently 'detached' as submarine fans (Steel *et al.* 2000).

Previous biostratigraphic works have considered the study interval (i.e. from the base of the Hollendardalen Formation upwards into the Aspelintoppen Formation, Fig. 2) to range in age from latest Paleocene to Middle Eocene (Manum & Throndsen 1986). As part of a larger sequence stratigraphic study, shale samples from the Gilsonryggen Member and the Battfjellet Formation were analysed for biostratigraphically significant dinocysts. Based on dinocyst zonation, all samples analysed to date are interpreted to be Early Eocene in age (J. Powell, pers. comm. 2001). The Paleocene/Eocene Boundary lies near the base of the Gilsonryggen Member (Manum & Throndsen 1986). The stratigraphically highest samples analysed to date were collected from the Battfjellet Formation, just above clinothem 16 on the mountain Storvola (Fig. 2B). This level in the Battfjellet Formation is stratigraphically equivalent to the lower part of the Aspelintoppen Formation on the adjacent mountain Brogniartfjellet. Based on Manum & Throndsen's (1986) assignment of a middle Eocene to Oligocene age to the Aspelintoppen Formation, we propose that the base of the middle Eocene lies above clinothem 16 on Storvola, within several tens of metres. We are currently testing this idea with continued biostratigraphic work.

The above preliminary assignment of the studied stratigraphic interval entirely to the Early Eocene (c. 6 Myr duration) provides a simple average of c. 300 000 years for the development of each of the mapped sequences. For this reason we refer to these sequences as fourth-order. The observed stacking of such sequences, in aggradational to progradational sets, produces larger-scale cyclic successions that could be referred to as third-order sequences.

Relationship of fans to their host clinoforms

Some of the Lower Eocene clinoforms are sandy only on their topset or shelf segments, so that their slope and basin-floor segments are entirely shale prone. This situation arises in the following way. Despite the relatively narrow width of the Early Eocene shelf (commonly less than 20 km), the deltas and strandplains that were the main sand delivery systems to the deepwater areas were sometimes unable to regressively transit the entire shelf. In such cases they turned around and retreated landwards again without delivering any significant sand volume across the shelf edge into deepwater (Steel & Olsen 2002). Clinoforms of this type are difficult to see, though siltstones or colour changes on their slope segments sometimes allow them to be mapped. Other clinoforms, where the sand-delivery delta systems did succeed in reaching the shelf margin, do have sandy slopes and/or sandy basin-floor segments. Such clinoforms are eye-catching from a distance, particularly because of sand-prone slopes (Fig. 2A).

At first glance the sand-prone clinoforms have two-dimensional sand continuity along their entire length (Fig. 2A), but closer inspection shows that the sand below the shelf-slope break is distributed unevenly, both in space and time, between five sites from the shelf edge to the basin floor. The terms *early* and *late* are used below with reference to the visible stratigraphic position of sites with respect to each other. The basin-floor fan is only one of these sites, the others being: (2) the incised, early shelf edge; (3) the early slope channels or canyons that fed the fans; (4) the later (retrograding) channel–levee system on the lower slopes and (5) the late shelf-edge deltas. These five sites of turbidite deposition tend to have a specific time relationship to each other, as indicated above and shown in Figure 4, because they develop in relation to a falling-to-rising relative sea-level cycle, below the shelf edge. Together, these different sites of turbidite accumulation make up the conventional lowstand systems tract (Posamentier *et al.* 1991) (the combined falling-stage and lowstand systems tracts of other researchers).

Because of the above uneven or discontinuous distribution of deepwater sand below the shelf edge, the basin-floor fan is often portrayed as disconnected from the sandy slope system. There is, of course, always a connection back up to the shelf edge, via channels, gullies or canyons, but these may be mud filled and sinuous, making the connection tenuous at best. In our database from Spitsbergen the basin-floor fans are always 'early' in the above sequence of events, presumably because the early sediment discharge from the shelf-edge area (?the falling-stage discharge) was the most efficient. The fans are also 'early' in the sense that they are (mappably) partially older than the lower slope channel–levee systems. The fans are also older than the latest component of the lowstand complex, the late deltas that re-established at the shelf edge

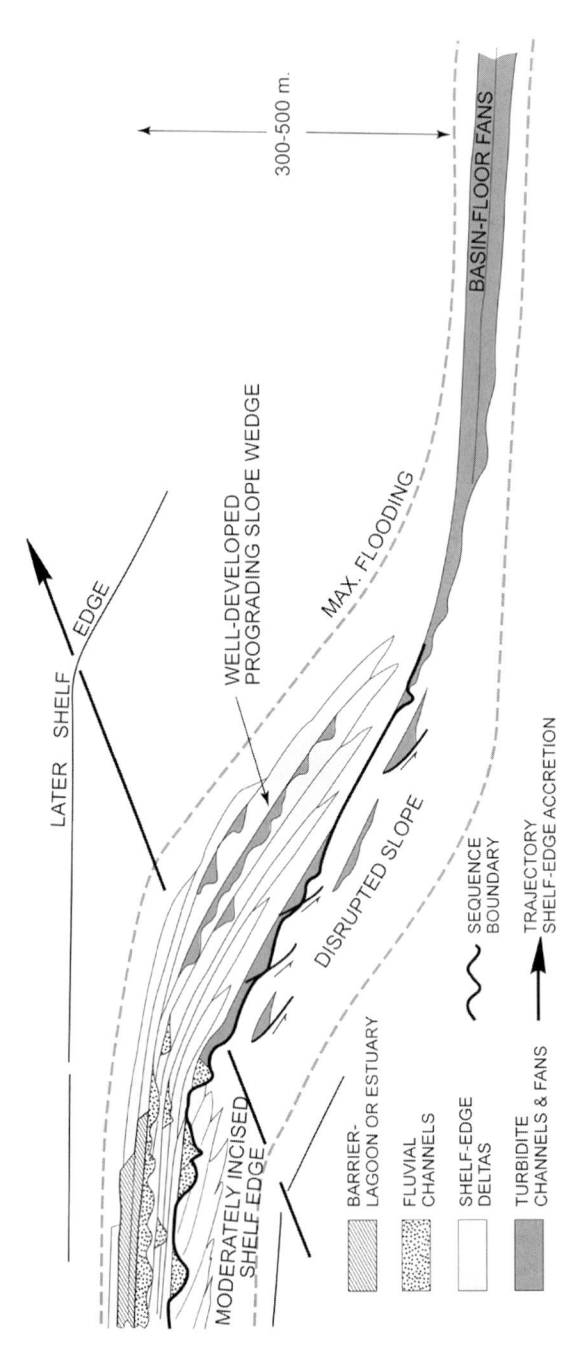

Fig. 4. Schematic illustration of the time–space relationship of the five main components of the deepwater, lowstand complex on some clinoforms in the Van Keulenfjorden transect: incised shelf edge, disrupted upper-slope canyons/gullies, basin-floor fan, lower-slope channel–levee system (sited just slopeward of the basin-floor fan), late prograding slope wedge from shelf-edge deltas.

when sea-level rose back up the shelf platform. These deltas are part of the lowstand complex because they downlap the slope with thin-bedded turbidites, onto all older lowstand components (Fig. 4), and are themselves overlain by a thick mudstone succession that contains the maximum flooding surface of the sequence.

Because the late-stage deltas are commonly extensive (tens of kilometres) along strike on the shelf edge, and they drape across the shelf margin at least down to the middle slope (in the Spitsbergen dataset), they are the most prominent lowstand element in normal two-dimensional views (Fig. 4). However, they are commonly separated stratigraphically from the older, lower-slope channels and the fans by 10 m or more of mudstones, so they typically lack communication with the underlying coarser sands.

The above picture of the space–time relationships between the basin-floor fan and the other sandy lowstand components, within the hosting, large-scale clinoform, is common in many of the Eocene sequences on Spitsbergen, and is probably relevant and analogous for other systems where water depth below the shelf edge is less than 1000 m. Shelf margins with greater water depths, a kilometre or more, tend to have more complex and segmented slopes, have much more extensive (but not necessarily thicker) fans, and do not form the type of clinoformed successions that are described here.

The relationship described above between the basin-floor fans and the other components of sand-prone clinoforms are analogous to the relationships described within deepwater lowstand systems tracts by other researchers (e.g. Vail 1987; Posamentier *et al.* 1991; Kolla 1993), though these workers generally assigned third-order status to their deepwater complexes. The lowstand fan, early lowstand wedge and late lowstand wedge of Posamentier & Vail (1988), or the early, middle and late lowstand systems tract of Normark *et al.* (1993) correspond with the basin-floor fan, lower-slope channel–levee complex and late shelf-edge deltas respectively of this study. Also, the evolutionary succession of Type 1 to Type 3 fans of Mutti (1985) resembles the change from sand-prone to mud-prone components seen here within individual clinoforms.

Scale and dimensions of the Eocene shelf-slope clinoforms

The Spitsbergen clinoforms are scaled with basin-water depth, and are similar in size to many other prograding shelf margins that build out from the edges of various basin types (Steel & Olsen 2002), including Lower Cretaceous clinoformed successions on the Alaskan North Slope (D. Houseknecht, pers. comm. 2001), and Eocene clinoformed successions in the North Atlantic Porcupine Basin (E. Johannessen, pers. comm. 2001). Because accretionary slopes beyond shelf breaks are normally no steeper than 7°, slope length is related mainly to water depth. Fans described from continental margins today, however, commonly occur in water depths of several kilometres, and are linked to much longer slopes than seen in this study. The size of modern fans is also commonly much greater than seen here, mainly because many of them have continent-large drainage systems.

The Eocene clinoforms on Spitsbergen had fairly narrow shelf widths, generally 20 km or less. Because basin water depth below the shelf-slope break was only 200–400 m and slopes were 3–4° steep, slope lengths were usually no more than 3–5 km. Basin-floor fans (or the sand-rich parts of them) appear to extend beyond the base of slope by no more than 12 km.

The basin-floor fans described here are relatively small, and would fall into the 'sand-rich submarine fan' category of Stow *et al.* (1996). Their sand-rich parts (?inner to mid fan) are generally less than 12 km in length, and so the entire length of the typical fan is likely to be less than 30 km. They are on a scale of some of the smaller, California margin fans of the present day (Normark & Piper 1985; see also Reading & Richards 1994), North Sea Jurassic fans (Garland 1993) or Eocene fans such as Frigg (Heritier *et al.* 1979), or Eocene Hecho Basin fans in northern Spain (Mutti 1977).

Recognition of fans versus fan lobes on basin floor

When only a basin-floor succession of stacked turbidite units is exposed in outcrop, it can be difficult to distinguish individual fans from component fan lobes. We use the term *lobe* in the sense of Stow *et al.* (1996) to refer to a lobe-shaped, low-relief depositional mound, usually generated at the distal ends of channels, and produced by the stacking of individual turbidite beds (see also Mutti & Normark 1991).

The clinoformal setting of the Spitsbergen succession, however, allows a resolution of this problem. As seen in Figure 5, a vertical section on Hyrnestabben displays almost 100 m of turbidites, stacked in four sand-bodies, each separated by 5–15 m of mudstones. Without knowledge of

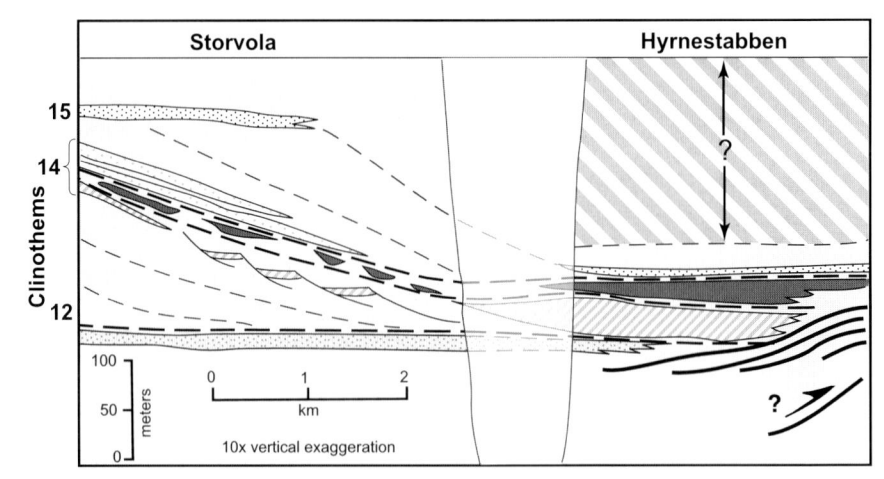

Fig. 5. Clinoforms 12, 14 and 15 on Storvola correlated with stacked basin-floor fan bodies on Hyrnestabben. Panel is oriented NW (left) to SE (right). Note that there are two basin-floor sand-bodies relating to Clinoform 14. Note also the slopeward-thickening mudstone succession that separates Clinoforms 12 from 14, and 14 from 15. Fan bodies (stratigraphically lowest to highest) are named green, orange, yellow and pink.

how these sand-bodies behave in a shelfward (landward) direction, it would be difficult to tell whether these turbidite bodies were all part of a single fan complex (for example as stacked lobes) or were individual fans. Despite the valley (gap) between Hyrnestabben and Storvola (Fig. 5) it is possible to trace and correlate these bodies towards their respective lower slope areas, and to demonstrate that they belong to three different fans. The critical evidence comes from the slopeward increase in mudstone thickness between successive fans. The clinoformal character of the succession causes the mudstones between individual clinoforms to thin on the basin floor (therefore only 5–15 m of mudstones between fans), to thicken drastically slopewards (>100 m thick), and to thin again towards and across the shelf edge (Fig. 5). We therefore use the term 'fan' to refer to a basin-floor sandstone body or complex that is genetically related (slopewards) to an individual sand-prone clinoform. Because clinoforms become sand-prone only occasionally, there can be a great thickness of muddy section between them on the slope, despite there being only a minimal thickness of time-equivalent mudstone on the shelf and on the basin floor. The tracing of a basin-floor body slopewards and monitoring the thickness of overlying and underlying mudstones thus allows discrete fans to be recognized. Fans in the Lower Eocene succession here (away from their termination area) vary in thickness from 15 to 60 m, whereas lobes within fans are usually less than 10–12 m thick.

Database: fans 12 and 14

Logged stratigraphic sections through sand-bodies at and beyond the toes of Clinoforms 12 and 14, combined with aerial- and ground-based photography, and detailed field examination of lateral facies relationships, provide a database for the facies analysis and architectural reconstruction of the studied basin-floor fan systems. The studied fans extend between the mountains Storvola and Hyrnestabben (Fig. 6A). Figure 5 shows that the lowest sandstone bench present on Storvola is the basin-floor fan of Clinoform 12. The next (younger) sandy clinoform on Storvola (Clinoform 14) has its basin-floor fan exposed as two turbidite bodies to the southeast on the mountain Hyrnestabben.

Six measured sections on Hyrnestabben (i.e. H-1 to H-6) and three measured sections on Storvola (i.e. S-1 to S-3) are located in Figure 6A. Note that whereas sections H-1, H-2 and H-3 are positioned along the same general trend of logged sections in clinoforms exposed on Storvola, sections H-4, H-5 and H6 are positioned at increasingly greater distance from this trend. This provides nearly three-dimensional exposure of the deepwater sandstone bodies.

Localized normal faults on Hyrnestabben (green lines) are related to modern outcrop failure (Fig. 6B). Red lines highlight the locations of logged stratigraphic sections H-1, H-2 and H-3.

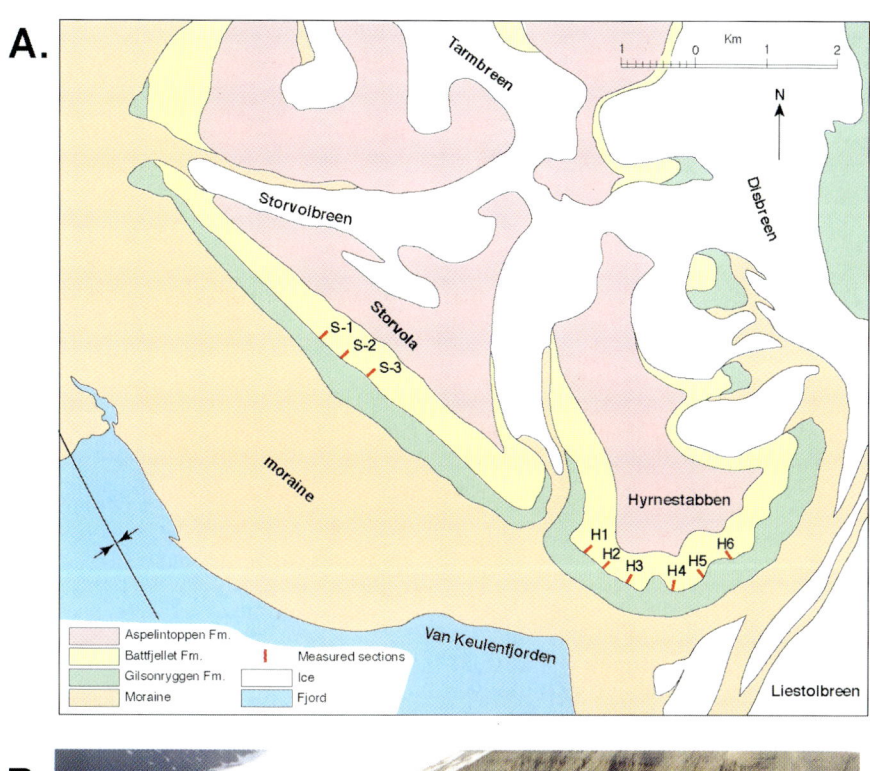

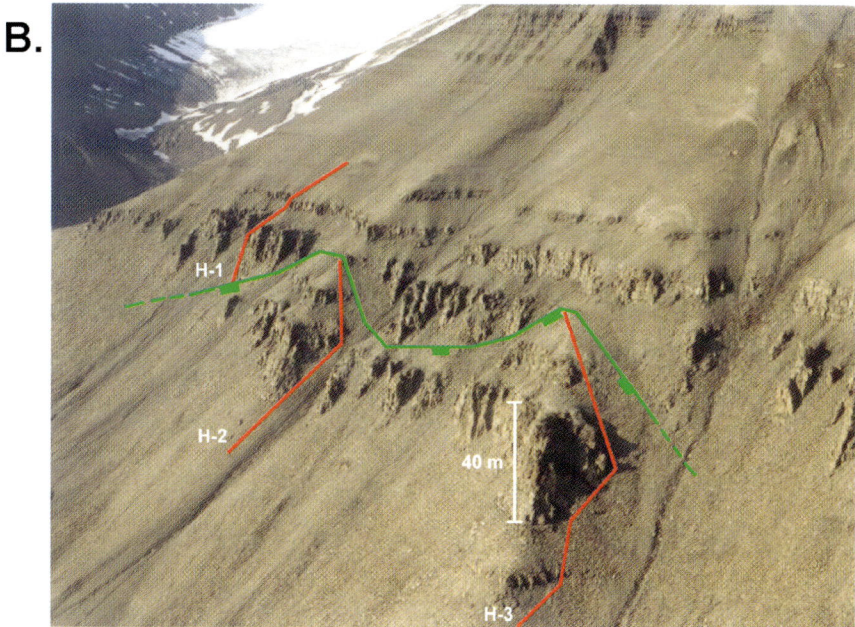

Fig. 6. (**A**) Location of measured sections (basin-floor fans of Clinoforms 12 and 14, Figs 8 & 9) on Storvola and on Hyrnestabben. (**B**) Aerial photograph of the southwest face of Hyrnestabben, showing location of measured profiles H1–H3 (see Figs 6A & 10A). Three prominent sandstone benches (basin-floor fan bodies) can be seen above trace of the fault (green line). The lower two relate to Clinoform 14, whereas the uppermost thin bench may belong to Clinoform 15. In the repeated section below the fault trace, a poorly exposed fourth bench (fan of Clinoform 12) can be seen at base of profile H3. Note the mudstone separation between the sandstone benches.

Turbidite facies

Structureless to parallel-laminated, thick bedded sandstone

Thick sandstone beds (>50 cm, and up to 5 m thick) typically lack internal stratification or are parallel-laminated, and frequently overlie an erosion surface (Fig. 7A; Fig. 8, S2) with basal relief rarely exceeding 30–60 cm over lateral distances of 50 m. Beds of this facies are typically composed of very fine- to medium-grained sandstone (Fig. 8), may be normally graded and display scattered toolmarks and flute casts on their bases. Coal clasts 0.5–3.0 cm in diameter occur most often at the base of units but are also frequently found scattered throughout thick beds (Figs 7B & 8). The flute casts are directed mainly to the east and southeast in Clinoform 12 fan (Fig. 8), but show a subtle upsection change in paleocurrents from E–SE to N–NE in Clinoform 14 (Fig. 9; H-1, 10–45 m).

Interpretation. These thick beds of structureless sandstone are interpreted as the product of deposition by continuous aggradation or collapse fallout from high-density turbidity currents (Stow *et al.* 1996; Kneller & Branney 1995). Evolution of turbidity currents into sandy debris flows cannot be ruled out as a mechanism of deposition for this facies. We interpret the common occurrence of laterally extensive, low-relief erosion surfaces beneath this facies as evidence of broad shallow channels that developed in a 'constructional' manner.

Parallel-laminated to ripple-laminated, thin-bedded sandstones

Thin sandstone beds (<50 cm thick), typically very fine- to fine-grained (Fig. 7A, level 29–31 m; Fig. 8: S1, 2–6 m) , are dominated by an internal motif that shows a vertical change from plane-parallel- up to ripple-lamination. Individual beds fine upwards to siltstone (0.3–3 cm). Capping mudstone is rare, and when present is only a few millimetres thick. Bed bases may be slightly erosional, but most commonly are simply sharp. A subordinate percentage of thin-bedded sandstones are structureless (Fig. 8, S1). There are rare occurrences of 1–3 m thick packages of beds, showing a coarsening- and thickening-upward trend (Fig. 8, section S1, metres 2–5). Bedding planes are most often approximately parallel within successions of these thin-bedded sandstones. However, rare occurrences of beds with non-erosive, divergent bedding planes (i.e. wedge-shaped beds) were documented. Basal surfaces of these beds do not truncate laminations within underlying units, but instead sets of ripple-lamination pinch out laterally within beds.

Bed bases occasionally display toolmarks. Paleocurrents from toolmarks and ripple lamination (Figs. 8 & 9) show a wide distribution, with many toolmark lineations oriented southeast–northwest to northeast–southwest, and ripple lamination directed south to northeast. The structureless beds have a narrower palaeocurrent dispersion, directed mainly between south and east.

Interpretation. The thin-bedded sandstone successions are interpreted as laterally extensive sheet sandstones. Individual beds show clear evidence of waning flow, and are interpreted as surge-type turbidites. They resemble Bouma-type turbidites but commonly lack a muddy top of any significant thickness.The spatial relationship between these turbidite sheets and the thick-bedded, channelized turbidites suggests that the sheets developed both laterally and (especially) distally to channelized fairways. The sheets probably represent both channel-mouth lobes and proximal overbank deposits, the latter including thin-bedded levee deposits and thicker bedded crevasse sheets (see discussion of thin-bedded turbidite environments by Mutti 1977). The wide spread in paleocurrent vectors in this facies supports the idea that these sheets were fed laterally from channel overbank flow as well as from the distal terminus of channels. The subordinate structureless, thin-bedded sandstones probably represent the transition from down-channel directed, high-density turbidity flows to the low-density turbidity currents that fed the thin, parallel- to ripple-laminated sheets at the distal (and lateral) margins of channels. Packages of thin-bedded turbidite sheets are almost as common as channelized turbidites in the proximal reaches of the Eocene fan bodies, and become dominant farther downfan, consistent with the notion that such sheetlike units are mainly channel-mouth depositional lobes.

Convolute bedded sandstones facies

Some thin bedded sandstone units display convolute lamination with amplitudes ranging from 5 cm to 50 cm (Fig. 8, S1). Flame structures sometimes occur within these beds lateral to the convolute zones. Thick-bedded sandstones can also be deformed in convolutions, and are

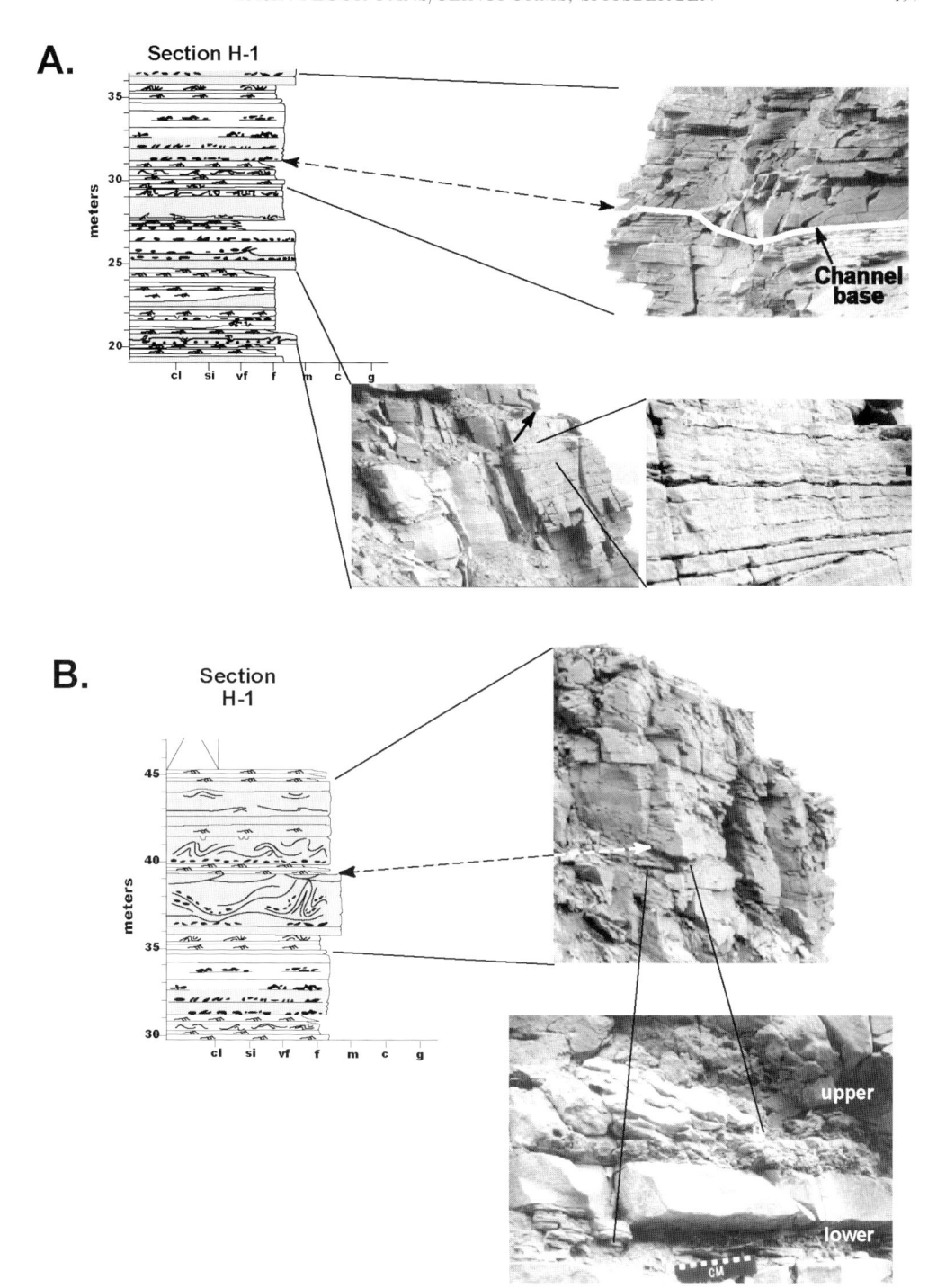

Fig. 7. Examples of some of the main basin-floor facies. (**A**) Interval 19–36 m in section H1 (Fig. 9) showing channel base with overlying structureless sandstones (upper photo), and underlying thin-bedded, ripple-laminated sandstones (lower photos). (**B**) Interval 30–45 m in section H1 (Fig. 9) showing some 5 m of structureless, partly deformed beds (upper photo), and a detailed view of associated upper and lower recessed intervals with coal clasts (lower photo).

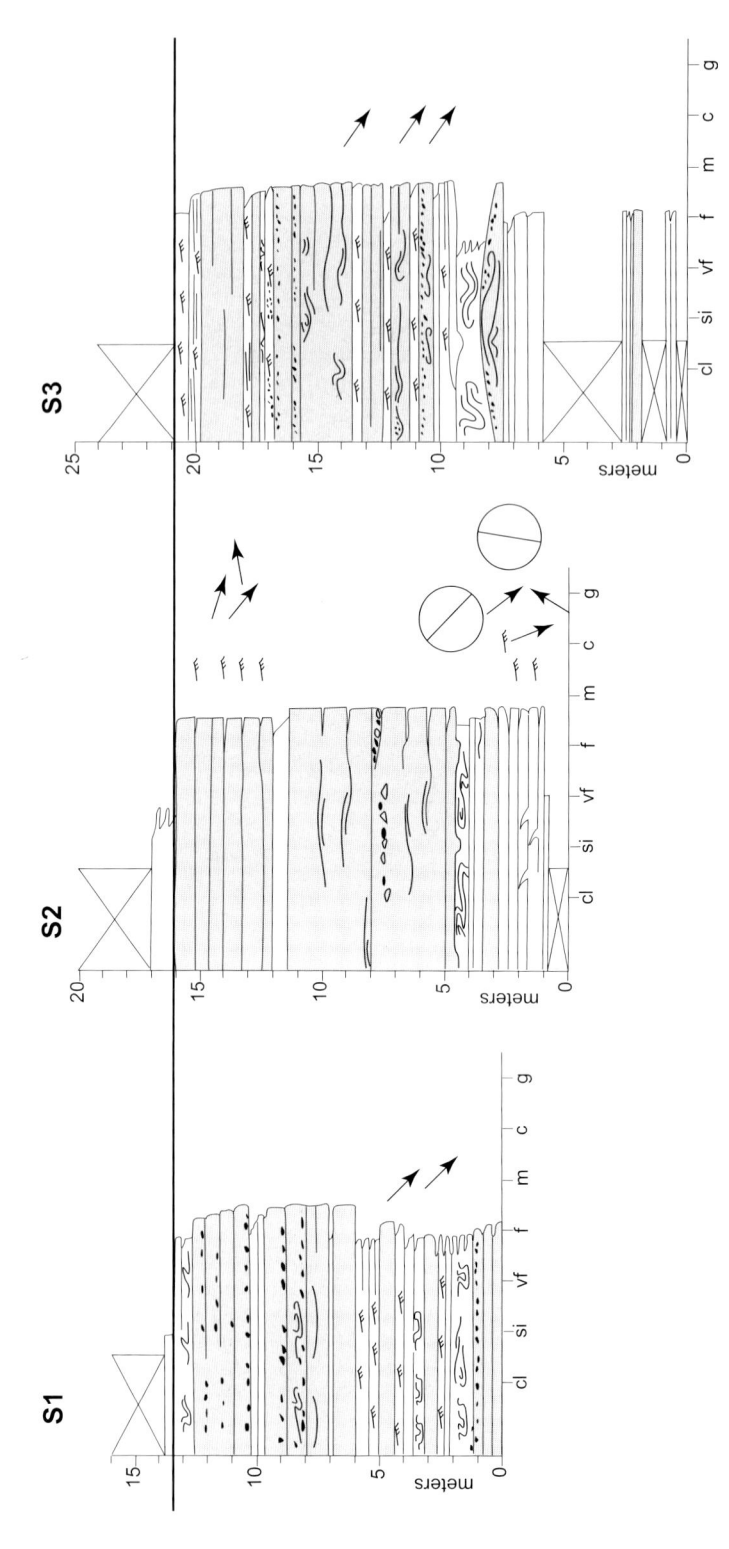

Fig. 8. Measured sedimentary sections through the basin-floor fan deposits of Clinoform 12 on Storvola. Thick (>50 cm), structureless beds are highlighted by shading. Note the crude upward change from thin to thick-bedded facies. Map locations of S1, S2 and S3 shown in Fig. 6A.

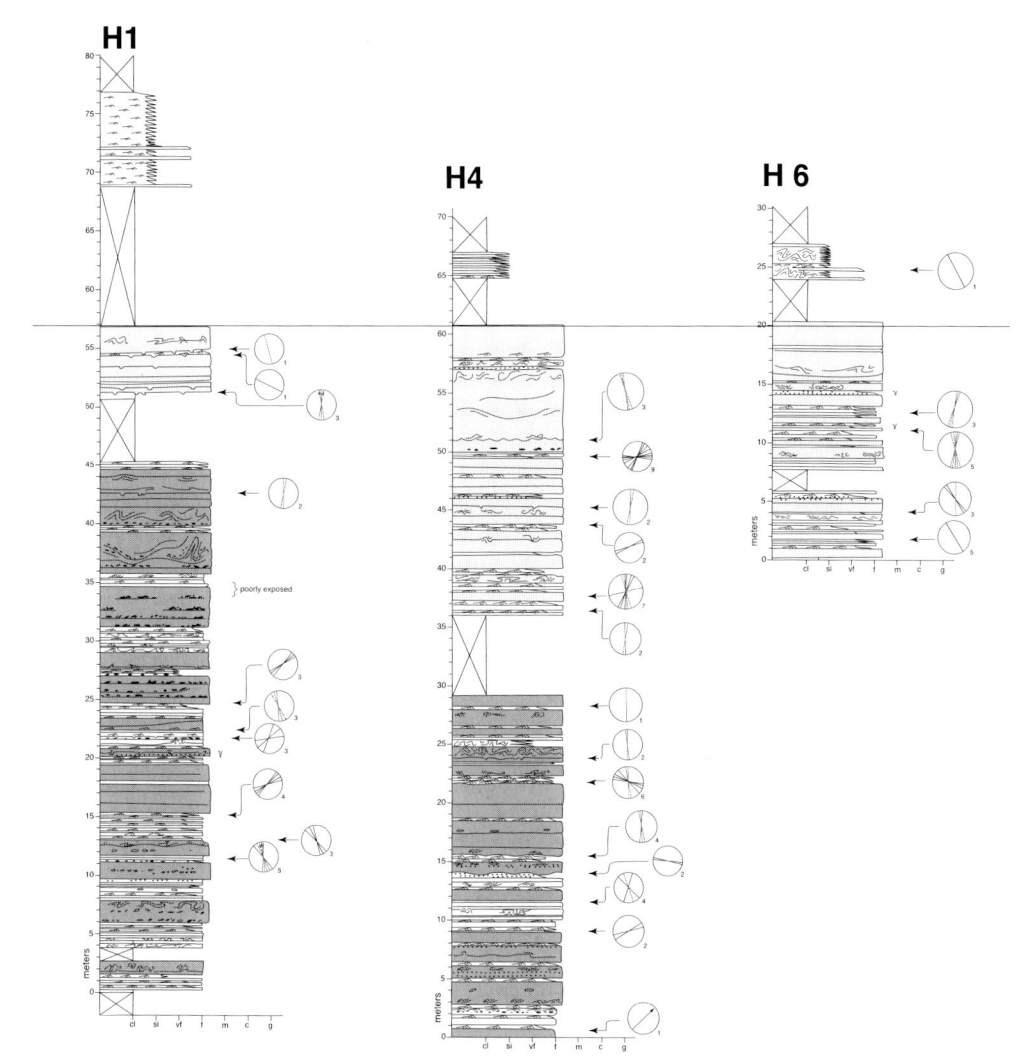

Fig. 9. Three sedimentary sections measured through Clinoform 14 (Orange and Yellow bodies of Fig. 10A) and Clinoform 15 (Pink body) on Hyrnestabben. Thick (>50cm), structureless beds are highlighted by shading. Palaeocurrent readings are mainly from flute casts and toolmarks. There is a crude upward increase in N-S oriented palaeocurrents within both Orange and Yellow bodies. Vertical scale in metres.

typically 0.3–1.0 m in height (Fig. 8, S1 and S3). Very large-scale convolute bedding (up to 8 m amplitude) occurs in the upper part of some fan bodies, particularly the lower fan body in Clinoform 14 (Fig. 9: H1, 36–39 m). In a few occurrences of large-amplitude convolutions (4–8 m in height), the ratio of amplitude to wavelength is so large that the convolutions are isoclinal folds.

Interpretation. Convolute bedding, load casts, and flame structures all attest to the liquefaction and fluidization processes that commonly accompany very high rates of sedimentation in basin-floor fans. Of particular interest is the large-scale deformation seen in the upper parts of the Clinoform 14 fan. This fan-capping instability occurs in the coarsest and thickest-bedded sandstones of the fan system, near the time of maximum progradation of the fan. The rapidity of deposition at this time, coupled with the thick mudstone cap that flooded back across the fan top immediately after fan abandonment, may have caused pore pressure build-up and liquefaction of the sands in the upper part of the fan.

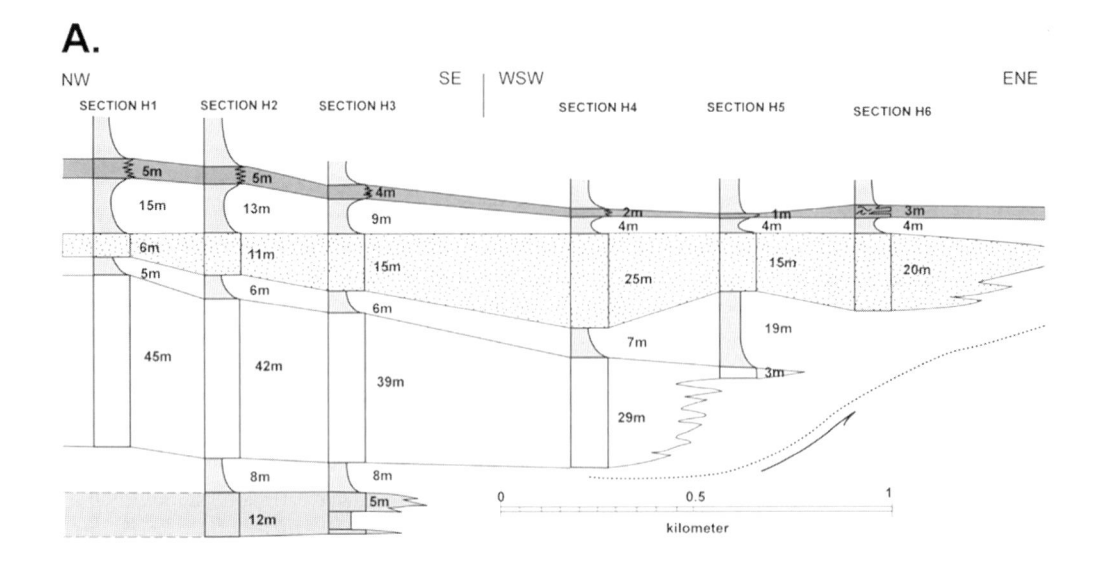

Fig. 10. (**A**) Correlation panel showing geometry of the four fan bodies on Hyrnestabben, at sections H1 to H6 (see Fig. 6A for map locations). The fan bodies are named (from stratigraphically lowest to highest) Green, Orange, Yellow and Pink. Note the relatively short lengths and abrupt pinchout of the Orange and Yellow bodies of Clinoform 14. It is argued in the text that this termination is not a distal pinchout but a lateral one, caused by a northward deflection of the fan on the basin floor. Thrust-created basin-floor topography is the likely cause of the fan deflection. (**B**) Upper sand-body (Yellow) in Clinoform 14 fan shows a relatively abrupt change from thin- to thick-bedded turbidites vertically. Location is 500 m southeast of section H1. Person for scale.

Siltstone and mudstone facies

Stratigraphic intervals of interbedded siltstone and mudstone (5–15 m in thickness) occur between the main fan-sandstone bodies (Figs 9 & 10). Rare thin, silty very fine-grained sandstones occur interbedded with siltstone and mudstone, resulting in this facies being poorly exposed.

Interpretation. This facies is interpreted as representing ambient background sedimentation on the basin floor. This style of sedimentation takes place away from the focus of basin-floor fan sedimentation. These 5–15 m thick units form continuous fine-grained sheets across individual basin-floor sandstone bodies, and often are documented as thickening slopewards (Figs 2A & 5). Therefore, based on the large-scale correlations that are possible within the clinothems of Van Keulenfjorden, it can be demonstrated that lithosomes such as these record periods of shelf and shelf-margin flooding (Steel *et al.* 2000).

Large-scale architecture

The stratigraphic panel of the southwest face of Hyrnestabben (Fig. 10A), keyed to the photograph in Figure 6B, shows the four separate turbidite–sandstone bodies on Hyrnestabben. For convenience, the bodies are colour-coded, also because one fan can include two or more sandstone bodies. The uppermost body (pink in Fig. 10A) is relatively thin, possibly belongs to Clinoform 15 (see also Fig. 5) and will not be dealt with further here. The two underlying sandstone bodies (yellow and orange in Fig. 10A) are well exposed and belong to Clinoform 14. The basinward pinchout of a fourth basin-floor sandstone body (green in Fig. 10A) is poorly exposed below the three other bodies (see base of profile H3 in Fig. 6B), and is the basin-floor fan of Clinoform 12 (Fig. 5). Sandstone percentages within these reaches of the basin-floor fans approach 90%.

Several geometric relationships are apparent in Figure 10A. The yellow sandstone body can be seen to thicken to the southeast, but then abruptly thins northeastwards from section H4. The green and orange sandstone bodies also pinch out abruptly in a southeasterly and easterly direction respectively. The orange and yellow bodies show thicknesses up to 45 m and 25 m respectively, and an apparent basinward extension of only 5 km. However, the palaeocurrent vectors in the thick-bedded turbidites for these bodies suggest a north–northeasterly transport, so that the abrupt easterly thinning noted above, beyond locations H5 and H6 (Fig. 6A), represents the lateral and not the distal edge of the fan system. The shale-prone interval between yellow and orange bodies shows a relatively constant thickness from section to section. However, the shale intervals between the yellow and pink, and the orange and green sandstone bodies thicken greatly toward the northwest (Fig. 5).

The green sandstone body extends across Storvola and pinches out to the southeast on the eastern end of Hyrnestabben (Fig. 5). This fan averages 15 m in thickness for much of its length (Fig. 8), and thins to 12 m at a location 250 m from its southeastern pinchout on Hyrnestabben. Its extension from toe-of-slope (between Storvola and Brogniartfjellet, Fig. 2B) out into the basin is about 10 km. Palaeocurrent directions show a dominant transport to the southeast (Fig. 8).

Vertical facies trends within individual sandstone bodies

The proximal reaches of the fan systems on Hyrnestabben show the following large-scale vertical trends (Fig. 9). The orange and yellow basin-floor bodies of Clinoform 14 show a preferred vertical succession of bed types. The lower portion of each body is dominated by rippled to parallel-laminated thin beds (<50 cm thick), interbedded with a subordinate proportion of structureless and parallel-laminated, thick-bedded units ($\geq$50 cm thick). These intervals display sole marks, indicating that the palaeoflow direction was dominantly east and southeastwards, i.e. in the same direction as the dip of the nearby slope.

A change, sometimes gradual (e.g. yellow body at H4, Fig. 9), sometimes abrupt (e.g. Fig. 10B), takes place upwards to units dominated by structureless and parallel-laminated thick beds. Intervals characterized by thick beds are 4–6 m in thickness, and display sole marks indicating north–northeasterly transport. Thick-bedded successions display lateral wedging, particularly at the top of a succession capped by thin beds, and commonly contain soft-sediment deformation structures. Erosion is common at the base of thick beds, but seldom exceeds 30–60 cm in relief over lateral distances of 50 m.

The Clinothem 12 fan shows an alternation of thin-bedded and thick-bedded units, with the former dominated by current-ripple lamination, and the latter by massive beds. However, there is also a crude but overall upward increase in the frequency of units with thick, massive beds (Fig. 8).

The above trends are crudely expressed on the scale of entire bodies (tens of metres), but also on a smaller scale (metres) within bodies.

Lateral trends within sandstone bodies

Figure 11A is a photomosaic of the Clinoform 14 fan bodies at site H-1, and Figure 11B is a tracing of dominant bedding surfaces on this photomosaic. The dominant feature of Figure 11 is the alternation of thick-bedded units that are slightly erosionally based with thin-bedded, recessed intervals in the cliff. Five thin, recessed intervals can be traced across the 150 m of lateral exposure in the lower, main body (orange in Fig. 10A). Laterally continuous, recessed intervals consist typically of thin-bedded, ripple-laminated sandstone and siltstone containing floating coal clasts. Coal clasts within these fine-grained

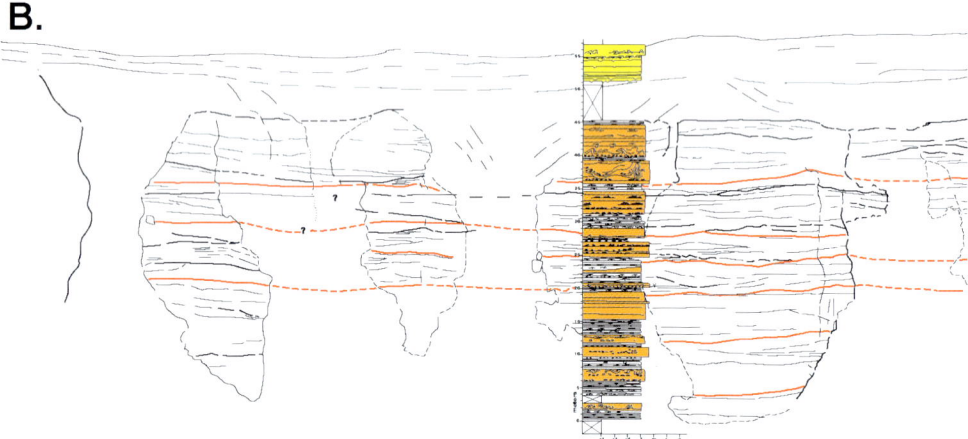

Fig. 11. (**A**) Outcrop photo, 150 m wide, of the two main sandstone bodies (orange and yellow of Fig. 10A) in the basin-floor fan of Clinoform 14 on Hyrnestabben. Location of measured section H1 is shown by white line. Red vertical bar is 5 m. (**B**) Tracing of dominant bedding surfaces on the photo of Fig. 11A with section H1 superimposed. Note, in the main sand-body: (1) the through-going, thin recessed intervals (highlighted by red lines) that are related to the finest-grained and thinnest-bedded horizons; and (2) the erosion surfaces related to the bases of thick-bedded (channelized) units.

Fig. 12. Turbidites in the upper sandbody (Yellow) of Clinoform 14 on Hyrnestabben. Note the lateral passage of thick-bedded (labeled A) to thin-bedded intervals (labeled B). Cliff is some 13 metres high.

units occur towards the top of fining-upward units. However, some recessed intervals are the result of coal-clast enrichment in the basal lag of thick-bedded, structureless sandstones. The number of thick-bedded units occurring between any two laterally continuous, recessed intervals highlighted in Figure 11B ranges between one and three.

The upper sandstone body in Figure 11A (yellow in Fig. 10A) is thin in its northwestern reaches but thickens significantly eastwards (Fig. 10A), and the proportion of thick-bedded, channelized units increases greatly. Lateral facies changes from thick-bedded to thin-bedded units are common between updip section H-1 and downdip section H-3 (e.g. Fig. 12). Basinward thickening and facies variability between sites H1 and H3 suggest that the northwestern end of this body on Hyrnestabben represents channelized, proximal fan deposits. The middle to southeastern reaches of the body thicken and contain a significant proportion of rippled-sheet sandstones.

The uppermost sandstone body that is barely visible in Fig. 11A (pink in Fig. 10A) contains a facies succession dominated by rippled, thin-bedded units with very rare, 5–10 cm thick turbidites that increase in number to the east.

Interpretation of lateral and vertical trends

Thin ripple- and parallel-laminated beds are interpreted as sheetlike turbidites that accumulated in crevasse splays, low-relief sandy levee complexes, and especially in channel-mouth depositional lobes (e.g. Mutti 1977; Mutti & Normark 1991). Erosively based, thick-bedded, structureless and parallel-laminated sandstones are interpreted as the fill of channels. The absence of associated muddy levee deposits suggests that the channels were of low sinuosity, and probably laterally mobile. The erosional relief at the base of these channels seldom exceeds 1 m. Lateral transition of erosively based thick-bedded units to conformably based thin bedded units, and the low relief of erosion surfaces, suggest that these channels were shallow, broad and of the 'constructional' type (Fig. 13). The shallow channels fed sheet complexes that preferentially built in distal reaches of the system. Despite the apparently mobile nature of the channels, they still produced sheet-like lobes in the distal areas.

Vertical facies trends of individual fan systems generally imply a progradational growth of channelized turbidite facies out over off-axis turbidite sheets. However, in Clinoform 14 sandstone bodies there is additionally implied a transition in which basin-axial transport eventually dominates over downslope transport (Fig. 14). This shift away from the dominance of downslope sediment transport during the construction of a fan-system implies a gathered drainage from multiple slope-entry points (Fig. 14). An additional aspect of the conceptual model for the youngest basin-floor fan evolution includes continued longitudinal transport, along the strike of Clinoform 14, and not revealed within Hyrnestabben outcrops.

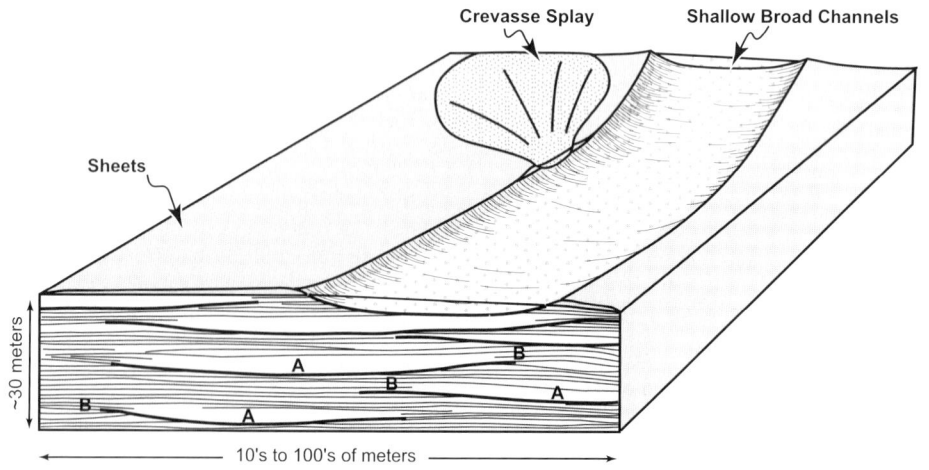

Fig. 13. Schematic representation of the likely lateral relationship between erosively based, thick-bedded (channelled) and conformably based, thin-bedded (mainly lobe sheets) turbidite facies in the basin-floor fans. The low relief of the channel erosion surfaces, lack of muddy levees and general sand-rich character of the fans suggest broad, low-sinuosity, mobile constructional channels.

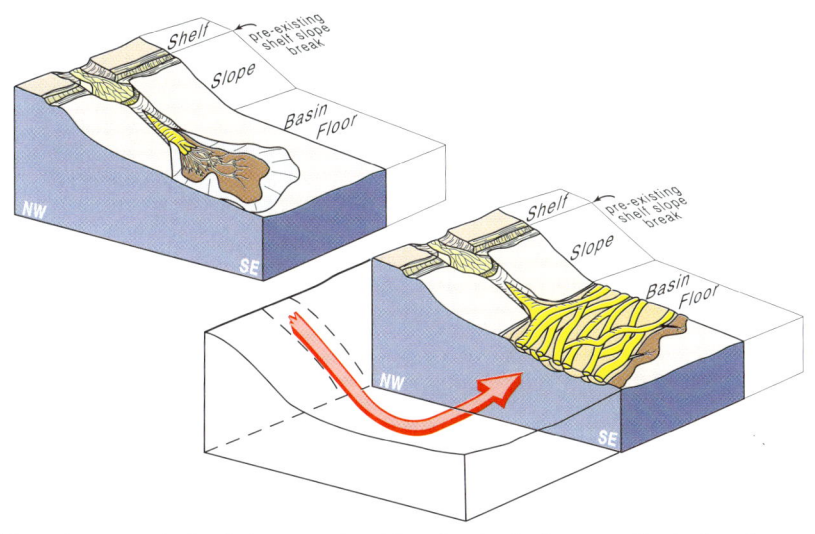

Fig. 14. Schematic portrayal of early (transverse) and late (longitudinal) stages of basin-floor fan construction for the youngest fans on Hyrnestabben. At an early stage, the shelf edge becomes deeply eroded by the distributary channels, allowing direct access between fluvial system and slope channels. At late stage, the longitudinal system on the basin floor has gathered the along-strike drainage from multiple slope channels.

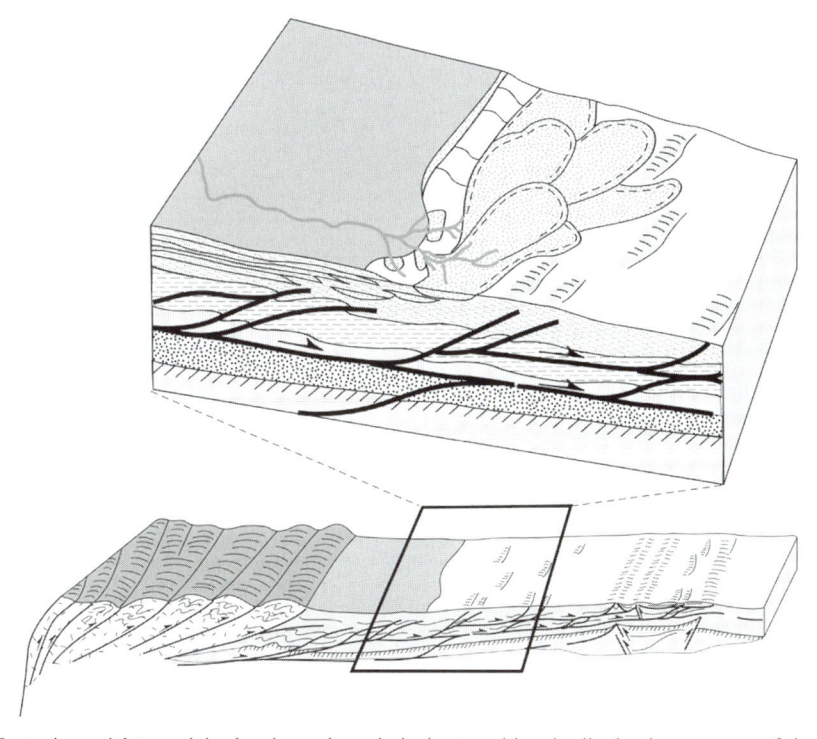

Fig. 15. Tectonic model to explain the abrupt lateral pinchout and longitudinal palaeocurrents of the youngest fan bodies. Thrust-generated basin-floor topography controlled deflection of fan drainage on the basin floor. Model is superimposed on the results of surface mapping (Braathen *et al.* 1999) and seismic reflection studies (Nottvedt *et al.* 1988), which show that the Lower Tertiary succession was involved in the deformation of the Spitsbergen Orogenic Belt.

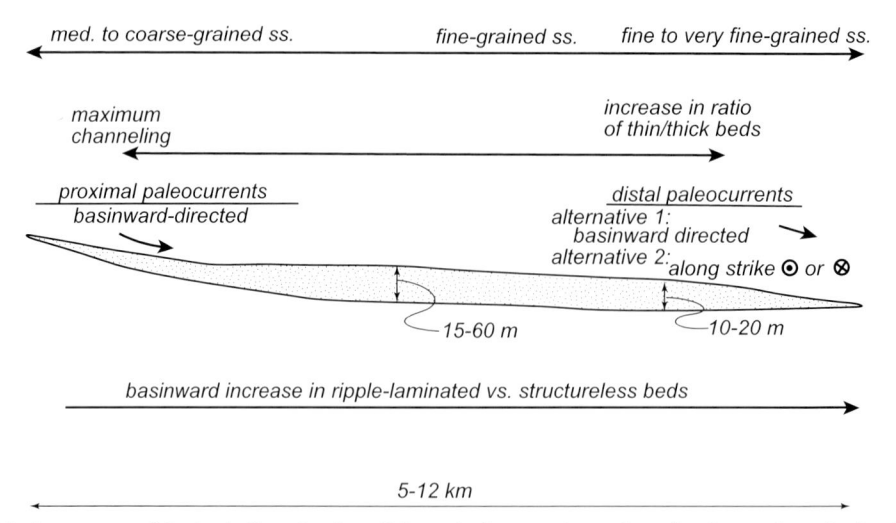

Fig. 16. A summary of the basin-floor fan data. Schematic diagram shows downdip changes in grain size, bed geometry and thickness, palaeocurrents, dominant sedimentary structures in beds, and the range of fan lengths on the basin floor.

Tectonic influence on sedimentation

It is likely that the above drainage shift for the toe of Clinoform 14 was caused by tectonically generated basin-floor topography. Surface mapping (Braathen et al. 1999) and seismic reflection studies in the western fjords of Spitsbergen (Nottvedt et al. 1988) demonstrate structural involvement of the Lower Tertiary section in the West Spitsbergen Orogenic Belt thrusting and folding (Fig. 1B). Support for synsedimentary structural control on fan deposition includes (1) the apparent short basinward extension of the fans on Hyrnestabben, and (2) the prominent shift from downslope to longitudinal direction of paleocurrent lineations. It is suggested that thrust-generated anticlinal structures on the basin floor deflected downslope sediment transport to a more slope-parallel trajectory once sediment reached the basin floor (Fig. 15).

A two-dimensional model of fan deposition

Although there is some variability in the character of the Lower Eocene submarine fans exposed in Van Keulenfjorden, there are also clear similarities between individual fan complexes. Figure 16 is an attempt to summarize some parameters characterizing these fans. Maximum grain size ranges from coarse sand in the proximal areas of a basin-floor fan to fine to very fine sand in the distal reaches. The maximum degree of channeling occurs at or near the base of slope, whereas the ratio of thin to thick beds increases in a basinward direction. Additionally, a basinward increase is seen in the ratio of ripple-laminated to structureless beds. Individual fan thicknesses range from 15–60 m at positions near the base of slope to mid-fan, and thin to 10–20 m within a kilometre of basinward pinch-out.

Some fans, such as those represented by Clinoform 12, record relatively simple basinward growth toward the southeast. Other fans, such as from Clinoform 14, show a more complex growth history. In these fans, distal paleocurrents demonstrate basin-axial sediment transport, caused by structurally induced topography on the basin floor. This hypothesis is consistent with the occurrence of synsedimentary folds documented from recent structural mapping in this area.

We thank the sponsors of the WOLF consortium (BP, Conoco, Norsk Hydro, PDVSA and Phillips) for continued support and enthusiastic discussion on the Spitsbergen clinoforms. D. Mellere is thanked for an early draft of Figure 11B, and T. Olsen (Statoil) for for much of the work that went into the construction of Figure 2B. Reviewers J. Ineson and R. Hiscott are thanked for encouraging us to make major changes to an early draft of this paper.

References

BLYTHE, A. E. & KLEINSPEHN, K. L. 1998. Tectonically versus climatically driven Cenozoic exhumation of the Eurasian plate margin, Svalbard: fission track analyses. *Tectonics*, **17**, 621–639.

BRAATHEN, A., BERGH, S. G. & MAHER, H. D. JR 1999. Application of a critical wedge taper model to the Tertiary transpressional fold-thrust belt on Spitsbergen, Svalbard. *Geological Society of America Bulletin*, **109**, 1468–1485.

CRADDOCK, C. E. C., HAUSED, H. D., MAHER, A., SUN, Y. & GUO-QIANG, Z. 1985. Tectonic evolution of the West Spitsbergen fold belt. *Tectonophysics*, **114**, 193–211.

ELDHOLM, O., SUNDVOR, E., MYHRE, M. A. & FALEIDE J. L. 1984. Cenozoic evolution of the continental margin off Norway and western Svalbard. *In*: SPENCER, A. M. (ed.) *Petroleum Geology of the North European Margin*. Graham & Trotman, London, 3–18.

GARLAND, C. R. 1993. Miller Field: reservoir stratigraphy and its impact on development. *In*: PARKER, J. R. (ed.) *Petroleum Geology of NW Europe: Proceedings of the 4th Conference*. Geological Society, London, 404–414.

HARLAND, W. B. 1969. Contribution of Spitsbergen to understanding of the tectonic evolution of the North Atlantic Gerion. *In*: KAY, M. (ed.) *North Atlantic Geology and Continental Drift*. American Association of Petroleum Geologists Memoirs, **12**, 817–851.

HELLAND-HANSEN, W. 1990. Sedimentation in a Paleocene foreland basin, Spitsbergen. *American Association of Petroleum Geologists Bulletin*, **74**, 260–272.

HELLAND-HANSEN, W. 1992. Geometry and facies of Tertiary clinothems, Spitsbergen. *Sedimentology*, **39**, 1013–1029.

HERITIER, F. E., LOSSEL, P. & WATHNE, E. 1979. Frigg Field: large submarine fan trap in Lower Eocene rocks of North Sea, Viking Graben. *American Association of Petroleum Geologists Bulletin*, **63**, 1999–2020.

KELLOGG, H. E. 1975. Tertiary stratigraphy and tectonism in Svalbard and continental drift. *American Association of Petroleum Geologists Bulletin*, **59**, 465–485.

KNELLER, B. C. & BRANNEY, M. J. 1995. Sustained high-density turbidity currents and the deposition of thick, massive sands. *Sedimentology*, **42**, 607–616.

KOLLA, V. 1993. Lowstand deepwater siliciclastic depositional systems: characteristics and terminology in sequence stratigraphy and sedimentology. *Bulletin Centres Recherche Exploration-Production Elf Aquitaine*, **17**, 67–78.

LOWELL, J. D. 1972. Spitsbergen Tertiary orogenic belt and the Spitsbergen fracture zone. *Geological Society of America Bulletin*, **83**, 3091–3102.

MANUM, S. B. & THRONDSEN, T. 1986. Age of tertiary formations on Spitsbergen. *Polar Research*, **4**, 103–131.

MELLERE, D., PLINK-BJORKLUND, P. & STEEL, R. J. 2002. Anatomy of shelf-edge deltas at an Eocene shelf margin, Central Basin Spitsbergen. *Sedimentology*, **49**, 1181–1206.

MUTTI, E. 1977. Distinctive thin-bedded turbidite facies and related depositional environments in the Eocene Hecho Group (south-central Pyrenees, Spain). *Sedimentology*, **24**, 107–131.

MUTTI, E. 1985. Turbidite systems and their relations to depositional sequences. *In*: ZUFFA, G. G. (ed.) *Provenance of Arenites*. Reidel Publishing Company, Dortrecht, 65–93.

MUTTI, E. & NORMARK, W. R. 1991. An integrated approach to the study of turbidite systems. *In*: WEIMER, P. & LINK, M. E. (eds) *Seismic Facies and Sedimentary Processes of Submarine Fans and Turbidite Systems*. Springer, New York, 75–106.

MYHRE, A. M., ELDHOLM, O. & SUNDVOR, E. 1982. The margin between Senja and Spitsbergen fracture zones: implications from plate tectonics. *Tectonophysics*, **89**, 33–50.

NORMARK, W. R. & PIPER, D. J. W. 1985. Navy Fan, Pacific Ocean. *In*: BOUMA, A. H., NORMARK, W. R. & BARNES, N. E. (eds) *Submarine Fans and Related Turbidite Systems*. Springer, New York, 87–94.

NORMARK, W. R., POSAMENTIER, H. W. & MUTTI, E. 1993. Turbidite systems, state of the art and future. *Reviews in Geophysics*, **31**, 91–116.

NYBERG, K., SOLHEIM, B., HELLAND-HANSEN, W. & NOTTVEDT A. 1995. Linked coastal and basin-floor fan systems, Battfjellet Formation, Spitsbergen: punctuated progradations during overall sea level rise. Abstract volume, Norwegian Petroleum Society, Stavanger 1995, 36.

NOTTVEDT, A., LIVBERG, F. & MIDBO, P. S. 1988. Tertiary deformation on Svalbard: Various models and recent advances. *In*: DALLMANN, W. K., OHTA, Y. & ANDRESEN, A. (eds) *Tertiary Tectonics of Svalbard*. Norsk Polarinstitutt Rapportserie, **46**, 79–84.

PLINK-BJORKLUND, P., MELLERE, D. & STEEL, R. J. 2001. Architecture and turbidite variability of sand-prone deepwater slopes: Eocene clinoforms in the Central Basin of Spitsbergen. *Journal of Sedimentary Research*, **71**, 895–912.

POSAMENTIER, H. W. & VAIL, P. R. 1988. Eustatic controls on clastic deposition II: Sequence and systems tract models. *In*: WILGUS, C. K., HASTINS, C. K., KENDALL, S. G. ST. C., POSAMENTIER, W. H., ROSS, C. A. & VAN WAGONER, J. C. (eds) *Sea Level Changes: An Integrated Approach*. Society of Economic Paleontologists and Mineralogists, Special Publications, **42**, 125–154.

POSAMENTIER, H. W., ERSKINE, R. D. & MITCHUM, R. M. JR 1991. Models for submarine-fan deposition within a sequence stratigraphic framework. *In*: WEIMER, P. E. & LINK, M. H. (eds) *Seismic Facies and Sedimentary Processes of Submarine Fans and Turbidite Systems*. Springer, New York, 127–136.

RICH, J. L. 1951. Three critical environments of deposition and criteria for recognition of rocks deposited in each of them. *Geological Society of America Bulletin*, **62**, 1–20.

READING, H. G. & RICHARDS, M. T. 1994. The classification of deepwater siliciclastic depositional systems by grain size and feeder system. *American Association of Petroleum Geologists Bulletin*, **78**, 792–822.

SPENCER, A. M., HOME, P. C. & BERGLUND, L. T. 1984. Tertiary structural development of the western Barents Shelf, Tromso to Svalbard. In: SPENCER, A. M. (ed.) Petroleum Geology of the North European Margin. Norwegian Petroleum Society, London, Graham & Trotman, 199–209.

STEEL, R. J., DALLAND, A., KALGRAFF, K. & LARSEN, V. 1981. The Central Tertiary Basin of Spitsbergen: sedimentary development of a sheared margin basin. In: KERR, J. W. & FERGUSON, A. J. (eds) Geology of the North Atlantic Borderland. Canadian Society of Petroleum Geology Memoirs, 7, 647–664.

STEEL, R. J. & WORSLEY, D. 1984. Svalbard's post-Caledonian strata: an atlas of sedimentational patterns and paleogeographic evolution. In: SPENCER, A. M. (ed.) Petroleum Geology of the North European Margin. Norwegian Petroleum Society, London, Graham & Trotman, 109–135.

STEEL, R. J., GJELBERG, J., HELLAND-HANSEN, W., KLEINSPEHN, K., NOTTVEDT, A. & LARSEN, M. R. 1985. The Tertiary strike-slip basins and orogenic belt of Spitsbergen. In: BIDDLE, K. T. & CHRISTIE-BLICK, N. (eds) Strike-Slip Deformation, Basin Formation and Sedimentation. Society of Economic Paleontologists and Mineralogists Special Publications, 37, 339–359.

STEEL, R. J., CRABAUGH, J. P., SCHELLPEPER, M., MELLERE, D., PLINK-BJORKLUND, P., DEIBERT, J. & LOESETH, T. 2000. Deltas vs rivers on the shelf edge: their relative contributions to the growth of shelf-margins and basin-floor fans (Barremian and Eocene, Spitsbergen). Proceedings of the GC-SEPM 20th Annual Perkins Research Conference, Deepwater Reservoirs of the World (CD-ROM), Houston, 981–1009.

STEEL, R. J. & OLSEN, T. 2002. Clinoforms, clinoform trajectories and deepwater sands. Proceedings of the GCS-SEPM Foundation 21st Annual Perkins Research Conference: Sequence Stratigraphy (CD-ROM), Houston.

STOW, D. A. V., READING, H. G. & COLLINSON, J. D. 1996. Deep seas. In: READING, H. G. (ed.) Sedimentary Environments: Processes, Facies and Stratigraphy. Blackwell Science, Oxford, 395–453.

VAIL, P. R. 1987. Seismic stratigraphy interpretation procedure. In: BALLY, W. (ed.) Seismic Stratigraphy Atlas. American Association Petroleum Geologists, Studies in Geology, 27, 1–10.

Turbidite systems influenced by structurally induced topography in the multi-sourced Welsh Basin

RU SMITH

A/S Norske Shell, PO Box 40, 4098 Tananger, Norway
Present address: Shell International E & P, Volmerlaar 6, PO Box 60,
2280 AB Zijswijk ZH, Netherlands (e-mail: Ru.Smith@shell.com)

Abstract: The latest Ordovician and lower Silurian fill of the Welsh Basin contains a range (in terms of scale, sediment texture, stratigraphic architecture and supply configuration) of deep-water depositional systems that record the influence of basin-floor topography on sediment distribution patterns. Systems supplied from the eastern basin margin at a time of broadly rising relative sea-level are interpreted to have initially filled an inboard base of slope depression lying above a tilted basement fault block (Cerig Gwynion Grits System). An opposing slope is thought to have caused deflection of turbidity currents to run parallel with the strike of the slope. Following this fill phase, a channel-fed lobe system (Caban-Ystrad Meurig System) extended further basinwards. After flooding of the eastern basin margin, voluminous, texturally immature axial systems, supplied from the southern basin margin, developed. These systems exhibit evidence for deflection of flows to run parallel to tectonically induced slopes and probable local flow reflections in areas of obliquity between bounding slopes and incident flows. The architecture of the lateral-slope to basin-axis profile has been plausibly modelled using a geometric forward model with a low-gradient lower basin slope and basinward higher aggradation rates and sand percentage.

The fill of the deep-water Welsh Basin has been a classic study area for deep-water facies and processes (Wood & Smith 1959; Kelling & Woollands 1969; Woodcock 1976). The Late Ordovician and early Silurian fill of the basin contains a succession of turbidite systems with both lateral and axial supply configurations (Smith *et al.* 1990). Unravelling basin history is aided by high-resolution biostratigraphy (Zalasiewicz 1990; Loydell 1991, 1993), mapped biofacies distributions (Smith 1987c, 1999 and references therein) and structural studies (Woodcock *et al.* 1988 and references therein; Davies *et al.* 1997).

Evidence for physiographic confinement of sediment gravity flow deposits was recognized early (James & James 1969; James 1972). An early fill-and-spill model was proposed for the Caban Conglomerate Formation (Smith 1987a; Smith *et al.* 1990) with the Cerig Gwynion Grits system at the base being confined to the southeast of a structurally induced submarine ridge along the Central Wales Lineament and the younger Caban-Ystrad Meurig system extending beyond this ridge. Younger axial systems contain evidence for lateral flow confinement and, locally, probable flow reflections off bounding slopes (Smith 1987b; Smith 1990; Smith & Anketell 1991).

This paper reviews the evidence for interactions between sediment gravity flows and palaeotopography in the basin, and extends and generalizes earlier discussions. Previously unpublished data are presented on a number of the systems, and the controls on stratigraphic architecture at bounding slopes are discussed with the aid of a two-dimensional geometric model.

The major themes of the paper are as follows:

(1) sequence stratigraphic and palaeobathymetric context;
(2) lateral supply of sediment gravity flows and evidence for deflections and fill-and-spill;
(3) axial supply of sediment gravity flows and evidence for flow reflections and deflections at bounding slopes;
(4) the controls on stratal architecture at bounding slope-basin transitions; and
(5) potential applications of the Welsh Basin dataset in subsurface prediction.

Structural context

The Ordovician-Silurian Welsh Basin lies on late Precambrian continental crust. In Late Ordovician times the basin occupied a back-arc setting (e.g. Kokelaar *et al.* 1984) and experienced widespread volcanism. Early Silurian volcanic rocks in south Wales have dominantly within-plate chemical characteristics with minor evidence for contemporaneous or earlier subduction (Thorpe *et al.* 1989). This evidence led to the suggestion that eruptions probably took place following the termination of subduction in the area. The

From: LOMAS, S. A. & JOSEPH, P. (eds) 2004. *Confined Turbidite Systems.* Geological Society, London, Special Publications, **222**, 209–228. 0305-8719/04/$15.00 © The Geological Society of London 2004.

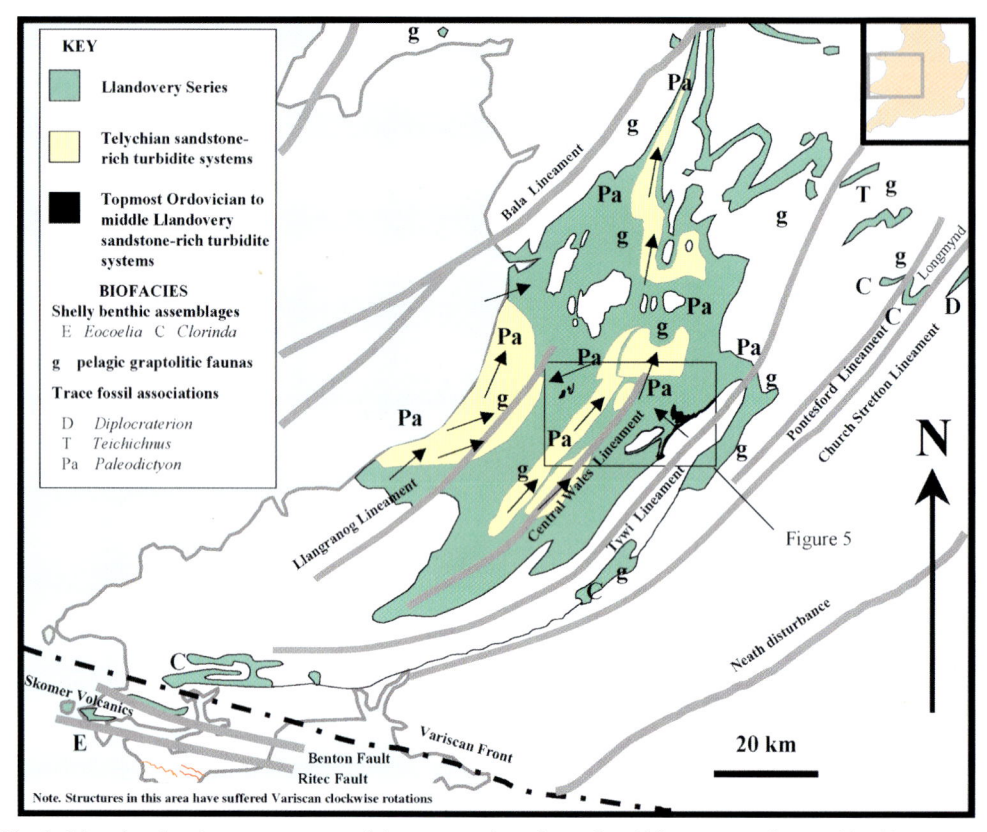

Fig. 1. Map showing the outcrop areas of the preserved portions of turbidite systems discussed in this paper. Structures south of the Variscan Front have experienced post-Acadian clockwise rotations (McClelland-Brown 1983). Arrows are average palaeocurrent directions. Biofacies information is from Smith (1999).

basin was affected by early Silurian extension/transtension, which was succeeded by transpressional inversion in the late Silurian to early Devonian Acadian phase of mountain building (e.g. Woodcock et al. 1988; Woodcock & Strachan 2000). A sigmoidal array of basement lineaments was repeatedly reactivated throughout the basin history.

Syndepositional structures active in late Ordovician and early Silurian time, with a strong component of extension across them have been inferred (Smith 1987a, b; Woodcock et al. 1988). The tilt block model of Smith (1987a) was subsequently confirmed by further mapping work (Woodcock 1990; BGS 1993, 1994; Davies et al. 1997) and supported by estimates of tectonic subsidence (Woodcock et al. 1996). The occurrence of extension/transtension in the basin at the same time as uplift on the southern basin margin is compatible with the view promulgated by Woodcock (1984) that the basin may have developed within a regional strike-slip context. The present deformation style of the basin-fill is dominated by open to tight asymmetric folds formed during the Acadian deformation phase that inverted the basin.

Stratigraphy

A high-resolution graptolite biostratigraphy is available for the basin-fill (e.g. Zalasiewicz 1990; Loydell 1991, 1993). Graptolite zones average a little over 1 Ma in duration, using the new timescale of Gradstein & Ogg (1996), with some parts of the record being more finely subdivided by subzones.

An inferred eustatic sea-level curve has been developed by Johnson et al. (1991) based on comparing palaeowater depth histories on a number of continents (Fig. 2). This shows a Rhuddanian rising limb followed by a mid Aeronian fall, a *sedgwickii* Zone high succeeded by an earliest Telychian fall and then a rise

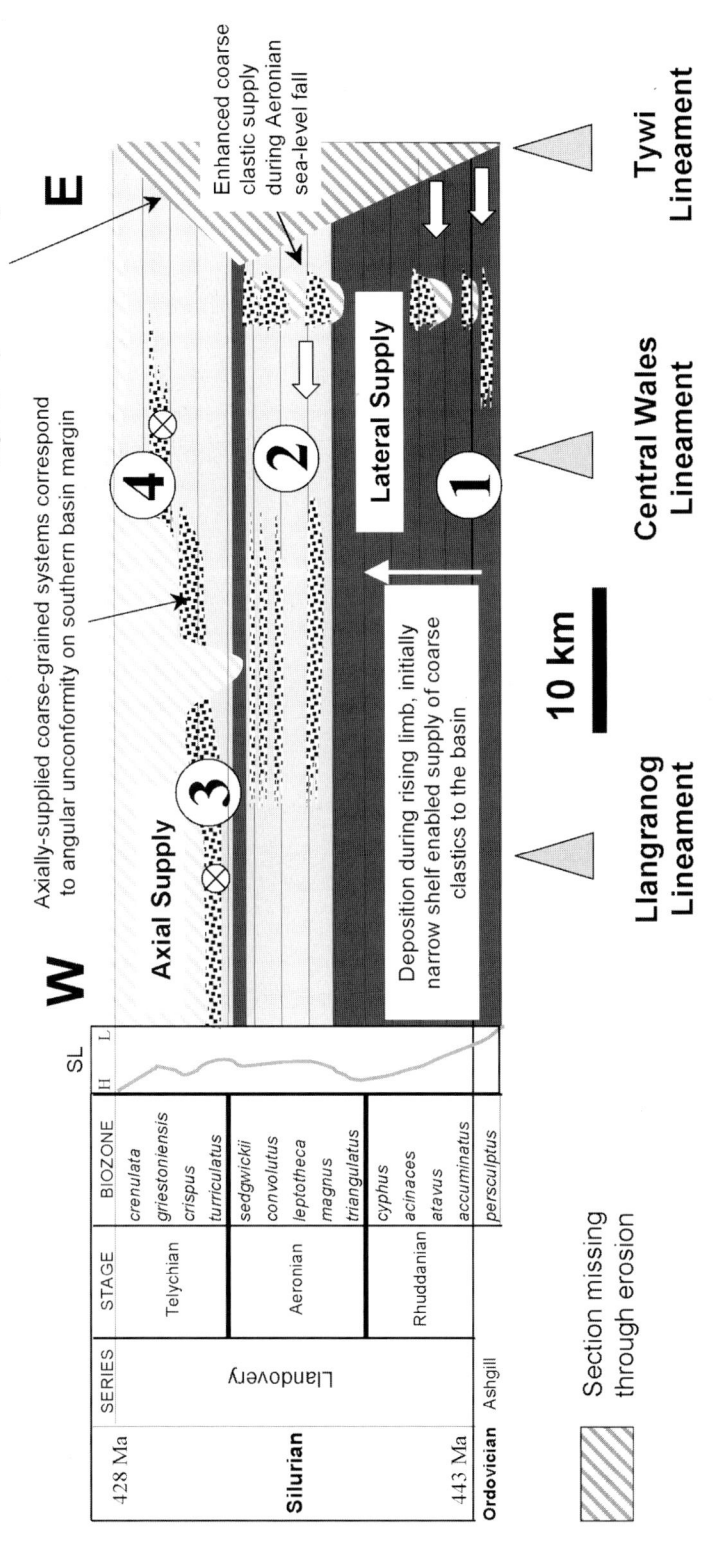

Fig. 2. West–east chronostratigraphic chart illustrating graptolite biostratigraphy of the uppermost Ordovician and Lower Silurian, distribution of coarse-grained turbidite systems and inferred eustatic sea-level history (Johnson *et al.* 1991). Ages of Caban Conglomerate main channel complexes and Ystrad Meurig lobes are taken from Davies & Waters (1995). Circled numbers indicate the four turbidite systems discussed in the text and shown in map view in Figure 4. Darker shade of background mudrocks indicates a preponderance of non-bioturbated anoxic hemipelagites.

towards a sea-level peak in the latest Telychian (*crenulata* Zone). The combination of such a eustatic curve with different subsidence histories on different margins of a basin will clearly result in differing relative sea-level (accommodation) histories (e.g. Smith 1987b). Figure 3 illustrates this principle by combining the Johnson *et al.* (1991) eustatic sea-level curve with illustrative total subsidence histories for representative locations on the southern (Marloes block, Pembrokeshire) and eastern (Midland Platform) margins of the Welsh Basin. For a given relative sea-level history, water depths and patterns of progradation/aggradation/retrogradation are then determined by the rate at which sediment supply exceeds/maintains pace with/fails to maintain pace with rates of creation of accommodation. The relative sea-level curves have been adjusted to match changes inferred from the platform areas (e.g. Bridges 1975), discussed below.

During the latest Ordovician and Rhuddanian sea-level rise (post Hirnantian glaciation) a very narrow shelf enabled supply of coarse-grained material to the basin. The Aeronian relative sea-level fall is expressed in an unconformity in the Garth District at the margin of the Midland Platform and probably led to an increase in the volumes of coarse clastic supply to the basin in the Caban-Ystrad Meurig System (Fig. 2). Unconformities are also found in the Berwyn Hills and Derwen High at this time (Wills & Smith 1922; Campbell 1983).

Later Aeronian sea-level rise is recorded on the western part of the Midland Platform by the first shallow marine deposits fringing the Longmynd-Shelve topographic high (Bridges 1975). The

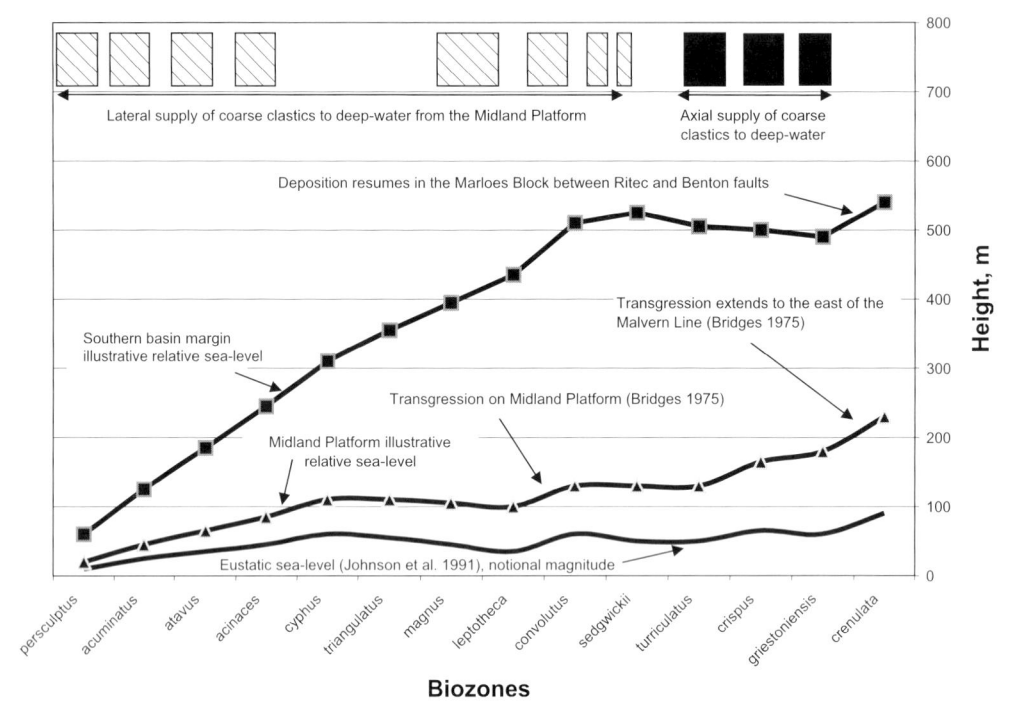

Fig. 3. Illustrative eustatic and relative sea-level curves for both eastern and southern basin margins together with timing of coarse sediment supply to the basin-fill. The eustatic curve is based on Johnson *et al.* (1991). Possible subsidence histories for the southern basin margin (between the Ritec and Benton faults) and Midland Platform margin have been combined with this eustatic curve to show relative sea-level (accommodation) curves. The high total (i.e. tectonic plus sediment load induced) subsidence rates contained in the southern basin margin curve are roughly consistent with the thickness of the poorly dated Skomer Volcanic Group. Water depth history depends on the rate at which this available space tends to be filled. For example, sediment supply rates on the southern basin margin were largely able to keep pace with or exceed rates of creation of accommodation prior to the Telychian relative sea-level fall. The palaeotopographic map of Bridges (1975) indicates that palaeorelief on the Midland Platform exceeding 200 m had been flooded by late Telychian (approximately *griestoniensis* Zone) times. High-frequency alternations of sandstone-rich and sandstone-poor packets in the axial turbidite systems (e.g. Figs 10 & 11) record higher-frequency oscillations of relative sea-level than depicted here.

latest Aeronian peak sea-level is recorded in the basin-fill by the *Monograptus sedgwickii* black shale. Major sediment sources for coarse clastic material were removed from the eastern margin of the basin at this time and voluminous supply from the south began in the succeeding *turriculatus* Zone.

Evidence for an earliest Telychian reduction in water depth at May Hill and the southern Longmynd on the Midland Platform has been noted by Johnson *et al.* (1991) and interpreted as a relative sea-level fall, discounting explanations related to increased sediment supply rate (Cocks & Rickards 1969). Later Telychian transgression above a hard-substrate shelf with relief in the order of 250 m is well documented (Bridges 1975). This extended the shelf sea some 120 km to the east of its earliest Silurian coastline.

On the southern basin margin a prolonged Telychian relative sea-level low associated with tectonic uplift is recorded by an angular unconformity in the Marloes Block between the Ritec and Benton Faults (Sanzen-Baker 1972; Bassett 1982). The pronounced alternation between sandstone bodies and thin-bedded turbidite background intervals (Smith 1995a) is strongly suggestive of higher-frequency (tens of thousands of years) relative sea-level changes. Deposition resumed in this area with the latest Telychian Coralliferous Series, which indicates that rates of rise of eustatic sea-level exceeded rates of tectonic uplift (Smith 1987b).

In addition to the angular unconformity in the Marloes Block of southwest Wales, evidence consistent with a widespread pulse of tectonic activity in earlier Telychian times occurs in the form of (a) stratigraphic growth in *turriculatus-crispus* Zone times across the Central Wales Lineament (Smith 1987a), and (b) pebbly mudstone debrites and slides derived from a stepped eastern basin margin (Smith 1987b; Smith *et al.* 1990; Woodcock *et al.* 1996; Davies *et al.* 1997).

Biofacies

Trace fossil assemblages of Telychian age have been used to define ichnofacies that change in character along a palaeobathymetric profile from the Midland Platform to the deep Welsh Basin (Smith 1999, Table 1). These have been combined with earlier studies of benthic assemblages across

Table 1. *Distribution of trace fossil ichnogenera along a Late Llandovery (Telychian) shelf to basin palaeobathymetric gradient (after Smith 1999)*

Ichnogenera	Locations			
	Basin-fill	Distal shelf (Buttington Quarry)	Mid shelf (Hughely Shales)	Proximal shelf (Gullet Quarry)
Monocraterion				•
Rosselia				•
Skolithos		•	•	•
Chondrites	•	•	•	•
Palaeophycus	•	•	•	•
Diplocraterion		•	•	
Diplichnites			•	
Scolicia			•	
Walcottia			•	
Cruziana/Rusophycos		•		
Muensteria		•		
Teichichnus		•		
Paleodictyon/Squamodictyon	•			
Megagrapton	•			
Acanthorhaphe	•			
Dictyodora	•			
Saerichnites	•			
Spirorhaphe	•			
Glockeria	•			
Gordia	•			
Cochlichnus	•			
Planolites	•			
Trichichnus	•			

In the basin-fill the only body fossils recorded are graptolites and very rare sponge spicules. Locality details can be found in Smith (1999).

the shallow marine domain of the Midland Platform to provide a biofacies context for the basin. Within the basin-fill trace fossil assemblages combined with the character of interturbidite hemipelagite character have been used to track temporal variations in bottom-water oxicity (Smith 1987c, 1988). Zalasiewicz (1990) provided a summary of the distribution of anoxic hemipelagites throughout the Llandovery Series. He highlighted two useful stratigraphic markers: (a) the mid Aeronian (*magnus* Zone) switch from underlying laminated non-bioturbated hemipelagites upwards into bioturbated hemipelagites; and (b) the anoxic interval in the *sedgwickii* Zone represented by the dark *sedgwickii* Shale.

Turbidite systems

The turbidite systems are discussed in terms of the following four systems (indicated in Figs 2 & 4), from oldest to youngest:

(1) Cerig Gwynion Grits system (*persculptus* Zone);
(2) Caban-Ystrad Meurig system (latest *persculptus*/*acuminatus* to *sedgwickii* Zones);
(3) Aberystwyth-Rhuddnant systems (*turriculatus* to *crispus* zones);
(4) Pysgotwr System (*griestoniensis* Zone) and the lateral bounding slope.

The first two of these were laterally supplied from the eastern basin margin, whereas the latter two were supplied axially from the southwest. Laterally supplied and axially supplied systems have contrasting mineralogical and textural characteristics, recorded in both the fine-grained material (Ball *et al.* 1992) and the coarse fraction (Morton *et al.* 1992). Turbidite sandstones supplied from the southern basin margin are far more texturally and mineralogically immature than those supplied earlier from the east.

Cerig Gwynion Grits System

The Cerig Gwynion Grits System maps out as an externally sheet-like sandstone body reaching approximately 100 m in thickness and with a northeast–southwest length of 15 km and northwest–southeast width of at least 5 km (Fig. 5; Kelling & Woollands 1969; Smith *et al.* 1990; Davies & Waters 1995). The system is composed of sandstones and conglomerates with low matrix content. Dish structures are common. It occurs within the latest Ordovician *persculptus* Zone, which represents a time of bottom water

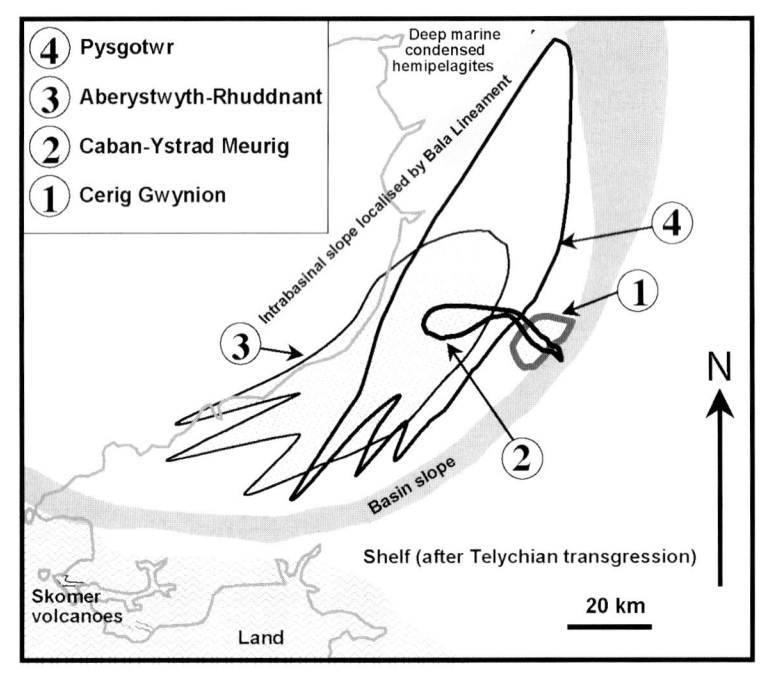

Fig. 4. Location map showing approximate outlines of the four turbidite systems discussed in this paper. Note that the proximal portions of the axial systems are shown schematically.

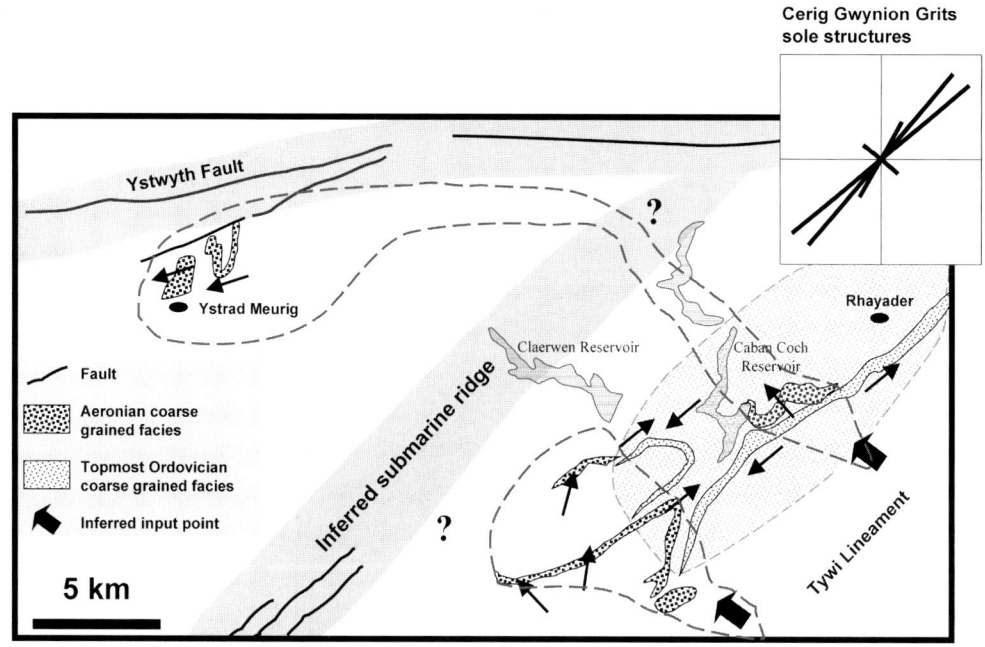

Fig. 5. Map showing outcrop patterns, inferred extent of coarse-grained sediment bodies and palaeocurrent patterns. Trends of palaeocurrents measured from sole structures in the Cerig Gwynion Grits Member of the Caban Conglomerate Formation (Davies & Waters 1995) are plotted in order to emphasize the absence of a radial palaeocurrent pattern. Detailed maps of the Lower Silurian succession in the vicinity of Rhayader have been published by Davies & Waters (1995). See Figure 1 for the location of this area.

anoxia associated with a glacioeustatic sea-level rise (Brenchley 1988).

Palaeocurrents from sole structures trend parallel to the base of the slope whereas sparse palaeocurrents from cross-lamination record palaeoflow towards the northwest, at a high angle to the strike of the slope. Palaeoflow directions inferred from sole structures run both to the northeast and the southwest. There is no indication of a radial palaeoflow pattern. Two main input points can be inferred (Fig. 5), one southeast of the location of vertically stacked Caban Conglomerate channel storeys, the other up depositional dip of the southwestern portions of the system.

Two interpretations of the Cerig Gwynion Grits turbidite system exist in the literature:

- a depositional sandstone lobe with no influence of palaeobathymetry (Davies & Waters 1995);
- a depositional sandstone lobe confined to the northwest by a south-east-facing slope localized by the Central Wales Lineament (Smith 1987a). This interpretation was based on the combination of abundant palaeocurrents recorded from sole structures that run parallel to the slope and the mapped elongate form of the sandstone body parallel to slope strike, together with structural mapping of the Central Wales Lineament to the northwest. The no-confinement model above is not consistent with the observed palaeocurrent patterns.

A possible analogue for the Cerig Gwynion system is that of the Pliocene fill of the San Diego Trough, inner California Continental Borderland (Teng & Gorsline 1989). Here sediment supplied from the La Jolla canyon is deflected to flow along the axis of a depression that is roughly 10 km across. Similar deflections are documented for turbidite systems of similar scale lying offshore eastern Corsica (Gervais et al. this volume). A comparable subsurface example is that of the Cretaceous Lucina Formation, offshore Gabon (Smith 1995b).

Caban-Ystrad Meurig System

The Caban-Ystrad Meurig System (Cave 1979, 1992; Smith et al. 1990) lies above the Cerig

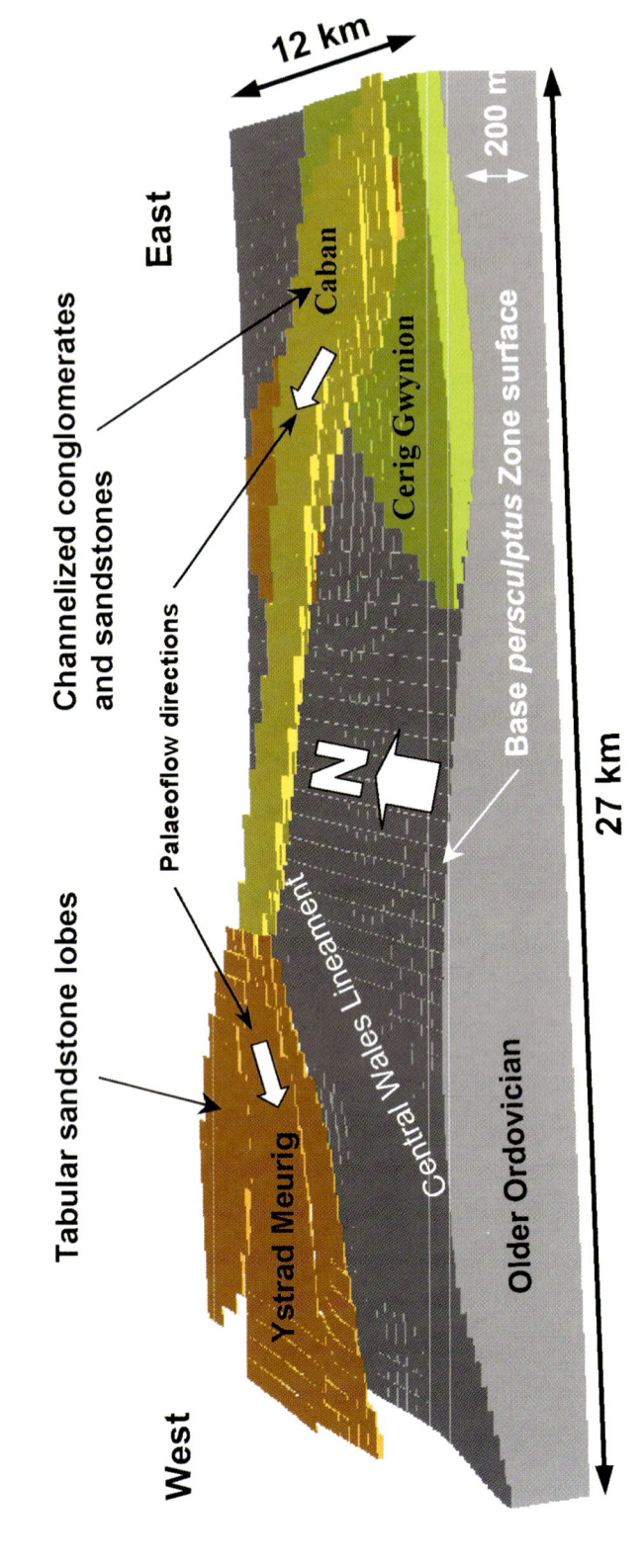

Fig. 6. Illustration of Caban–Ystrad Meurig architecture in a quantitative three-dimensional model. Model dimensions are 27 km east–west by 12 km north–south and 1 km vertically. Cell sizes in the model are 500 × 500 × 1 m. Post-*persculptus* Zone mudrocks have been made transparent in order to make visible the modelled architecture of the coarse-grained stratigraphic elements.

Gwynion Grits system. It consists of two palaeogeographic domains: (a) a stack of coarse-grained channel complexes in the vicinity of Rhayader; and (b) a system of tabular depositional lobes to the west in the vicinity of Ystrad Meurig (Figs 5 & 6). Detailed maps of these elements have been published by Davies & Waters (1995).

The oldest of the Caban conglomerate channel-fills occur within the uppermost *persculptus* to *acuminatus* zones (Davies & Waters 1995) directly above the Cerig Gwynion Grits. The gravel and sand fills of these channels exhibit deep scouring and associated steep depositional dips. Mudstones are discontinuous through erosional truncation.

The Ystrad Meurig Grits are composed of near-tabular thin to thick, fine- to coarse-grained sandstones with local conglomerates at the base of the succession (Smith *et al.* 1990; Cave 1992). Beds are typically normally graded and contain abundant parallel and convolute lamination. Davies & Waters (1995) distinguished four discrete Ystrad Meurig lobes. The three oldest of these are of (a) *magnus-leptotheca* Zone age (mid Aeronian), (b) early *convolutus*, (c) late *convolutus*, and are not documented to the north of the Ystwyth Fault. A very thin younger interval of *sedgwickii* Zone age is recorded from the Cwm Rheidol borehole to the north of the Ystwyth Fault (Davies & Waters 1995). A northward side-lapping of tabular lobes from pre-*leptotheca* to *sedgwickii* Zone times has been inferred. Smith *et al.* (1990) reported soft-sediment faults with strong components of downthrow to the south. These were considered compatible with an influence on depositional patterns and palaeocurrents by an ancestral Ystwyth Fault. There is a close match between the trend of the present Ystwyth Fault and measured palaeocurrents, and there is no record of a radial palaeocurrent pattern (Davies & Waters 1995).

Figure 6 shows a three-dimensional quantitative computer model constrained by measured sections and the detailed biostratigraphy recorded in Davies & Waters (1995). It has been constructed by interpolating the tops and bases of correlated sediment bodies with anisotropy values being guided by the measured palaeocurrents. The model illustrates the basal topography-filling (in this interpretation) Cerig Gwynion Grits system overlain by vertically stacked storeys of Caban channels. The higher of these channel storeys are linked to approximate age equivalents in the depositional lobe domain close to Ystrad Meurig. Smith (1987a) conjectured the possibility (although impossible to test without subsurface control) that erosion and bypass of the palaeoridge took place at the crossing point of the Central Wales Lineament.

Aberystwyth-Rhuddnant systems

Extensive axial turbidite systems developed during the Telychian stage (Fig. 7). Basal ages of the intervals containing sandstone lobe packets become younger from west (*utilis* Subzone of the *turriculatus* Zone, Cave & Lloydell 1997) to east (*griestoniensis* Zone, Smith 1987b). Together, they provide an excellent example of a depositional sequence (or sequence set) developed in response to relative sea-level fall caused by local tectonic uplift at source (Smith 1987b, 1988; Smith *et al.* 1990). Sandstones in the axial systems contrast with those supplied from the eastern basin margin in commonly having very high matrix contents (commonly 15–40%), much of which is probably primary. Variations in matrix content are readily visible in the field through the greater tendency of mud-rich sandstones to take up tectonic cleavage.

Bimodal palaeocurrents associated with the basal Aberystwyth Grits. Thin-bedded turbidites underlying and interfingering with the basal Aberystwyth Grits (Fig. 8) exhibit strikingly bimodal palaeocurrent patterns, with sole structures indicating flow towards present northeast and ripple cross-lamination indicating flow towards the northwest (Anketell & Lovell 1976; Smith & Anketell 1992). Following the experimental results of Kneller *et al.* (1991), Smith & Anketell (1992) interpreted this pattern in terms of oblique flow reflection off an intrabasinal slope lying to the southeast of the coastal belt of exposures and facing towards the northwest.

Palaeoflow deflections and confinement of the Aberystwyth Grits. There has been some controversy over the nature of lateral facies changes to the east of the coastal exposures of the Aberystwyth Grits. The view of Wilson *et al.* (1992) that the sand-rich Aberystwyth Grits system is abruptly bounded by a fault system to the east was refuted by Cave & Loydell (1997) on the evidence of detailed biostratigraphic work. Cave and Loydell did, however, suggest that a certain degree of confinement could have been provided by depositional dips associated with the underlying Devil's Bridge Formation. This interval of thin-bedded fine-grained turbidites they attributed to a 'slope apron' setting despite having been deposited

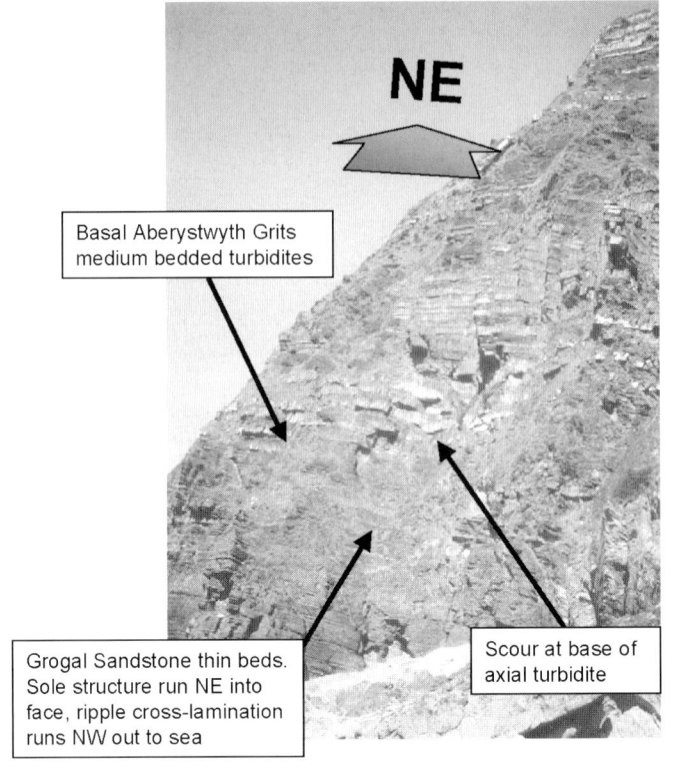

Fig. 7. Block diagram summarizing lateral shifts in the eastern edge of the axial coarse-grained turbidite systems and the relations between palaeocurrents and structurally induced bounding slopes. AG, Aberystwyth Grits; RG, Rhuddnant Grits; PG, Pysgotwr Grits. The schematic surface beneath the turbidite systems represents the base Telychian surface.

over 40 km basinwards of the coeval slope base.

McCann & Pickering (1989) documented south-directed palaeocurrents from ripple cross-lamination in thin-bedded Telychian turbidites close to the northern termination of the Aberystwyth Grits. Smith (1990) suggested that the Bala Lineament localized a syndepositional slope in this area that deflected distal turbidity currents and acted as a slope down which northerly derived flows could travel, or alternatively off which flows might reflect. Palaeocurrent patterns in the Aberystwyth and Rhuddnant Grits are summarized in Fig. 9.

Structural confinement of the Rhuddnant Grits. Smith (1987a) noted a dramatic facies change in rocks of *turriculatus* and *crispus* Zone age across the Central Wales Lineament. Packets of medium-bedded sandstone lobe facies occur only to the northwest of the lineament, and thickness changes are dramatic (Woodcock 1990 estimated a change from 1500 m or more in the northwest to about 500 m in the southeast).

There is, however, no evidence, in the form of debrites or slide deposits, for fault scarps having been exposed at surface. Thus instantaneous structurally induced slopes were most probably never steep and were probably associated with monoclines above blind faults (unlike the reconstructions in Wilson *et al.* 1992). Mapping by the British Geological Survey (Davies *et al.* 1997) suggests that structural confinement was greatly reduced by *crispus* Zone times relative to that recorded in the underlying *turriculatus* Zone section (Fig. 7). Clayton (1993) investigated the palaeocurrent record from this succession to test for flow reflection off this bounding slope. He found that there was no record of reflection off the Central Wales Lineament, an observation that is consistent with the palaeocurrent record for the succeeding Pysgotwr Formation (Smith 1987b). It is here suggested that flows had been turned to run parallel to bounding slopes in a more proximal position, and hence there was no obliquity between flow path and strike of bounding slope.

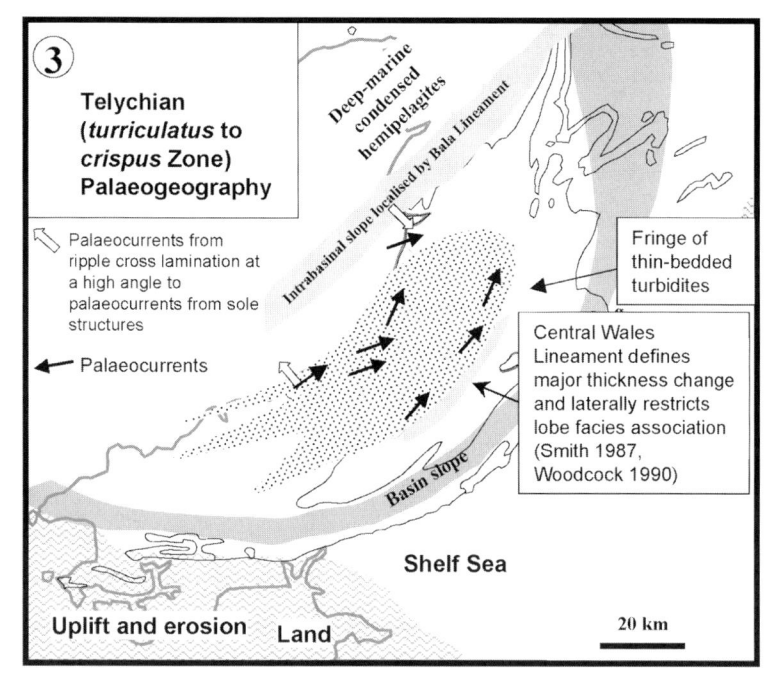

Fig. 8. Grogal Sandstones and basal Aberystwyth Grits at Craig Grogal. The view is towards the northeast. Sole structures indicate flow directly into the outcrop face whereas ripple migration directions indicate flow seawards towards the left (northwest). A section through a scour cut by a flow travelling towards the northeast is visible in the middle of the field of view. Approximately 70 m of section is visible in the photograph.

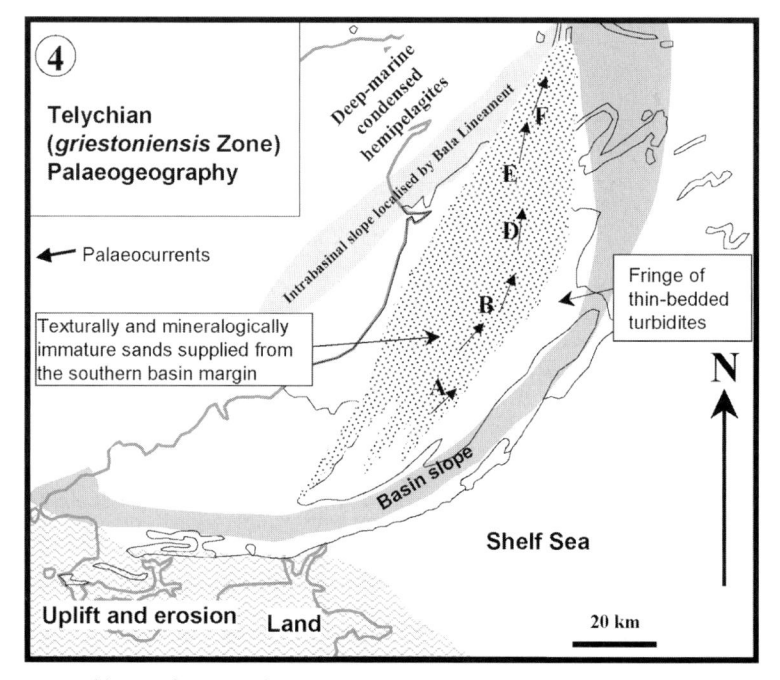

Fig. 9. Palaeogeographic map for *turriculatus* to *crispus* Zone times. Note the palaeocurrents showing a predominant parallelism with bounding tectonic structures.

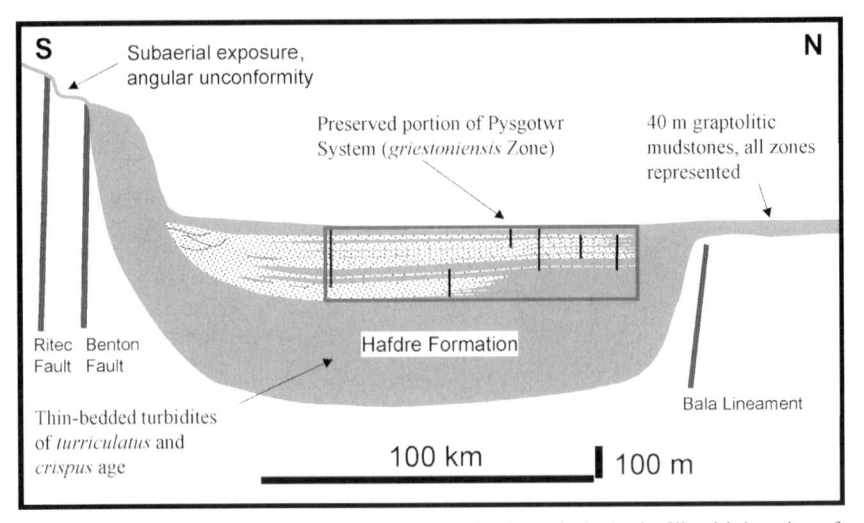

Fig. 10. Reconstructed section from the southern basin margin through the basin-fill, with location of measured sections shown in Figure 11 indicated by black vertical lines.

Pysgotwr Formation and the Lateral Basin Slope

The Pysgotwr Formation is the most completely preserved of the axially supplied turbidite systems. Its preserved portions reach a down-palaeoflow extent of approximately 100 km (Figs 10 & 11). Four facies associations have been recognized in this system and are described below.

(1) *'Channel-lobe transition'* (Smith 1987b) or *'Amalgamated conglomerates and sandstones'* (Smith 1995a). Amalgamated packages of sandstones and lenticular conglomerates are recorded from the most proximal preserved portions of the Pysgotwr Formation (Smith 1987a). Dune-scale cross-stratification also occurs in this proximal area (Smith 1995a), but not further basinwards.

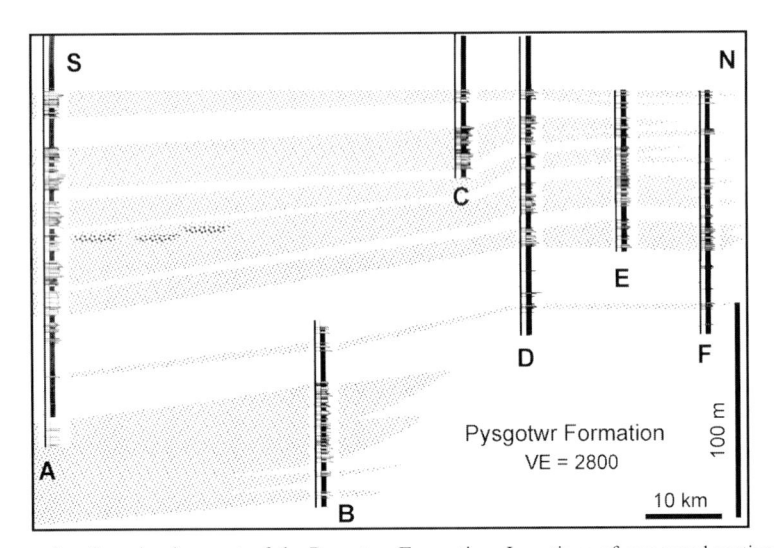

Fig. 11. Down-palaeoflow development of the Pysgotwr Formation. Locations of measured sections are shown in Figure 13. For detailed correlations in the proximal Craig Twrch area see Smith (1995). Systematic down-palaeoflow changes in percentage amalgamation, maximum grain size and maximum bed thickness are documented in Smith (1987b).

(2) *Sandstone lobe.* Packages of thin- to thick-bedded sandstones punctuate the dominant thin-bedded turbidite background facies of the Telychian basin-fill. This is the preponderant sandstone-rich facies association in the Aberystwyth Grits, Rhuddnant Grits and Pysgotwr Grits formations (Cave & Hains 1986; Smith 1987b, 1995; Clayton 1994). Bed thickness patterns show no prevalence of thickening-up over thinning-up tendencies and have been argued to be random through the application of runs tests (Smith 1987, 1988). Bed thickness populations for sandstone beds having non-laminated sandstone divisions at their bases record a clear proximal to distal change (Fig. 12). Matrix-rich slurried beds (Wood & Smith 1959; Hiscott & Middleton 1979) occur in this facies association, reaching a proportion of 20% close to the eastern limit of the depositional sandstone lobe area (e.g. in the Talerddig area), but lower towards the west. These were interpreted as debris flow deposits, possibly transformed from turbulent flows during deceleration.

(3) *Thin-bedded turbidites.* Thinly interbedded laminated coarse siltstones/fine sandstones and mudstones are the background facies to the sandstone lobe packets in the Aberystwyth Grits and Cwmystwyth Grits groups. Climbing ripple cross-lamination and convolute lamination are common features. Coarse fraction bed thicknesses range from less than 1 cm up to 25 cm. Thin-bedded fine-grained turbidites are separated by hemipelagites, which may be bioturbated or organic-rich and non-bioturbated (Cave 1979; Smith 1987c).

(4) *Muddy slope.* This facies association occurs in the vicinity of Rhayader and includes mud turbidites, hemipelagites and deformed slid mudstones. Local pebbly mudstone debrites occur in the Cwmbarn Formation (Smith *et al.* 1990).

The Telychian lateral basin slope. Mapping suggests that something close to true onlap occurred against a fault-induced non-depositional slope in the Rhayader area, at least during Telychian times (Roberts 1929) and possibly earlier (James 1992). However, the eastern edges of the medium- to thick-bedded lobe sandstones lie many kilometres to the west of this onlap area (Figs 13 & 14), leading Smith (1987b) to suggest that there was no discrete single slope break, but an upwards-concave lower slope. Clear lateral variations in percentage sand, the degree of amalgamation and thickness within the sandstone lobe facies association (Smith 1997b, 1988) are also consistent with such an inference.

A step in the slope associated with a fault further to the east of the onlap termination mapped by Roberts (1929) and others was inferred by Smith (1987b, 1988). The theme of fault-related steps was further developed by other authors (Smith *et al.* 1990; Waters *et al.* 1992; Woodcock *et al.* 1996; Davies *et al.* 1997), who documented local debrites consistent with the presence of exposed fault scarps.

A quantitative two-dimensional forward geometric model is used to illustrate plausible stratal architectures of the Telychian bounding slope. Input per time or depth increment consists of slope gradients and aggradation rates along the modelled basin to slope profile. In the simplest slope–basin model, rate of lateral overlap (i.e. horizontal component of the movement of the

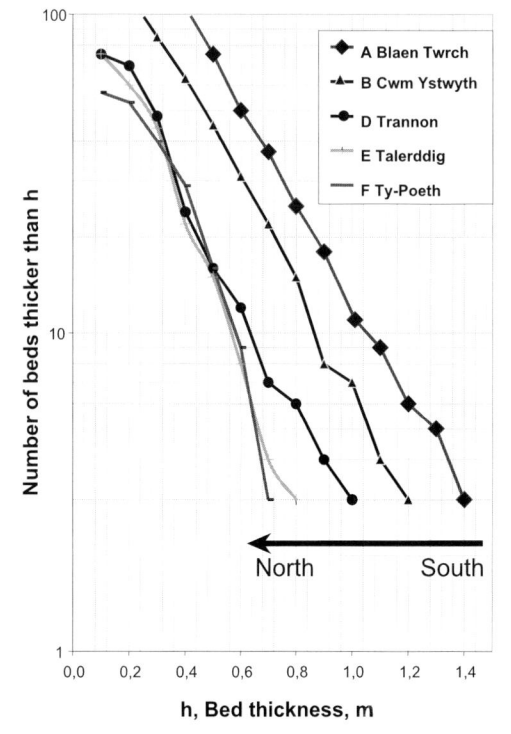

Fig. 12. Bed thickness distributions for beds having non-laminated basal intervals from sections in the Pysgotwr Formation along a 65 km proximal to distal transect (data from Smith 1988). Data are plotted as the log of number of beds thicker than a given bed thickness against bed thickness. The data are consistent with a negative exponential distribution (e.g. see Drummond 1999). The arrow shows proximal to distal direction. Locations are noted on the correlation panel shown in Figure 11.

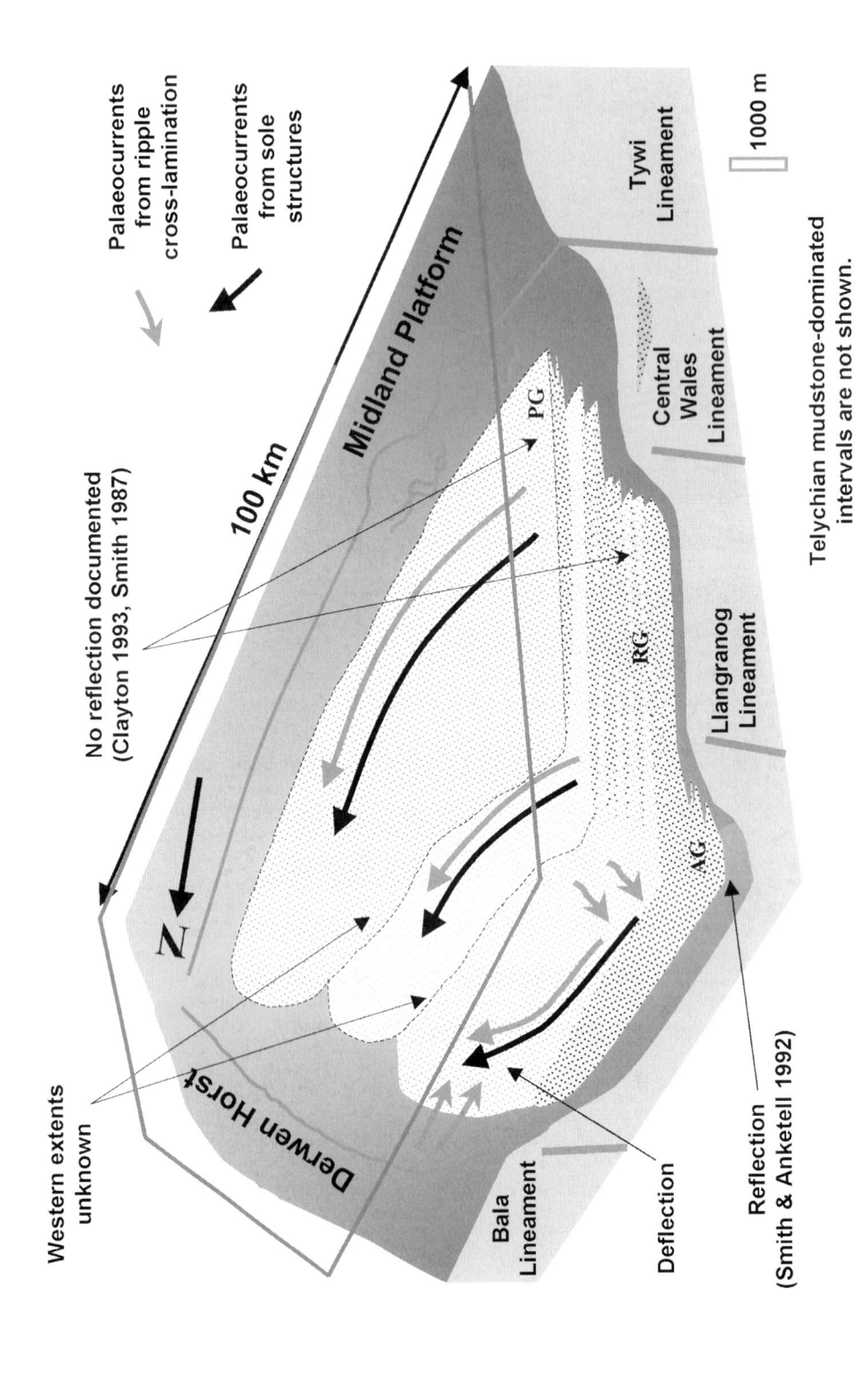

Fig. 13. Palaeogeographic map for *griestoniensis* Zone times. Locations of measured sections shown in Figure 11 are indicated.

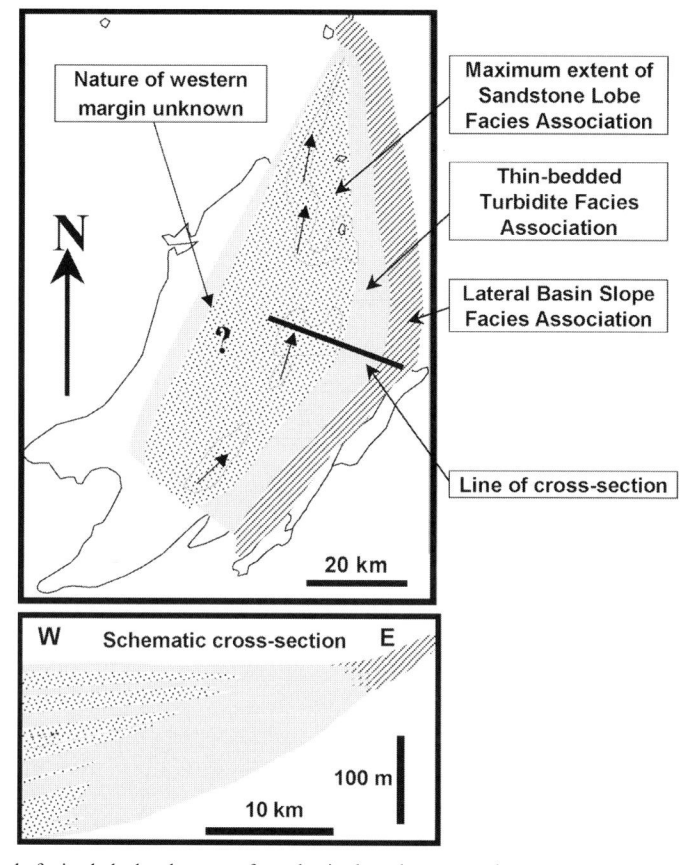

Fig. 14. Large-scale facies belt development from basinal to slope areas in *griestoniensis* Zone times.

slope-base to basin break in an upslope direction) equals the difference between the aggradation rate in the basin and the aggradation rate on the slope divided by the tangent of the slope (Smith 1988). Thus pure onlap occurs when aggradation rate on the slope is zero, and a 'feathering' or aggradational onlap occurs if aggradation rates on the slope are higher (Smith 1988; Smith & Joseph 2004). Such a model can be extended to more closely match the concave-up slope envisaged by Smith (1987b) by adding further slope segments. Figure 15 shows such an illustrative model of stratal architectures associated with the slope laterally bounding the Pysgotwr turbidite system. The input assumptions are summarized in Figure 16. Slope gradients and aggradation rates relative to the basin-floor succession have been adjusted in order to match thicknesses measured for the belt of thin-bedded turbidites and adjacent muddy slope and the observed widths of these belts. The model shows how the trajectory of successive slope break positions is governed by lateral variations in aggradation rates along the bathymetric profile. This style of aggradational onlap stands in marked contrast to well-known abrupt onlap configurations such as those in parts of the Grès d'Annot, southern France (Smith & Joseph 2004). The modelled stratal architecture corresponds, at a very large scale, to the convergent thinning seismic facies used in subsurface seismic interpretation (Prather *et al.* 1998).

Potential application of Welsh Basin analogues to subsurface prediction

The characteristics of the Welsh Basin turbidite systems discussed above could be applied to subsurface problems of reservoir prediction as follows:

(1) The basin-fill provides an excellent illustration of the very different reservoir architecture and mineralogical and textural

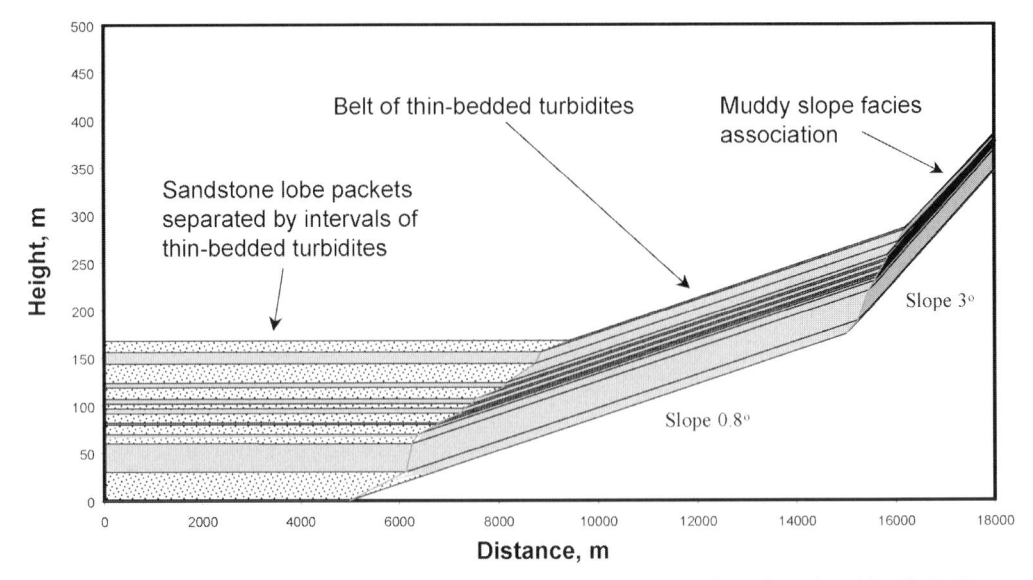

Fig. 15. Model approximating an upward concave slope. Bold lines indicate the trajectories of breaks in slope through time. The basinal succession is that recorded in the Craig Twrch area (Smith 1995a). Input to this model is summarized in Figure 16.

characteristics (hence reservoir properties in the subsurface) that can occur in sands supplied from different margins of a basin. Comparisons can be made with the North Sea (den Hartog Jager *et al.* 1993) and Norwegian Sea (Vergara *et al.* 2001) basins.

(2) Slurried beds (in the sense of Wood & Smith 1959) are common in lateral and distal positions in turbidite systems.

(3) Caution may be needed in interpreting palaeocurrents from ripple cross-lamination, as previously noted by Kneller *et al.* (1991) and others. Interpretations must be made in full awareness of the context of potential bounding slopes.

(4) Although large angular differences between sole structures and ripple migration directions can be a clear indicator of the presence

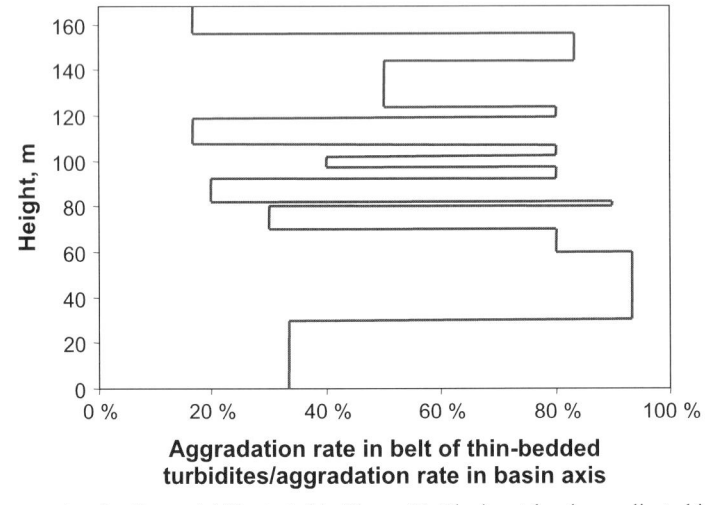

Fig. 16. Input parameters for the model illustrated in Figure 15. The input has been adjusted in order to match observed width of belt of thin-bedded turbidites and mapped thicknesses.

of a nearby bounding slope, the absence of such features does not preclude nearby abrupt terminations of sand-rich sections associated with fault-induced differential subsidence. In the latter case, the lack of obliquity between flows and slopes (perhaps due to upflow complete turning to slope-parallelism) may explain the lack of reflections.

(5) Close facies and thickness matches can be made between the Telychian axial turbidite systems of the Welsh Basin and the Nise Formation of the Norwegian Sea Basin (e.g. Høgseth *et al.* 1999). The very different Cerig Gwynion Grits system has several features of similarity with Paleocene turbidite systems offshore mid Norway (e.g. Gjelberg *et al.* 2001; Smith & Møller in press).

(6) Sediment gravity flow sandstones with textural and mineralogical properties favourable for good reservoir quality at depth, prior to the onset of intense quartz cementation (e.g. Cerig Gwynion Grits), may be associated with systems fed through small canyons during times of relative rises of sea-level. Greater textural immaturity may be associated with direct supply from deltaic sources, especially if associated with a high relief non-vegetated hinterland, at times of low relative sea-level (e.g. the Telychian axial systems).

(7) The outcrop observations related to the Telychian basin to slope profile, and the geometric model used to explain them, illustrate how high percentage sandstone bodies may occur in the basinal portions of large-scale convergent thinning stratal architectures.

Conclusions

The uppermost Ordovician and Lower Silurian fill of the Welsh Basin provides a rich natural laboratory in which to study a range of turbidite system styles, in terms of supply configuration, scale, sandstone texture, sediment body architecture and responses of sediment gravity flows to palaeoslopes. The following inferred examples of interactions between slopes and gravity flows have been discussed:

(1) deflections of flows supplied from the eastern basin margin to run parallel with bounding slopes in the Cerig Gwynion Grits system;
(2) basinward-stepping deposition of channel-fed lobes following fill of inboard depression;
(3) parallelism between palaeocurrents and the Ystwyth Fault in the Ystrad Meurig sheet portion of the Caban-Ystrad Meurig system;
(4) inferred flow reflections in parts of currents moving coarse silt and very fine sand as ripples in the Grogal sandstones and lowest portions of the Aberystwyth Grits;
(5) deflections to run parallel to bounding slopes in the younger Telychian systems. However, in these cases, no evidence for reflections off laterally bounding slopes has been encountered, and this has been attributed to an absence of flow obliquity in the preserved portions of these systems.

Helpful reviews from N. Woodcock, K. Pickering and A. Bell are gratefully acknowledged.

References

ANKETELL, J. M. & LOVELL, J. P. B. 1976. Upper Llandoverian Grogal Sandstone and Aberystwyth Grits in the New Quay area, central Wales: a possible upwards transition from contourites to turbidites. *Geological Journal*, **11**, 101–108.

BALL, T. K., DAVIES, J. R., WATERS, R. A. & ZALASIEWICS, J. A. 1992. Geochemical discrimination of Silurian mudstones according to depositional process and provenance within the Southern Welsh Basin. *Geological Magazine*, **129**, 567–572.

BASSETT, M. G. 1982. Silurian rocks of the Marloes and Pembroke Peninsulas. *In*: BASSETT, M. G. (ed.) *Geological Excursions in Dyfed, South-West Wales*. 103–122.

BRENCHLEY, P. J. 1988. Environmental changes close to the Ordovician-Silurian boundary. *In*: COCKS, L. R. M. & RICKARDS, R. B. (eds) *A Global Analysis of the Ordovician-Silurian Boundary*. Bulletin of the British Museum (Natural History), Geology Series, **43**, 377–385.

BRIDGES, P. H. 1975. The transgression of a hard substrate shelf: the Llandovery (Lower Silurian) of the Welsh Borderland. *Journal of Sedimentary Petrology*, **45**, 79–94.

BRITISH GEOLOGICAL SURVEY. 1993. *Rhayader*, England & Wales Sheet 178.

BRITISH GEOLOGICAL SURVEY. 1994. *Llanilar*, England & Wales Sheet 179.

CAMPBELL, S. D. G. 1983. *The geology of an area between Bala and Betws y Coed, North Wales*. PhD thesis, University of Cambridge.

CAVE, R. 1979. Sedimentary environments of the basinal Llandovery of mid-Wales. *In*: HARRIS, A. L., HOLLAND, C. H. & LEAKE, B. E. (eds) *The Caledonides of the British Isles, Reviewed*. Geological Society, London, Special Publications, **8**, 517–526.

CAVE, R. 1992. The Ystrad Meurig Grits, an early Silurian turbiditic sandstone lobe in mid-Wales. *Journal of the Earth Science Teachers' Association*, **17**, 87–95.

CAVE, R. & HAINS, B. A. 1986. Geology of the country between Aberystwyth and Machynlleth. *Memoir of the British Geological Survey*, Sheet 163 (England and Wales).

CAVE, R. & LOYDELL, D. K. 1997. The eastern margin of the Aberystwyth Grits Formation. *Geological Journal*, **32**, 37–44.

CLAYTON, C. 1993. Deflection versus reflection of sediment gravity flows in the late Llandovery Rhuddnant Grits turbidite system, Welsh Basin. *Journal of the Geological Society, London*, **150**, 819–822.

CLAYTON, C. 1994. Contrasting sediment gravity flow processes in the late Llandovery Rhuddnant Grits turbidite system, Welsh Basin. *Geological Journal*, **29**, 167–181.

COCKS, L. R. M. & RICKARDS, R. B. 1969. Five boreholes in Shropshire and the relationships of shelly and graptolitic facies in the Lower Silurian. *Quarterly Journal of the Geological Society, London*, **124**, 213–238.

CRIMES, T. P. & CROSSLEY, J. D. 1980. Inter-turbidite bottom current orientation from trace fossils with an example from the Silurian Flysch of Wales. *Journal of Sedimentary Petrology*, **50**, 821–830.

DAVIES, J. R. & WATERS, R. A. 1995. The Caban Conglomerates and Ystrad Meurig Grits Formation: nested channels and lobe switching on a mud-dominated latest Ashgill to Llandovery slope-apron, Welsh Basin, UK. *In*: PICKERING, K. T., HISCOTT, R. N., KENYON, N. H., RICCI LUCCHI, F. & SMITH, R. D. A. (eds) *An Atlas of Deep-Water Environments: Architectural Style in Turbidite Systems*. Chapman & Hall, London, 184–193.

DAVIES, J. R., FLETCHER, C. J. N., WATERS, R. A., WILSON, D., WOODHALL, D. G. & ZALASIEWICZ, J. A. 1997. Geology of the country around Llanilar and Rhayader. (ed.) *Memoir, British Geological Survey*, Sheets 178 & 179.

DORÉ, A. G., LUNDIN, E. R., BIRKELAND, Ø., ELIASSEN, P. E. & JENSEN, L. N. 1997. The NE Atlantic Margin: implications of late Mesozoic and Cenozoic events for hydrocarbon prospectivity. *Petroleum Geoscience*, **3**, 117–131.

DRUMMOND, C. N. 1999. Bed-thickness structure of multi-sourced ramp turbidites: Devonian Brallier Formation, Central Appalachian Basin. *Journal of Sedimentary Research*, **69**, 115–121.

GJELBERG, J. G., ENOKSEN, T., KJAERNES, P., MANGERUD, G., MARTINSEN, O. J., ROE, E. & VÅGNES, E. 2001. The Maastrichtian and Danian depositional setting, along the eastern margin of the Møre Basin (Mid-Norwegian Shelf): implications for reservoir development of the Ormen Lange Field. *In*: MARTINSEN, O. J. & DREYER, T. (eds) *Sedimentary Environments Offshore Norway: Palaeozoic to Recent*. NPF Special Publications, **10**, 421–440.

GRADSTEIN, F. M. & OGG, J. 1996. A Phanerozoic time scale. *Episodes*, **19**, 3–4.

HARTOG JAGER, D. G. DEN, GILES, M. R. & GRIFFITHS, G. R. 1993. Evolution of Paleogene submarine fans of the North Sea in space and time. *In*: PARKER, J. R. (ed.) *Petroleum Geology of Northwest Europe: Proceedings of the 4th Conference*. Geological Society, London, 59–71.

HISCOTT, R. N. & MIDDLETON, G. V. 1979. Depositional mechanics of thick-bedded sandstones at the base of a submarine slope, Tourelle Formation (Lower Ordovician), Quebec, Canada. SEPM Special Publications, **27**, 307–326.

HØGSETH, K., VAGLE, G. B., BERGFJORD, E., GRANHOLM, P. G. & SKJERVOLD, R. 1999. The Cretaceous depositional systems of the frontier Vøring Basin: evidence from the Nyk High well (6707/10-1) and the Vema Dome well (6706/11-1). *Sedimentary Environments Offshore Norway, Grieghallen, Bergen, Norway, 3–5 May 1999*, extended abstract.

JAMES, D. M. D. 1972. Sedimentation across an intrabasinal slope: the Garnedd-wen Formation (Ashgillian), West Central Wales. *Sedimentary Geology*, **7**, 291–307.

JAMES, D. M. D. 1992. Discussion of 'A late Ordovician/early Silurian non-depositional slope and perched basin along the Tywi Anticline, Mid Wales' by D. M. D. James. Reply. *Geological Journal*, **27**, 289–294.

JAMES, D. M. D. & JAMES, J. 1969. The influence of deep fractures on some areas of Ashgillian–Llandoverian sedimentation in Wales. *Geological Magazine*, **106**, 562–582.

JOHNSON, M. E., KALJO, D. & RONG, J.-Y. 1991. Silurian eustasy. *Special Papers in Palaeontology, London*, **44**, 145–163.

KELLING, G. & WOLLANDS, M. A. 1969. The stratigraphy and sedimentation of the Llandoverian rocks of the Rhayader district. *In*: WOOD, A. (ed.) *The Pre-Cambrian and Lower Palaeozoic Rocks of Wales*. University of Wales Press, Cardiff, 255–282.

KNELLER, B., EDWARDS, D., MCCAFFREY, W. & MOORE, R. 1991. Oblique reflection of turbidity currents. *Geology*, **19**, 250–252.

KOKELAAR, B. P., HOWELLS, M. F., BEVINS, R. E., ROACH, R. A. & DUNKLEY, P. N. 1984. The Ordovician marginal basin of Wales. *In*: KOKELAAR, B. P. & HOWELLS, M. F. (eds) *Volcanic and Associated Sedimentary and Tectonic Processes in Modern and Ancient Marginal Basins*. Geological Society, London, Special Publications, **16**, 245–269.

LOYDELL, D. K. 1991. The biostratigraphy and formational relationships of the upper Aeronian and lower Telychian (Llandovery, Silurian) formations of western mid-Wales. *Geological Journal*, **26**, 209–244.

LOYDELL, D. K. 1993. Upper Aeronian and lower Telychian (Llandovery) graptolites from western mid-Wales. Part 2. *Palaeontographical Society Monograph*, 56–180.

MCCANN, T. & PICKERING, K. P. 1989. Palaeocurrent evidence of a northern structural high to the Welsh Basin during the Late Llandovery. *Journal of the Geological Society, London*, **146**, 211–212.

MCCLELLAND-BROWN, E. 1983. Palaeomagnetic studies of fold development and propagation in the Pembrokeshire Old Red Sandstone. *Tectonophysics*, **98**, 131–149.

MORTON, A. C., DAVIES, J. R. & WATERS, R. A. 1992. Heavy minerals as a guide to turbidite provenance in the Lower Palaeozoic Southern Welsh Basin: a pilot study. *Geological Magazine*, **129**, 573–580.

PRATHER, B. E., BOOTH, J. R., STEFFENS, G. S. & CRAIG, P. A. 1998. Classification, lithologic calibration, and stratigraphic succession of seismic facies of intraslope basins, deep-water Gulf of Mexico. *American Association of Petroleum Geologists Bulletin*, **82**, 701–728.

ROBERTS, R. O. 1929. The geology of the district around Abbey-cwmhir (Radnorshire). *Quarterly Journal of the Geological Society, London*, **85**, 651–676.

SANZEN-BAKER, I. 1972. Stratigraphical relationships and sedimentary environments of the Silurian–early Old Red Sandstone of Pembrokeshire. *Proceedings of the Geologists' Association*, **83**, 139–164.

SAVOYE ET AL. this volume, submitted

SMITH, R. D. A. 1987a. Structure and deformation history of the Central Wales Synclinorium, NE Dyfed: evidence for a long-lived basement structure. *Geological Journal*, **22**, 183–198.

SMITH, R. D. A. 1987b. The *griestoniensis* Zone Turbidite System, Welsh Basin. *In*: LEGGETT, J. K. & ZUFFA, G. G. (eds) *Marine Clastic Sedimentology: Concepts and Case Studies*. Graham & Trotman, London, 89–107.

SMITH, R. D. A. 1987c. Early diagenetic phosphate cements in a turbidite basin. *In*: MARSHAL, J. M. (ed.) *Diagenesis of Sedimentary Sequences*. Geological Society, London, Special Publications, **36**, 141–156.

SMITH, R. D. A. 1988. *A sedimentological analysis of the Late Llandovery Welsh Basin*. PhD thesis, University of Cambridge.

SMITH, R. D. A. 1990. Discussion on palaeocurrent evidence of a northern structural high to the Welsh Basin. *Journal of the Geological Society, London*, **147**, 886–887.

SMITH, R. D. A. 1995a. Sheet-like and channelized sediment bodies in a Silurian turbidite system, Welsh Basin, UK. *In*: PICKERING, K. T., HISCOTT, R. N., KENYON, N. H., RICCI LUCCHI, F. & SMITH, R. D. A. (eds) *An Atlas of Deep-Water Environments: Architectural Style in Turbidite Systems*. Chapman & Hall, London, 250–254.

SMITH, R. D. A. 1995b. Reservoir architecture of lacustrine turbidite systems, Lower Cretaceous, offshore Gabon. *In*: LAMBIASE, J. (ed.) *Hydrocarbon Habitat in Rift Basins*. Geological Society, London, Special Publications, **80**, 197–210.

SMITH, R. D. A. 1999. Late Llandovery trace fossil assemblages of the Welsh Basin and Midland Platform. *In*: BOUCOT, A. J. & LAWSON, J. (eds) *Ecostratigraphy*. Cambridge University Press, 438–443.

SMITH, R. D. A. & ANKETELL, J. M. 1992. Welsh Basin 'contourites' reinterpreted as fine-grained turbidites: the Grogal Sandstones. *Geological Magazine*, **129**, 609–614.

SMITH, R. D. A. & JOSEPH, P. 2004. Onlap stratal architectures in the Grès d'Annot: geometric models and controlling factors. *In*: JOSEPH, P. & LOMAS, S. (eds) *Deep-Water Sedimentation in the Alpine Foreland Basin of SE France: New Perspectives on the Grès d'Annot and Related Systems*. Geological Society, London, Special Publications.

SMITH, R. D. A., WATERS, R. A. & DAVIES, J. 1990. *Upper Ordovician and Lower Silurian Turbidite Systems of the Welsh Basin*. Field Guide for Trip A18, International Sedimentological Congress, Nottingham, England, August 1990.

TENG, L. S. & GORSLINE, D. S. 1989. Late Cenozoic sedimentation in California Continental Borderland basins as revealed by seismic facies analysis. *Geological Society of America Bulletin*, **101**, 27–41.

THORPE, R. S., LEAT, P. T., BEVINS, R. E. & HUGHES, D. J. 1989. Late-orogenic alkaline/subalkaline Silurian volcanism of the Skomer Volcanic Group in the Caledonides of south Wales. *Journal of the Geological Society, London*, **146**, 125–132.

VERGARA, L., WREGLESWORTH, I., TRAYFOOT, M. & RICHARDSEN, G. 2001. *Petroleum Geoscience*, **7**, 395–408.

WATERS, R. A., DAVIES, J. R., FLETCHER, C. J. N. & WILSON, D. 1992. Discussion of 'A late Ordovician/early Silurian non-depositional slope and perched basin along the Tywi Anticline, Mid Wales' by D. M. D. James. *Geological Journal*, **27**, 285–289.

WILLS, L. J. AND SMITH, B. 1922. The Lower Palaeozoic rocks of the Llangollen district with special reference to the tectonics. *Quarterly Journal of the Geological Society, London*, **78**, 176–226.

WILSON, D., DAVIES, J. R., WATERS, R. A. & ZALASIEWICZ, J. A. 1992. A fault-controlled depositional model for the Aberystwyth Grits turbidite system. *Geological Magazine*, **129**, 595–607.

WOOD, A. & SMITH, A. J. 1959. The sedimentation and sedimentary history of the Aberystwyth Grits (Upper Llandoverian). *Quarterly Journal of the Geological Society, London*, **114**, 163–195.

WOODCOCK, N. H. 1976. Ludlow Series slumps and turbidites and the form of the Montgomery Trough, Powys, Wales. *Proceedings of the Geologists Association*, **87**, 169–182.

WOODCOCK, N. H. 1984. Early Palaeozoic sedimentation and tectonics in Wales. *Proceedings of the Geologists Association*, **95**, 323–335.

WOODCOCK, N. H. 1990. Transpressive Acadian deformation across the Central Wales Lineament. *Journal of Structural Geology*, **12**, 329–337.

WOODCOCK, N. H. & STRACHAN, R. (eds) 2000. *Geological History of Britain and Ireland*. Blackwell Science, Oxford.

WOODCOCK, N. H., AWAN, A., JOHNSON, T., MACKIE, A. & SMITH, R. D. A. 1988. Acadian tectonics in Wales during Avalonia/Laurentia convergence. *Tectonics*, **7**, 483–495.

WOODCOCK, N. H., BUTLER, A. J., DAVIES, J. R. & WATERS, R. A. 1996. Sequence stratigraphical analysis of late Ordovician and early Silurian depositional systems in the Welsh Basin: a critical assessment. *In*: HESSELBO, S. P. & PARKINSON, D. N. (eds) *Sequence Stratigraphy in British Geology*. Geological Society, London, Special Publications, **10**, 197–208.

ZALASIEWICZ, J. A. 1990. Silurian biostratigraphy in the Welsh Basin. *Journal of the Geological Society, London*, **147**, 619–622.

Submarine fans within small basins: examples from the Tertiary of New Zealand

CHRISTOPH ZINK[1] & RICHARD J. NORRIS[2]

[1] *Badley Ashton & Associates Ltd, Winceby House, Winceby, Horncastle, Lincolnshire, LN9 6PB, UK (e-mail: christoph-zink@web.de)*
[2] *Geology Department, Otago University, PO Box 56, Dunedin, New Zealand*

Abstract: This study uses the example of two small sedimentary basins to draw lessons for the stacking patterns of submarine fan facies in restricted basins. The Te Anau and Waiau basins are located in southwest New Zealand. They developed from the Middle Eocene onwards close to and under the direct influence of the evolving Australia–Pacific plate boundary. Although never more than a few tens of kilometres wide, both basins accumulated sedimentary successions more than 6 km thick. We discuss the scale, nature and composition of submarine fans within the basins. Sediment stacking patterns are characterized by rapid lateral and vertical changes within the full range of facies predicted by most submarine fan models. Despite the small basin size, some submarine fans contain sandstone bodies of impressive thickness (50 m) and many kilometres lateral extent. Such bodies of well-sorted, highly permeable sandstone are excellent potential hydrocarbon reservoirs. The occurrence of complete Bouma sequences and well-developed couplets of clastic- and hemipelagic mudstones indicates that such features are not restricted to large-scale fans (covering thousands of square kilometres). Even in small sedimentary basins, submarine fans can develop that show an impressive range of sedimentary facies and architectural elements, commonly described only from their large-scale counterparts.

During the Cretaceous, New Zealand separated from Gondwana with the opening of the Tasman Sea (Fig. 1). Following this phase of spreading, the Eocene to Miocene epochs saw the reorganization of the Australia–Pacific plate boundary. The Te Anau and Waiau basins of Western Southland evolved in close proximity and under the direct influence of this newly developing plate boundary (Carter & Norris 1976; Norris & Turnbull 1993). Basin evolution began in the Middle Eocene with an extension-dominated phase as the leading edge of an evolving rift (marked on Fig. 1 as the Emerald Basin), extended northwards into New Zealand continental crust. An initially small component of dextral displacement became more and more important into the Miocene while the extensional component diminished. By mid–late Miocene times, movements along the palaeoplate boundary had changed into a dextral strike-slip-dominated regime. This phase subsequently inherited an increasingly transpressional component of relative plate movements resulting in basin eversion (Norris & Turnbull 1993). The described plate tectonic history of the region is well studied and is reasonably well constrained, based on seafloor data (Weissel *et al.* 1977; Walcott 1978; Stock & Molnar 1982; Sutherland 1995). These studies provide independent evidence, constraining the amount of extension that might have occurred during the opening of the basins. Studies of the basin-fill sediments can be compared with these models, and can also produce a more detailed picture of the basin evolution.

Basin geometry

In their present form, the Te Anau and Waiau basins are located to the east of the Fiordland Complex (Fig. 2), a body of Palaeozoic and Mesozoic metamorphic and plutonic rocks. The Fiordland Complex also separates the basins from the Alpine Fault, the present main locus of the Australia–Pacific plate boundary, some 80 km to the west. Both basins have been important depocentres since the Middle Eocene in which 6–8 km of sediment have been accumulated.

The Te Anau Basin has a roughly triangular shape (Fig. 2). It is delineated by the north-trending Hollyford Fault System to the east, and by faults of the NNE-trending Te Anau Fault System to the northwest. The NE-trending Moonlight Fault System separates it from the Waiau Basin to the southeast. The original shape of the basin was destroyed during basin eversion. The southern part of the basin is shortened by faulting, whereas the northern part is folded into a broad syncline plunging south. However, an original half-graben

From: LOMAS, S. A. & JOSEPH, P. (eds) 2004. *Confined Turbidite Systems.* Geological Society, London, Special Publications, **222**, 229–240. 0305-8719/04/$15.00 © The Geological Society of London 2004.

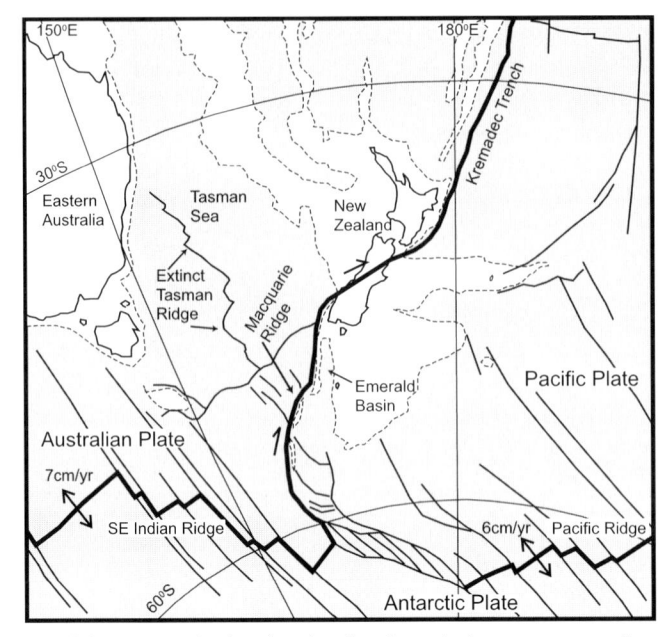

Fig. 1. Tectonic setting of the New Zealand region showing the main fracture zones and spreading ridges, traced from gravity anomaly maps of Sandwell & Smith (1999). Grey-shaded areas indicate a water depth of more than 2000 m. This indicates roughly the distribution of oceanic (grey) and continental (white) crust.

geometry is implied for the Te Anau Basin by sedimentary successions thickening towards the west, where the more proximal sedimentary facies are observed.

The Waiau Basin (Fig. 2) can also be described as a half-graben with sedimentary successions thickening and coarsening towards the west–northwest, where the basin is delineated by the Moonlight Fault System (Norris et al. 1978; Norris & Carter 1980). Along the eastern and southeastern borders, sedimentary rocks of the Waiau Basin unconformably overlie or are faulted against Triassic–Permian basement rocks of the Takitimu Mountains and Longwood Ranges. Offshore to the south, the basin is terminated by the Mid Bay High (Norris & Carter 1980; Bishop et al. 1992).

Most authors described the two basins as having developed adjacent to the eastern margin of the Fiordland Complex. The Waiau Basin is thought to have developed to the south of the Te Anau Basin (Kamp 1986; Norris & Turnbull 1993; Turnbull et al. 1993), with which it was connected at times. Zink (2000), however, proposed that the two basins developed as two sub-parallel, roughly north–south-trending half-grabens, subsequently displaced and distorted along the bounding faults. In this way, the Waiau Basin would have developed to the east of the Te Anau Basin, separated from it by a basement ridge. In either case, the basins are regarded as being of similar scale, only a few tens of kilometres wide.

Basin-fill sediments

The sediment fill of both the Te Anau and the Waiau basins is characterized by sediment bodies of great thickness (single bodies within the succession often reaching more than 1000 m), showing rapid lateral and vertical facies changes (Fig. 3). The sediments indicate a wide range of depositional environments. Facies range from very proximal rock-fall and debris flow deposits, including clasts several metres in size, through thick sand-dominated fluvial and submarine fan successions, to hemipelagic mudstones (Waicoe Formation of Fig. 3).

Among the terrestrial facies, early rift successions comprising alluvial fan and fluvial deposits are dominant. The highly carbonaceous fluvial successions in particular form sediment bodies of great thickness (up to 1800 m) and many square kilometres lateral extent (Earl Mountains Sandstone as well as parts of the Hope Arm and Point Burn formations of Fig. 3). Among the marine sediments, submarine fan facies are most common. More than 10 submarine fan bodies are described from the two basins

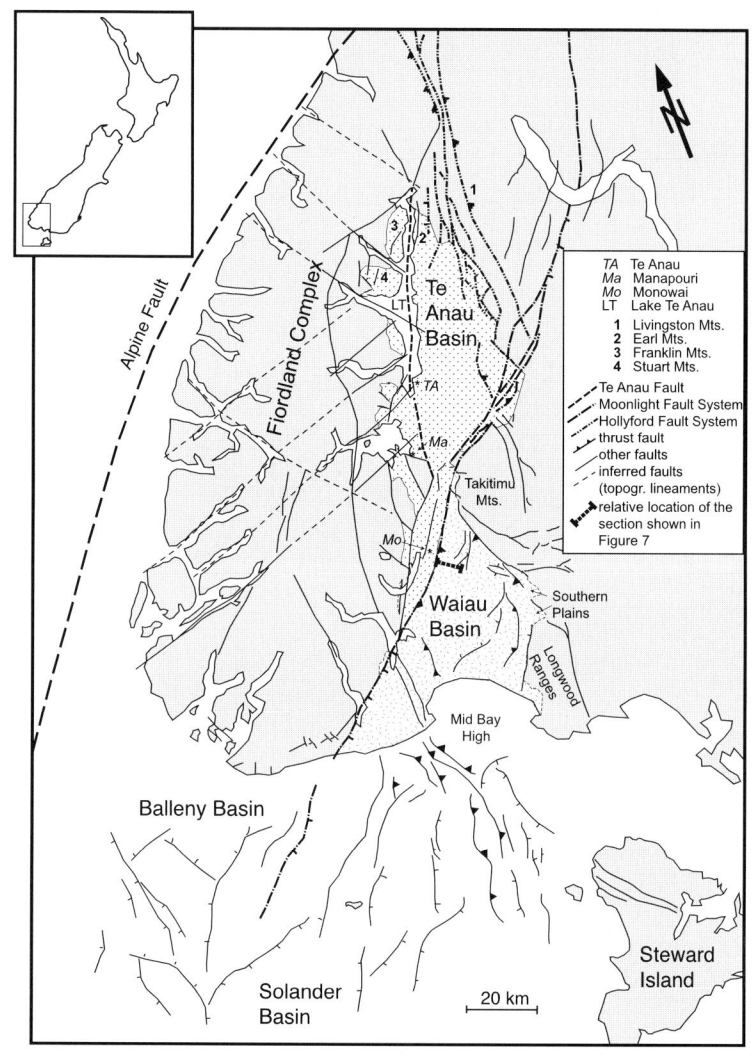

Fig. 2. Structure and location map of southwest New Zealand showing the Te Anau and Waiau basins. The major controlling structures and the locations referred to in the text are illustrated. Inset map shows the location of the study area in the southwest of New Zealand. Changed after Zink (2000), based on Norris & Turnbull (1993) and Norris & Carter (1980).

(Norris & Turnbull 1993; Turnbull et al. 1993; Zink 2000). Out of these, we pick three well-exposed examples, which allow us to highlight and describe the facies diversity and fan geometries present within the two basins. Proximal submarine fan facies are best exposed within the northernmost, uplifted parts of the Te Anau Basin. From this area, we describe the Oligocene Turret Peaks Formation. Further south, the Waiau River exposes a partial cross-section through the Waiau Basin. Here, more distal, mud-rich facies are exposed, which are examined in the Blackmount and MacIvor formations.

The Turret Peaks Formation

The easternmost Stuart Mountains (Fig. 4) provide fantastic large-scale cliff exposures of the sand-dominated Turret Peaks Formation deposits. Some 1800 m of sandstone-dominated fan deposits have infilled a major pull-apart depocentre within the northern Te Anau Basin, some 20 by 30 km in size. Controlled by N–S and NW–SE trending sets of faults, the depocentre opened up in Early to Middle Oligocene times, as the unit is dated Whaingaroan throughout (Turnbull et al. 1993), a time of increasingly

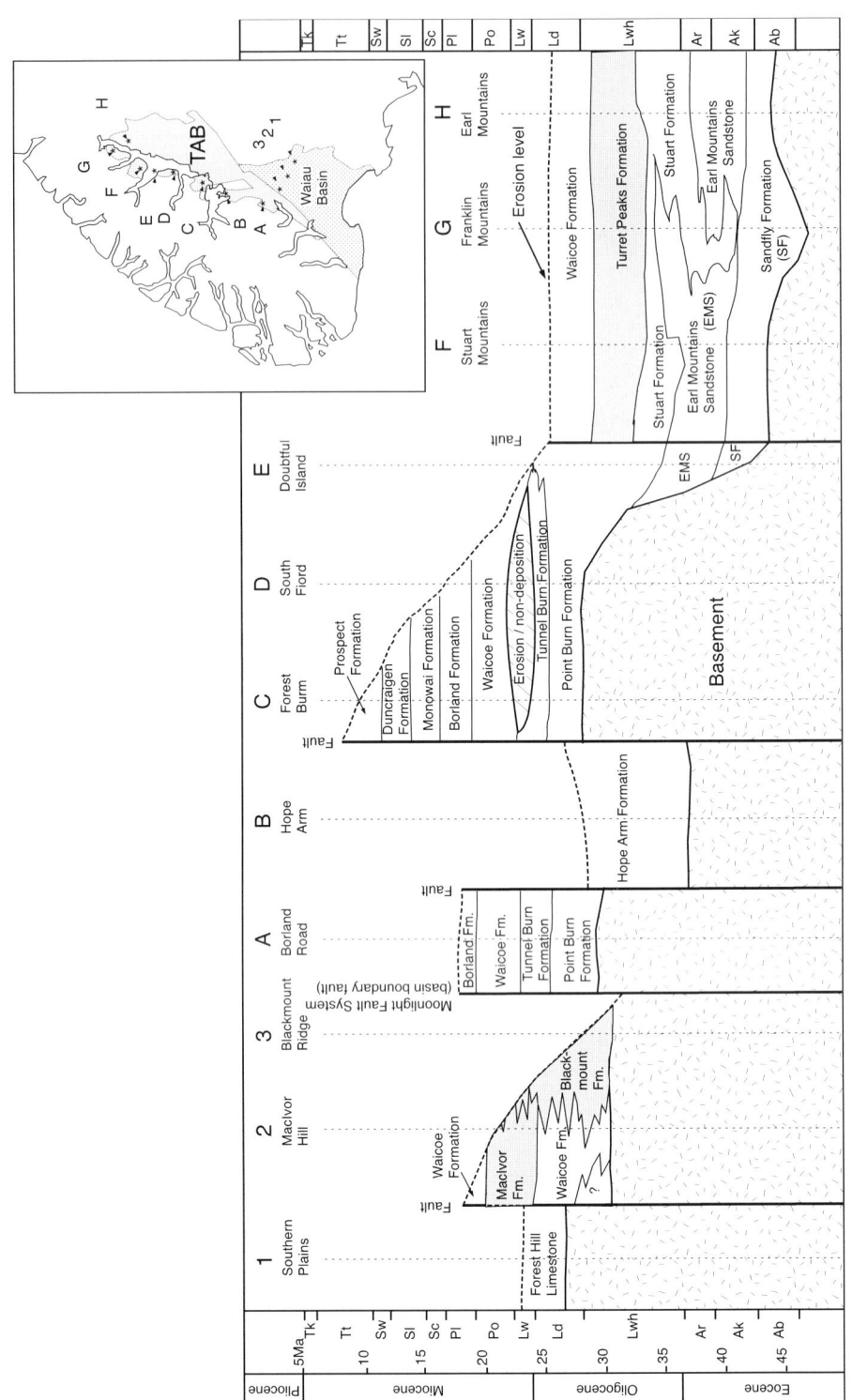

Fig. 3. Stratigraphic correlation chart for the Te Anau (TAB) and Waiau basins. The figure is based in parts on Turnbull *et al.* (1993). Formations described in this paper are shaded grey; for information on all other formations, please refer to Turnbull *et al.* (1993) or Zink (2000).

Fig. 4. Large-scale cliff exposure of the basal 500 m of Turret Peaks Formation strata within the western Stuart Mountains. The base of the fan deposits is located at the top of the white slips (see arrow) developed within underlying shallow marine strata. View looking northeast over Lake Te Anau towards the Livingston Mountains seen in the far background.

oblique extension. We use the term *submarine fan* in its most commonly used way, to describe a deepwater–clastic sediment body, of sediment gravity flow origin. We do not imply an actual fan geometry. In the case of the Turret Peaks Formation, such a geometry has never been established.

The basal 700–800 m of this succession, as observed in the Stuart Mountains, is made up of metre-bedded, often amalgamated sandstone beds (Fig. 5). On kilometre-scale cliff outcrops, broad channelized sand-bodies, reaching up to 50 m in thickness, can be observed within this basal unit (Fig. 4). Most sand beds appear internally unstructured; the sand is medium-coarse grained, sometimes gritty and generally well sorted. Centimetre-sized clasts and rare shell fragments appear floating throughout some of the beds. Graded beds are very rare and thin muddy interbeds even less common.

Vertically up-section, the average bed thickness decreases and normally graded beds become the dominant facies. From about 1000 m onwards within the succession, bedding tends to be on decimetre-scale. Also at around 1000 m, three conglomerate beds are found within the succession. Separated from each other by only a few decimetres of sandstone, the conglomerate beds are between 0.5 and 3 m thick. Each bed shows well-developed inverse grading as small pebbles of 2–5 cm at the base of each bed grade into a pebble–cobble conglomerate with clasts up to 25 cm in diameter at the top. Clast a(p)a(i) imbrication is well developed in all beds. The sand matrix content of the conglomerates is highest at the base, where they are matrix-supported in places, and decreases towards the tops to around 25–30%.

Above 1300–1500 m, massive sandstone beds are very rare, and are replaced by mudstone intercalated with graded sandstone of a more classical flysch facies assemblage (Fig. 5). The Turret Peaks Formation as a whole is interpreted as a ponded deepwater massflow system, filling a pull-apart type depocentre, receiving sediment from several sides. Recurring sandy debris flows are a likely transport and depositional process for the massive sandstone beds (Hiscott & Middleton 1980). Amalgamation of the deposits of several flows can produce the metre-thick, massive-looking beds described. Debris flows are also capable of keeping the clasts and shell fragments, found within the lithofacies, in suspension. Turbidity currents provide another possible transport and depositional mechanism. If each of the structureless beds represents a series of turbidity currents, or at least the waning stages of an individual flow, then it is possible to explain the outsized clasts throughout the beds. Reading & Richards (1994) propose a 'Sand-Rich Ramps' model, fed by turbidity currents, which produces deposits very similar to the ones described here. It features amalgamated sand-bodies formed in channels and as more distal sheet sandstones. The sand deposits of this ramp type are described as well-sorted,

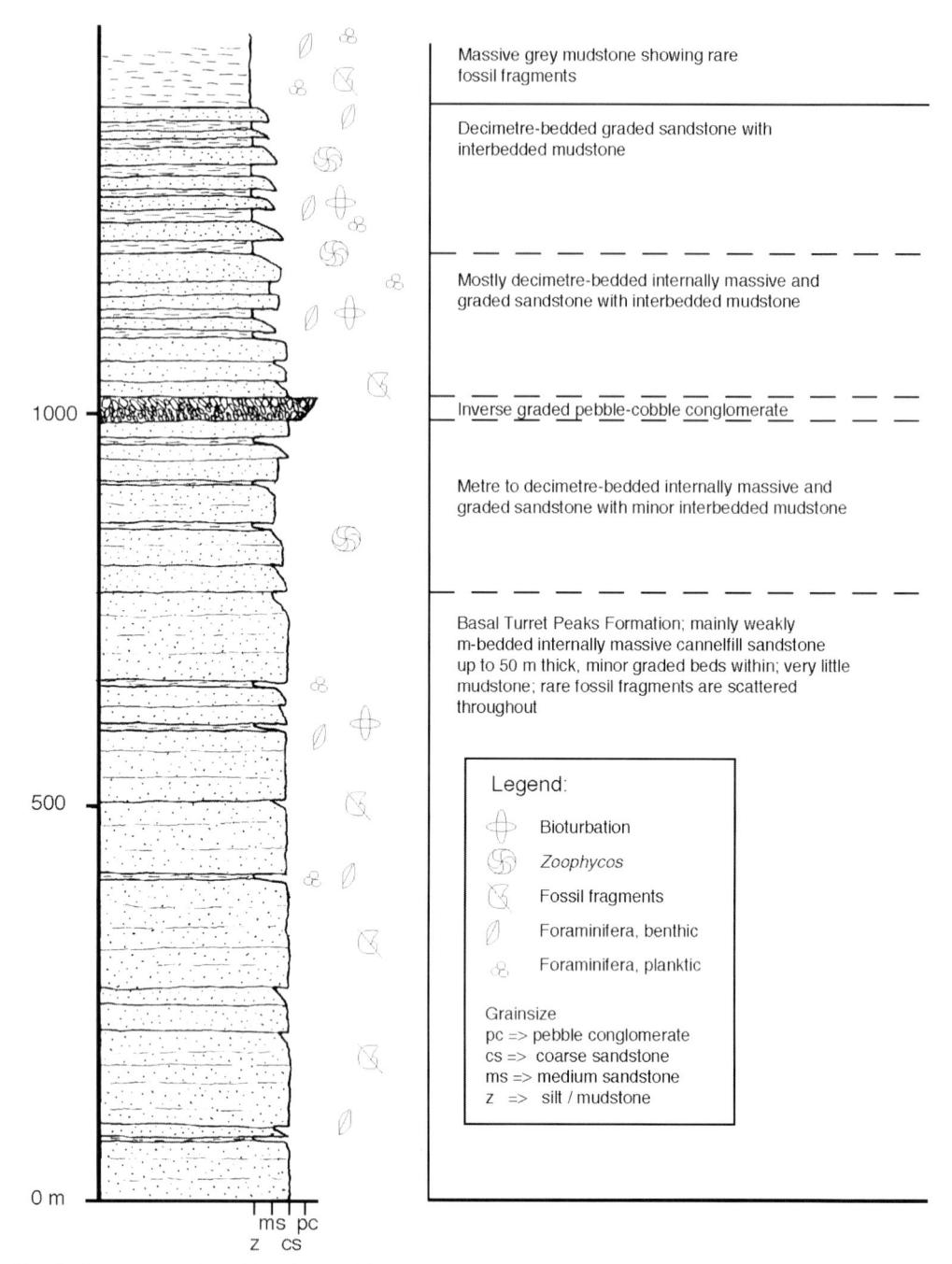

Fig. 5. Summary stratigraphic column of the Turret Peaks Formation taken along the ridge crest of the Stuart Mountains (simplified after Zink 2000).

ungraded, structureless, and frequently amalgamated sandstones, and interpreted as the deposits of sand-rich low-efficient turbidites (Mutti and Ricci Lucchi, 1981).

For the conglomerates, inverse grading, a high sand-matrix content and bimodal grain size together with the a(p)a(i) clast imbrication, suggest a cohesionless debris flow (grainflow)

(Nemec & Steel 1984) origin. Kinematic sieving causes the inverse grading within the flow (Lowe 1982). Walker (1975), on the other hand, interprets similar imbricated conglomerates as turbidite deposits. Walker states that turbulence could not have been the force that kept pebble-sized clasts suspended, and suggests dispersion of the flow instead. The deposits would therefore represent the basal traction carpets of high-concentration turbidity currents. Unfortunately, the outcrop situation does not allow the examination of lateral facies changes. A conclusive interpretation of the actual flow mechanism is therefore not possible. The normally graded sandstone beds with mudstone interbeds higher up in the succession are more easily explained. They are interpreted as turbidite deposits (Bouma 1962). Normal size-grading is widely regarded as a reliable criterion to interpret a deepwater sandstone as a turbidite (e.g. Shanmugam & Moiola 1997).

Because of uplift and erosion, less resistant mud-rich facies are rarely exposed within the Te Anau Basin. The Waiau River valley, however, provides good exposures through such facies in the northern Waiau Basin, east of the township of Monowai (Carter & Norris 1977). From this section we describe the facies and facies architecture of the Blackmount and MacIvor formations.

The Blackmount Formation

The Blackmount Formation is Oligocene to Miocene in age and demonstrates exceptionally well the rapid lateral and vertical facies changes that are common to all fans within the two basins. The fan succession starts right at the basin-margin fault with a locally derived, near-monomict mass-flow breccia. Clasts of the poorly sorted breccia tend to reflect the underlying crystalline basement rocks (diorite, gneiss). The breccia varies in thickness between 5 and 100 m; its top is marked by a rapid transition into a sandstone–mudstone flysch facies (Fig. 6). However, breccia material can be found slumped into the sandstone–mudstone succession even several tens of metres higher in the succession. The lower Blackmount Formation interval

Fig. 6. Base of the Blackmount Formation deposits, exposed by the Waiau River some 2 km southwest of the township of Monowai. Note the rapid change from the basal breccia (to the right of the solid line) in the foreground and far right of the picture, into the proximal flysch deposits exhibiting sand-dominated packets separated by muddy intervals (highlighted with dotted lines for the first three intervals). The distinct sand packet seen in the centre of this picture is some 12 m thick.

Fig. 7. Distal Blackmount Formation deposits exposed along the Waiau River some 4 km west of the recent expression of the basin boundary fault. Bedding is highlighted by colour banding of clastic (dark) and hemipelagic (light) mudstone.

(Fig. 6) is characterized by 5–15 m thick packets of amalgamated sandstone beds, with individual beds ranging in thickness from 0.2 to 1 m. Beds are normally graded, exhibiting Bouma A and sometimes B divisions. These sand-dominated packets are separated by thin-bedded (0.05–0.3 m), mudstone-dominated alternating sandstone–mudstone facies, also in packets of similar thickness. Where the thick sandstones can be traced laterally, they can in many cases be observed to thin out over tens or a few hundreds of metres. We interpret these sandstone beds as sand-rich feeder channels migrating over the surface of the proximal fan. Sole-marks on these proximal sandstone bodies (in agreement with slump geometry) indicate a palaeoslope towards the east–southeast. Up section and away from the controlling fault, discrete sandstone-channel packets are still clearly visible. Sandstone beds, however, are thinner, laterally more extensive,

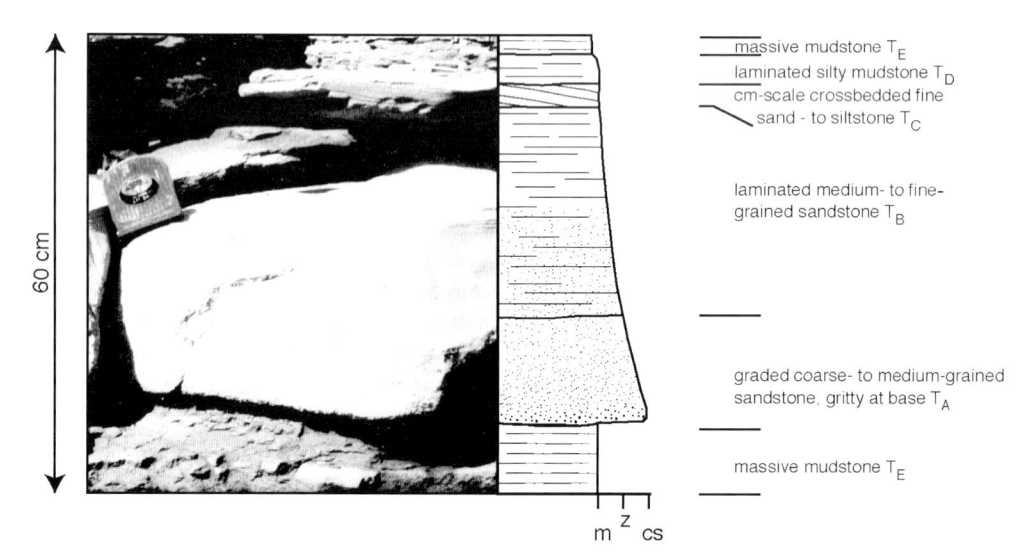

Fig. 8. A complete Bouma sequence developed in the MacIvor Formation strata.

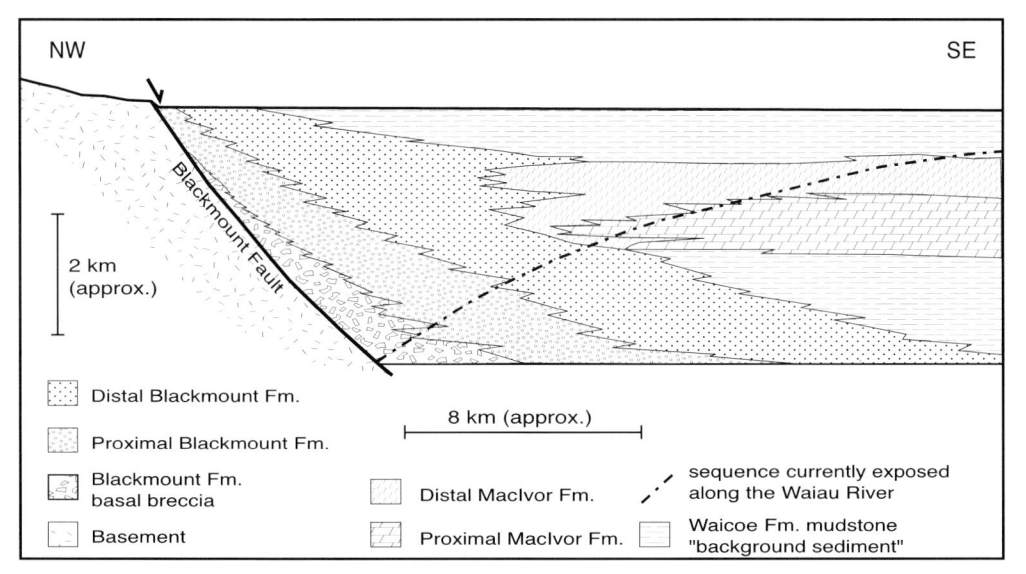

Fig. 9. Diagrammatic representation of the relationships between the Blackmount and McIvor submarine fan bodies with the effects of later reverse faulting and folding removed. The Blackmount Fault is interpreted to have been a normal fault active during sedimentation and is shown as such. The given scale is approximate!

and the average grain size is generally finer. The most distal facies of the Blackmount Formation can be observed some 4 km east of the present-day expression of the basin-margin fault. The sediment here is mudstone-dominated, and only rarely do the bases of thicker turbidite beds reach silt or sand grade. Bedding is highlighted by magnificent examples of colour-banded turbidites, created by couplets (typically 0.1–0.5 m thick) of sharp-based dark grey clastic mudstone grading into light grey hemipelagic mudstone rich in planktonic foraminifera (Fig. 7). This most distal facies is found interfingering with the easterly derived McIvor Formation facies described below.

The MacIvor Formation

This Miocene fan is clearly sourced from a limestone shelf area as, petrographically, the beds are bryozoan biosparites with various amounts of terrigenous sands (Carter & Norris 1977). Only the more distal parts of the MacIvor Formation are well exposed. Towards the east, the deposits are buried under a thick succession of younger, mostly molass-type deposits. The MacIvor Formation sections often comprise well-developed flysch successions including complete Bouma sequences (Fig. 8). Well-preserved flute casts indicate east-to-west flow directions for the MacIvor turbidity currents. An extensive belt of shelf bryozoan limestones of similar age is found on the Southern Plains some 20 km to the southeast (Hyden 1979), and is the likely source area.

The fact that westerly derived Blackmount Formation facies interfinger with easterly derived MacIvor Formation facies indicates that these deposits may have been laid down near the centre of the basin and therefore present a clear limitation on basin width (Fig. 9).

Discussion

Size of the basins

Although there is some leeway regarding the original size and width of the Te Anau and Waiau basins, both regional plate tectonic reconstructions and the sedimentary record restrict the amount of extension and therefore the basin width to the order of tens of kilometres. Based on plate tectonic reconstructions, Zink (2000) established palaeographic reconstructions assuming an Eocene to Miocene extension value of around 50 km, followed by some 90 km of shortening thereafter, while the continuous dextral component of lateral displacement added up to a combined 190 km. These values were compared against estimates based on the study of the sediment fill within the two basins, and appeared realistic. Although the Te Anau

Basin might have reached a maximum width of around 80 km at 25 Ma (Zink 2000), much of this width was occupied by extensive shallow-water shelf areas that existed at the time (Zink 2000). The actual deepwater clastics were more likely to have been deposited within a basin with a width closer to 30 km. The interfingering Blackmount and MacIvor Formation deposits, which are derived from opposite sides of the basin, also constrain the width of the Waiau Basin to little more than 25 km (Fig. 9). However, reconstructing all normal and reverse offsets along the basin margin faults is problematic, and additional offsets along smaller parallel structures might be easily overlooked within a rather monotonous flysch succession. Such offsets could well add up to a considerable amount. Nevertheless, even including these uncertainties, it seems unlikely that the basin centre, marked by the interfingering fan strata, was located more than 10–12 km east of the western basin margin fault.

The basins during their early development may have resembled early basins developed in the Gulf of California (Stock & Hodges 1989; Legg et al. 1991; Zanchi 1994).

Tectonics and facies diversity

This study highlights the facies diversity that can develop within a small-scale basin. Facies changes, both laterally and vertically, are abrupt and intense (Fig. 3). These changes, however, cannot simply be attributed to the size of the basins. They also appear to reflect the basins' complex tectonic evolution. The early rift-dominated phase resulted in uplifted shoulders, steep topography and intense proximal sediment input. During the transition to a more strike-slip-dominated regime, pull-apart depocentres developed. At the same time, however, the topographic gradient of the source areas appears to have diminished and the basins became starved of sediment. This is reflected by the abundance of muddy deeper-water, distal sedimentary facies throughout the basins (Norris & Turnbull 1993). With the onset of compression and basin inversion, basin flanks and basin strata were uplifted and eroded. Remaining depocentres were infilled quickly by proximal deposits, commonly comprising reworked older basin sediments (Turnbull et al. 1993; Zink 2000).

Similar sedimentary characteristcis have been described from small 'pull-apart' basins in California, such as the Los Angeles and Ventura basins (Hsu et al. 1980; Biddle 1991; Sorlien et al. 2000). In western Southland however, active tectonism may have been more intense throughout the basin history. Globally such basins are important repositories of hydrocarbons, suggesting that the New Zealand basins may also warrant evaluation as potential hydrocarbon prospects.

Hydrocarbon prospectivity

This study demonstrates that smaller-scale basins should not be underestimated regarding their hydrocarbon potential. The early rift sediments in both basins comprise thick and laterally extensive fluvial successions rich in organic matter, which have to be regarded as a good potential hydrocarbon source. Based on vitrinite reflectance, it has been established that these deposits were buried down to a depth within the oil-window (Turnbull et al. 1993). Overlying this potential source unit, sandstone bodies, such as the Turret Peaks Formation deposits, make potentially excellent reservoirs. They comprise well-sorted, highly permeable sandstone bodies (reservoirs) of great thickness in addition to being capped by distal, mud-rich facies (seals). All ingredients to produce a commercial oil play are therefore present. For the Te Anau Basin however, basin eversion and related faulting were too intense for the preservation of any undisturbed larger-scale traps. The less intensely everted Waiau Basin, however, is still seen as a potential hydrocarbon prospect; it has indeed been the subject of limited exploration and is currently under investigation. The most important lesson to be learned is that even small basins may have the necessary ingredients for good hydrocarbon prospectivity.

Conclusions

Size doesn't matter! Even in small sedimentary basins, submarine fans can develop that show an impressive range of sedimentary facies and architectural elements, commonly described only from their large-scale counterparts. Facies diversity is not just related to the size of a basin; factors such as basin tectonism and source area evolution are also important. This example highlights the fact that, under the right circumstances, even small-scale basins might prove to have good hydrocarbon potential.

CZ would like to thank C. Landis and I. Turnbull for many stimulating discussions and also acknowledge the support received, working at the Department of

Geology and Petroleum Geology at Aberdeen University, Scotland. RJN acknowledges the work of Bob Carter with whom he jointly carried out most of the original fieldwork in the Blackmount area and established the stratigraphy of the Blackmount and McIvor Formations. Financial support from University of Otago research grants is acknowledged. Financial assistance to CZ was provided by an Otago University Scholarship, an Otago University Bridging Grant, and two awards of a Jack Bradshaw Scholarship.

References

BIDDLE, K. T. 1991. The Los Angeles Basin: an overview. *In*: BIDDLE, K. T. (ed.) *AAPG Memoir, Active Margin Basins*. American Association of Petroleum Geologists, 5–24.

BISHOP, D. G., KOONS, P. O., REAY, A. & TURNBULL, I. M. 1992. The composition and significance of Mid Bay and Mason Bay reefs, Foveaux Strait, Southland. *New Zealand Journal of Geology and Geophysics*, **35**, 109–112.

CARTER, R. M. & NORRIS, R. J. 1976. Cainozoic history of southern New Zealand: an accord between geological observations and plate-tectonic predictions. *Earth and Planetary Science Letters*, **31**, 85–94.

CARTER, R. M. & NORRIS, R. J. 1977. Blackmount, Waiau Basin (Fieldtrip Guide), Geological Society of New Zealand Conference. Geological Society of New Zealand, Queenstown.

HISCOTT, R. N. & MIDDLETON, G. V. 1980. Fabric of coarse deep-water sandstones Tourelle Formation, Quebec, Canada. *Journal of Sedimentary Petrology*, **50**, 703–722.

HSU, K. J., KELTS, K. & VALENTINE, J. W. 1980. Resedimented facies in Ventura Basin, California, a model of longitudinal transport of turbidity currents. *American Association of Petroleum Geologists Bulletin*, **84**, 1034–1051.

HYDEN, F. M. 1979. Mid-Tertiary temperate shelf limestones, Southland, New Zealand. PhD thesis, University of Otago, Dunedin.

KAMP, P. J. J. 1986. The mid-Cenozoic Challenger Rift System of western New Zealand and its implications for the age of Alpine Fault inception. *Bulletin of the Geological Society of America*, **97**, 255–281.

LEGG, M. R., WONG, O. V. & SUAREZ, V. F. 1991. Geologic structure and tectonics of the inner continental borderland of northern Baja California. *In*: DAUPHIN, J. P. & SIMONEIT, B. R. T. (eds) *The Gulf and Peninsular Province of the Californias*. American Association of Petroleum Geologists, Tulsa, OK, Memoirs, **47**, 145–177.

LOWE, D. R. 1982. Sediment gravity flows: II. Depositional models with special reference to the deposits of high-density turbidity currents. *Journal of Sedimentary Petrology*, **52**, 279–297.

MUTTI, E. & RICCI LUCCHI, F. 1981. Introduction to the excursions on siliciclastic turbidites, *2nd International Association of Sedimentologists European Regional Meeting*, Bologna, 1–3.

NEMEC, W. & STEEL, R. J. 1984. Alluvial and coastal conglomerates: Their significant features and some comments on gravelly mass-flow deposits. *In*: KOSTER, E. H. & STEEL, R. J. (eds) *Sedimentology of Gravels and Conglomerates*, Canadian Society of Petroleum Geologists Memoirs, **10**, 1–31.

NORRIS, R. J. & CARTER, R. M. 1980. Offshore Sedimentary Basins at the Southern End of the Alpine Fault, New Zealand. International Association of Sedimentologists, Special Publications, **4**, 237–265.

NORRIS, R. J. & TURNBULL, I. M. 1993. Cenozoic basins adjacent to an evolving transform plate boundary, southwest New Zealand. *In*: BALANCE, P. F. (ed.) *South Pacific Sedimentary Basins. Sedimentary Basins of the World*. Elsevier Science, Amsterdam, 251–270.

NORRIS, R. J., CARTER, R. M. & TURNBULL, I. M. 1978. Cainozoic sedimentation in basins adjacent to a major continental transform boundary in southern New Zealand. *Journal of the Geological Society of London*, **135**, 191–205.

READING, H. G. & RICHARDS, M. 1994. Turbidite systems in deep-water basin margins classified by grain size and feeder system. *American Association of Petroleum Geologists Bulletin*, **78**, 792–822.

SANDWELL, D. T. & SMITH, W. H. F. 1999. Marine gravity anomaly from Geosat and ERS1 satellite altimetry. *Journal of Geophysical Research*, **102**, 10039–10054.

SHANMUGAM, G. & MOIOLA, R. J. 1997. Reply: Reinterpretation of depositional processes in a classic flysch sequence (Pennsylvanian Jackfork Group), Ouachita Mountains, Arkansas and Oklahoma. *American Association of Petroleum Geologists Bulletin*, **81**, 476–491.

SORLIEN, C. C., GRATIER, J.-P. *ET AL*. 2000. Map restoration of folded and faulted late Cenozoic strata across the Oak Ridge fault, onshore and offshore Ventura basin, California. *Bulletin of the Geological Society of America*, **112**(7), 1080–1090.

STOCK, J. M. & HODGES, K. V. 1989. Pre-Pliocene extension around the Gulf of California, and the transfer of Baja California to the Pacific Plate. *Tectonics*, **8**(1), 99–115.

STOCK, J. & MOLNAR, P. 1982. Uncertainties in the relative positions of the Australia, Antarctica, Lord Howe, and Pacific plates since the Late Cretaceous. *Journal of Geophysical Research*, **87**, 4697–4717.

SUTHERLAND, R. 1995. The Australia–Pacific boundary and Cenozoic plate motions in the SW Pacific: some constraints from Geosat data. *Tectonics*, **14**, 819–831.

TURNBULL, I. M., URUSKI, C. I. *ET AL*. 1993. Cretaceous and Cenozoic sedimentary basins of Western Southland, New Zealand. *New Zealand Geological Survey Basin Studies*, **4**, 86.

WALCOTT, R. I. 1978. Present tectonics and late Cenozoic evolution of New Zealand. *Geophysical Journal of the Royal Astronomical Society*, **52**, 137–184.

WALKER, R. G. 1975. Generalized facies models for resedimented conglomerates of turbidite association. *Geological Society of America Bulletin*, **86**, 737–748.

WEISSEL, J. K., HAYES, D. E. & HERRON, E. M. 1977. Plate tectonic synthesis: the displacements between Australia, New Zealand and Antarctica since the late Cretaceous. *Marine Geology*, **25**, 231–277.

ZANCHI, A. 1994. The opening of the Gulf of California near Loreto, Baja California, Mexico: from basin and range extension to transtensional tectonics. *Journal of Structural Geology*, **16**(12), 1619–1639.

ZINK, C. 2000. *Middle Eocene to Middle Miocene evolution of the Te Anau Basin, western Southland, New Zealand*. PhD thesis, Otago University, Dunedin.

Down-channel variations in stratal patterns within a conglomeratic, deepwater fan feeder system (Miocene, Adana Basin, Southern Turkey)

N. SATUR[1*], B. CRONIN[1], A. HURST[1], G. KELLING[2] & K. GÜRBÜZ[3]

[1]*Department of Geology and Petroleum Geology, King's College, Aberdeen AB24 3UE, UK*
[2]*Department of Earth Sciences, Keele University, Keele, Staffs ST5 5BG, UK*
[3]*Çukurova University, Geology Department, 01330 Balcali, Adana, Turkey*
[*]*Present address: Statoil, N-4035 Stavanger, Norway (e-mail: NISA@Statoil.no)*

Abstract: The proximal, channelized section of a deepwater fan in the Adana Basin, southern Turkey, provides an opportunity to study down-channel changes in stratal patterns. This is a multisourced, bypass fan with at least four feeder channels. The downdip changes within one of these channels are described along a 10 km transect. Down-channel changes in grading and organization of clasts are observed for non-grading disorganized facies in the most proximal locations to more organized, inverse-normal and normal graded conglomerates midway down the transect. Bedding style changes from scoured and more sheet-like beds updip into a pattern of small channels and bars midway along the transect, and into thick-bedded and structureless conglomerates in the more distal locations within the channel. These changes in stratal patterns coincide with observed changes in depositional gradient of the channel. The gradient changes are interpreted to be a contributing factor in controlling hydrodynamics within the gravity flows and thus the depositional stratal patterns.

The Lower Miocene Cingöz Formation in the northeastern area of the Adana Basin provides an opportunity to investigate down-channel variations in stratal patterns within the fill of deepwater channels (Fig. 1). This study aims to provide a qualitative and quantitative description of the stratal patterns of the channel-fill and illustrates how these vary down-channel. Some possible causes of these changes will be addressed in terms of transport and depositional processes along with some reasoning as to why these changes occur. With increased quality of datasets required for hydrocarbon exploration and production, the importance of changes in sediment transport processes down-channel over distances of tens of kilometres has become vital in efforts to predict economically significant variations. Investigation, quantification and qualification of down-channel stratal pattern variations in well-exposed examples can thus yield information on changes in depositional processes that have predictive values.

Study area

The Neogene Adana Basin of SE Turkey was formed as a foreland basin following the late Eocene to Oligocene thrust emplacement of an ophiolite complex and associated mélange of the Tauride Belt, situated to the north (Williams *et al.* 1995; Kelling *et al.* 1987). Extensional faulting and a marine invasion accompanied a widespread early Miocene phase of subsidence.

This caused gradual infilling and shoaling upwards within the basin during middle to late Miocene times (Gürbüz & Kelling 1993; Williams *et al.* 1995; Yetis *et al.* 1995; Fig. 2). Sediment was transported by turbidity currents and debris flows across a narrow (1–3 km wide; Satur 1999), contemporaneous, carbonate-siliciclastic shelf and slope (Karaisalı and Kaplan-kaya Formations, Figs 1 & 2) and accumulated on the deeper basin floor as two deepwater clastic systems that had conical planform geometry with radial palaeoflow directions in their proximal reaches. These were interpreted as deepwater, turbidite fans (Schmidt 1961; Yetis 1988; Gürbüz 1993).

Investigation of planktonic foraminifera indicated that these two deepwater turbidite fans, called the Eastern and Western Fans (Gürbüz 1993), were deposited during the Late Burdigalian to the Late Serravallian interval (Nazik & Gürbüz 1992; Fig. 2). This formation has previously been interpreted as two coeval deepwater fan systems (Gürbüz 1993). Subsequent work has revealed that these two fans merged to form one deepwater system during times of maximum progradation, separating into two distinct geometric features as sea-level rose (Satur 1999; Satur *et al.* 2000). Here data are present from the northeastern area, referred to as the proximal part of the Eastern Fan, immediately to the south of Meydan Yayla village (Fig. 1).

Despite having a similar tectonic setting, being coeval and sourced from a similar region, the two fans display important differences (Gürbüz &

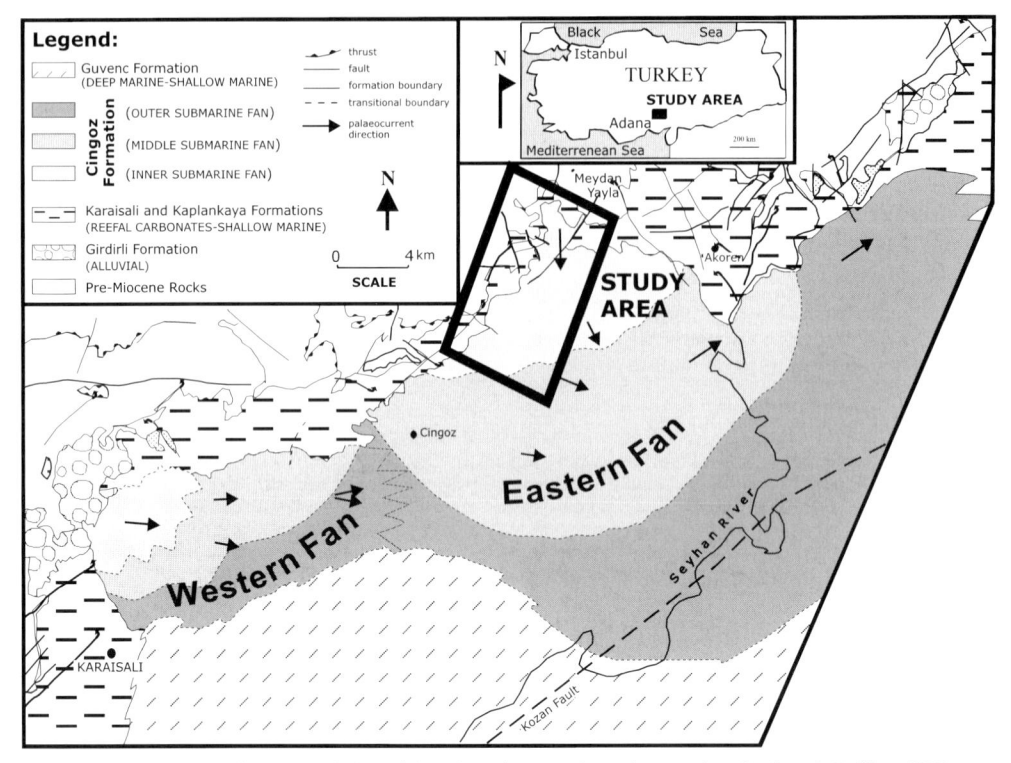

Fig. 1. Geological map of Miocene facies within the Adana Basin (redrawn after Gürbüz & Kelling 1993).

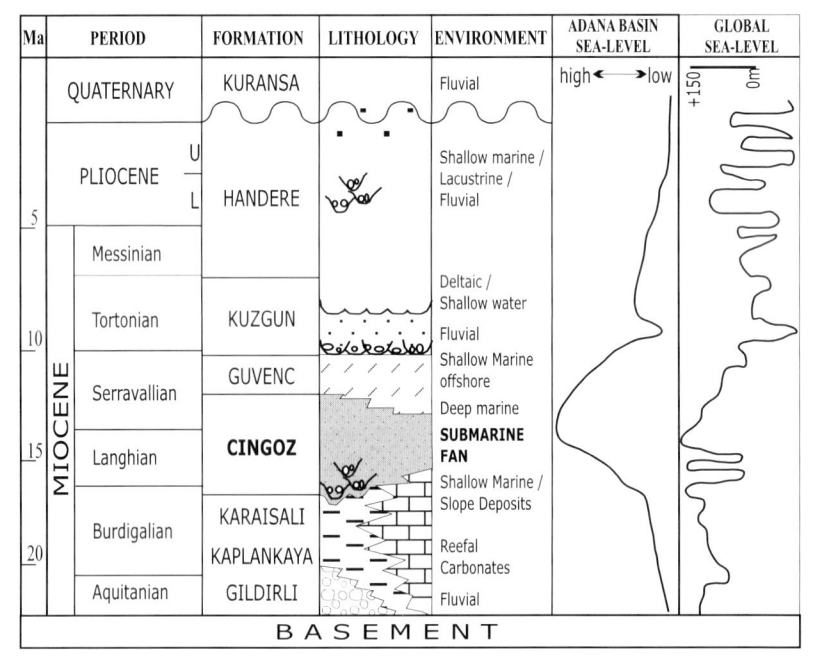

Fig. 2. General Cenozoic stratigraphy for the study area with interpreted sea-level curve for the Adana Basin (redrawn from Nazik & Gürbüz 1992).

Kelling 1993). The point-sourced Western Fan is elongate, with a downdip length of 25 km and thickness of approximately 1000 m, with predominant W–E palaeocurrent directions. The channelized facies were deposited within an asymmetric fairway, which was influenced by seafloor topography (Satur *et al.* 2000). Granular and pebbly sandstones with occasional conglomerates dominate the fairway-fill. These have accompanying sheet sandstones and siltstones in the more distal sectors forming depositional tongue- and lobe-shaped features (Satur *et al.* 2000). The Western Fan accords well with the low-efficiency, type II turbidite system of Mutti & Normark (1987) with sand deposited within channel-fill and lobe packages.

The Eastern Fan has a radius of approximately 25 km and is sourced from the north (Gürbüz & Kelling 1993). Southerly palaeocurrent directions are recorded in the upper fan, trending towards the E in the middle fan and SE in the outer fan (Fig. 1). Recent mapping has revealed this fan to have multiple sources, with at least four feeder channels (Fig. 3). The conglomeratic channel-fill package of the largest fairway, channel 1, is 85 m thick and represents a

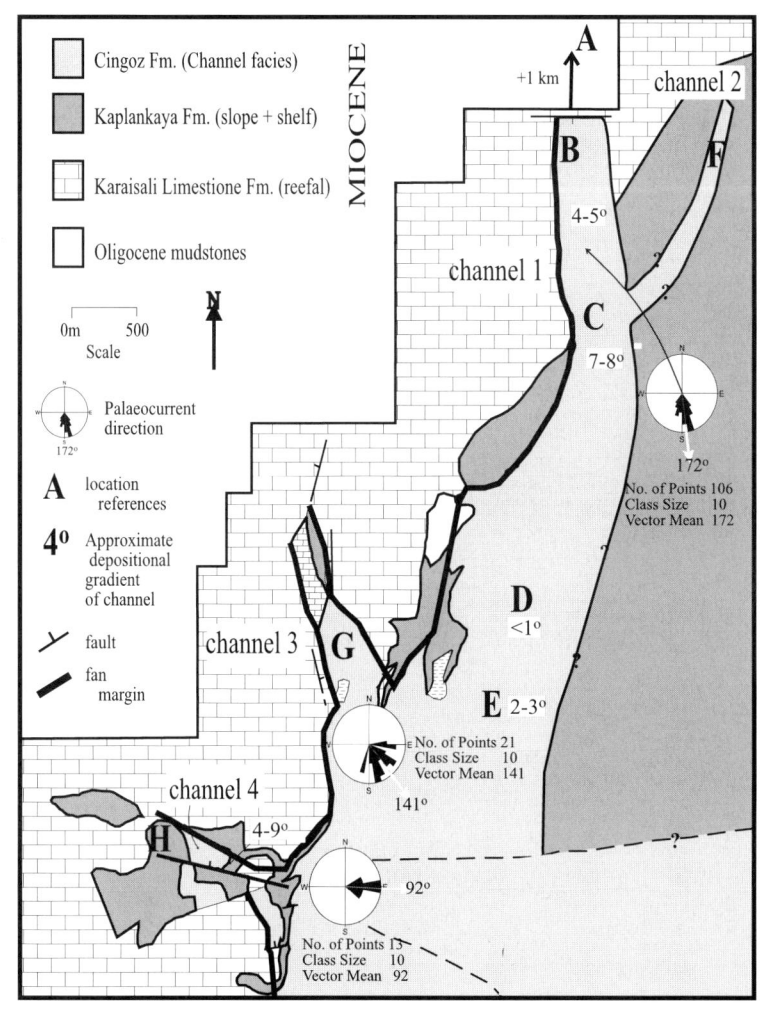

Fig. 3. Map of the multisourced feeder system in the northeastern parts of the Cingöz Fan. This map illustrates the distribution of basin margin facies (Karaisalı and Kaplankaya formations) and the channelized sections of the Cingöz Formation. The distributions of the overlying lobe sandstones of the Cingöz Formation are omitted for clarity. Letters are location points referred to in the text. Rose diagrams illustrate palaeocurrent directions measured predominantly from imbricated clasts and groove casts. Channel-floor depositional gradients are illustrated, estimated from the dip of beds.

small fraction of the 3000 m thick succession of this fan (Gürbüz 1993). In this fan system sand is deposited predominantly in non-channelized lobes with the feeder channels acting as bypass conduits, thus broadly according with the highly efficient, type I fan system as defined by Mutti & Normark (1987). The channel system that is preserved today did not act as a bypass system throughout the lifespan of the fan as it is overlain by more distal parts of the fan system, comprising sandy lobes. The younger feeder system that would have been coeval to these more distal facies is not preserved.

The aim of this study is to investigate the downdip variations in facies and architecture of the fills of the channels within the upper reaches of the Eastern Fan (Fig. 3). The study is based on the use of detailed vertical logs, geological maps and photomosaics of a series of closely spaced depositional dip and strike sections that allow some interpretation of three-dimensional stratal geometry in a well-exposed, relatively undeformed succession. Palaeocurrent directions are predominantly measured from the orientation of imbricated clasts and less frequent groove marks. Lithofacies are classified according to the facies scheme of Ghibaudo (1992) (Table 1), and analysed using the thickness of each facies as a function of the total thickness of the measured section. Clast type ratios are measured by counting the number of clasts (up to a maximum of 100) and their rock type within a

Table 1. *Lithofacies scheme used in this study (after Ghibaudo 1992)*

Facies	Facies code	Graphical representation
Gravelly mud	GyM	
Muddy gravel	MyG	
Gravel	G	
Gravel–sand couplet	GS	
Gravelly sand	GyS	
Sand	S	
Mud	M	

CONGLOMERATIC FAN FEEDER SYSTEM, TURKEY 245

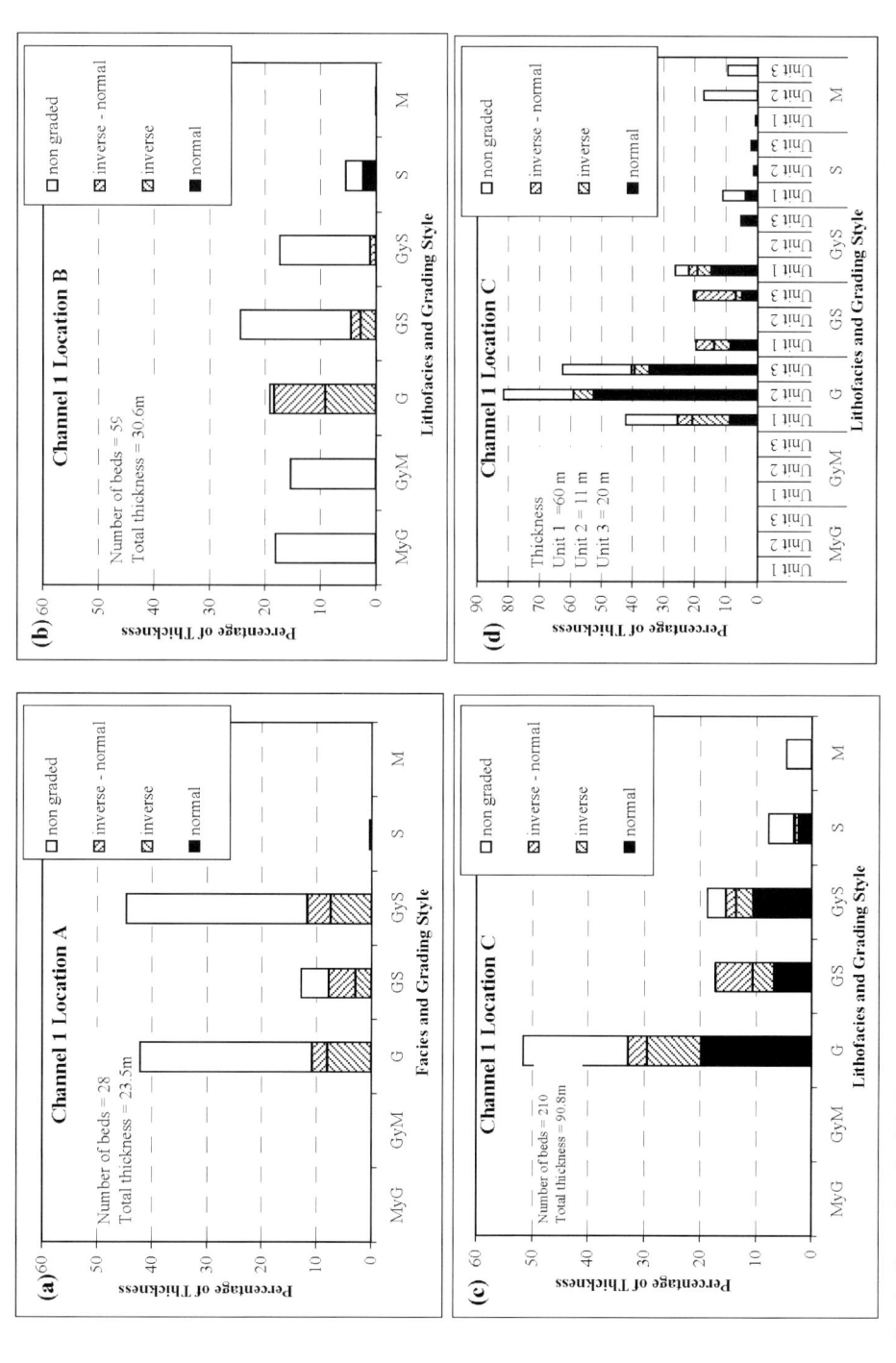

Fig. 4. Graphs illustrating the lithofacies and grading types recorded within Channel 1: (a) most proximal location within the channel; (b) 1 km to the south of location A; (c) 3.5 km south of location A; (d) subdivision of the fill of Channel 1, at location C, into three units.

1 × 1 m square. The depositional gradient within the feeder channels is estimated by calculating the difference in the structural dip of the channel base and the sandstone beds that overlie the conglomeratic channel-fill. This estimation is valid with the assumption that the sandstone beds (lobe deposits, Gürbüz 1993) were deposited on very low gradients and there is no differential compaction or subsidence.

The feeder system

The northern margin of the Adana Basin is well exposed. Although post-depositional faulting has made some modification to this margin, the spatial relationships between the uplifted hinterlands, the shelf, slope and deepwater fan are readily observed (Fig. 1). This allows the basin margin physiography to be explored with ease.

The Eastern Fan was sourced from the uplifted Tauride orogenic belt to the north (Gürbüz 1993). Sediment was transported across a narrow shelf (Satur 1999) supplying at least four feeder channels (Fig. 3). Channel 1, with its small tributary (channel 2), was incised into the shelf and slope facies, creating steep channel walls and limiting lateral migration of the channel. Normal faulting on the shelf, oblique to the coastline, formed smaller conduits (channels 3 and 4) for sediment to be transported basinwards (Fig. 3; Satur 1999). Again, the steep limestone walls of the channels prevented lateral migration of the channels.

Channel 1, location A

Channel 1 is a N–S trending feeder incised into the underlying slope shales (Kaplankaya Formation) and shelfal limestones (Karaisalı Formation). At the most northerly exposures of this channel (location A, Fig. 3), less than 1 km from the shelfal limestones of the Karaisalı Formation, the fill is dominated by disorganized and complex graded gravels and gravelly sandstones (Fig. 4a) that are typically 1–2 m thick. Mesozoic ophiolite clasts are predominant within the fill (Fig. 5) and are often randomly oriented. Here the channel incises into Kaplankaya shales and is approximately 30 m thick and 100–150 m wide (aspect ratio 3–5:1). This location is interpreted as representing the most proximal exposures of channel 1 due to the close proximity to the shelfal limestones and the disorganized nature of the conglomerates (cf. Walker 1975). Complex grading is common and probably reflex surging or waxing and waning currents (Kneller 1995) within the channel. Disorganized conglomeratic beds were deposited within a debris flow, or alternatively, also considering the bed thickness, may have

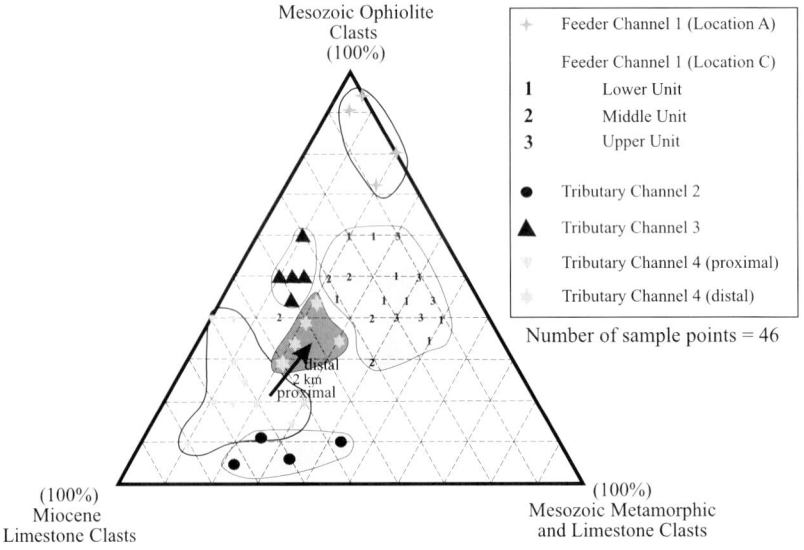

Fig. 5. Clast-type ratios of the four feeder channels to the eastern area of the Cingöz Fan. This graph plots the ratio of Mesozoic ophiolitic clasts, Mesozoic metamorphic clasts and Miocene limestone clasts. The triangular plot of clast-type ratios shows that the channels have slightly different source areas. Clasts of the distal locations of Channel 4 are different from more proximal areas within the same channel and may indicate some reworking of the older channel-fill of Channel 1.

been deposited as a lag deposit during sediment bypass (cf. Walker 1975).

Channel 1, location B

One kilometre down-channel (location B, Fig. 3), non-graded and inverse-graded gravels and non-graded muddy gravels and gravelly mudstones persist (Fig. 4b). The gravels display frequent cross-stratification, internal and basal scouring. The top of the section is composed of a structureless muddy gravel, greater than 6 m thick with randomly oriented clasts. Metre-scale limestone blocks are present within a gravelly mudstone bed midway up the section and also within the slope shales of the Kaplankaya Formation lateral to the channel-fill. The gravelly mudstone bed at this location has a wedge shape in cross-section with a thickness change from 5.4 to 0.55 m over a distance of 10 m (Fig. 6). This unit is interpreted as a muddy debrite, and with no evidence of post-depositional deformation the geometry suggests that this unit infilled some seafloor topography, probably an unfilled scour within the channel. The disorganized muddy gravel at the top of the section is interpreted as a pebbly debrite and the majority of the non-graded and inverse-graded stratified gravels as being dominated by traction and erosive processes.

Channel 1, location C

At 3.5 km down-system from location A, the Karaisalı Formation limestone is seen to confine the steep (40–70° from horizontal) western margin of the channel, which is well exposed (location C, Fig. 3). Here the channel is 500–600 m wide and contains an 85 m thick fill succession (aspect ratio ~6:1; log C, Fig. 7). Palaeocurrent direction, measured from clast imbrication, ranges from 135° to 215° (number of measurements = 172) with mean southerly palaeocurrent directions of 178° (location C, Fig. 3). The base of the channel had a dip of 19° and the sandstones overlying the channel-fill of 11°: thus the channel had a depositional gradient of approximately 8°, if the sandstones are assumed to have been deposited horizontally. Clast-supported conglomerates with a coarse sandy matrix and well-rounded clasts dominate the channel-fill. Normal and non-graded gravel and normal-graded gravelly sandstone are predominant with minor sandstones and mudstones (Fig. 4c). Typical clast ratios are Miocene shelfal limestone (15%) and basement Mesozoic metamorphic (30%), carbonates (10%) and ophiolitic material (45%) (Fig. 5). The channel-fill can be assigned to three fill phases on the basis of bed geometry and facies, forming the lower, middle and upper units (log C, Fig. 7), which record changes in deposition and erosion within the channel as it filled (Fig. 8).

Lower unit. The lower unit of the channel-fill is 50.5 m thick and consists of lenticular, laterally discontinuous beds in two-dimensional dip-section with scoured bases (Fig. 9a, log C; Fig. 7). It is composed predominantly of normal, inverse to normal and non-graded, clast-supported conglomeratic units up to 5 m thick with minor sandstone intervals (Fig. 4d). Clasts have a mean diameter of 0.05 m and are frequently imbricated. Two types of imbrication occur, first where clasts have their long axis (a-axis) dipping to the N and NW and second where clasts have their b-axis dipping to the N and NW and their a-axis horizontal in an E–W direction. The latter case is more prevalent where stratification is observed within this unit. At the top of the lower unit (log C, 36–44 m, Fig. 7), two thick conglomeratic beds occur. The lower bed is a 2.5 m thick, scoured, clast-supported conglomerate. It has inclined (up to 20°) planar, sandy cross-stratified layers that dip to the south (down-channel). This unit is more continuous than the beds below and can be traced for 20 m. Immediately above this are two thin silt layers and a 4 m thick unit comprising 3 m of 27 repeated, 9–15 cm thick horizontal layers of inversely graded conglomerates and sandstones that marks the top of the lower unit (Fig. 9b).

The thick, more laterally continuous, conglomeratic unit with down-channel dipping cross-stratification is ascribed to large migrating barforms. The length of these bars is unknown, but their height would be up to 2.5 m. Alternatively these surfaces could represent lateral migration of a channel thalweg. The overlying, horizontally stratified conglomerate and sand intervals are similar to those described by Lowe (1982) and attributed to deposition from traction carpets (Dzulynski & Sanders 1962; Fig. 9b). This traction carpet state of flow must have been sustained for a substantial time period in view of the 27 sand and pebble intervals that make up this 4 m thick bed. This interpretation is supported by the orientation of imbricated clast axes, which have their a-axis dipping up-channel and b-axis transverse to the channel. This indicates that the clasts were transported as bedload by tractional current (Piper 1970; Walker 1975). This is in contrast to clasts that

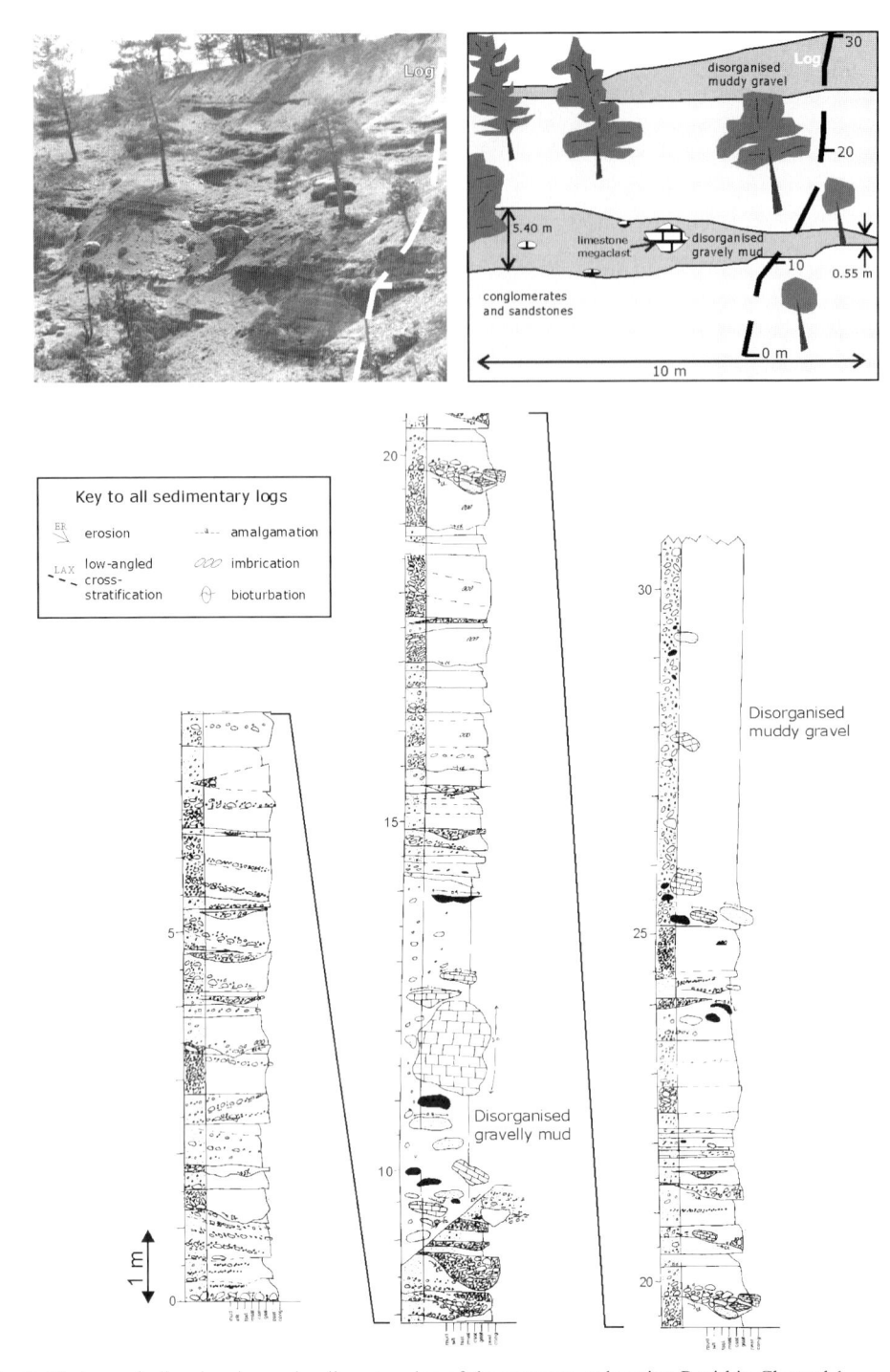

Fig. 6. Photograph, line drawing and sedimentary log of the exposure at location B within Channel 1. Conglomerates and sandstones deposited by tractional processes are predominant with a lower and upper parts. The disorganized gravelly mudstone contains randomly distributed limestone clasts, typically 0.1–0.2 m in diameter with occasion metre-scale limestone blocks. The disorganized muddy gravel consists of 0.02–0.06 m diameter, randomly oriented rounded clasts (clast-supported) with a muddy matrix.

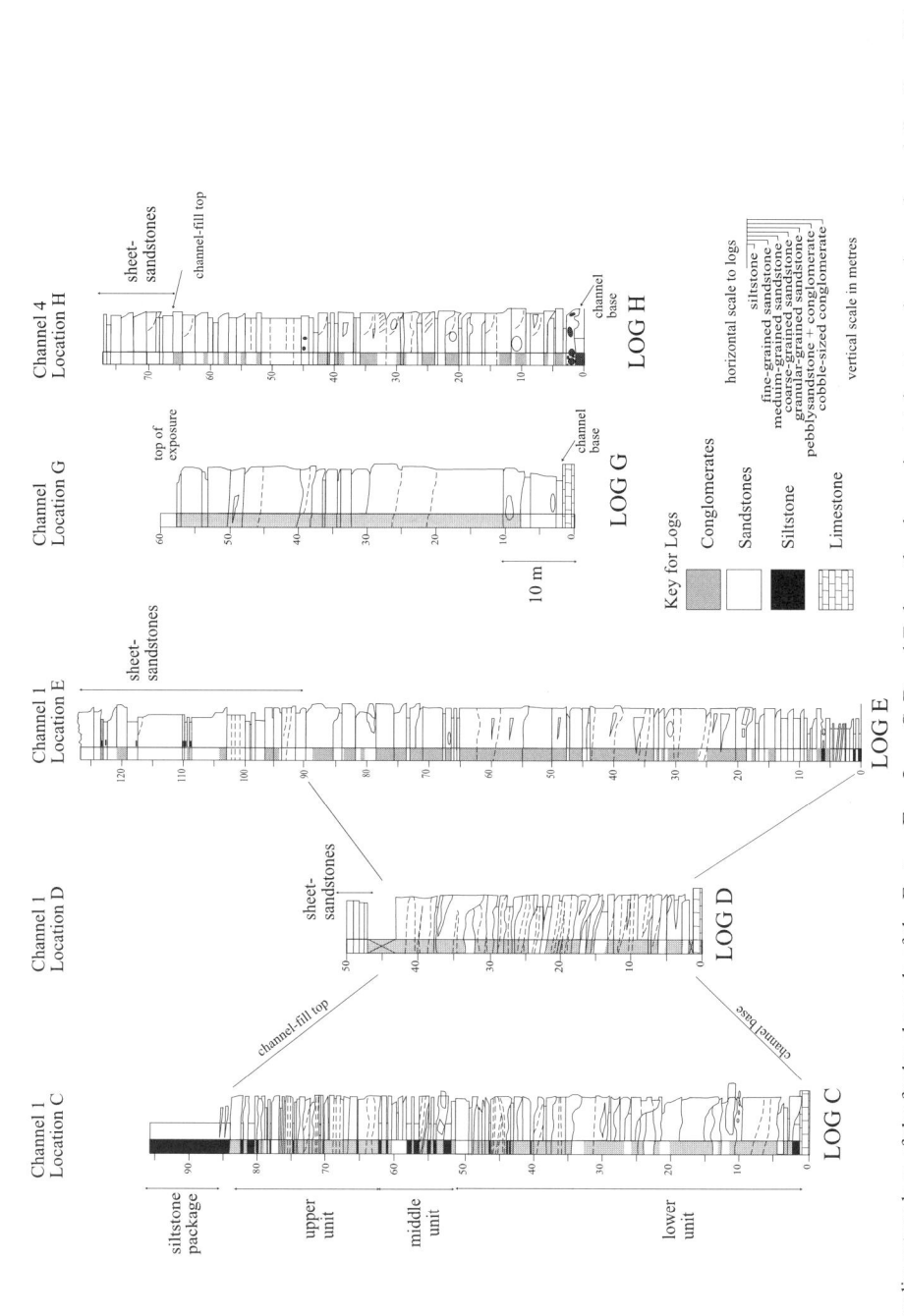

Fig. 7. Summary sedimentary logs of the feeder channels of the Eastern Fan. Logs C, D and E show the change in statal patterns in a down-channel direction within Channel 1. Sheet sandstones that overlie parts of Channels 1 and 4 are interpreted as depositional lobes by Gürbüz (1993). Vertical scale of logs in metres. See Figure 3 for locations.

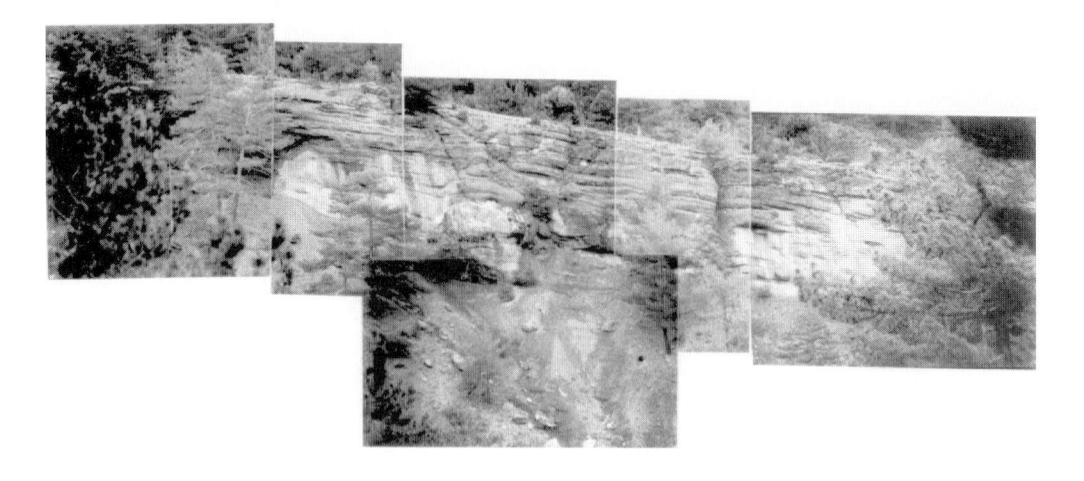

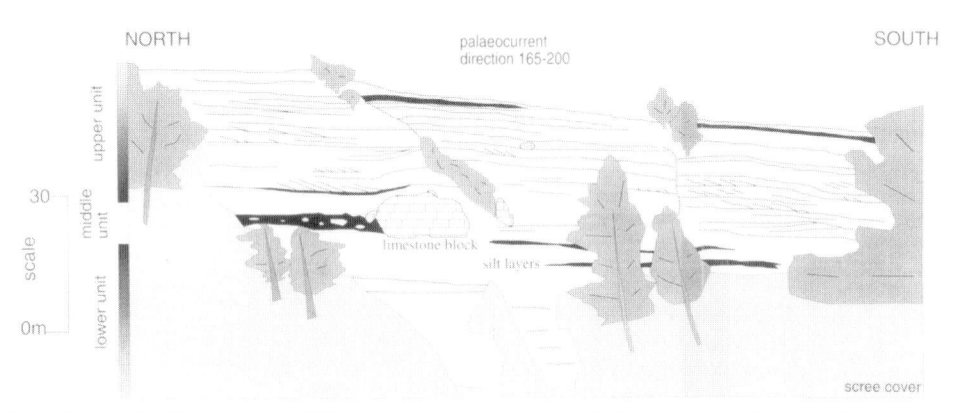

Fig. 8. Part of the 90 m thick fill of Channel 1, location C. Channel-fill architecture illustrates three-stage channel-fill, lower scoured discontinuous conglomerates, hemipelagic and disorganized conglomerates with occasional large (>10 m) blocks of limestone in the middle and sheet-like, cross-stratified conglomerates forming the upper unit.

have their b-axis dipping up-channel and a-axis transverse to the channel and are interpreted as having been deposited out of suspension (Davies & Walker 1974; Walker 1975).

Middle unit. The middle unit is 7 m thick and is composed of five 0.30–1.50 m thick siltstone beds that are laterally continuous down depositional dip for over 300 m (log C 52–59 m, Fig. 7). They are ungraded, parallel laminated, and have sparse traces of chondrites bioturbation (Fig. 9b). Interbedded with the siltstones are disorganized (non-graded) and normally graded, clast- and matrix-supported conglomerates that contain undeformed mud clasts up to 1 m in diameter and outsized, subangular limestone blocks up to 3 m in diameter (Fig. 8). These are interpreted as muddy and pebbly debrites with the large limestone blocks either transported the 3–5 km from the shelf within the debris flow or fell into the channel from the channel margin, tens of metres away. The siltstone beds are interpreted as being of hemipelagic origin.

This unit is interpreted as representing a quiescent period in sedimentation within the channel, reflecting a reduction in sediment supply from the source area, during which hemipelagic processes prevailed and were interrupted only by occasional debris flows resulting from failure of the steep channel margins. The deposition of the debrite allowed the hemipelagic layers to be preserved, protected from erosion by subsequent turbidity currents. The accumulation and preservation of hemipelagic sediments indicates that activity within the channel was reduced as the channel effectively shut down.

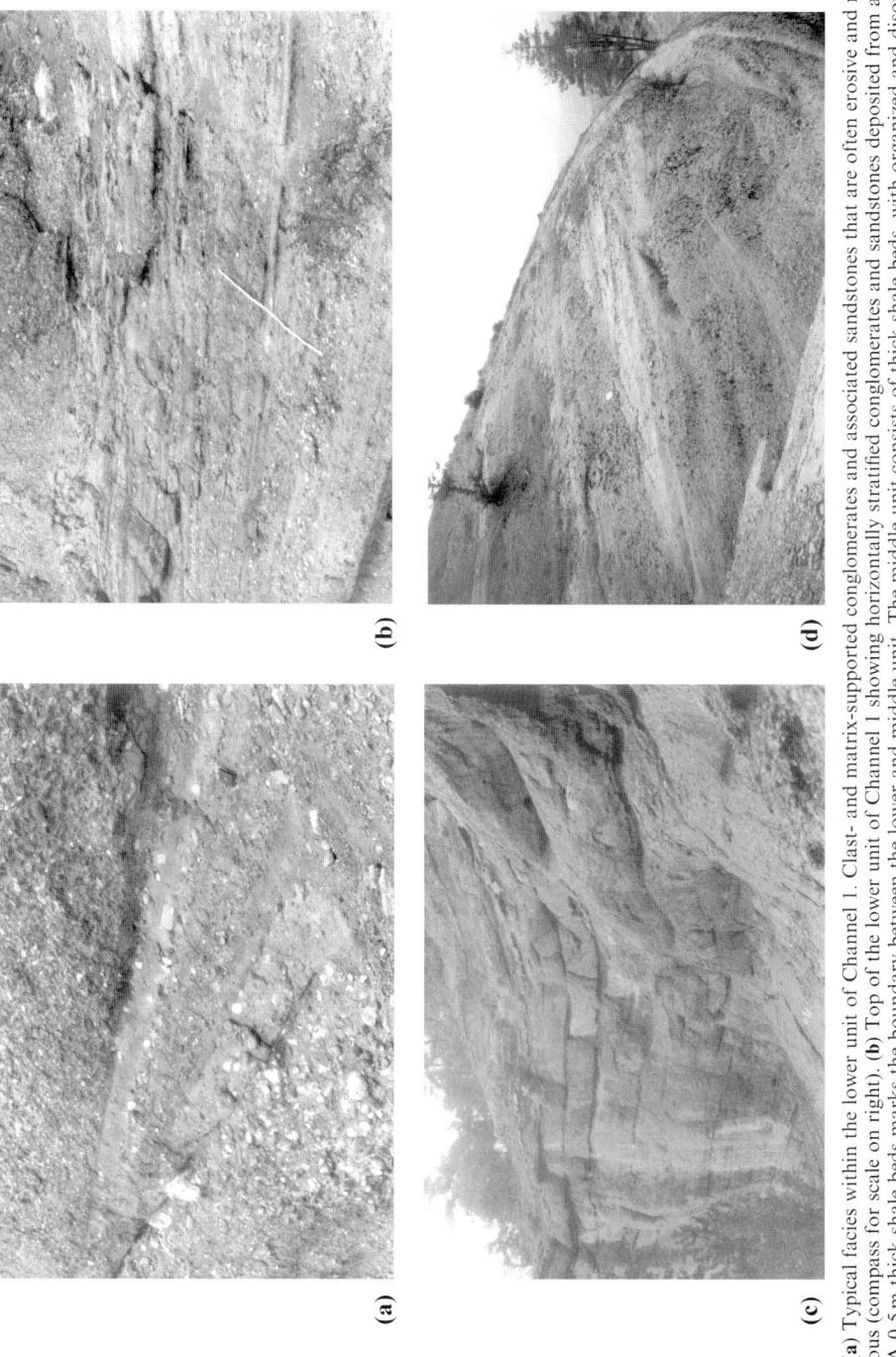

Fig. 9. (a) Typical facies within the lower unit of Channel 1. Clast- and matrix-supported conglomerates and associated sandstones that are often erosive and non-continuous (compass for scale on right). (b) Top of the lower unit of Channel 1 showing horizontally stratified conglomerates and sandstones deposited from a traction carpet. A 0.5 m thick shale beds marks the boundary between the lower and middle unit. The middle unit consists of thick shale beds, with organized and disorganized conglomerate beds (1 m ruler for scale). (c) Upper unit of Channel 1 illustrating the conglomeratic and sheet-like nature of the fill. Towards the top of the channel-fill, conglomeratic beds become thinner and shale beds thicker, passing upwards into a thick packages of claystones (cliff section 32 m high). (d) Scouring conglomerates and lenticular sands forming a braided stream morphology from reworking of the channel floor (water can centre 0.5 m high).

Upper unit. The upper unit is 24 m thick and comprises laterally continuous over the exposed sections (Figs 8 & 9c), 0.30–0.70 m thick sheets of both normally graded and ungraded, clast- and matrix-supported conglomerates (log C 59–83 m, Fig. 7). These conglomeratic beds can be traced downdip for over 400 m. They exhibit only localized scouring at their bases (Figs 8 & 9c). Large-scale (2–5 m long), low-angle cross-bedding (5–15°) is observed within many of the beds dipping to the south (down-channel). Also frequent centimetre-thick, horizontally stratified conglomerate and sandstone intervals are observed with clasts having their b-axis dipping up-channel and horizontal a-axis oriented across the channel. The top section of the upper unit shows an increase in frequency and thickness of siltstone beds (Fig. 9c) with a 1 m thick disorganized muddy gravel marking the top of the channel-fill succession. Above the channel-fill, a 30 m thick package of horizontally laminated, non-bioturbated siltstones and claystones occur, and this package pinches out gradually over 4 km in a down-channel direction. This package does not contain any material coarser than siltstone-size grains.

The cross-bedding within the sheet conglomerates is envisaged to have formed by the down-channel migration of bedforms (Sanders 1965) that, given the scale of cross-bedding, must have been in the order of several metres in amplitude. The orientations of the a- and b-axes of the clasts suggest that tractional processes were predominant within the channel, with clasts rolling along the channel floor (Piper 1970; Walker 1975). Tractional processes would also permit the migration of conglomeratic bedforms, and may be a response to gravity currents bypassing the channel, reworking the channel-floor deposits. The siltstone and claystone package that overlies the coarse clastic fill of channel 1 at this location is interpreted as being part of the prograding slope. A regional sequence stratigraphic interpretation cannot be interpreted, as exposures are lost below ground to the south. However, the lack of any coarse material within the package, its location close to the basin margin and the rapid nature of its downdip pinchout suggest that it is not a channel abandonment package. Overlying this slope package is the thick succession of sheet sandstones that represents the sandy lobate sector of the Eastern Fan, backstepping over the channel-fill and slope package (Gürbüz 1993).

Channel 1, location D

Some 7 km down-channel from the headward part of the feeder channel 1 (location D, Fig. 3), several stacked packages occur that consist of lenticular-shaped, granular to medium-grained gravelly sandstones that thin and fine upwards. These are typically 1–3 m thick, are ungraded or normally graded, and cross- or planar stratified (log D, Fig. 7). Between the sandy lenses there are units of clast- to matrix-supported conglomerates, which have sandy intrabedded lenses and are typically 4–8 m in thickness (Fig. 9d). These conglomerates often display a basal inverse grading interval, followed by a thicker, normally graded section. These conglomerates have broad (10–20 m), deep (2–4 m), scoured bases that incise into the sandy units. The true width of these erosive features is not recorded in this laterally limited strike-section. Both the sandstones and conglomerates have planar cross-stratification and along with frequent clast imbrication indicate a general flow direction to the S and SE. On the basis of the dip of the channel base and the sandstones beds overlying the channel-fill, the depositional gradient of the channel at this location is estimated to be <1° (location D, Fig. 3).

The stacked conglomerate lenses are interpreted as representing the fills of shallow channels (furrows) that are separated laterally and vertically by sandy bedforms. This exposure is limited to a depositional strike cross-section across the channel-fill, but these stratal patterns are interpreted as forming a braided stream morphology in planform. Alternatively these channel features could be part of a megaflute/scoured surface (cf. Normark *et al.* 1979; Normark & Piper 1991). The former interpretation is preferred here as explaining the vertical separation of conglomeratic scoured features and more sandy bar-like features, given that in other locations within the channel the fill is composed of gravel and gravelly sandstone and lacks sandstone beds (Fig. 4).

Channel 1, location E

Some 2 km down-channel from these facies (location E, Fig. 3), a 70–90 m thick package of sediment comprising several 2–5 m thick and structureless conglomerate beds occurs (log E, Fig. 7): they pass stratigraphically upwards into coarse sandy turbidites and then sheet sandstones that represent lobe deposits (Gürbüz 1993). These deposits represent the coarsest material seen in the feeder system, with well-rounded clasts with a mean diameter of 0.07–0.1 m, but clasts often exceed 0.5 m in diameter. Where clast imbrication is preserved, clasts often have their a-axis dipping to the north (up-channel). Most beds are normal- and non-graded and display little evidence of erosion. The thick-bedded and structureless nature of these conglomerates suggests that they were deposited rapidly with clast deposited out of suspension, given the orientation of their a-axis (Walker 1975).

Channel 2, location F

Channel 2 is situated 2 km east of Channel 1 (location F, Fig. 3) and is at a similar stratigraphic level. It is 30 m wide with a 6 m thick fill in its uppermost reaches (aspect ratio of 5:1) (Fig. 10). It has a lenticular cross-section incised into and overlain by a 40–50 m thick, horizontally laminated siltstone succession. The fill of Channel 2 is composed of cross-stratified, clast-supported conglomerates and pebbly sandstones with thin silt layers near the top of the channel-fill (Fig. 10). Clast types are typically dominated by Miocene limestone clasts (60%) with Mesozoic metamorphic and limestone clasts (35%) and minor amounts of Mesozoic ophiolite clasts (5%) (Fig. 5). These conglomerate beds display some clast imbrication with a-axis horizontal and oriented across the channel and b-axis dipping towards the north (up-channel). Towards the top of the channel-fill, low-angle ($<5°$), planar cross-bedded surfaces are observed dipping towards the east.

This channel is located within a package of siltstones, of which the upper parts can be correlated with the siltstone package overlying the fill of Channel 1 at location C. This siltstone package is interpreted as forming part of the slope succession. Palaeocurrents from clast imbrication are towards the SSW and, given the similar stratigraphic location of Channel 2 to Channel 1, would suggest that this channel was a tributary to Channel 1. Low-angled cross-bedded surfaces are interpreted as being lateral accretionary surfaces within the channel, interpreting migration of a channel thalweg across the channel towards the east.

Channel 3, location G

Channel 3 is fault-bounded and incised into the underlying limestone and shale of the shelf (location G, Fig. 3). It has a NNW–SSE trend with measured palaeocurrent directions of 120–165°, a width of up to 400 m and a fill of at least 55 m, with aspect ratio of 8:1, and is tributary to Channel 1. Poorly sorted, angular-grained sandstones that contain reworked shallower-water, shelly fauna from the lowest units of the channel-fill (log G 0–7 m, Fig. 7). Towards the SW margin these shelly sandstones are overlain and eroded into by thick-bedded (up to 5 m) clast-supported conglomerates that comprise 90% of the 55 m thick channel-fill at this location within the channel. Some of these conglomerates show inverse-normal grading, but are otherwise structureless with occasional scoured bases and isolated megaclasts up to 3 m in diameter. The conglomerates are typically composed of 50% ophiolitic material, 38% Miocene Karaisalı Limestone, 12% Mesozoic limestone and metamorphic clasts (Fig. 5). From the SW to the NE margin of the channel conglomerates grade laterally into turbidite sandstones. This relationship is probably associated with faulting at the SW margin of the channel (location G, Fig. 3). This forms a half-graben feature in which the coarsest material is deposited in a channel thalweg within the deepest sectors, in close proximity to the faulted SW margin. Towards the shallower NE margin sand is deposited away from the channel thalweg.

Channel 4, location H

Channel 4 trends in a W–E direction with easterly palaeocurrent directions (from clast imbrication), and is 150 m wide with a 60 m thick fill package (aspect ratio of 2.5:1) (location H, Fig. 7). The base of the channel can be traced over several hundred metres, incising into the underlying Kaplankaya shales and more distally into the channel-fill of Channel 1. At the base of the fill succession there is a 5 m thick, laterally thinning, disorganized, clast-supported conglomerate with large cobble-size clasts of limestone, sandstone and siltstone (log H, Fig. 7). The clasts have no preferred alignment, and there is only limited deformation of the siltstone clasts. Texturally mature sands with clast and matrix-supported, cross-stratified gravels and gravelly sandstones fill the remainder of this channel, similar to the facies observed in Channels 1 and 3. Channel 4 is composed of Karaisalı Limestone (62%), basement limestone and metamorphic clasts (18%) and ophiolitic clasts (20%) in its more proximal locations (Fig. 5). The channel-fill conglomerates of Channel 4 are dominated by shelfal and basement carbonates with a lower content of ophiolitic material compared with those within Channels 1, 2 and 3. In a 2 km more distal location the channel-fill is typically composed of Karaisalı Limestone (42%), basement limestone and metamorphic clasts (21%) and ophiolitic clasts (37%) (Fig. 5).

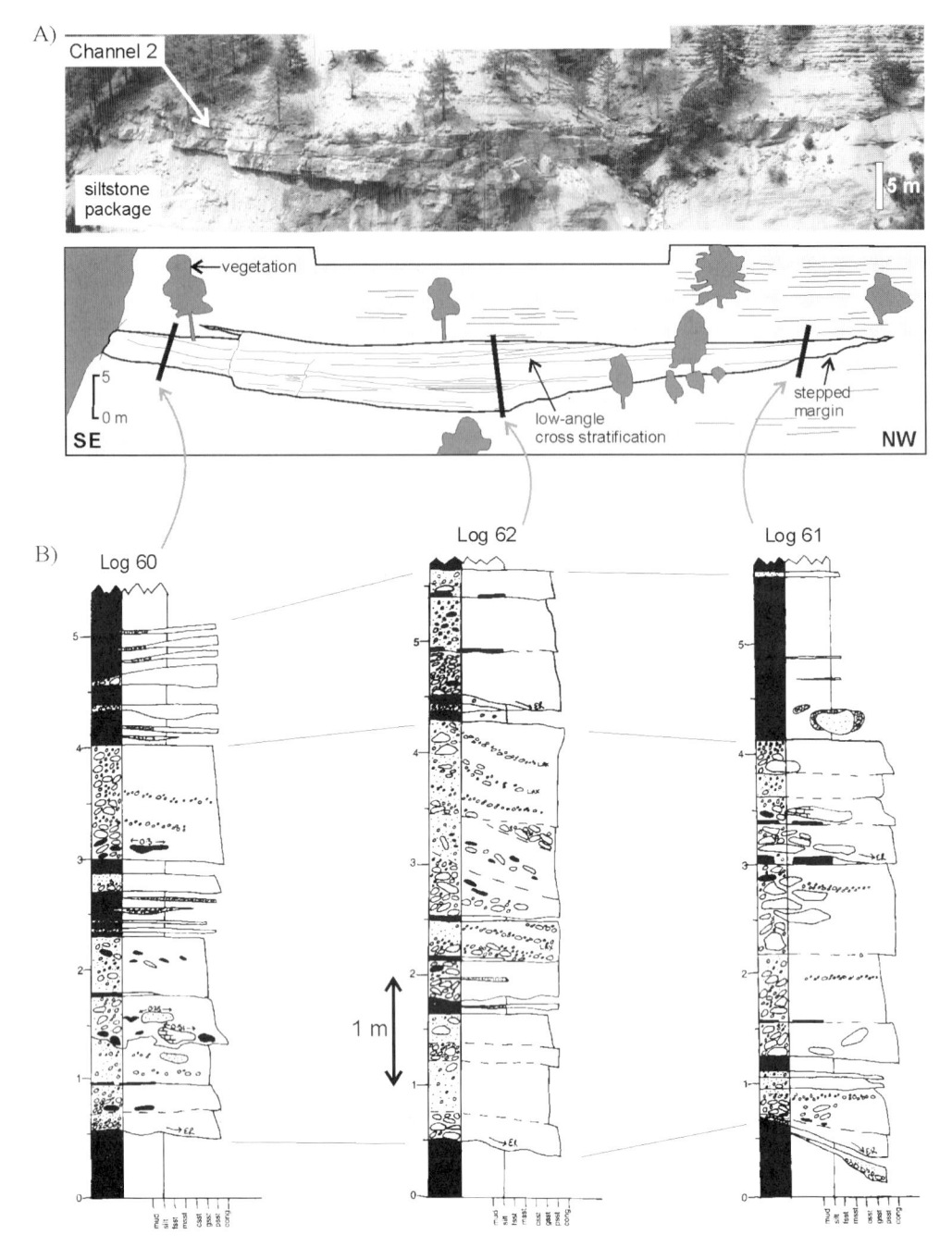

Fig. 10. (a) Photomosaic, line drawing and (b) sedimentary logs of Channel 2 (location F). This small lenticular channel is a tributary of Channel 1, incising into the Kaplankaya slope facies. Net to gross is high at the centre of the channel, decreasing towards both margins.

The disorganized conglomerate at the base of the channel-fill has a texture that is typical of a cohesive debris flow (Fisher 1971; Postma 1986). The lateral thinning of this unit implies that it has infilled a previously scoured hollow within the fine-grained slope succession. However, at some stage the flow that deposited this unit must have been erosive to incorporate the

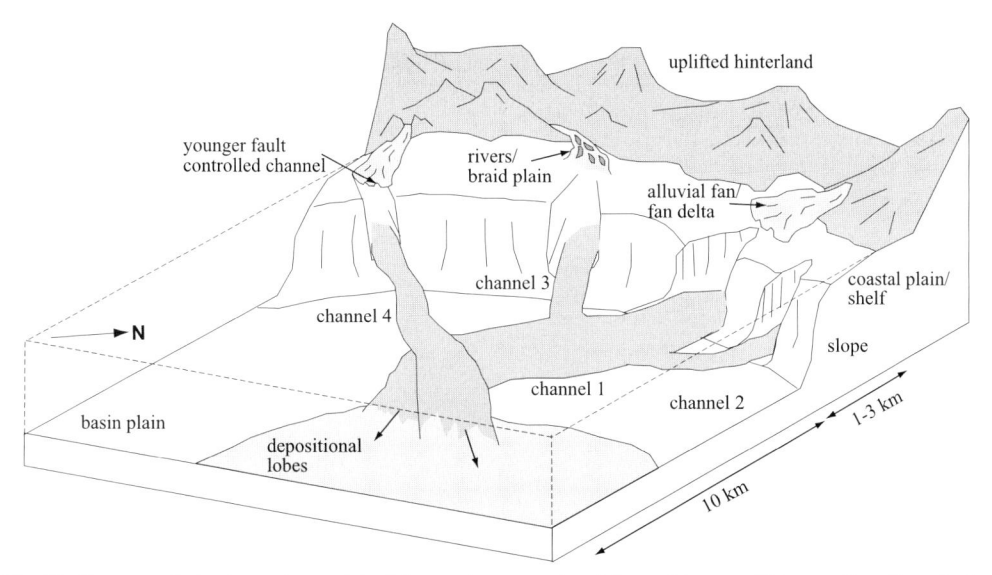

Fig. 11. Conceptual physiography model of the northern margin of the multi-sourced Eastern Fan. Four feeder channels incise the basin margin and allow sediment to bypass the shelf and slope and be deposited basinwards to form a 3 km thick and 25 km radial fan package consisting predominantly of sandstone lobes.

siltstone clasts of the underlying formation. Incision of Channel 4 into the fill of Channel 1 implies that it is younger in age than the fill of Channel 1. Analysis of the clast-type ratios within the separate channel-fills (Fig. 5) indicates that Channels 1 and 3 have similar clast types, with the proximal part of Channel 4 being distinctly different with a high percentage of Miocene age limestone clasts. At 2 km distally within Channel 4, clast-type ratios are a mixture of those seen within all the channels. Ophiolitic material and Cingöz turbidite sandstone clasts occur at this location, which are not recognized in the upper reaches of Channel 4. Clast-type ratios in the distal portion of Channel 4, and the observation that Channel 4 erodes into Channel 1, indicate reworking of Channel 1 fill facies in the more distal locations of Channel 4.

Ratios of clast types in the most proximal areas of Channel 1 (location A) show them to be dominated by Mesozoic ophiolite clasts (Fig. 5). The clast types contrast with those observed in the same channel approximately 2.5 km distally. This is not fully understood, but if the channel fill deposits at location A are interpreted as channel lag facies, deposited from bypass currents, then they may not be coeval with the more distal fill of Channel 1. At location C within Channel 1, the clast-type ratios for the three units are as shown in Fig. 5. There is no clear distinction between the three units, which suggests that there was no significant change in source area during the time of channel infilling.

Channel 2 has a very distinctive clast type with a high percentage of Miocene Limestone clasts and low proportion of Mesozoic ophiolite clasts (Fig. 5). This may indicate that this channel had a different source area from that of Channel 1 despite it being its tributary.

The proximal area of the Eastern Fan, Cingöz Formation, is interpreted as being multisourced with at least four feeder channels (Figs 3 & 11). These channels incise into the slope and limestone shelfal deposits of the Kaplankaya and Karaisalı formations (Fig. 12). Channels 1, 2 and 3 appear to be active at the same time. Given the change in clast types within the fill of Channel 4, it would appear to be younger, with gravity flows within this channel reworking some of the deposited within Channel 1.

Discussion

Down-channel changes in facies and bedding styles allow some comparison with previously defined fan models. Walker (1975) recognized that, in a conglomeratic rich fan system, there was a systematic change in a proximal to distal direction in style of the beds ungraded-disorganized into inverse-normal and normal grading. A similar change in the style of beds in the conglomeratic fill of Channel 1 has been observed. Proximally, complex grading and non-graded gravel and gravelly sandstone beds are prevalent, with individual clasts displaying a chaotic a-axis orientation (Fig. 4a). Facies at location B within Channel 1 are dominated by non-graded, inverse and inverse-to-normal grading conglomerates (Fig. 4b) that are generally unstratified. Location C within Channel 1 has a higher proportion of normally graded conglomerate beds (Fig. 4c). The association between these three locations within Channel 1 is known to be a proximal-distal relationship from field mapping and interpretation of palaeocurrent data. If the processes for the changes in bedding are taken from Walker (1975), then the depositional gradient and distance in which the flow travelled appear to be the main controlling factors and appear

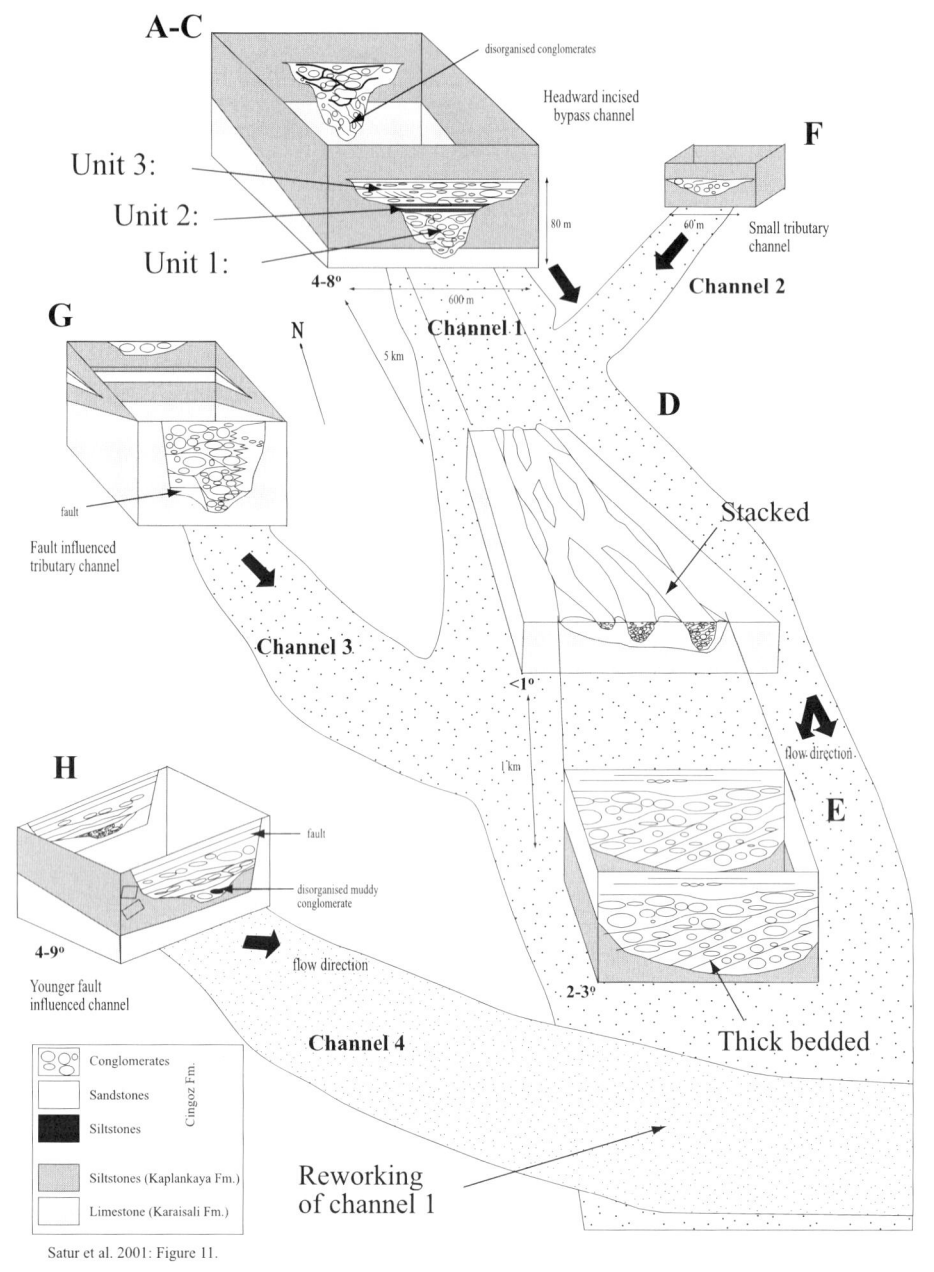

Fig. 12. Schematic block diagram illustrating down-channel variations in the facies organization within the upper reaches of the Eastern Fan. (Bold letters refer to location markers, Fig. 3)

to be supported by this study. The identification of a systematic change in facies is important and can help build predictive models.

Bedforms are interpreted as having existed within the fill of Channel 1. However, there are very few descriptions of bedforms from studies of ancient exposures. High-angled cross-bedding in ancient deepwater conglomeratic systems has been interpreted as bedforms in other studies (Piper 1970; Winn & Dott 1979; Hein & Walker 1982). However, studies from modern deepwater systems often describe bedforms within channels. These features are regularly identified on high-resolution, deep-towed, side-scan sonar images of modern deepwater channel systems, where they are described as elongate lineaments oriented perpendicular to the channel axis. For example, Belderson et al. (1972) described bedforms in the Naples Canyon, Malinvero et al. (1988) in the Var channel system, Piper et al. (1988) in the Larentian Fan, and Masson et al. (1995) in the Monterey Fan Channel. However, the analogy between modern and ancient examples is unclear, because of the contrast in scale of the bars, but examples from modern systems clearly indicate that they can form in deepwater systems and that they should be recognized more frequently in ancient examples if they can be preserved.

The most proximal exposures of Channel 1 indicate that it had a high gradient, and was relatively narrow and erosive (lower unit), with presumably high sediment flux rates in the early phases of infilling (Fig. 8 and locations A–C in Fig. 12). A quiescent period then allowed the deposition of fine-grained hemipelagic material along with pebbly and muddy debrites (Middle Unit). Sheet conglomerates formed in the final phase of filling probably represent a decrease in sediment supply rates and/or widening of the channel, causing the gravity flows to be less well confined (Upper Unit). This vertical change in bedding style and facies demonstrates how channel processes can change through time in response to changes in shape of the channels and/or sediment supply rates. One unanswered question is how these temporal changes in channel-fill manifest themselves down-channel, as timelines between the different localities within the channel-fill are not resolvable.

The down-channel changes in stratal patterns within the proximal channelized sector of this turbidite fan are interpreted in terms of variations of flow processes and hydrodynamic conditions during deposition. These changes in flow conditions are here primarily attributed to the observed changes in gradient within the channel. Possible additional influences from changes in confinement of the flow as the channel with increased down-system are unknown and cannot be ruled out. The relative gentle and uniform tectonic dip of the beds allows the reconstruction of channel gradient. Steep gradients (4–8°) are recorded in the proximal locations of the channel (locations A–C, Fig. 13), decreasing in gradient ($<1°$) at location D before steepening (2–3°) again at location E. This formed a 1.5 km long, relatively flat area within the channel profile. Similar features are described in modern canyon systems by Le Pichon & Renard (1982), who attributed its formation to structural control or increased erosion at the confluence to two channels. Within Channel 1 the increase in gradient at location E is close to the confluence between Channels 1 and 3 and may have been a contributing factor in the formation of this flat area. However, faulting is observed close to this location and may also play a role in controlling channel gradients.

Some 7 km down-channel from the head of the fan (location D), a break in slope at the upper reaches of the flat area caused conditions within flows to change and led to the formation of a braided stream network within the channel (Fig. 13). In this example the formation of a braided stream morphology is not a result of sediment deposition and choking of the channel, as in fluvial examples. This is because the coarsest fractions of the flows are deposited down-channel from the braided stream deposits. It is envisaged that, at the first break in slope, a hydraulic jump from supercritical to subcritcal flow conditions occurred (Fig. 13). This resulted in reworking of the channel floor by tractional processes as a consequence of increased turbulence. This allowed coarse-grained sediment from the channel floor to be entrained within the flow, producing a braided stream morphology.

This braided morphology is recognized in modern and ancient deepwater systems. The Lower Var Fan Channel (Malinvero et al. 1988; Piper & Savoye 1993; Mulder et al. 1997), the Orinoco Fan (Beldersen et al. 1984) and the Labrador Sea Fan (Hesse et al. 2001) are examples of modern systems where braided streams have been interpreted. In ancient examples, Winn & Dott (1979), Hein & Walker (1982) and Buatois & Angriman (1992) all interpreted facies in their respective study areas as deepwater braided channels. The scale of channels and bars varies between systems, but typical dimensions from Hein & Walker (1982) indicate braided channels to be approximately 1/50th the size of the main channel in which the braided morphology is located. A similar relationship between the size (depth) of braided channels and that of Channel

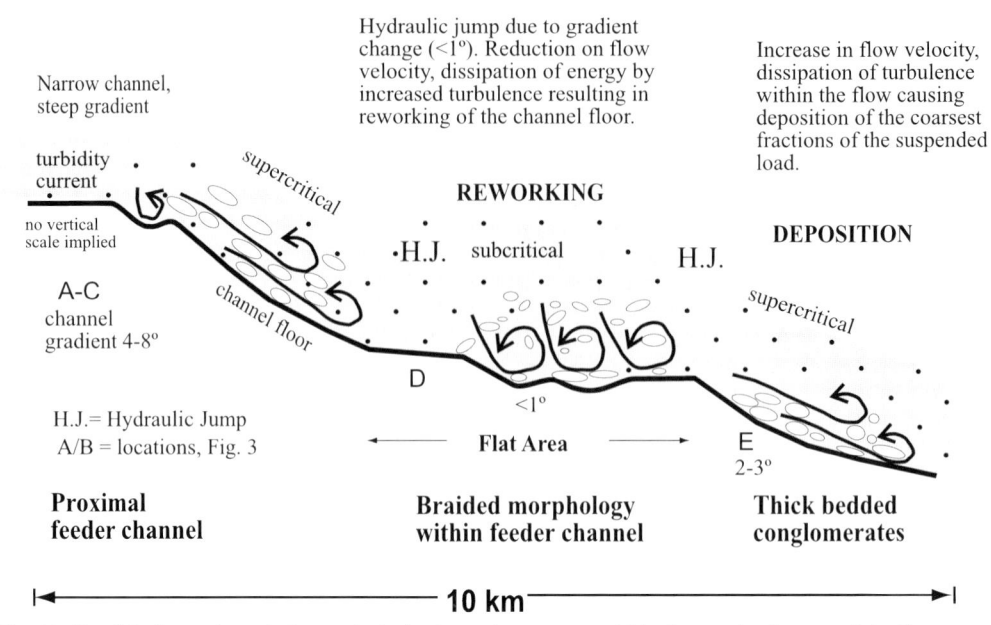

Fig. 13. Possible down-channel changes in hydrodynamic processes within the proximal sector of the Eastern Fan to explain the observed variations in stratal patterns. Changes in slope as the flow enters a relatively flat area within the channels initiate a hydraulic jump. The consequent increase in turbulence causes reworking of the channel floor, producing a braided stream morphology in planform. A second hydraulic jump occurs downslope as the flow exits the flat area, reducing the flow's turbulence and capacity, causing deposition of the coarsest fraction of the flow with the finer-grained material transported further into the basin.

1 is seen in the Adana Basin, but true widths of the braided channels are not recorded because of limited exposure.

As turbidity currents continued downslope, flows that had increased in density as they entrained sediment reworked from the channel floor could not keep coarse sediment in suspension. This may be a result of gradual increase in density of the flow, or more likely a result of a second hydraulic jump from subcritical to supercritical flow at the break in slope as the flow exited the flat area (Fig. 13). This would have caused reduced turbulence and therefore competence and capacity of the flow, thus initiating deposition (Menard 1964; Komar 1971). This would explain the thick-bedded conglomerates deposited down-channel from the stratal patterns associated with sub-aerial braided streams.

A similar downdip morphological change from a single channel to a braided network and back to a single channel is recorded in the Orinoco Deepsea Fan (Beldersen et al. 1984). In this example the change from a single channel into a braided network is attributed to a reduction in seafloor gradient and a widening in the sediment pathway. The change from a braided network back to a single channel downdip is due to an increase in gradient and also a narrowing of the sediment pathway due to a structural feature. This study by Beldersen et al. (1984) does not describe changes within the gravity currents that would result in these morphological changes, but seafloor gradient and confinement of the flow appear to be important factors, similar to those interpreted in this study.

Conclusions

The proximal part of this turbidite fan shows at least four separate feeder channels that allowed sediment to bypass the shelf and slope and be deposited basinwards. Analysis of the fill of the channel system shows distinct and hydrodynamically consistent down-channel variations in facies architecture. Within Channel 1 there is a predictable down-channel change in bedding pattern from more disorganized facies with ungraded beds into more organized and inverse-normal and normal graded beds. The three-stage infilling of Channel 1 (location C) indicates that this channel evolved from an erosive into a depositional channel, with a quiescent period of current activity between these two phases of infilling. These temporal changes in stratal patterns may illustrate changes in sediment supply and/or sediment discharge rates in response to changes in channel cross-sectional shape and channel gradient.

As turbidity current flowed down-channel, a hydraulic jump occurred as the flow entered the area of reduced depositional channel-floor gradient (location D), resulting in reworking of the channel floor and the formation of a braided stream network within Channel 1. Sediment was resuspended within the flow and transported down-channel, where channel-floor gradients increased, causing increased velocity and reduced turbulence within the flow. This resulted in the deposition of the coarsest fractions of the suspended load to form thick-bedded conglomerates, as the turbidity current did not have sufficient energy to maintain this material in suspension. The channelized feeder system of this fan demonstrates a wide variety of facies changes down-channel and most probably results from a combination of factors including variations in slope gradient, sediment supply, confinement of the turbidity currents and the influence of the tributary feeder channels. It is important to recognize and account for such variations in other modern and ancient deepwater fan systems to promote better understanding of subsurface (seismic and particularly borehole) data from turbidite fans. Such knowledge has considerable economic significance, for example in hydrocarbon exploration and production, both in enhancing generic understanding of facies distribution within turbidite fan feeder systems and potentially in providing a predictive tool for subsurface investigations in specific cases.

The authors would like to thank the Mesostratigraphy of Deepwater Sandstones Consortium (Amerada Hess, BP-Amoco, Conoco, TotalFinaElf and Enterprise) for funding this study. S. Lomas and G. Cornamusini are thanked for their useful comments when reviewing this paper, and Y. Topak, O. Çelik and I. Naz from Çukurova University for their assistance in the field.

References

BELDERSON, R. H., KENYON, N. H., STRIDE, A. H. & STUBBS, A. R. 1972. *Sonographs of the Sea Floor*. Elsevier, Amsterdam.
BELDERSON, R. H., KENYON, N. H., STRIDE, A. H. & PELTON, C. D. 1984. A 'braided' distributary system on the Orinoco deep-sea fan. *Marine Geology*, 56, 195–206.
BUATOIS, L. A. & ANGRIMAN, A. O. L. 1992. The ichnology of a submarine braided channel complex: the Whisky Bay Formation, Cretaceous of James Ross Island, Antarctica. *Palaeogeography, Palaeoclimatology, Palaeoecology*, 94, 119–140.
DAVIES, I. C. & WALKER, R. G. 1974. Transport and deposition of resedimented conglomerates: the Cap Enragé Formation, Gaspé, Quebec. *Journal of Sedimentary Petrology*, 44, 1200–1216.
DZULYNSKI, S. & SANDERS, J. 1962. Current marks on firm mud bottoms. *Connecticut Academic Arts and Sciences Transactions*, 42, 57–96.
FISHER, R. V. 1971. Features of coarse-grained, high-concentration fluids and their deposits. *Journal of Sedimentary Petrology*, 41, 916–927.
GHIBAUDO, G. 1992. Subaqueous sediment gravity flow deposits: practical criteria for their field description and classification. *Sedimentology*, 39, 423–424.
GÜRBÜZ, K. 1993 *Identification and evolution of Miocene submarine fans in the Adana Basin, Turkey*. PhD thesis, University of Keele.
GÜRBÜZ, K. & KELLING, G. 1993. Provenance of Miocene submarine fans in the northern Adana Basin, southern Turkey: a test of discriminate function analysis. *Geological Journal*, 28, 277–293.
HEIN, J. H. & WALKER, R. G. 1982. The Cambro-Ordovician Cap Enragé Formation, Quebec, Canada: conglomeratic deposits of a braided submarine channel with terraces. *Sedimentology*, 29, 309–329.
HESSE, R., KLAUCKE, I., KHODABAKHSH, S., PIPER, D. J. W., RYAN, W. B. F. & NAMOC STUDY GROUP. 2001. Sandy submarine braid plains: potential deep-water reservoirs. *American Association of Petroleum Geologists Bulletins*, 85, 1499–1521.
KELLING, G. G., GÖKÇEN, S., FLOYD, P. & GÖKÇEN, N. 1987. Neogene tectonics and plate convergence in the eastern Mediterranean: new data from southern Turkey. *Geology*, 15, 425–429.
KNELLER, B. 1995. Beyond the turbidite paradigm: physical models for deposition of turbidites and their implication for reservoir prediction. *In*: HARTLEY, A. & PROSSER, J. (eds) *Characterization of Deepmarine Clastic systems*. Geological Society, London, Special Publications, 31–50.
KOMAR, P. D. 1971. Hydraulic jumps in turbidity currents. *American Association of Petroleum Geologists Bulletin*, 82, 1477–1488.
LE PICHON, X. & RENARD, V. 1982. Avalanching; a major process of erosion and transport in deep-sea canyons: evidence from submersible and multi-narrow beam surveys. *In*: SCRUTTON, R. & TALWANI, M. (eds) *The Ocean Floor*. John Wiley, New York, 113–128.
LOWE, D. 1982. Sediment gravity flows: II. Depositional models with special reference to the deposits of high-density turbidity currents. *Journal of Sedimentary Petrology*, 52, 279–297.
MALINVERO, A., RYAN, W., AUFFRET, G. & POUTOT, G. 1988. Sonar images of recent failure events on the continental margin off Nice, France. *Geology Society of America Special Papers*, 299, 59–75.

MASSON, D. G., KENYON, N. H., GARDNER, J. V. & FIELD, M. E. 1995. Monterey Fan: channel and overbank morphology. *In*: PICKERING, K. T., HISCOTT, R. H., KENYON, N. H., RICCI LUCCHI, F. & SMITH, R. D. A. (eds) *Atlas of Deep-water Environments: Architectural Styles in Turbidite Systems*. Chapman & Hall, London, 74–79.

MENARD, H. W. 1964. *Marine Geology of the Pacific*. McGraw-Hill, New York.

MULDER, T., SAVOYE, B. & SYVITSKI, J. 1997. Numerical modelling of a mid-sized gravity flow: the 1979 Nice turbidity current (dynamics, processes, sediment budget and sea-floor impact). *Sedimentology*, **44**, 305–326.

MUTTI, E. & NORMARK, W. 1987. Comparing examples of modern and ancient turbidite systems: problems and concepts. *In*: LEGGETT, F. & ZUFFA, G. (eds) *Marine Clastic Sedimentology*. Graham & Trotman, London, 1–38.

NAZIK, A. & GÜRBÜZ, K. 1992. Karaisalı-Çatalan-Egner yöresinin (KB-Adana) alt-orta Miyosen istifinin planktonik foraminifer giyostratigrafisi. *Turkiye Jeoloji Bulteni*, **35**(1), 67–80.

NORMARK, W. R. & PIPER, D. J. W. 1991. Initiation processes and flow evolution of turbidity currents: implications for the depositional record. *In*: OSBORNE, R. H. (eds) *From Shoreline to Abyss: Contributions in Marine Geology in Honor of Francis Parker Shepard*. Economic Paleontologists and Mineralogists, Special Publications, **46**, 207–230.

NORMARK, W. R., PIPER, D. J. W. & HESSE, G. R. 1979. Distributary channels, sand lobes and mesotopography of Navy submarine fan, California Borderland, with application to ancient fan sediments. *Sedimentology*, **26**, 749–744.

PIPER, D. J. W. 1970. A Silurian deep sea fan deposit in western Ireland and its bearing on the nature of tubidity currents. *Journal of Geology*, **78**, 509–522.

PIPER, D. J. W. & SAVOYE, B. 1993. Processes of Late Quaternary turbidity current flow and deposition on Var deep-sea fan, north-western Mediterranean Sea. *Sedimentology*, **40**, 557–582.

PIPER, D. J. W., SHOR, A. N. & HUGHES CLARKE, J. E. 1988. *The 1929 Grand Bank earthquake, slump and turbidity current*. Geological Society of America, Special Paper, **229**, 77–92.

POSTMA, G. 1986. Classification for sedimentary gravity-flow deposits based on flow conditions during sedimentation. *Geology*, **14**, 291–294.

SANDERS, J. 1965. *Primary sedimentary structures formed by turbidity currents and related resedimentation mechanisms*. Society of Economic Paleontologists and Mineralogists, Special Publications, **12**, 192–219.

SATUR, N. 1999. *Internal architecture, facies distribution and reservoir modelling of the Cingöz deep-water clastic system in southern Turkey*. PhD thesis, University of Aberdeen.

SATUR, N., CRONIN, B., HURST, A., KELLING, G. & GÜRBÜZ, K. 2000. Sand body geometry in a sand-rich, deep-water clastic system, Miocene Cingöz Formation of southern Turkey. *Marine and Petroleum Geology*, **17**, 239–252.

SCHMIDT, G. 1961. Stratigraphic nomenclature for the Adana region petroleum district VII. *Petroleum Administration Bulletin*, **6**, 47–63.

WALKER, R. G. 1975. Generalized facies models for resedimented conglomerates of turbidite association. *Geological Society of America Bulletin*, **86**, 737–748.

WILLIAMS, G., ÜNLÜGENÇ, U., KELLING, G. & DEMIRKOL, Ç. 1995. Tectonic controls on stratigraphic evolution on the Adana Basin, Turkey. *Journal of the Geological Society, London*, **152**, 873–882.

WINN, R. D. & DOTT, R. H. JR 1979. Deep-water fan-channel conglomerates of late Cretaceous age, southern Chile. *Sedimentology*, **26**, 203–288.

YETIS, C. 1988. Reorganization of the Tertiary stratigraphy in the Adana Basin, southern Turkey. *Newsletters Stratigraphy*, **20**(1), 43–58.

YETIS, C., KELLING, G., GÖKÇEN, S. & BAROZ, F. 1995. A revised stratigraphic framework for Late Cenozoic sequences in the north-eastern Mediterranean region. *Geology Rundsch*, **84**, 794–812.

Sand-rich turbidite system of the Late Oligocene Northern Apennines foredeep: physical stratigraphy and architecture of the 'Macigno costiero' (coastal Tuscany, Italy)

GIANLUCA CORNAMUSINI

Dipartimento di Scienze della Terra, Università di Siena, Via Laterina, 8, I-53100 Siena, Italy (e-mail: cornamusini@unisi.it)

Abstract: The 'Macigno costiero' turbidite system characterized the oldest foredeep clastic wedge of the Northern Apennines during the Late Oligocene collisional phase. The cropping-out thickness is about 500 m. The features of the 'Macigno costiero' indicate a sand-rich, low-efficiency turbidite system. The system developed within a partially confined basin, which was part of a complex foredeep system. The stacking pattern of the turbidite system was determined through the analysis of facies and physical stratigraphy. It consists of a succession organized in sedimentary units, which are characterized by particular associations of facies linked to distinct depositional environments. Several architectural elements are seen: (1) unchannelized and channelized lobes; (2) distributary channels with channel-fill, overbank and channel-margin deposits; (3) main channel with channel-fill, channel-margin and interchannel deposits. Five turbidite stages were identified. From the bottom up they consist of four lobe stages and one proximal channel stage. The lobe stages are characterized by thickening–coarsening upward trends, from distal lobes to proximal lobes up to the channel–lobe transition zone. The uppermost, fifth stage is linked to a main channel complex with stacked channel-fill, channel-margin and interchannel deposits. This final stage also marks the maximum progradation of the system up to its closure due to the synsedimentary overthrusting of the orogenic wedge.

The Northern Apennines (NA) collisional clastic wedges display a great variety of turbidite infill (Ricci Lucchi 1990). The sedimentation developed starting from the Late Oligocene up to the Pliocene. Wide-foredeep, narrow and confined foredeep, and thrust-top basins have been identified (Ricci Lucchi 1986; Argnani & Ricci Lucchi 2001) and linked to the migrating collisional wedge. The younger outer (with respect to the chain) NA successions have been well constrained from a sedimentological and stratigraphical point of view thanks to magnificent exposures in the field. Examples include the Miocene Marnoso-arenacea Formation (Ricci Lucchi 1986 and references therein), deposited in the wide foredeep, and the Epiligurian successions, deposited in thrust-top basins (Mutti *et al.* 1995 and references therein). In contrast, the older inner NA successions, except for the northern ones (Bortolotti *et al.* 1970; Ghibaudo 1980; Andreozzi & Di Giulio 1994; Bruni *et al.* 1994), have been poorly constrained. These deposits are represented by the early synorogenic clastic wedges of the Macigno Formation and the Cervarola-Falterona Unit (Late Oligocene–Early Miocene).

This paper illustrates a detailed stratigraphical and architectural reconstruction of the 'Macigno costiero' (Late Oligocene), one of the oldest, innermost (with respect to the chain) syncollisional turbidite systems of the NA (Fig. 1).

The 'Macigno costiero' represents an important and particular depositional system of the NA foredeep. Whereas the main foredeep units were deposited in wide, migrating foredeep basins (Ricci Lucchi 1986), the 'Macigno costiero' developed in a prograding system located in a complex foredeep within a restricted and tectonically confined basin (see also Boccaletti *et al.* 1990). The main foredeep units are all characterized by high-efficiency turbidite systems, longitudinal with respect to the development of the basin and its feeder, whereas the 'Macigno costiero' developed transversally with respect to these elements. The 'Macigno costiero' represents the only turbidite system linked to the NA early synorogenic foredeep with a clearly prograding system and a transversal influx of sediments.

Previous sedimentological-stratigraphical works (Bracci *et al.* 1984; Ferrini & Pandeli 1985; Cornamusini & Costantini 1997; Cornamusini 2001; Cornamusini *et al.* 2002a) have helped define the system in general. This study presents the temporal and spatial evolution of sedimentation, and the stratigraphic organization of the sediments in architectural elements and depositional units (*sensu* Mutti & Normark 1987) within this sand-rich, low-efficiency turbidite system (*sensu* Mutti 1985). Research involved a detailed analysis of facies (genetic

From: LOMAS, S. A. & JOSEPH, P. (eds) 2004. *Confined Turbidite Systems.* Geological Society, London, Special Publications, **222**, 261–283. 0305-8719/04/$15.00 © The Geological Society of London 2004.

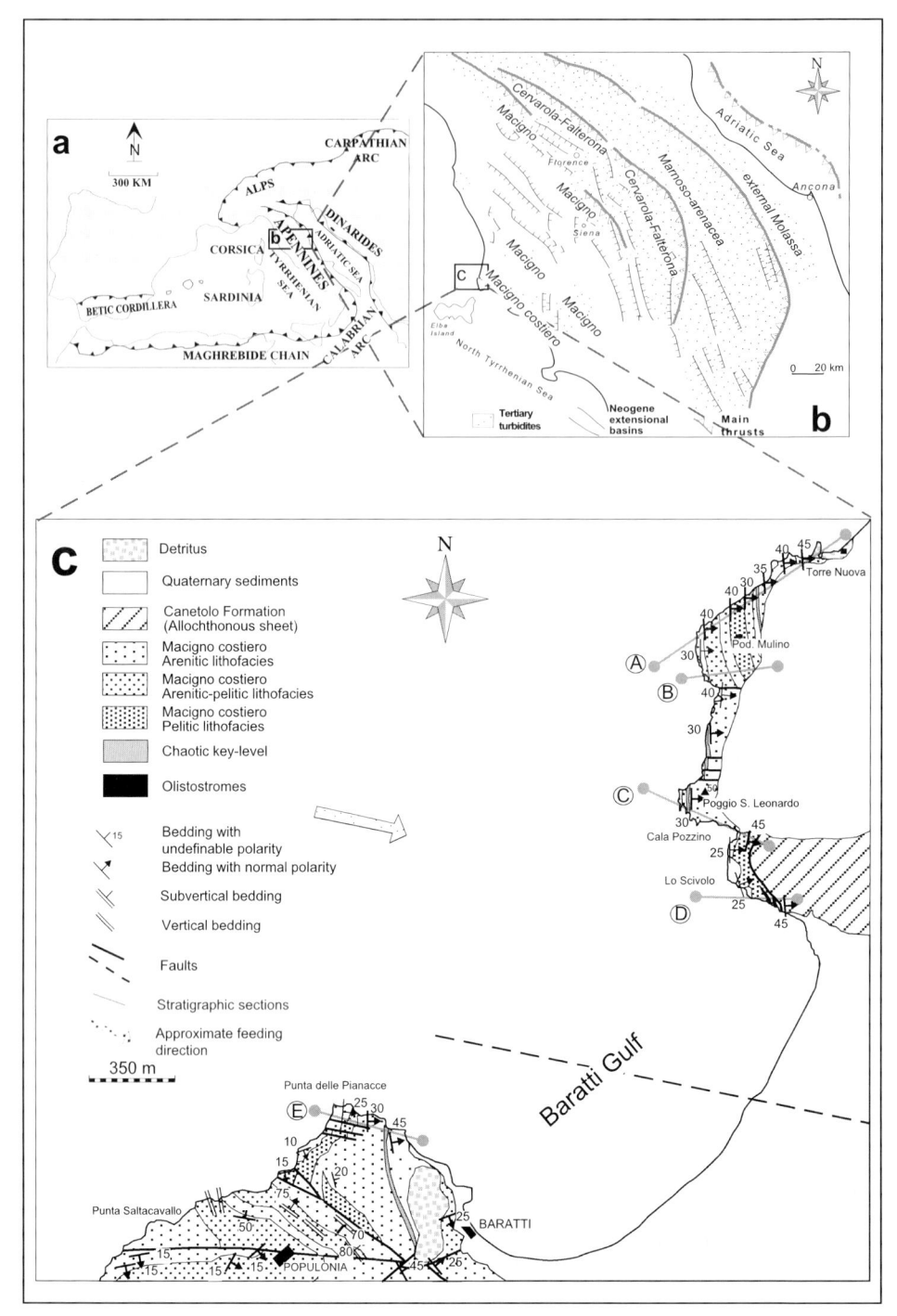

Fig. 1. Geological sketches. (**a**) structural sketch map of the circum-Mediterranean area; (**b**) structural sketch map of the Northern Apennines: main thrusts separating main turbidite systems (they are, from west to east: 'Macigno costiero', Macigno, Cervarola-Falterona, Marnoso-arenacea) are schematically represented; (**c**) schematic geological map of the study area, with the distribution of the 'Macigno costiero' log sections.

facies related to depositional processes), the physical stratigraphy of stacking and correlation patterns. The study is based on five logs, laterally correlated through key-beds and roughly aligned in a north–south direction. The log alignment is transversal to the approximately eastward influx of sediments into the basin (Gandolfi & Paganelli 1992; Cornamusini 2001; Cornamusini et al. 2002a).

Several depositional stages linked to a prograding turbidite system were identified. Numerous architectural elements (and their stacking pattern) typical of a low-efficiency turbidite system were also identified.

Geological-regional setting

The 'Macigno costiero' represents a minor but distinct clastic wedge, west of and inner with respect to that of the larger 'Macigno appenninico' or Macigno s.s. (Bortolotti et al. 1970; Sestini et al. 1986; Boccaletti et al. 1990; Argnani & Ricci Lucchi 2001; Cornamusini 2001; Biserni & Cornamusini 2002).

The outcrops of the 'Macigno costiero' form a discontinuous belt along the northern Tyrrhenian Sea coast. The belt is parallel to the NA chain and is delimited to the west by the Tyrrhenian Sea (Fig. 1), which represents a set of basins developed during the Neogene extensional tectonic regime (Carmignani et al. 1995). The northern Tyrrhenian Sea sediments bury the Alps/Apennines suture and the innermost portion of the NA orogenic wedge (Bartole 1995; Cornamusini et al. 2002b). To the east (southern Tuscany) the 'Macigno costiero' is delimited by a complex tectonic stack, which developed during the collisional phases and was largely dismembered during the Neogene post-collisional-extensional phases that affected the hinterland of the chain (Carmignani et al. 1994; 2001).

The 'Macigno costiero' is a 500 m thick turbidite succession, which was involved in collisional tectonic phases and was overlain by the allochthonous thrust-sheet of the Ligurian–Subligurian units. The substratum of the 'Macigno costiero' is not well defined because the base does not crop out; the base could be the deepsea pelitic unit of the Scaglia toscana Formation (this formation forms the substratum of the more external Macigno Formation) or the deformed units of the orogenic wedge frontal-thrusts.

The 'Macigno costiero' is a deepsea siliciclastic turbidite succession characterized by a high sand/pelite ratio, interbedded with extremely rare, thin carbonate beds. The sandstones are arkose-lithic or arkose and show a basement-collisional orogen provenance (Valloni & Zuffa 1984; Gandolfi & Paganelli 1992; Costa et al. 1997; Cornamusini 2001, 2002; Biserni & Cornamusini 2002). Cornamusini et al. (2002a) suggested that the source area was in the Corsica–Sardinia Hercynian basement, located in the hinterland of the basin. Based on nannofossil associations, the turbidite succession was recently ascribed to the Chattian (Late Oligocene) (Costa et al. 1997; Cornamusini 2001).

Facies and facies associations

Each bed of turbidite sediments was examined and classified according to the genetic-interpretative criteria of depositional processes and hydrodynamic flow types (Lowe 1982; Fisher 1983; Mutti 1992).

The main features of facies and of facies associations, including their hydrodynamic interpretation, are shown in Tables 1 and 2 respectively. Facies were described in terms of texture and sedimentary structures, and interpreted in terms of depositional mechanisms (see also Cornamusini & Costantini, 1997).

In terms of the evolution of an ideal flow (Mutti 1992; Mutti et al. 1999), several facies were recognized in this sand-rich turbidite system. Five facies associations were also identified. The distinctive assemblage of facies and facies associations allows the identification of depositional elements (Mutti & Normark 1987, 1991).

The vertical and lateral distribution of facies revealed five facies associations. Linked to depositional environments, the facies associations represent distinctive units and architectural elements of the system and refer to different stages of growth (Mutti & Normark 1987).

Facies association I is characterized by a sharp lithological contrast in which coarser deposits represent the channel infill and the finer deposits represent the lateral channel sediments. The fine-grained, thin classical turbidites represent overbank deposits due to overflow and spilling from the channelled flow (Mutti & Ricci Lucchi 1972; Mutti 1977; Walker 1978, 1985; Hill 1984; Mutti & Normark 1987; Pickering et al. 1989; Bruhn & Walker 1997; Cronin et al. 1998, 2000). Thin-bedded turbidites represent channel-margin and interchannel deposits (Mutti 1977). Debris flow deposits and slumps often drape or infill the channel. The coarse sandstones are characterized by deep concave scours and non-tabular amalgamation surfaces, and by

Table 1 *Macigno costiero facies scheme. Flow types according to Mutti (1992) and Mutti et al. (1999)*

Facies	Sedimentary features	Flow type
Facies A: slumps and slides	Slumps, sliding structures, disrupted beds	Slumpings and slidings
Facies B: chaotic deposits	Matrix-supported, crude inverse-grading, structureless or crude internal organization	Cohesive debris flow
Facies C: pebbly sandstones with muddy matrix	Partly matrix-supported, massive, normal grading, load casts, mud-clasts	Hyperconcentrated flow or granular flow
Facies D: massive pebbly sandstones	Massive, structureless or cross- and plane-parallel lamination, water-escape structures, traction carpets, mud-clasts, scours	Gravelly high-density turbidity current or granular flow
Facies E: pebbly sandstones with traction carpets	Traction carpets, massive, inverse graded, cross- and plane-parallel lamination, scours, load casts, mud-clasts	Gravelly high-density turbidity current or granular flow
Facies F: graded pebbly sandstones	Normal graded, cross- and plane-parallel lamination, mud-clasts, flute casts, scours	Gravelly high-density turbidity current or granular flow
Facies G: massive sandstones	Massive, structureless or normal graded, inverse graded, thin traction carpets, plane-parallel and cross-lamination, water-escape structures, mud-clasts, scours, load and flute casts, Ta division	Sandy high-density turbidity current or turbulent flow
Facies H: graded and laminated sandstones	Normal graded, plane-parallel- and cross-lamination, convolute lamination, water-escape structures, mud-clasts, scours, load and flute casts, Ta-e division	Sandy high-low-density turbidity current or turbulent flow
Facies I: thin-bedded-turbidites	Thin bedding, low S/P ratio, normal graded, plane-parallel ondulate and convolute lamination, ripples, Tc-e, Td-e divisions	Low-density turbidity current or turbulent flow
Facies J: cross-laminated coarse-grained sandstones	Generally thin bedded, cross-laminated coarse-middle sandstones	Bypassing flow

scours with rip-up mud clasts typical of channels (Mutti 1992; Mattern 2002). Thin-bedded turbidites are sometimes characterized by convergent stratification, climbing ripples and trace fossils; they are typical of the channel-margin and channel–levee environments (Hesse & Dalton 1995).

Facies association II is characterized by massive, coarse sediments, with slightly lenticular geometry, deep scours, amalgamation surfaces, dewatering structures, traction carpets and impact-injection structures with associated small sedimentary dykes and sills, along with low- to high-angle cross-laminated sediments

Table 2 *Facies associations scheme: depositional elements according to Mutti & Normark (1987)*

Facies association	Facies	Depositional elements
Facies association I: disorganized deposits and thin-bedded turbidites	Mainly A, B, C and I, subordinately D, J and E	Channel, channel margin, interchannel
Facies association II: coarse-grained sandstones	Mainly D, subordinately G, H, J and I	Channel-fill, channel–lobe transition, very proximal lobe
Facies association III: medium- to coarse-grained sandstones	Mainly G and subordinately D	Proximal lobe
Facies association IV: medium- to fine-grained sandstones	Mainly G and subordinately H and I	Intermediate-distal lobe
Facies association V: fine-grained sandstones and thin-bedded turbidites	Mainly H and I	Distal lobe, lobe/fan fringe, interlobe

(bypassing facies). This association is characteristic of the transition between channel and attached lobe environments (Mutti 1985; Mutti & Normark 1987, 1991), or of very proximal lobe environments (Hurst et al. 1999) and of channelized fan-lobes (Mattern 2002).

Facies association III is characterized by massive or slightly normal-graded sandstones with tabular scours and rip-up mudstone clasts, injection structures, tabular amalgamations and shale-clast levels. It is typical of proximal lobe environments (Mutti & Normark 1987).

Facies association IV is characterized by tabular beds of massive to normal-graded sandstones representing the whole Bouma (1962) sequence, and by small tabular scours. The association is typical of intermediate-distal lobe environments (Mutti & Normark 1987).

Facies association V is characterized by thin-bedded turbidites with beds representing the upper portion of the Bouma (1962) sequence, by high tabularity, low frequency of sole erosive structures, and hemipelagic portions with bioturbation and trace fossils. There is no high-grade convolution of the laminations. This association of facies is referred to distal lobe, lobe–fan fringe or interlobe environments (Mutti 1977; Mutti & Normark 1987; Shanmugam & Mojola 1988).

Architecture of the turbidite system

Research involved detailed facies studies and measurement of the five logs. They are from north to south (Fig. 1): log A (Porticciolo), log B (Podere Mulino), log C (Cala Pozzino), log D (Baratti beach) and log E (Punta delle Pianacce). Log A, illustrated in Figure 2, is the most complete succession.

The correlated sections (Fig. 3) show the stratal pattern with bed geometry and changes in thickness, facies and shape. These were used to define the architectural elements and depositional stacks of the system.

Several architectural elements (Mutti & Normark 1987, 1991; Mutti 1977, 1985, 1992; Shanmugam & Mojola 1988; Stow & Mayall 2000; Cronin et al. 2000; Satur et al. 2000; Saito & Ito 2002) were identified in the 'Macigno costiero' turbidite system.

The analysed portion of the 'Macigno costiero' turbidite system is divided into five stages according to the hierarchical classification of Mutti & Normark (1987, 1991). The five stages were identified on the basis of facies and facies associations organized in vertical sequences and on their lateral correlation over 3 km. Every turbidite stage records a particular growth phase of the system, following Mutti & Normark (1987). The boundary of each stage was set where abrupt modifications in the vertical organization and abrupt changes in facies and lithology, reflecting variations in the depositional environment, were observed.

The following sections describe and interpret the five stages (starting from the lowest) and their relative architectural elements. Every stack of beds, which can be ascribed to a single facies association, is referred to as a unit.

Stage 1

The exposed portion of stage 1 (Figs 2a & b) is about 55 m thick (the lower part does not crop out). This stage is shown in logs A and B, and was traced laterally for about 100 m (Figs 3 & 4).

Unit 1.1

This lower portion of the section is characterized mainly by massive-coarse pebbly sandstones (facies D) in slightly lenticular beds (Figs 2b, 3 & 4). Massive sandstones contain traction carpets, dewatering structures and amalgamation surfaces. Thin, fine-grained fractions, separated from the massive fraction through mud-clast

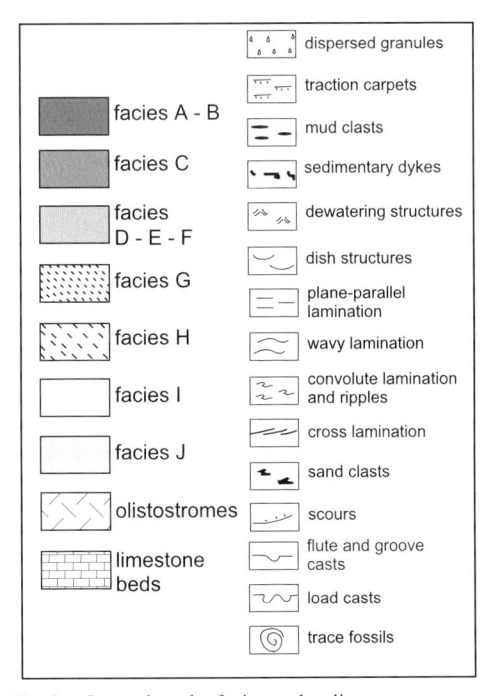

Fig. 2a. Legend to the facies and sedimentary structures.

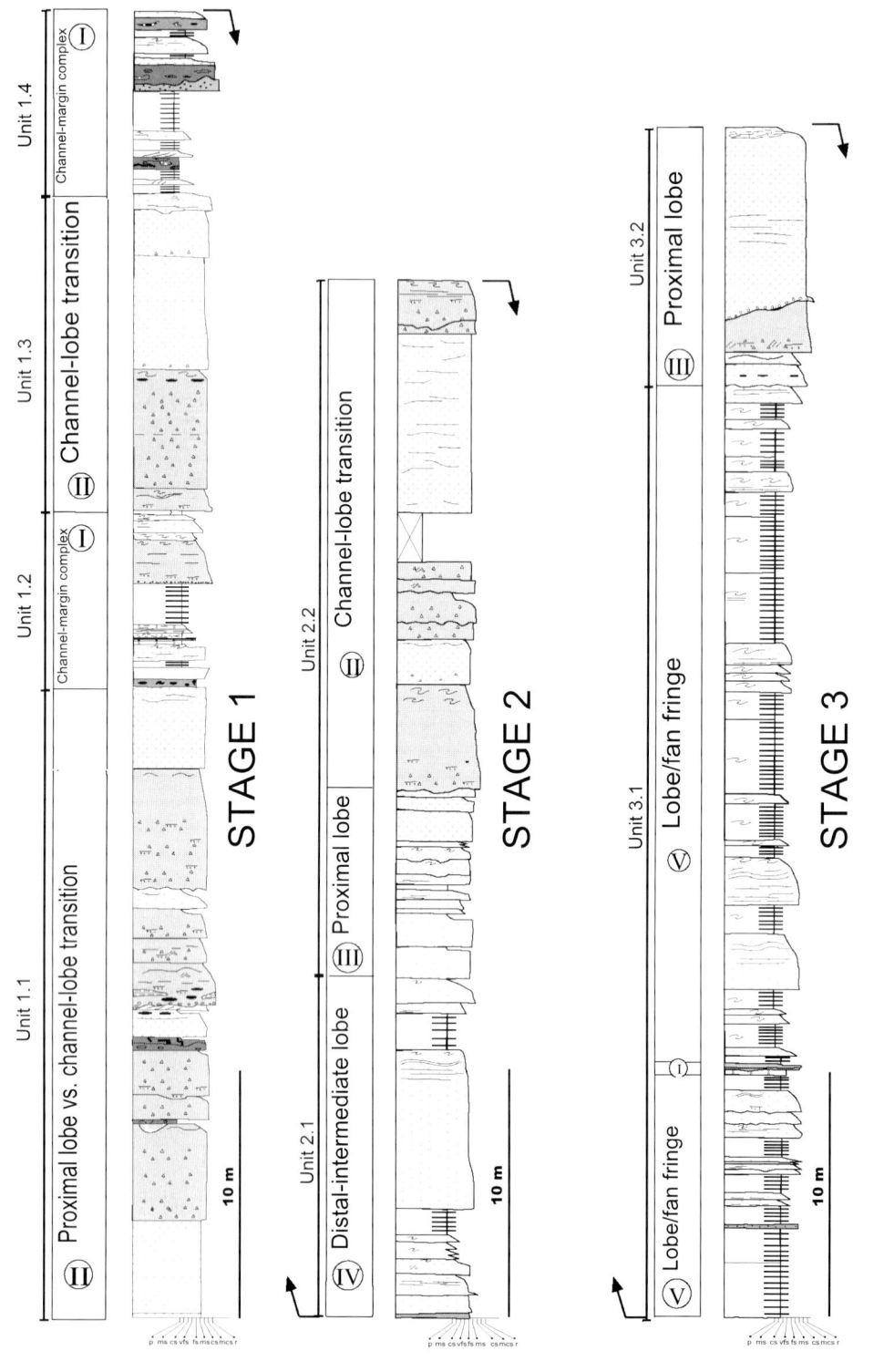

Fig. 2b. Log A. Stages 1, 2 and 3 are shown. Facies associations (with Roman numerals) and depositional elements are indicated on the left of each log.

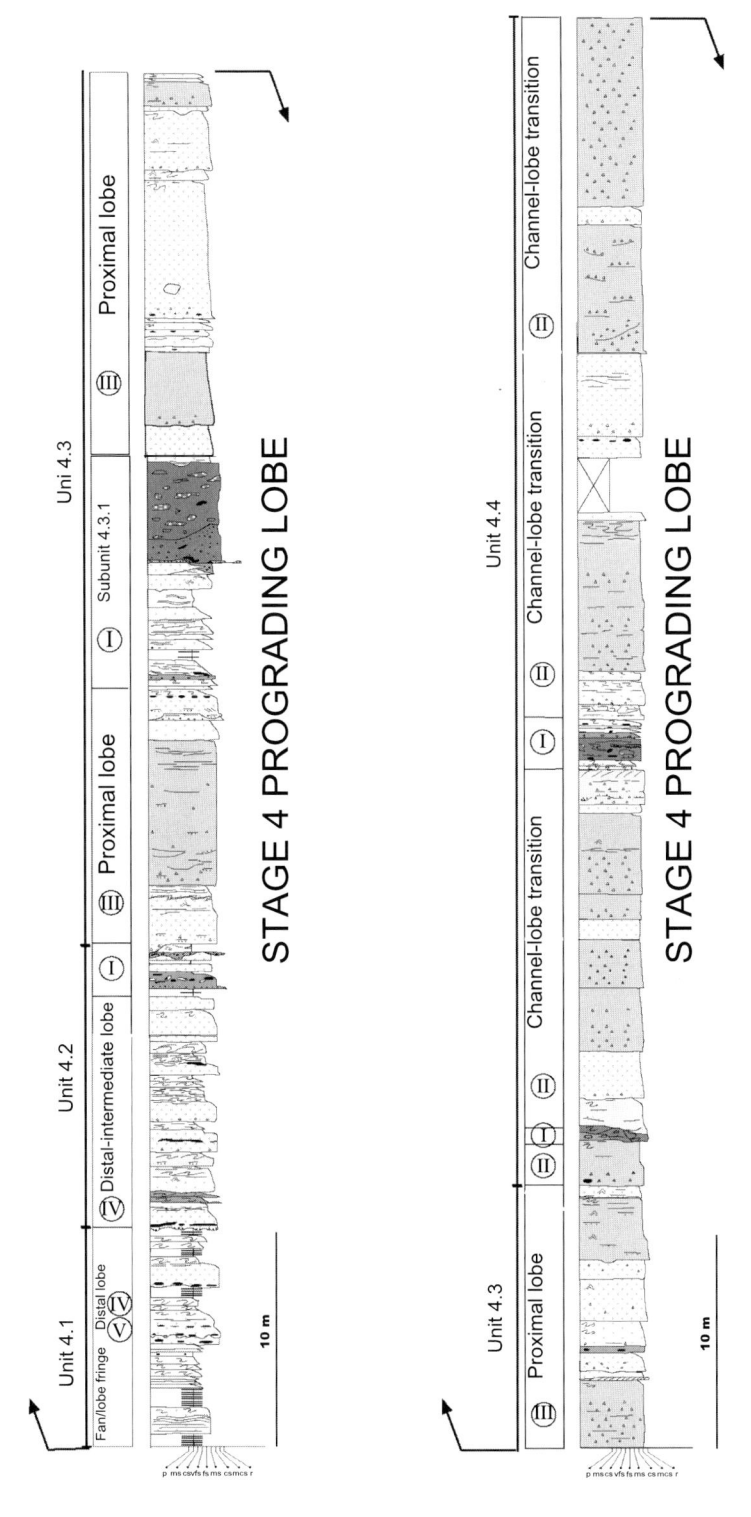

Fig. 2c. Log A. Stage 4 is shown. Facies associations (with Roman numerals) and depositional elements are indicated on the left of each log.

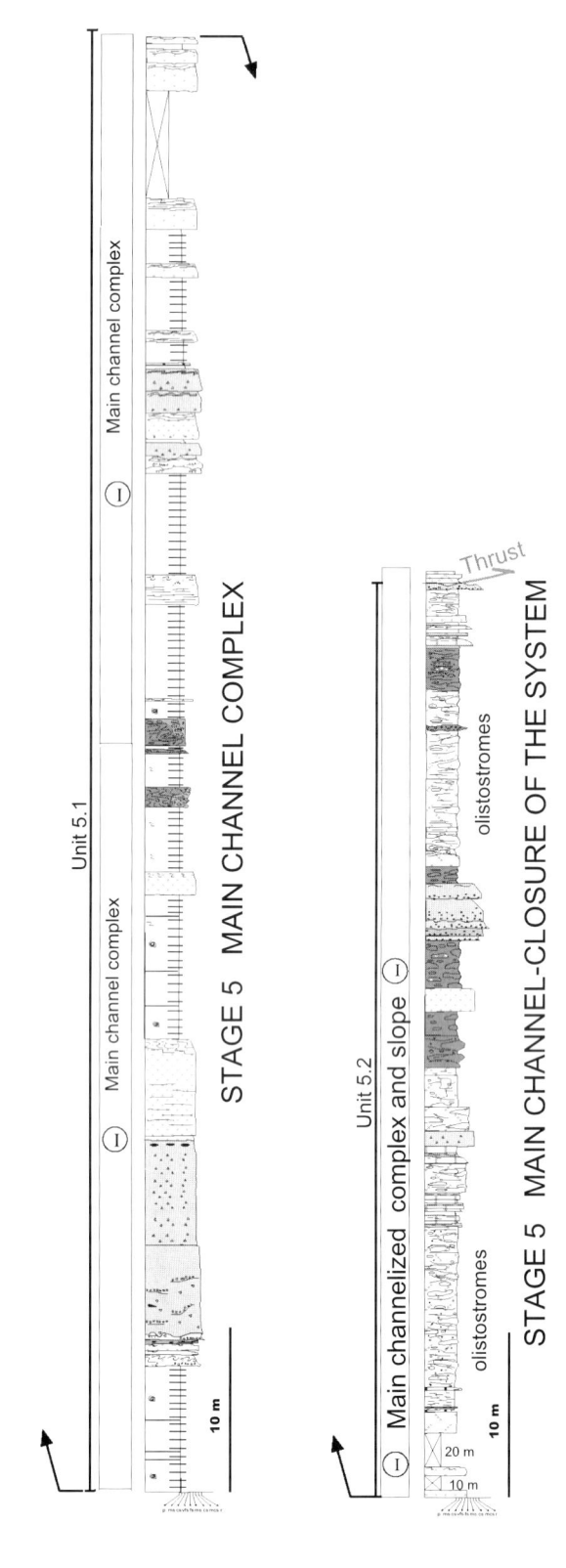

Fig. 2d. Stage 5 is shown (log A on the left and log D on the right). Facies associations (with Roman numerals) and depositional elements are indicated on the left of each log.

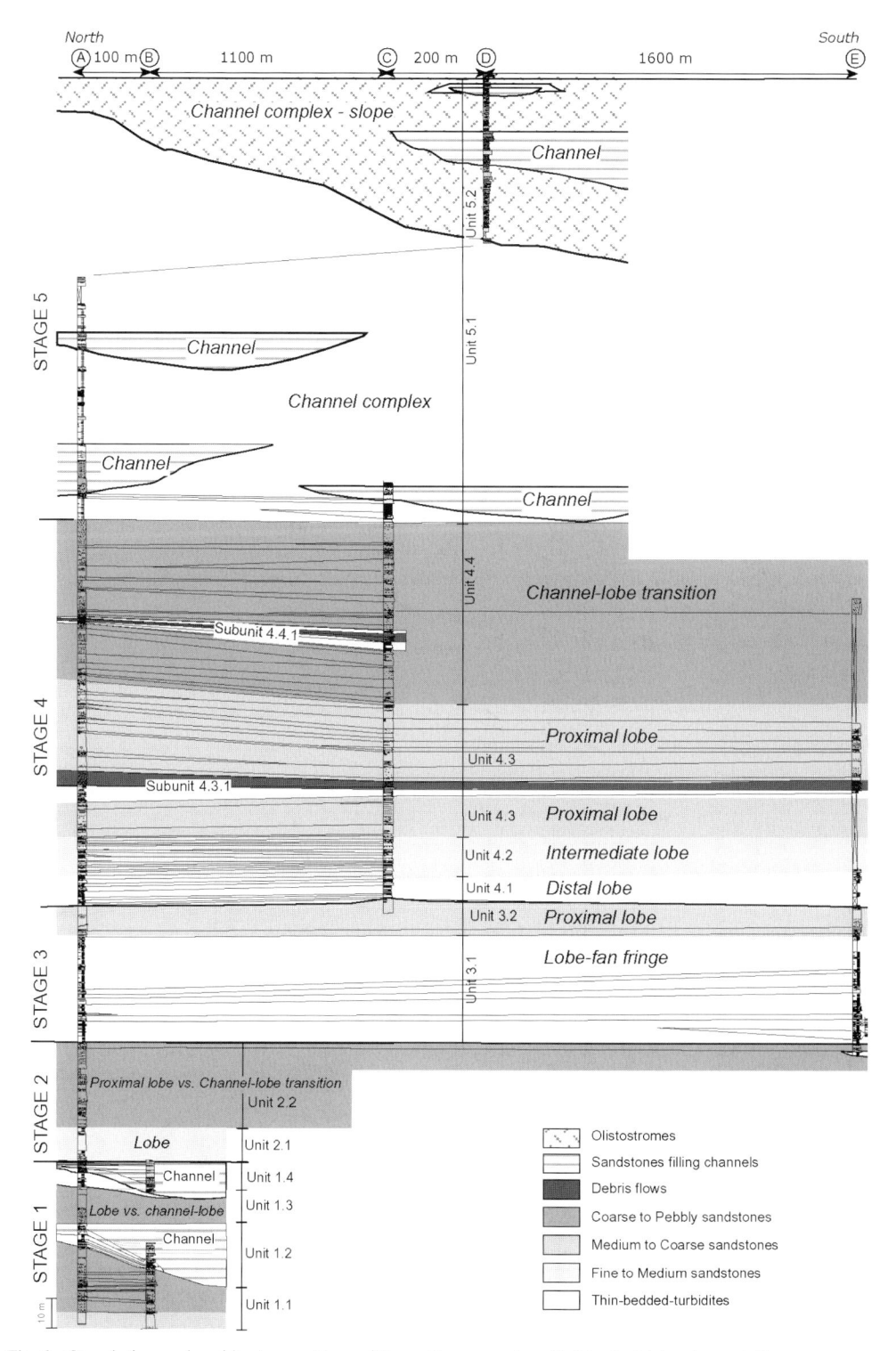

Fig. 3. Correlation and architecture pattern of the entire succession. Bed-by-bed lateral correlations are indicated with black lines.

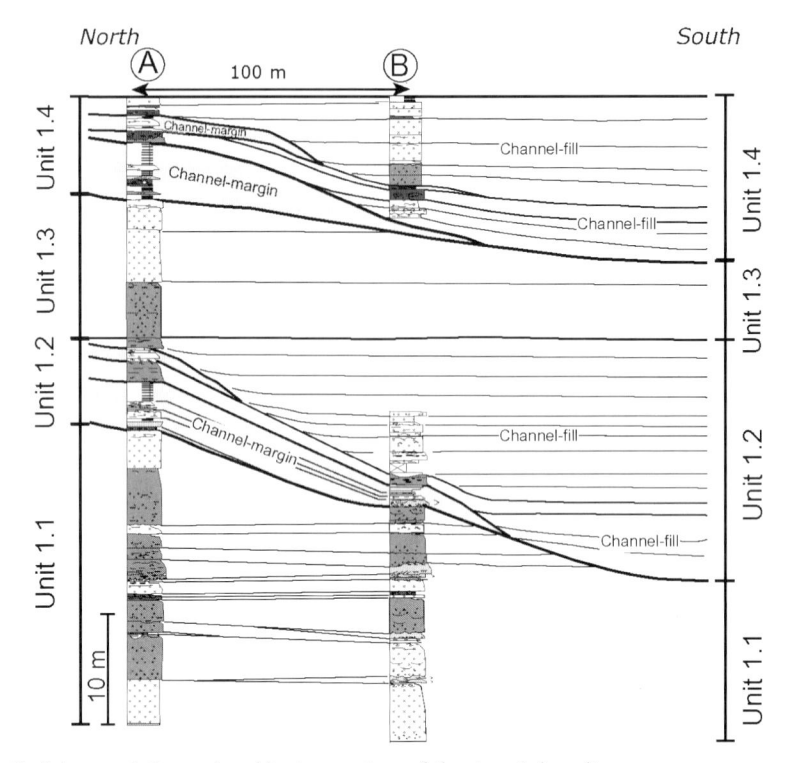

Fig. 4. Detail of the correlation and architecture pattern of the stage 1 deposits.

horizons, are frequent on top of the bed. Scours at the base of the massive beds are also frequent. Coarser sandstone beds sometimes contain a few huge, floating, limestone clasts (some dm in size) and spheroidal carbonate concretions. Finer, massive sandstone beds ascribed to the Ta Bouma (1962) division (facies G) are present.

Pebbly sandstone beds dominated by inverse grading, organized in traction carpets relative to facies E, are also present. There are other sediments (facies J) characterized mainly by thin-bedded cross-lamination and coarse sandstones, and by water-escape and traction-carpet structures (F6 facies of Mutti 1992). Thin beds of muddy sandstones with poor internal organization (facies B and C) are also present; small sedimentary dykes sometimes cut them.

Units 1.2, 1.3 and 1.4

A stack of thin-bedded turbidites (facies I) associated with slightly lenticular, cross-laminated sandstones (facies J), thin carbonate turbidites and minor, classical, thin turbidites (facies H) is represented in log A (Fig. 2b). The stacking pattern in log B is quite different (Fig. 4). The association of facies I and facies J sediments is restricted to the lower portion of the unit stack and is thinner than that of log A. The upper part of the stack in log B consists of massive and classical turbidites (facies G and H), which are absent in log A.

Unit 1.2 is covered by a package (unit 1.3) of coarse-massive, amalgamated pebbly sandstones and coarse sandstones (facies D and G). It is followed by a stack ascribed to facies association I, which consists of thin-bedded turbidites with subordinate muddy sandstones and cross-stratified sandstones (lower portion of unit 1.4).

The upper coarser sediments of unit 1.4 consist of very coarse sandstones—pebbly sandstones (facies F) in irregular beds. These beds are topped by a portion of chaotic deposits (facies B) in both logs and by an association of chaotic deposits and thin-bedded turbidites (log A), or coarse massive sandstones with a thinning-fining upward trend from facies D to facies G (log B). The latter deposits, characterizing log B (Fig. 4), show dewatering structures, scours and lenticular bedding.

Stage 2

Stage 2, about 40 m thick, shows sheet sandstone bodies in a thickening-coarsening upward

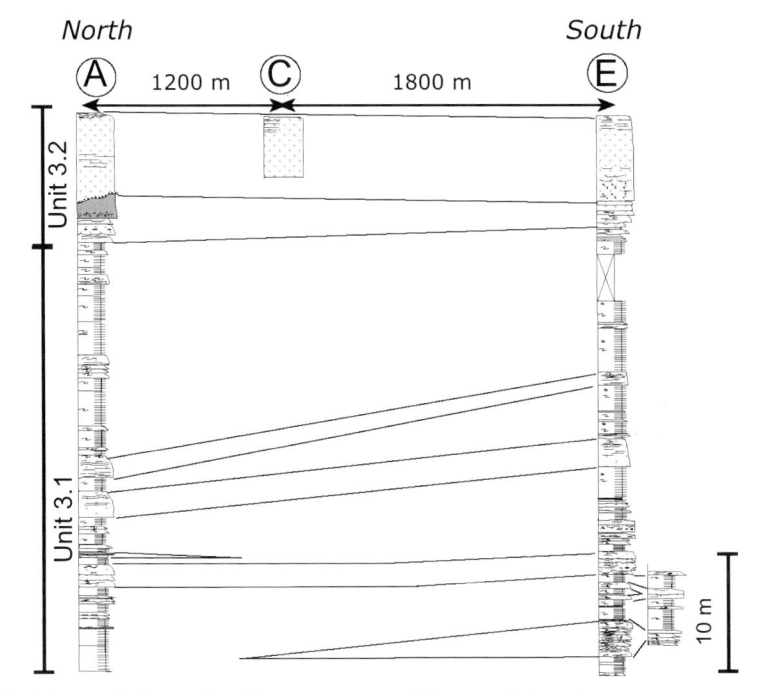

Fig. 5. Detail of the correlation and architecture pattern of the stage 3 deposits.

sequence (Figs 2a & b), from thin-bedded turbidites (facies I) to thick, massive pebbly sandstones and coarse sandstones (facies D and G). It can be seen in its entirety in log A. Log B shows only the basal portion, and log E shows only the top portion.

Units 2.1 and 2.2
From the bottom up the deposits of units 2.1 and 2.2, in a clear thickening–coarsening upward sequence, consist of thin-bedded turbidites (facies I), thin-bedded and classical turbidites (facies H), massive and classical turbidites (respectively facies G and H), and pebbly and coarse-massive sandstones (facies D). Thin-bedded turbidites are characterized by fine sandstone–siltstone couplets, sometimes with hemipelagic partitions. The Tc-e, Td-e Bouma (1962) intervals are frequent, as are current ripples and convolute laminations. Classical turbidites (*sensu* Walker 1992) are characterized by Ta-d, Tb-d intervals, whereas amalgamation surfaces, scours, traction carpets and tails at the top characterize massive turbidites.

The sediments of this stage, observed in log E (Fig. 3), can be ascribed to the top portion of stage 2. The sediments at the top of stage 2, as seen in log E, consist of thin-bedded turbidites associated with lenticular, thin-bedded pebble beds, massive sandstone and coarse, cross-stratified sandstones (facies J).

Stage 3
Stage 3 is about 50 m thick (Fig. 2b). It can be observed in logs A and E, and its upper portion is also seen in log C (Figs 3 & 5).

It presents a thickening–coarsening upward trend from thin-bedded turbidites up to massive sandstones, through a marked contrast in grain size.

Stage 3 is characterized by sandstone bodies. The sheet-like beds have constant thickness and facies for over 3 km (logs A and E). These sediments are topped, through an abrupt thickening–coarsening upward trend, by facies D and G sandstone beds with uniform thickness and facies (tracing on logs A, C and E).

Unit 3.1
This portion of the succession (characterizing the lower-middle part of stage 3) consists of sheet-like, thin-bedded turbidites (facies I) and classical turbidites (facies H) (Figs 6 & 7). The former have Tc-e Bouma (1962) divisions, whereas the latter have Tb-e divisions, and both often lack the Ta division. Beds are of

Fig. 6. Thin-bedded turbidite package of unit 3.1, measured on log E.

uniform thickness with great lateral continuity over a distance of at least 3 km (from log A to log E). Only a few bed stacks laterally pinch-out gently. The pelite portions of the thin-bedded turbidites are often bioturbated and contain trace fossils.

Unit 3.2

Unit 3.2 (a few metres thick) characterizes the top part of stage 3. The vertical transition from thin-bedded turbidites (facies I) to classical-amalgamated (tabular amalgamation surfaces; see Mutti 1992; Mattern 2002) and massive

Fig. 7. Thin-bedded turbidites (facies I) with interbedded classical turbidites (facies H) of unit 3.1.

sandstone beds (respectively facies H and G) is very swift. The latter give way above to thick beds of pebbly or coarse, massive sandstones (facies D) with non-tabular amalgamation surfaces, traction carpets, scours and water-escape structures (dish and pillar).

Stage 4

Stage 4, 130 m thick, is seen in logs A, C and E (Figs 3 & 8). It shows a prominent thickening and coarsening upward trend (Fig. 2c). Stage 4 is characterized by sandstone packages of regular thickness through logs A, C and E. The lower portion consists of thin-bedded turbidites and sheet-like, classical sandstone turbidites with good lateral continuity (Fig. 9), the lower-intermediate portion of massive and classical turbidites. There are also sediments characterized by association of chaotic facies and thin-bedded turbidites. One of these chaotic beds, due to its sedimentological peculiarity and lateral continuity, has been used as a key-bed for the whole succession (Fig. 8).

Unit 4.1

This unit has good lateral continuity with sheet-like beds and slight amalgamation surfaces (Figs 2 & 8). It is characterized by thin-bedded turbidites (facies I), classical turbidites (facies H) and medium-thin massive sandstones (facies G), which give way to thicker beds above (facies G). Classical and massive turbidites have erosive tabular bases with flute- and groove-casts. Levels of mud clasts are also frequent; they either separate the massive portion from the graded portion of every bed or are located close to the bottom of the massive beds.

Units 4.2 and 4.3

With respect to the underlying unit, unit 4.2 is characterized by a decrease in thin-bedded turbidites (facies I) and an increase in classical and massive turbidites (facies H and G). In this portion, thin levels of thin-bedded turbidites (facies I), massive scoured sandstones (facies D and G) and disorganized pebbly mud beds (facies B) are diffuse (Fig. 2c).

The overlying sediments (unit 4.3) consist of thick, coarse sandstone beds showing a clear continuation of the thickening–coarsening upward trend. Sandstones in the lower portion are finer, and a few contain Bouma (1962) divisions with scours (facies G and H). The upper portion prevalently consists of massive-coarse and pebbly sandstone beds (facies G, D and E) with amalgamation surfaces, traction carpets, crude lamination, water-escape structures and scours, and with abundant bigger dispersed clasts, especially in the top portion (Fig. 10). There are also some beds with planar and trough cross-lamination (facies J).

A 10 m thick package (sub-unit 4.3.1), consisting of thin-bedded turbidites (facies I) and thin, classical turbidites below and of a peculiar marker chaotic bed (facies B) above, is interbedded in unit 4.3. This package of beds is seen in logs A, C and E. It extends laterally and continuously for over 3 km, but with great variations in thickness (Figs 3 & 8). Both the thin-bedded turbidites and the chaotic portions thin out toward the southern sections. The thin-bedded turbidite portion has interlayered sandstone beds with low-angle cross-lamination (facies J). The chaotic key-bed, laterally continuous in all logs, is divided into three portions (see in Fig. 2c): a graded pebbly sandstone at the base (facies F), a pebbly mudstone portion (facies B) in the middle, and a slumped and disrupted portion consisting of semi-lithified, marly mudstone clasts in a predominant muddy-marly matrix with scattered limestone clasts and shallow water fossils (facies B and A) at the top (see Postma 1984). The thickness of the chaotic bed is very irregular and reduces towards the southern sections. The upper portion has a convex shape. The pebbly sandstone portion is particularly irregular in thickness. The thickness of the chaotic upper portion is more regular but increases to the north.

Unit 4.4

This portion of the stage consists of massive, medium-coarse and pebbly sandstones (mainly facies D, subordinately facies G) in thick beds characterized by water-escape structures, traction carpets, erosive structures (scours), tabular and non-tabular amalgamation surfaces and, sometimes, cross-lamination on top of the bed (Fig. 2c). The geometry of beds gently pinches out, with irregularities due to scours (Fig. 11) and large-scale bedforms. Beds with cross-laminated sandstones and traction carpets (facies J), and chaotic, thin beds (facies C and B) are also present.

Unit 4.4 shows a thickening and coarsening upward trend from middle- to coarse-grained massive, slightly graded sandstones (facies G) to pebbly, very coarse-grained, completely massive sandstones with traction carpets (facies D). There is an upward increase in the frequency and width of scours. The lenticularity of beds also increases toward the top.

Sub-unit 4.4.1 is interbedded in unit 4.4 (Figs 3 & 8). The bottom and top portions of this

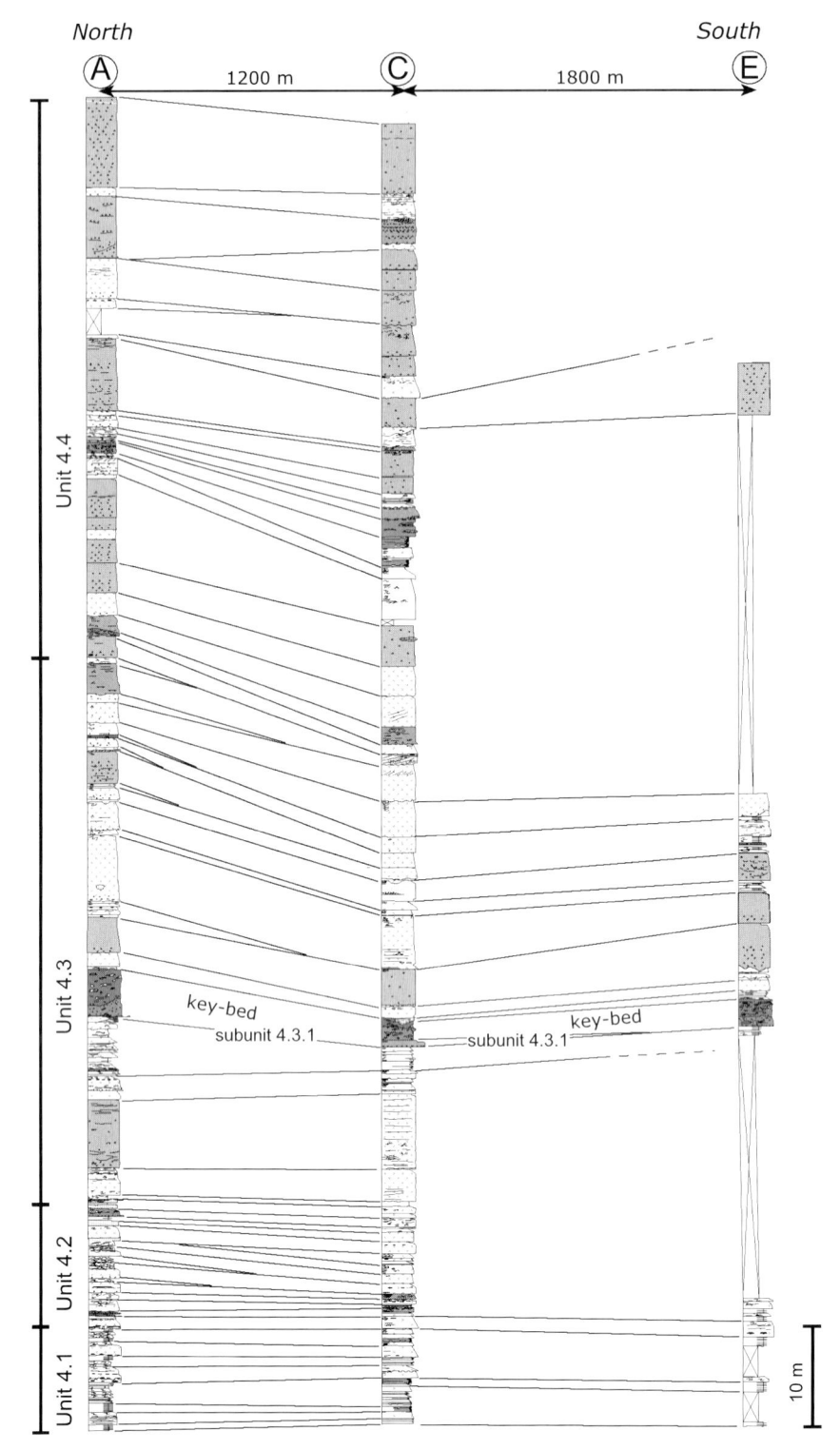

Fig. 8. Detail of the correlation and architecture pattern of the stage 4 deposits.

Fig. 9. Lower portion of stage 4. Thin-bedded turbidites of unit 4.1, giving way above to classical and massive turbidites of units 4.2 and 4.3 respectively. This sequence is part of log C.

sub-unit consist of thin-bedded turbidites (facies I), whereas the middle portion consists of chaotic microconglomerates (facies B, C and F). This package pinches out towards log A and increases in depth towards log C.

Stage 5

This is the final stage of the turbidite system in the area. It is about 180 m thick, and is divided into two units (Figs 2d & 3). The lower one, 85 m thick can be seen in log A and partly in log C (Fig. 3). The upper one, 95 m thick, is seen in log D. The lateral extent of this body cannot be observed, but is deduced on the basis of the regional setting.

Unit 5.1
This package of sediments (Fig. 2d) consists of thin-bedded turbidites (facies I) with interbedded massive, coarse and pebbly sandstones (facies G, H and D). The thin-bedded turbidites are characterized by Tc-e Bouma (1962) divisions, bioturbations and trace fossils, CCC-turbidites (Walker 1985, 1992), and a few thin levels of black shale. Thin, intermediate-chaotic beds (facies B and C) are also interlayered in the thin-bedded turbidites. Slump (facies A) typical of the facies *slumped thin-bedded turbidites* of Walker (1992) are also present. The sandstones are characterized by lenticular beds with erosional surfaces. The sedimentary structures include deep non-tabular scours, pebbly sandstone levels, water escapes, traction carpets, crude laminations and diffuse, bigger dispersed clasts.

Unit 5.2
Unit 5.2 consists mainly of chaotic slides, generally strongly deformed limestones and shales stacked in chaotic levels. Rare carbonate turbidites are interlayered with the chaotic levels. In the middle and upper part of this unit a thick stack of conglomerate and chaotic beds (facies F and B) is interlayered with the allochthonous slides and olistostromes. Conglomerates are characterized by a roughly normal gradation, strong lenticular shape, erosional surfaces and, sometimes, wavy laminations on the overlying fine sediments.

Interpretation

About unit 1.1, the prominent presence of scoured massive, coarse sandstone and pebbly sandstone beds, which leads to irregularity in bedding, contemplated in the turbiditic model of Mutti & Normark (1987), indicates a transitional channel–lobe depositional environment (channel–lobe transition of the authors). This interpretation is also sustained by the presence of cross-stratified, coarse-grained sandstones, which characterize transitional and bypassing facies (Mutti 1992; Mutti *et al.* 1999). These deposits develop mainly at the transition between high gradients with channelized flows and in the more regular, open morphologies with non-channelized flows (Mutti & Normark 1987,

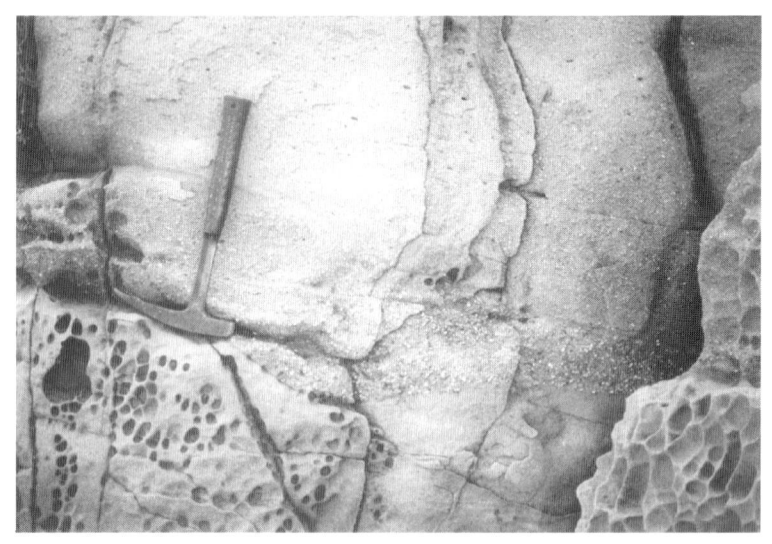

Fig. 10. Small scour separating a pebbly sandstone bed (facies D) above and a massive, slightly laminated, coarse sandstone bed (facies G) below.

Fig. 11. A scour is observed on the left side of the photo. The upper, thick bed consists of structureless, massive pebbly sandstones with dishes (facies D), whereas the lower, thick bed consists of massive or crudely plane-parallel laminated, coarse sandstones (facies G) with cross-lamination on top. Panel for scale is 15 cm wide.

1991). Moreover, the presence of muddy sandstone beds, ascribed to debris flows and hyperconcentrated flows, and of limestone floating clasts (the latter interpreted as allochthonous elements such as olistoliths), is compatible with such a depositional environment. According to Hurst *et al.* (1999), the analogous sedimentary structures and similar geometry identified in the Big Monster unit of the Annot Sandstones, indicate a purely lobe depositional environment situated in a very proximal position. Nevertheless, based on a study by Mattern (2002) on the sand-rich system of the Reiselsberger Sandstein, the high frequency of sedimentary structures such as coarse-grained and pebbly sandstones, fluid-escape structures, amalgamation and scour structures, and the very thick bedding of the examined unit, can be ascribed to a channelized fan-lobe environment.

The whole package of unit 1.2 sediments is considered a channel-fill and channel margin ascribed to facies association I (Mutti 1977, 1992). The channel deepens southward, and channel-margin facies characterized by thin-bedded turbidites and bypass deposits (channel-wing deposits of Elliott 2000) are well developed in log A. Typical facies transformation deposits (facies J) are present along the channel, both on the margin and in the deep portions. The latter are characteristic of hydraulic jumps and flow transformations that accompanied the high influx of sediments flowing and bypassing in the channel (Mutti 1977, 1992). The bypassing beds may also indicate overflow and spill-channel turbidites

(Hurst et al. 1999). In the deeper portion of the channel, illustrated in log B (Fig. 4), the channel-margin deposits diminish, whereas coarser infill deposits are well developed. This last package of sandstones has a planar geometry and onlaps the channel-margin deposits. These sandstone beds are interpreted as fill and channel-margin infill facies, and (for the upper part of log A) as overbank deposits (Hurst et al. 1999).

Unit 1.3 is in relation to a channel–lobe transition zone. Unit 1.4 is a distributary channel, which deepens southward, as seen in logs A and B (Fig. 4). Log A shows a channel-margin facies association, with thin-bedded turbidites, thin debris flow, pebbly deposits and overbank deposits (Mutti 1992; Hesse & Dalton 1995). In contrast, log B shows a classical channel infill consisting of debris flow deposits at the base, which can be correlated with channel-margin deposits, and the pebbly sandstone and massive sandstone beds that infill the margins of the channel. The infill represents the aggrading onlap of Clark et al. (2001). In this case the debris-flow deposit drapes the surface of the margin-channel and fills in the axial position, whereas the coarse-massive sandstone beds onlap the margin of the channel and laterally give way to the channel-margin thin-bedded turbidites facies.

Sediments of units 2.1 and 2.2 are interpreted as a series of lobes. In particular, the lower deposits are linked to an intermediate-distal lobe (facies association IV), the intermediate deposits to a proximal lobe (facies association III), and the upper ones to a channel–lobe transition zone (facies association II) (Mutti & Normark 1987; Mutti 1992). The beds of log E, ascribed to facies association I, are laterally discontinuous and single out a small channel complex at the top of stage 2 (Fig. 3). In particular, they are ascribed to channel-margin deposits with typical bypass facies of a distributary channel within the channel–lobe transition zone (Mutti & Normark 1987).

Unit 3.1 features are relative to facies associations V and IV (Fig. 5). In particular, they can be referred to distal lobe or lobe–fan fringe environments for the abundance of thin-bedded and classical turbidites, the good lateral continuity and planar geometry of the beds, the lack of channelization and the low sand/pelite ratio (Mutti 1977; Mutti & Normark 1987; Shanmugam & Mojola 1988; Swart 1992; Saito & Ito 2002). The fairly good lateral continuity (in terms of both facies and sedimentary structures) of unit 3.2 sediments, relative to facies association III (seen in logs A, C and E), the high sand/pelite ratio and the presence of scours indicate a proximal lobe deposition zone (Mutti & Normark 1987; Hurst et al. 1999; Satur et al. 2000).

The deposits of unit 4.1 are ascribed to facies association V and give way above to the deposits of facies association IV. Their main characteristics are a dominant sheet-like geometry (Fig. 8) and low sand/mud ratio, which indicate a distal lobe or lobe/fringe environment (Mutti 1977, 1992; Mutti & Normark 1987).

The thickening–coarsening upward trend of units 4.2 and 4.3 (Figs 2c & 8) indicates a gradual transition from intermediate lobe (facies association IV) to proximal lobe facies (facies association III). These sediments are characterized by an upward decrease in compensation cycles (*sensu* Mutti & Sonnino 1981), and an increase in the sand/pelite ratio, scouring features and amalgamations (see Mutti & Normark 1987; Walker 1992; Hurst et al. 1999; Clark et al. 2001).

The chaotic key-bed, which is part of sub-unit 4.3.1 (Figs 2c & 8), shows features of a bipartite, cohesive, debris flow deposit associated with an upper slide and slump portion (Cornamusini & Costantini 1997; Cornamusini 2001). All of sub-unit 4.3.1 contains sediments of facies association I. Their general upward convex shape, combined with the wide-basinal distribution and the features of the cohesive debris-flow bed, indicates that they deposited in an unchannelized position. The marker chaotic bed is interpreted as the product of destabilization of the inner foredeep slope sediments (e.g. Agirrezabala & Garcia-Mondéjar 1994; Haughton 2001), probably in relation to the tectonic thrust of the advancing orogenic wedge (Fielding et al. 1997), or the decreasing sea-level (Shanmugam & Mojola 1988).

Unit 4.4 deposits, characterized mainly by very high sand/pelite ratios, abundant scours and cross-lamination, belong to facies association II. They are believed to be linked to a proximal lobe vs. channel–lobe transition zone (*sensu* Mutti & Normark 1987), or to a very proximal channelized lobe (Shanmugam & Mojola 1988; Hurst et al. 1999; Mattern 2002).

The deposits of sub-unit 4.4.1 can be referred to facies association I and are interpreted as wide, gentle channel infill close to the lobe. The package of sediments, seen in logs A and C, increases in thickness in southern log C and pinches out with onlap geometry in northern log A (Hurst et al. 1999). The channel margin is represented in log A, whereas a channel-fill portion is represented in log C (Fig. 8). The very coarse sandstone/pebbly-sandstone bed (middle in sub-unit 4.4.1) shown in log C characterizes the channel axis and is not present along the margin.

In unit 5.1, the thin-bedded turbidites (facies I) are linked to channel margins and channel–levees (see Hesse & Dalton 1995; Bruhn & Walker 1997; Cronin *et al.* 2000) and interchannel deposits, with interlayering of a few lenticular, classical turbidite beds (facies H) in relation to overbank deposits. Thin debris-flow beds and slump could be interpreted as deposits linked to the instability of the nearby slope (Haughton 2001). Three channels cutting and lateral to the interchannel deposits are represented (Fig. 3). They are infilled by massive, coarse, pebbly sandstones with marked erosional features and by bypass sediments.

Unit 5.2 is characterized by a complex and chaotic stack of allochthonous olistostromes (derived from the substratum of the slope, which consists of allochthonous material from the orogenic wedge) with only two reduced sets of autochthonous deposits. The deposits of this unit consist of a major channel-fill of conglomerates and cohesive debris flows and a minor channel-fill with coarse sandstones and cohesive debris flow sediments; both are overlain by channel-margin and channel–levee, thin-bedded turbidites (Mutti 1977). The turbidite system is closed by the synsedimentary thrust of the allochthonous wedge; the front of the wedge formed the internal slope of the basin.

Evolution of the turbidite system

The vertical organization of the sediments and of the facies association indicates five stages of growth in a progradational turbidite system. The progradation of the whole system developed through at least four lobe stages (stages 1 to 4) and ended with channel stage 5 (Fig. 12). Every lobe stage is characterized by a clear thickening–coarsening upward trend, from a distal-intermediate lobe to a proximal lobe or channel–lobe transition zone (*sensu* Mutti & Normark 1987).

There are differences among the four lobe stages. Stage 1 (only the upper portion is partially exposed) shows important channelization of the lobes. Two channel fills with channel-margin deposits are present; they are interpreted as distributary channels close to the proximal lobe or the channel–lobe transition zone (Mutti & Normark 1987).

Stages 2 and 3 are prograding lobes with similar thickness but great differences in development. Stage 2 starts with intermediate-distal lobe sediments (unit 2.1) and ends with a large package (unit 2.2) of proximal channel–lobe transition sediments; a small distributary channel lies on top. In contrast, stage 3 starts with a very thick, distal-intermediate lobe package of sediments (unit 3.1) and ends with a thin package (unit 3.2) of proximal lobe sediments with no channelization.

Stage 4 is much thicker than the others, but its vertical development is quite similar to that of stage 2. The lower portion (units 4.1 and 4.2) of the distal-intermediate lobe facies is thinner than the proximal lobe up to the upper portion of the channel–lobe transition (units 4.3 and 4.4).

In this case, stage 2 is anomalous. It records a more distal lobe environment (Fig. 12) linked to the temporary retreat of the system or to a shift in the lobe (Mutti & Ricci Lucchi 1972).

The key-level package relative to sub-unit 4.3.1, consisting of thin-bedded turbidites below and a cohesive debris flow and slumped bed above, indicates another interruption in the thickening–coarsening upward trend. The underlying thin-bedded turbidites may indicate a momentary reduction in the supply of sand in this portion of the basin (starved deposits of Ghibaudo 1980; Shanmugam 1980), followed by a coarse-mass flow. The cohesive debris-flow key bed, comprising slumped mud-draping slope marls, testifies to the destabilization of the slope. Normal turbidite deposition in the prograding lobe was then renewed.

The interruption of normal lobe deposition may have been determined by the following factors: (1) tectonic activity modified the source area and caused a temporary reduction in the sediment supply rate, with subsequent destabilization of the slope; (2) sea-level fluctuations determined a decreased supply of turbidites and destabilized the slope and shelf-edge; (3) a mix of both components.

The main channelization of stage 5 indicates the significant progradation of the system (Fig. 12) in the proximal fan environment, close to the slope, as indicated by the upward increase in slope-slides, until sedimentation ceased because of the overthrusting of the internal margin of the basin.

The depositional system inside the collisional framework

The 'Macigno costiero' succession can be interpreted as a mainly prograding turbidite system (*sensu* Piper & Stow 1991). This is indicated by the repeated upward thickening and coarsening of sequences, and by the substitution of lobe complexes (stages 1, 2, 3 and 4) by proximal-channel complexes (stage 5) (Mutti & Normark 1987; Shanmugam & Mojola 1988).

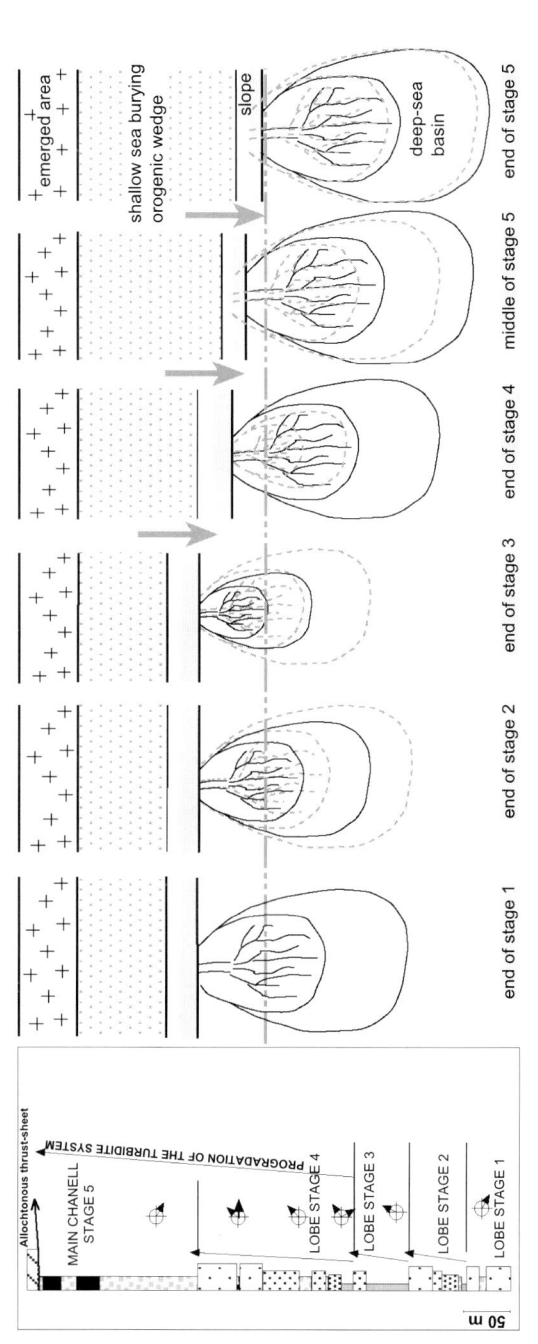

Fig. 12. Schematic evolution of the 'Macigno costiero' turbidite system. The various phases of the system/fan shifting/development are shown. The proximal channel, and the channelized and unchannelized fans, are illustrated for the turbidite system. Black lines indicate the position of the system at the time, whereas grey, dotted lines indicate its position in the previous stage. The straight grey, dotted line indicates the position of the respective measured sequence. The three grey arrows indicate the tectonic push and advancement of the orogenic wedge. A summary log of the whole turbidite system with the palaeocurrent patterns is shown.

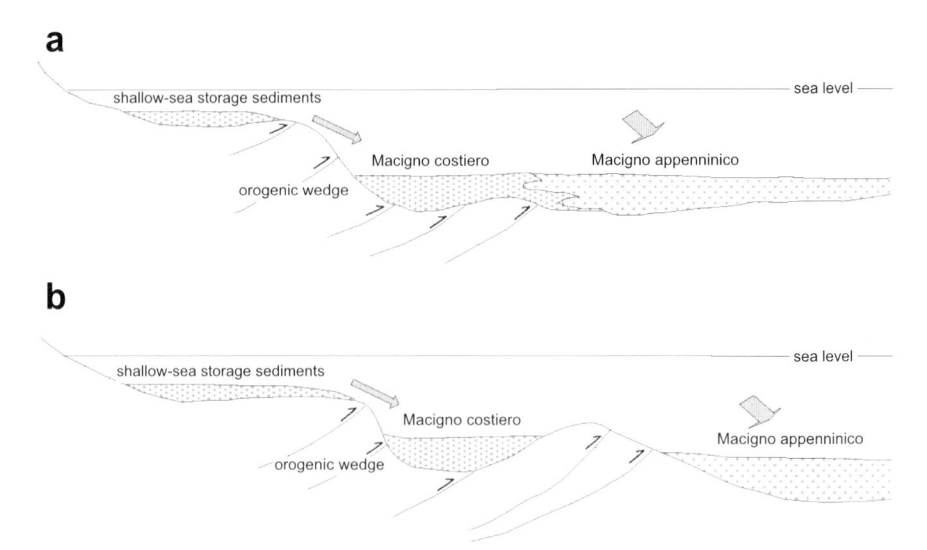

Fig. 13. Diagrams showing the two different phases of the depositional system: (**a**) open basin phase; (**b**) closed basin phase. The 'Macigno costiero' and 'Macigno appenninico' turbidite systems are represented; the grey

The internal organization and architecture of the turbidite system correspond to systems with a close connection between channels and lobes, e.g. the type II low-efficiency systems of Mutti (1985). These characteristics also accommodate the medium-sized systems of Mutti (1992), the Type III 'fan channel–lobe complex' of Stow & Johansson (2000) and the 'sand-rich systems' of Reading & Richards (1994).

The progradation of the 'Macigno costiero' was linked to the eastward migration of the orogenic wedge (progradational systems of Macdonald 1986). The final stages of progradation initially determined extremely proximal fan depositional environments, and successively resulted in closure of the basin due to the sliding of portions of the slope (olistostromes) and the overthrust.

The chaotic deposits interbedded mainly in the top portion of the turbidite succession were fed by the internal slope of the basin as a result of the tectonic activity of the orogenic front (Fielding et al. 1997).

The 'Macigno costiero' turbidite system has many sedimentological–architectural features that fit models for confined systems. The presence of thick, channel–lobe transition sediment packages and of lobe sand-bodies with a tabular geometry and a complex internal organization (presence of abundant concave and tabular scours, bypass sediments, injection structures, dewatering structures, etc.) all suggests a confined system (Lomas et al. 2000).

As for the geodynamic setting, the 'Macigno costiero' basin may have developed in a sub-basin of a complex foredeep (sensu Ricci Lucchi 1986). In particular, data seem to indicate a foredeep–frontal thrust-top basin (Fig. 13) that evolved from an initial open stage to a late closed stage (Evans & Elliott 1999; Argnani & Ricci Lucchi 2001). The vertical trend of the turbidite succession agrees with this setting. The first three stages of growth, with three independent thickening–coarsening upward trends, deposited in a portion of the foredeep basin that was marginal-proximal to the slope or in an open, frontal thrust-top basin (Fig. 13). During the open phase, the turbidite system interacted with the main foredeep basin deposits of the 'Macigno appenninico' (Biserni & Cornamusini 2002). The advancing thrust-sheet orogenic wedge determined the progressive and final progradation of the system (stages 4 and 5; see Fig. 12), and changed the physiography of the basin into a closed thrust-top basin with no important link to the main external foredeep (Fig. 13).

Palaeocurrent data collected for the whole succession show a radial pattern with a polymodal distribution (Fig. 12), indicating palaeoflows from the western, internal side of the basin towards the east. The basin was fed transversally, probably through several entry points (Cornamusini 2001; Cornamusini et al. 2002a).

The 'Macigno costiero' shows the architecture and evolution of a confined turbidite system, closed by the superpositioning of an active thrust-sheet linked to the Apennine collisional phase.

The source of this transversally fed, deepwater, sand-rich turbidite system was not distant. In such systems the shelf separating the basin from the source may be narrow (Bouma 2000). The shelf was occupied by storage sediments coming from fan-delta systems close to where the Hercynian basement crops out (Carmignani et al. 1995; Cornamusini 2001; Cornamusini et al. 2002a). These shelf sediments then spilled into the deepwater basin of the 'Macigno costiero' (Fig. 13).

Conclusions

The evolution of the 'Macigno costiero' turbidite system was marked by a complex pattern of spatial-temporal variations. These variations can be seen in the framework of a developing and prograding turbidite system, which was transversal to the basin. This system represents the oldest and innermost portion of the Tertiary NA foredeep complex, and developed as a sand-rich turbidite system through at least five stages of growth. The five stages indicate an evolution of sedimentation from lobes to a main channel complex close to the slope of the basin, up to closure of the basin by an allochthonous syn-sedimentary thrust-sheet.

Stages 1 to 4 represent a series of prograding lobes, from a distal-intermediate lobe environment to a proximal-lobe or channel–lobe environment. Particularly in stages 1 and 4, channelized lobes with distributary channels characterize the upper portions. Stage 5 marks the final depositional phase of the system, grading from a main channel complex, characterized by channel-fill and interchannel deposits, to the sliding of the approaching orogenic wedge. The latter was characterized by allochthonous tectonic units that finally overthrust the turbidite succession, interrupting sedimentation and determining the migration of the basin depocentre (migrating foredeep of Ricci Lucchi 1986).

In particular, stages 1, 2 and 3 represent thickening–coarsening upward stacking patterns typical of lobes. The deactivation of every lobe can be ascribed to tectonic-eustatic causes or to shifting of the lobes. In contrast, stage 4 marks the gradual and continuous progradation of the whole turbidite system until sedimentation ceased because of tectonic processes.

For the first three stages, the system was not definitively prograding, but: (1) it was in a more or less stable position in the basin, and was accompanied by deactivation and shifting of lobes, or by a partial or total momentary interruption of the influx of sediments due to tectonic or eustatic causes; (2) it underwent cyclic progradation and retrogression due to tectonic and/or eustatic causes that affected both the source area and the basin, changing the shape and geometry of the latter.

On the whole, the features and spatial/temporal evolution of the 'Macigno costiero' system were determined by the collisional tectonic regime, which controlled the system's development and end. The tectonic regime also controlled the shape and size of the basin and the degree of confinement of the turbidite system. Deposition developed in a thrust-top basin in front of the orogenic wedge, which partially interfered with deposition in the more external, wide foredeep basin. Interaction with deposition in the wide foredeep basin was controlled by tectonic highs, which partly separated the two sub-basins.

This paper is part of the author's PhD thesis. A. Lazzarotto, A. Costantini and F. Sandrelli are thanked for discussions in the field. Suggestions, comments and critical reviews by S. Lomas and by the referee M. Carr and another anonymous reviewer were extremely precious.

References

AGIRREZABALA, L. M. & GARCIA-MONDÉJAR, J. 1994. A coarse grained turbidite system with morphotectonic control (Middle Albian, Ondarroa, northern Iberia). *Sedimentology*, 41, 383–407.

ANDREOZZI, M. & DI GIULIO, A. 1994. Stratigraphy and petrography of the Mt. Cervarola Sandstones in the type area, Modena Province. *Memorie della Società Geologica Italiana*, 48, 351–360.

ARGNANI, A. & RICCI LUCCHI, F. 2001. Tertiary siliciclastic turbidite systems of the Northern Apennines. In: Vai, G. B. & Martini, I. P. (eds) *Anatomy of an Orogen: the Apennines and Adjacent Mediterranean Basins*. Kluwer Academic, Dordrecht, 327–350.

BARTOLE, R. 1995. The North Tyrrhenian-Northern Apennines post-collisional system: constraints for geodinamic model. *Terra Nova*, 7, 7–30.

BISERNI, G. & CORNAMUSINI, G. 2002. Sistemi torbiditici interagenti nell'avanfossa oligocenica nordappenninica: dati petrografico-stratigrafici del Macigno della Toscana meridionale costiera. *Bollettino della Società Geologica Italiana*, 121, 2, 253–263.

BOCCALETTI, M., CALAMITA, F., DEIANA, R., GELATI, R., MASSARI, F., MORATTI, G. & RICCI LUCCHI, F. 1990. Migrating foredeep-thrust belt system in the northern Apennines and southern Alps. *Palaeogeography, Palaeoclimatology, Palaeoecology*, 77(1), 3–14.

BORTOLOTTI, V., PASSERINI, P., SAGRI, M. & SESTINI, G. 1970. The Miogeosyclinal Sequences. In: Sestini, G. (ed) Development of the Northern Apennines Geosyncline. Sedimentary Geology, 4, 341–444.

BOUMA, A. H. 1962. Sedimentology of Some Flysch Deposits: A Graphic Approach to Facies Interpretation. Elsevier, Amsterdam.

BOUMA, A. H. 2000. Coarse-grained and fine-grained turbidite systems as end member models: applicability and dangers. Marine and Petroleum Geology, 17, 137–143.

BRACCI, G., DALENA, D. & BRACACCIA, V. 1984. Caratteristiche sedimentologiche dell'Arenaria di Calafuria (Toscana). Atti della Società Toscana di Scienze Naturali, Memorie, Serie A, 91, 189–202.

BRUHN, C. H. & WALKER, R. G. 1997. Internal architecture and sedimentary evolution of coarse-grained, turbidite channel–levee complexes, Early Eocene Regência Canyon, Espírito Santo Basin, Brasil. Sedimentology, 44, 17–46.

BRUNI, P., CIPRIANI, N. & PANDELI, E. 1994. New sedimentological and petrographical data on the Oligo-Miocene turbiditic formations of the Tuscan Domain. Memorie della Società Geologica Italiana, 48, 251–260.

CARMIGNANI, L., DECANDIA, F. A., FANTOZZI, P. L., LAZZAROTTO, A., LIOTTA, D. & MECCHERI, M. 1994. Tertiary extensional tectonics in Tuscany (Northern Apennines, Italy). Tectonophysics, 238, 295–315.

CARMIGNANI, L., DECANDIA, F. A., DISPERATI, L., FANTOZZI, P. L., LAZZAROTTO, A., LIOTTA, D. & OGGIANO, G. 1995. Relationships between the Tertiary structural evolution of the Sardinia–Corsica–Provençal Domain and the Northern Apennines. Terra Nova, 7, 128–137.

CARMIGNANI, L., DECANDIA, F. A. ET AL. 2001. Inner Northern Apennines. In: Vai, G. B. & Martini, I. P. (eds) Anatomy of an Orogen: The Apennines and Adjacent Mediterranean Basins. Kluwer, Dordrecht, 197–214.

CLARK, J., GARDINER, A. ET AL. 2001. Turbidite Sedimentation in Confined Systems. Field-trip Guide-book, Nice, France, September.

CORNAMUSINI, G. 2001. The early depositional phases of the Northern Apennine foredeep-thrust belt system: implications from the 'Macigno costiero' (Late Oligocene, Italy). Ofioliti, 26(2a), 263–274.

CORNAMUSINI, G. 2002. Compositional evolution of the Macigno Formation. of southern Tuscany along a transect from the Tuscan coast to the Chianti Hills. Bollettino della Società Geologica Italiana, Vol. spec. n.1, 365–374.

CORNAMUSINI, G. & COSTANTINI, A. 1997. Sedimentology of a Macigno turbidite section in the Piombino-Baratti area (northern Apennines, Italy). Giornale di Geologia, 59(1-2), 129–141.

CORNAMUSINI, G., ELTER, F. M. & SANDRELLI, F. 2002a. The Corsica–Sardinia Massif as source area for the early northern Apennines foredeep system: evidence from debris flows in the 'Macigno costiero' (Late Oligocene, Italy). International Journal of Earth Sciences, 91, 280–290.

CORNAMUSINI, G., LAZZAROTTO, A., MERLINI, S. & PASCUCCI, V. 2002b. Eocene–Miocene evolution of the North Tyrrhenian Sea. Bollettino della Società Geologica Italiana, Vol. spec. n.1, 769–787.

COSTA, E., DI GIULIO, A., PLESI, G., VILLA, G. & BALDINI, C. 1997. I flysch oligo-miocenici della trasversale Toscana meridionale-Casentino: dati biostratigrafici e petrografici. Atti Ticinensi di Scienze della Terra, 39, 281–302.

CRONIN, B. T., HURST, A., CELIK, H. & TÜRKMEN, I. 2000. Superb exposure of a channel, levee and overbank complex in an ancient deep-water slope environment. Sedimentary Geology, 132, 205–216.

CRONIN, B. T., OWEN, D., HARTLEY, A. & KNELLER, B. 1998. Slumps, debris flows and sandy deep-water channel systems: implications for the application of sequence stratigraphy to deep water clastic sediments. Journal of the Geological Society, 155, 429–432.

ELLIOTT, T. 2000. Megaflute erosion surfaces and the initiation of turbidite channels. Geology, 28(2), 119–122.

EVANS, M. J. & ELLIOTT, T. 1999. Evolution of a thrust-sheet-top basin: the Tertiary Barrême basin, Alpes-de-Haute-Provence, France. Geological Society of America Bulletin, 111(11), 1617–1643.

FERRINI, G. & PANDELI, E. 1985. Facies e sequenze verticali nel Macigno di Calafuria (Livorno). Bollettino della Società Geologica Italiana, 104, 445–458.

FIELDING, C. R., STEPHENS, C. J. & HOLCOMBE, R. J. 1997. Submarine mass-wasting as indicator of the onset of foreland loading: Late Permian Bowen Basin, Queensland, Australia. Terra Nova, 9, 14–18.

FISHER, R. V. 1983. Flow transformations in sediment gravity flows. Geology, 11(5), 273–274.

GANDOLFI, G. & PAGANELLI, L. 1992. Il Macigno costiero fra La Spezia e Grosseto. Giornale di Geologia, 54(1), 163–179.

GHIBAUDO, G. 1980. Deep-sea fans deposits in the Macigno Formation (Middle-Upper Oligocene) of the Gordana Valley, Northern Apennines, Italy. Journal of Sedimentary Petrology, 50(3), 723–742.

HAUGHTON, P. 2001. Contained turbidites used to track sea bed deformation and basin migration, Sorbas Basin, south-east Spain. Basin Research, 13, 117–139.

HESSE, R. & DALTON, E. 1995. Turbidite channel/overbank deposition in a Lower Devonian orogenic shale basin, Fortin Group of Gaspe Peninsula, Northern Appalachians, Canada. Journal of Sedimentary Research, B65(1), 44–60.

HILL, P. R. 1984. Sedimentary facies of the Nova Scotian upper and middle continental slope, offshore eastern Canada. Sedimentology, 31, 293–309.

HURST, A., VERSTRALEN, I., CRONIN, B. T. & HARTLEY, A. 1999. Sand-rich fairways in deep-water clastic reservoirs: genetic units, capturing uncertainty, and a new approach to reservoir modelling. American Association of Petroleum Geologists Bulletin, 83(7), 1096–1118.

LOMAS, S., CRONIN, B. T. ET AL. 2000. Characterisation of lateral heterogeneities in an exceptionally exposed turbidite sand-body, Grès d'Annot (Eocene–Oligocene), SE France. *In*: WEIMER, P. SLATT, R. M. *ET AL*. (eds) *Deep-Water Reservoirs of the World*. SEPM, Gulf Coast Section, 20th Annual Research Conference (CD-ROM).

LOWE, D. R. 1982. Sediment gravity flows: II. Depositional models with special reference to the deposits of high-density turbidity currents. *Journal of Sedimentary Petrology*, **52**(1), 279–297.

MACDONALD, D. I. M. 1986. Proximal to distal sedimentological variation in a linear turbidite trough: implications for the fan model. *Sedimentology*, **33**, 243–259.

MATTERN, F. 2002. Amalgamation surfaces, bed thicknesses, and dish structures in sand-rich submarine fans: numeric differences in channelized and unchannelized deposits and their diagnostic value. *Sedimentary Geology*, **150**, 203–228.

MUTTI, E. 1977. Distinctive thin-bedded turbidite facies and related depositional environments in the Eocene Hecho Group (South-central Pyrenees, Spain). *Sedimentology*, **24**, 107–131.

MUTTI, E. 1985. Turbidite systems and their relations to depositional sequences. *In*: ZUFFA, G. G. (ed) *Provenance of Arenites*. Reidel, Dordrecht, 65–93.

MUTTI, E. 1992. *Turbidite Sandstones*. Agip, Istituto di Geologia, Università di Parma, Milan.

MUTTI, E. & NORMARK, W. R. 1987. Comparing examples of ancient and modern turbidite systems: problems and concepts. *In*: LEGGETT, J. K. & ZUFFA, G. G. (eds) *Marine Clastic Sedimentology*. Graham & Trotman, London, 1–38.

MUTTI, E. & NORMARK, W. R. 1991. An integrated approach to the study of turbidite systems. *In*: WEIMER, L. & LINK, M. H. (eds) *Seismic Facies and Sedimentary Processes of Submarine Fans and Turbidite Systems*. Springer, New York, 75–125.

MUTTI, E. & RICCI LUCCHI, F. 1972. Le torbiditi dell'Appennino Settentrionale: introduzione all'analisi di facies. *Memorie della Società Geologica Italiana*, **11**(2), 161–199.

MUTTI, E. & SONNINO, M. 1981. Compensation cycles: a diagnostic feature of turbidite sandstone lobes. *In*: Abstracts Volume, 2nd IAS European Regional Meeting, Bologna, Italy, 120–123.

MUTTI, E., PAPANI, L., DI BIASE, D., DAVOLI, G., MORA, S., SEGADELLI, S. & TINTERRI, R. 1995. Il Bacino terziario epimesoalpino e le sue implicazioni sui rapporti tra Alpi ed Appennino. *Memorie di Scienze Geologiche*, **47**, 217–244.

MUTTI, E., TINTERRI, R., REMACHA, E., MAVILLA, N., ANGELLA, S. & FAVA L. 1999. *An Introduction to the Analysis of Ancient Turbidite Basins from an Outcrop Perspective*. AAPG Continuing Education Course Note Series, **39**.

PICKERING, K. T., HISCOTT, R. N. & HEIN, F. J. 1989. *Deep Marine Environments*. Unwin Hyman, London.

PIPER, D. J. W. & STOW, D. A. W. 1991. Fine grained turbidites. *In*: EINSELE, G., RICHEN, W. & SEILACHER, A. (eds) *Cycles and Events in Stratigraphy*. Springer, Berlin, 360–376.

POSTMA, G. 1984. Slumps and their deposits in fan delta front and slope. *Geology*, **12**, 27–30.

READING, H. G. & RICHARDS, M. 1994. Turbidite systems in deep-water basin margins classified by grain size and feeder system. *American Association of Petroleum Geologists Bulletin*, **78**, 792–822.

RICCI LUCCHI, F. 1986. The Oligocene to recent foreland basins of the northern Apennines. *In*: ALLEN, P. A. & HOMEWOOD, P. (eds) *Foreland Basins*. International Assocciation of Sedimentologists, Special Publications, **8**, Blackwell Scientific, Oxford, 105–139.

RICCI LUCCHI, F. 1990. Turbidites in foreland and on-thrust basins of the northern Apennines. *Palaeogeography, Palaeoclimatology, Palaeoecology*, **77**(1), 51–66.

SAITO, T. & ITO, M. 2002. Deposition of sheet-like turbidite packets and migration of channel–overbank systems on a sandy submarine fan: an example from the Late Miocene–Early Pliocene forearc basin, Boso Peninsula, Japan. *Sedimentary Geology*, **149**, 265–277.

SATUR, N., HURST, A., CRONIN, B. T., KELLING, G. & GÜRBÜZ, K. 2000. Sand body geometry in a sand-rich, deep-water clastic system, Miocene Cingöz Formation of southern Turkey. *Marine and Petroleum Geology*, **17**, 239–252.

SESTINI, G., BRUNI, P. & SAGRI, M. 1986. The flysch basins of the Northern Apennines: a review of facies and of Cretaceous–Neogene evolution. *Memorie della Società Geologica Italiana*, **31**, 87–106.

SHANMUGAM, G. 1980. Rhythms in deep sea, fine-grained turbidite and debris-flow sequences, Middle Ordovician, eastern Tennessee. *Sedimentology*, **27**, 419–432.

SHANMUGAM, G. & MOJOLA, R. J. 1988. Submarine fans: characteristics, models, classification, and reservoir potential. *Earth Sciences Reviews*, **24**, 383–428.

STOW, D. A. V. & JOHANSSON, M. 2000. Deep-water massive sands: nature, origin and hydrocarbon implications. *Marine and Petroleum Geology*, **17**, 145–174.

STOW, D. A. V. & MAYALL, M. 2000. Deep-water sedimentary systems: new models for the 21st century. *Marine and Petroleum Geology*, **17**, 125–135.

SWART, R. 1992. The sedimentology of the Zerrissene Turbidite System, Damara Orogen, Namibia. *Memoirs of the Geological Survey of Namibia*, **13**.

VALLONI, R. & ZUFFA, G. G. 1984. Provenance changes for arenaceous formations of the northern Apennines, Italy. *Geological Society of America Bulletin*, **95**, 1035–1039.

WALKER, R. G. 1978. Deep-water sandstone facies and ancient submarine fans: models for exploration for stratigraphic traps. *American Association of Petroleum Geologists Bulletin*, **62**, 932–966.

WALKER, R. G. 1985. Mudstones and thin-bedded turbidites associated with the Upper Cretaceous Wheeler Gorge conglomerates, California: a possible channel–levee complex. *Journal of Sedimentary Petrology*, **55**, 279–290.

WALKER, R. G. 1992. Turbidite and submarine fans. *In*: WALKER, R. G. & JAMES, N. P. (eds) *Facies Models: Response to Sea Level Change*. Geological Association of Canada, St. Johns, 239–263.

Spatial variability of Hurst statistics in the Castagnola Formation, Tertiary Piedmont Basin, NW Italy: discrimination of sub-environments in a confined turbidite system

FABRIZIO FELLETTI

Università degli Studi di Milano. Dipartimento di Scienze della Terra, Via Mangiagalli 34, 20133-Milano, Italy (e-mail: fabrizio.felletti@unimi.it)

Abstract: The purpose of this paper is to investigate the potential of the use of the Hurst test as a statistical tool for quantifying the degree of clustering of high and low values of field measurements in vertical turbiditic sections within a small and confined tectonically mobile basin. The paper (1) documents the fact that significant clustering occurs in almost all sections, (2) shows that there are systematic spatial variations in the degree of clustering such that different parts of the basin have different Hurst statistics, (3) discusses the origin of those spatial trends, and (4) compares Hurst H with Chen and Hiscott's Hurst K. The case study focuses on one well-exposed reservoir-scale turbiditic unit cropping out in the eastern part of the Tertiary Piedmont Basin (NW, Italy) and representing the lower portion of an Upper Oligocene–Lower Miocene turbidite system (Castagnola Formation). The sedimentological model that has been adopted is based on the description of 41 sedimentological logs (approximately 25–35 m apart) physically correlated bed by bed. Three variables were studied: thickness of sandstones and siltstones, grain size score, and sand and silt thickness percentage (i.e. the ratio of coarse division to overlying mudstones, in percent). The Hurst test has revealed the existence of common clustering for the three studied variables (only 2 of 41 logs failed to pass the test at the 10% significance level for all the studied variables). This clustering is associated with lateral and vertical facies variations in response to changes in depositional sub-environments within the Castagnola Basin. A geostatistical analysis has been performed in order to estimate the spatial distribution of the Hurst statistics within the studied area. The analysis makes possible predictions at unsampled locations.

Chen & Hiscott (1999) have recently proposed a methodology for quantifying the degree of clustering in turbiditic facies in vertical sections. This method, utilizing the *Hurst statistic* (Hurst 1951) or *rescaled range analysis* (Feder 1988; Plotnick & Prestegaard 1995), provides a means of determining the degree of clustering of low and high values in a time series. Chen & Hiscott (1999) showed that such clustering can be used as an index to distinguish submarine-fan sub-environments. Their results suggest that channel–levee deposits tend to exhibit strong clustering; lobe–interlobe deposits display moderate clustering and basin-floor sheet-sand systems tend to have weak clustering. Chen & Hiscott (1999) calculated the Hurst statistic for 19 turbidite sections that span a wide range of geological time, tectonic settings, facies characteristics and depositional environments.

The aim of this paper is to apply the statistical analysis proposed by Chen & Hiscott (1999) to a turbidite unit deposited within a small confined tectonically mobile basin, in order to get a better knowledge of the variability of this index and to refine the position of group boundaries. The paper records how the Hurst statistic changes laterally (parallel and perpendicular to the main palaeocurrent direction) and whether it can really be used to discriminate sub-environments within a geologically complex sand-body with facies heterogeneities at different scales. Geostatistical analysis allowed the prediction of Hurst statistics at unsampled locations within the studied area.

For all the measured stratigraphic sections, the same bed-by-bed measurements used by Chen & Hiscott (1999) were considered: coarse-division thickness (sandstone and siltstone), grain size score, and coarse- division thickness percentage (i.e. the ratio of the coarse division to overlying mudstones, in percent).

Methodology

The Hurst statistic has been used as a tool to determine the degree of clustering of low and high values in vertical stratigraphic sections. The Hurst method was applied for the first time in deep marine sedimentology by Chen & Hiscott (1999). Hurst (1951) derived the relationship $K = \log(R/S)\log_{10}(N/2)$ (R is the maximum range in cumulative departures from the mean over N observations, and S is the standard

From: LOMAS, S. A. & JOSEPH, P. (eds) 2004. *Confined Turbidite Systems*. Geological Society, London, Special Publications, **222**, 285–305. 0305-8719/04/$15.00 © The Geological Society of London 2004.

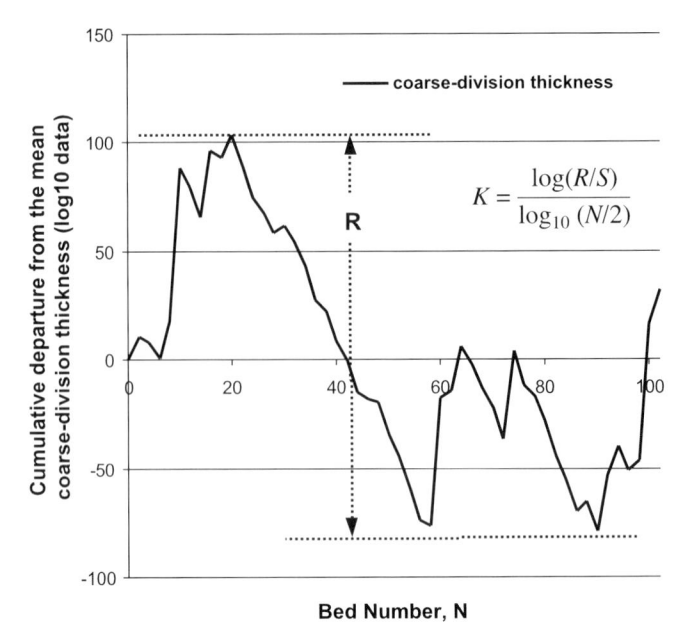

Fig. 1. This plot show the variables used in calculation of Hurst index K. The plotted data represent a series of 104 bed measurements (Section 36) of coarse-division thickness (sand and silt) within the sand-body. R is the maximum range in cumulative departures from the mean over N observation. S is the standard deviation for the N observations (after Chen & Hiscott 1999).

$$K = \frac{\log(R/S)}{\log_{10}(N/2)}$$

deviation) (Fig. 1) and showed the tendency for a natural sequence of data with a large N value ($N > 100$) to yield values of $K > 0.5$. By contrast, for purely random processes he predicted values of K close to 0.5.

For a small N value, an alternative estimator, H, is determined by dividing the series of observations (after $\log_{10}$ conversion) into a number (n) of subseries of different lengths (Wallis & Matalas 1970, 1971; Chen & Hiscott 1999). Separate values of R_n/S_n are calculated for each subseries and then $\log(R_n/S_n)$ is plotted against $\log n$. The slope of the straight line of points is H. The two estimators K and H are strongly correlated, but values of H are generally somewhat smaller than values of K, and for random data are closer to 0.5 than are values of K (Chen & Hiscott 1999). In the present study, Hurst H statistics have been computed because the number of beds in the sand-body towards the terminations of the unit is smaller than 100.

In order to determine an appropriate expected value of Hurst H for randomized turbidite succession, a Monte Carlo simulation technique has been used. From the original data, 300 randomly shuffled sequences are simulated. The expected value is then compared with the H values from the original field section to assess the likelihood that it belongs to the same population. A Fortran-77 program, HURST.FOR (Chen & Hiscott 1999), has been used to calculate the Hurst statistics and the Monte Carlo simulation. Known distribution parameters permit the use of hypothesis tests for values of H calculated from the original turbidite successions. If H_0 is the Hurst H value of a measured turbidite section, and the proportion of sequences with a $H \geq H_0$ in the 300 randomly shuffled sequences is smaller than a selected significance level α, then the hypothesis that the measured sequence is vertically random distributed is rejected at that significance α.

The estimation of the spatial distribution of the Hurst statistics within the studied area has been performed using a geostatistical approach. The term *geostatistics* is associated with a class of techniques used to analyse and predict values of a variable distributed in space or time (Matheron 1971; Armstrong 1998). The application of geostatistical methods in this case includes three main steps:

(1) exploratory data analysis;
(2) variogram structural analysis (calculation and modelling of variograms);
(3) making predictions (kriging).

Variogram analysis allows the quantification of the correlation between any two values separated by a distance h (usually called the *lag* distance), and uses this information to make predictions at unsampled locations (kriging). Variogram modelling is a prerequisite for kriging, or making predictions. Experimental variograms are first computed from the stratigraphic log data (one variogram for each variable), then fitted with a theoretical curve (theoretical variogram). The theoretical variogram represents the spatial variability of the variable under study, and it is considered fitted when it replicates the experimental variogram. Depending on the behaviour of the considered variables, the variogram reveals several spatial characteristics such as continuity (correlation length), anisotropy, zone of influence and trend as a result of different geological controls (Journel & Huijbregts 1978; Isaaks & Srivastava 1989; Goovaerts 1997). Variograms are usually described using two parameters. The *range* is the lag distance at which all successive values are independent of each other. The *sill* is the variogram value corresponding to the total variability. The results of the geostatistical analysis are shown by contour maps (one map per variable) representing the spatial distribution of the studied variable.

Geological model of the studied turbidite lithosome

This case study focuses on one well-exposed reservoir-scale turbiditic unit, cropping out in the eastern part of the Tertiary Piedmont Basin and representing the lower portion of an Upper Oligocene–Lower Miocene turbidite system (Castagnola Formation; Cavanna et al. 1989; Baruffini et al. 1994; Felletti, 1999, 2002) (Fig. 2).

The Tertiary Piedmont Basin (TPB) and the Castagnola Formation

The Tertiary Piedmont Basin (TPB) is a tectonically mobile episutural basin that has developed on the Adria microplate behind the thrust front of the meso-Alpine belt. The basin seals the junction between the Alpine and Apenninic orogenic chains (Biella et al. 1987, 1992; Gelati & Falletti, 1996).

The present paper focuses on a small area in the easternmost part of the TPB (Fig. 2). Here is an almost 3000 m thick turbidite succession, ranging from late Eocene to early Miocene in age (Baruffini et al. 1994). Several unconformities are present within the succession. The major ones correspond to regional tectonic events that induced important changes in basin size and configuration, as well as variations in the sediment dispersal pattern and facies distribution (Cavanna et al. 1989; Di Giulio & Galbiati 1993). The turbidite deposits record sea-bottom morphological variations due to the activity of a crustal-scale lineament (Villalvernia-Varzi Line, VVL) which played an important role of tectonic release during Oligo-Miocene times (Biella et al. 1992; Laubscher et al. 1992).

A Chattian–Aquitanian phase of transpressive motion produced a structural depression (Castagnola Basin) to the south of the VVL. It was progressively filled by the Castagnola turbidites (Mutti 1992; Baruffini et al. 1994; Di Giulio & Galbiati 1998). The axis of the basin is oriented ENE–WSW, approximately parallel to the VVL. This basin geometry exerted a structural and morphological control on the shape of turbidite bodies, facies assemblages and stacking patterns (Baruffini et al. 1994) during Aquitanian time. This structural confinement apparently becomes less important in the uppermost part of the turbiditic succession where the palaeocurrent directions are more dispersed. The Castagnola turbidite system (at least 800 m of turbidite sediments at present-day thickness) appears to pinch out with an onlap relationship to the underlying deformed slope marls of the Rigoroso Formation. Well-exposed onlap terminations occur at the basin margins, and can be traced across several hundred metres-wide exposures, parallel to the predominantly northeast palaeocurrent direction (Fig. 3).

According to Cavanna et al. (1989), Stocchi et al. (1992) and Baruffini et al. (1994), the Castagnola turbidite system was deposited by gravity flows of relatively large volume and charged with fine-grained sediment, evolving from high-density to low-density turbidity currents during their movement. The lower portion of the Castagnola Formation has been interpreted by Baruffini et al. (1994) as the record of turbidite deposition on a small basin-plain.

The studied turbidite unit

The studied succession (Felletti 2002) represents the lowermost turbiditic unit at the base of the lower member of the Castagnola Formation; it is the first lithosome deposited in the Castagnola Basin since its early subsidence (Fig. 3). The selected stratigraphic interval is bounded by one megabed (*Key-bed A*) at the top and by the hemipelagic marlstones of the Rigoroso Formation (middle-upper Oligocene) at the base (Fig. 4). Key-bed A is the first megabed occurring in the Castagnola Basin; it marks a change of

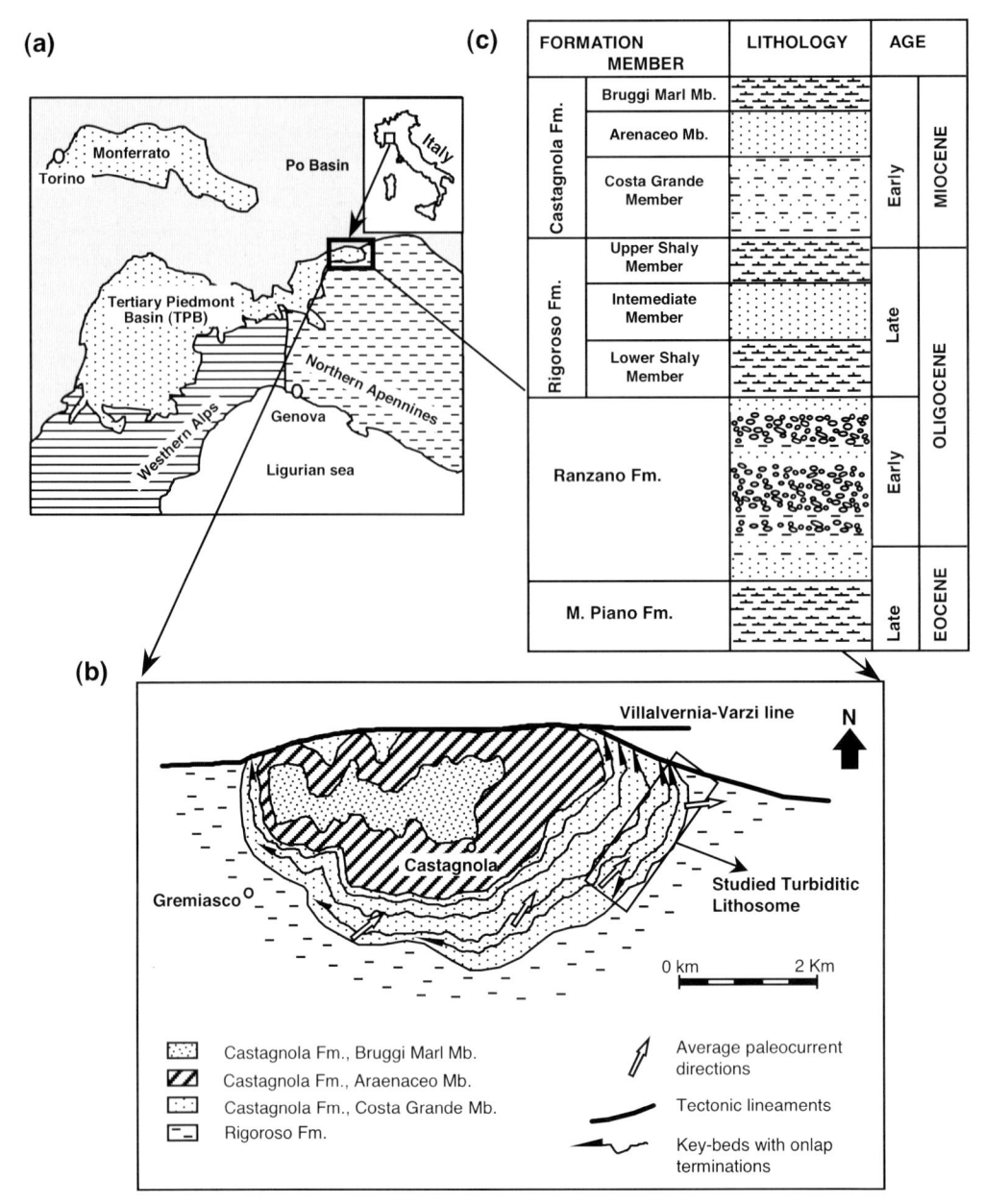

Fig. 2. (a) Location of Tertiary Piedmont Basin; (b) geological sketch map of the Castagnola basin redrawn and modified after Stocchi *et al.* (1992); (c) stratigraphic column for the eastern sector of the TPB.

depositional style from the sand-rich and complex studied sand-body (below) to the mud-rich and parallel-bedded succession (above). The selected sand-body represents a discrete turbidite succession that shows a peculiar history in the early evolutionary stage of the Castagnola basin. The studied outcrop is about 1.8 km wide; the thickness of the sand-body ranges from approximately 30 m in the depositional areas to 0 m at the northern and southern termination. Palaeocurrent data taken from basal flutes (184 measurements) and ripples (27 measurements) indicate that the currents flowed principally toward the NE. Close to the northern termination of the sand-body, a predominantly easterly directed trend is present at the base of

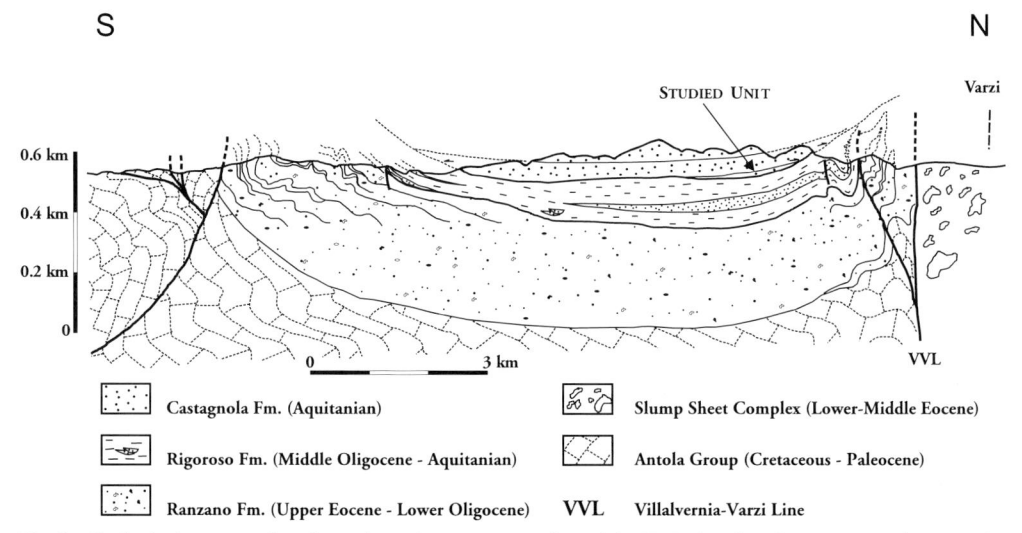

Fig. 3. Geological cross-section throughout the eastern portion of the TPB showing the gross stratal geometries related to the tectonic features (redrawn and modified after Di Giulio & Galbiati 1993).

the beds, whereas palaeocurrents from ripples higher in the bed (12 measurements) show a large spread and indicate a dispersal pattern, principally southerly directed, different from that deduced from the associated sole structures (Fig. 5).

The geometry of the northern onlap can be considered comparable with the *onlap type 1* (structural tilting and onlap) described by Sinclair (2000) and with the *infill geometry* described by Hurst *et al.* (1999), where the original depositional surface has significant relief. By contrast, the style of southern termination of the sand-body can be compared with the *onlap type 3* (gradual pinchout of sandstones suggesting depositional growth of a topographic high onto an original approximately horizontal depositional surface), described by Sinclair (2000).

The stacking pattern of the sand-body consists of an overall coarsening- and thickening-upward trend. The sand/mud ratio increases toward the northern termination of the sand-body. Conversely, a strong decrease of the sand/mud ratio occurs toward the southern lateral termination of the sand-body (Fig. 6). The frequency distribution of bed thickness is bimodal (Figs 7 & 8), as observed in many turbidite systems (Talling 2001). There are thick-bedded turbidites (with a well-developed massive base) and there are thin-bedded turbidites (comprising only Bouma division C, D and E). This bimodality is particularly evident in the northern part of the basin, where both modes are present.

The studied cross-section reveals the diachronous development of several sand packages. The correlation lines (timelines representing tops or bases of individual beds) clearly show that the sand-body is subdivided in several sedimentary sub-units, which show lateral thickness variations due to pinching and swelling of individual beds. Nine different sub-units that are vertically stacked and laterally juxtaposed have been recognized. They range in thickness from 0 cm to 8 m and have been named B, C, C_1, D, E, F, G, H, I. The stacked sub-units assemblage is shown in Figure 9 with a large vertical-scale exaggeration. Each bundle consists of a group of individual sandstone beds separated by muddy intervals, and shows different vertical stacking patterns and different horizontal trends in facies arrangement. These different internal architectures are related to the location of the sub-units relative to the basin margins (Felletti 2002).

Geometry, stratigraphic relationships, facies distribution and palaeocurrent directions indicate that the sand-body was deposited in a basin controlled by synsedimentary transpressive tectonic activity linked with the Villalvernia-Varzi Line to the north. Turbidite deposition during accumulation of most of the sand-body was controlled by:

(1) synsedimentary tilting of the slopes;
(2) the distribution of structural and depositional relief within the basin;
(3) the thickness and volume of the turbidite flows;
(4) the angle of impingement of turbidity currents against the confining slopes (that induced flow modifications).

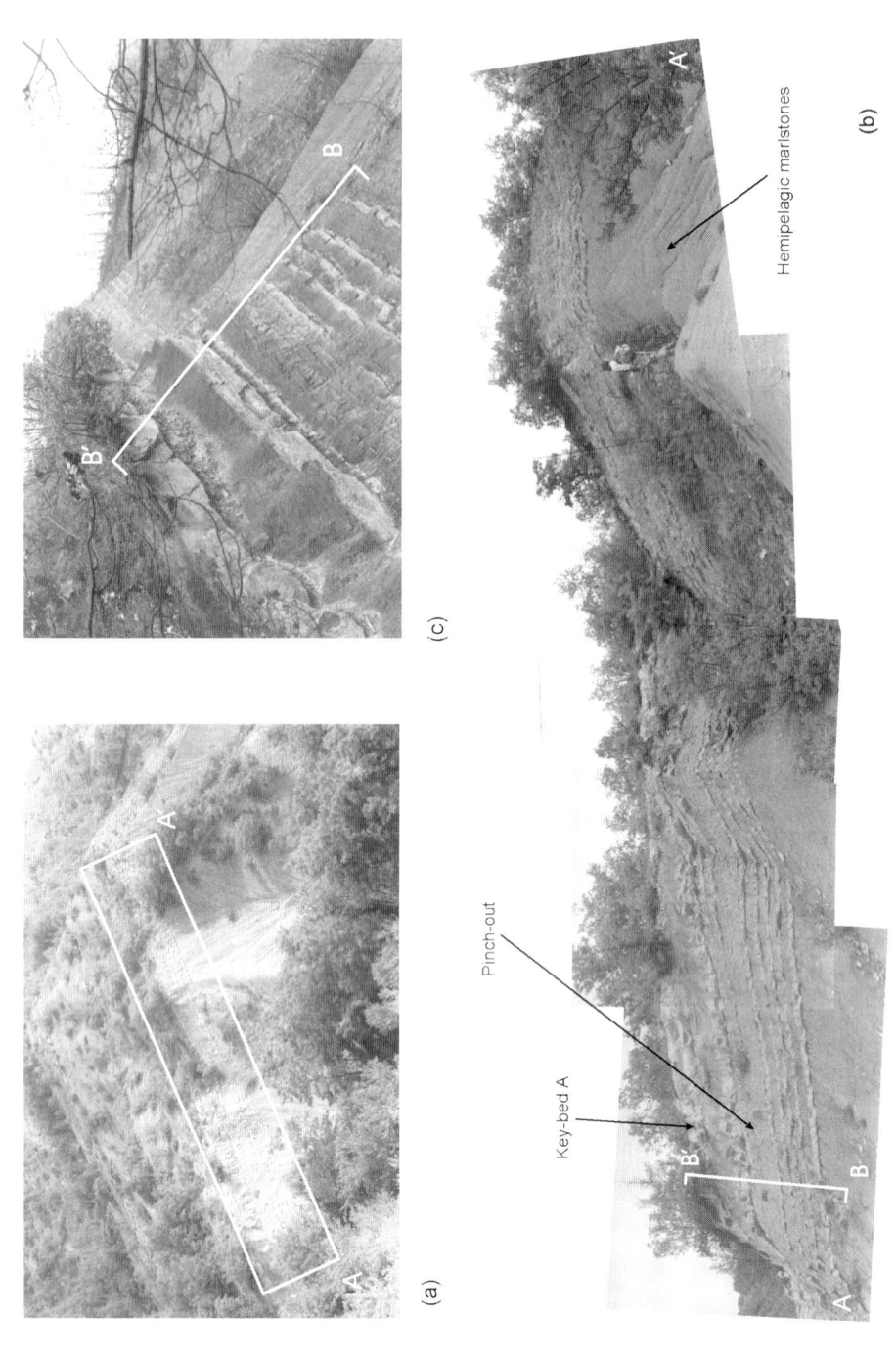

Fig. 4. (a) Panoramic view of the studied area from logs 15–17. (b) Photo-montage shows the selected stratigraphic interval, bounded by one megabed (key-bed A) at the top and by the slope hemipelagic marlstones at the base (Marne di Rigoroso Formation, middle–upper Oligocene). The progressive thinning of sandstone turbidites toward the south represents compensation of the entire sub-unit, caused by smoothing of the underlying irregular topography (sub-unit G). (c) Closer view of (b): the beds are bipartite, with a lower, fine- to medium-grained sandstone grading up into a thick silty mudstone top, which is moderately bioturbated.

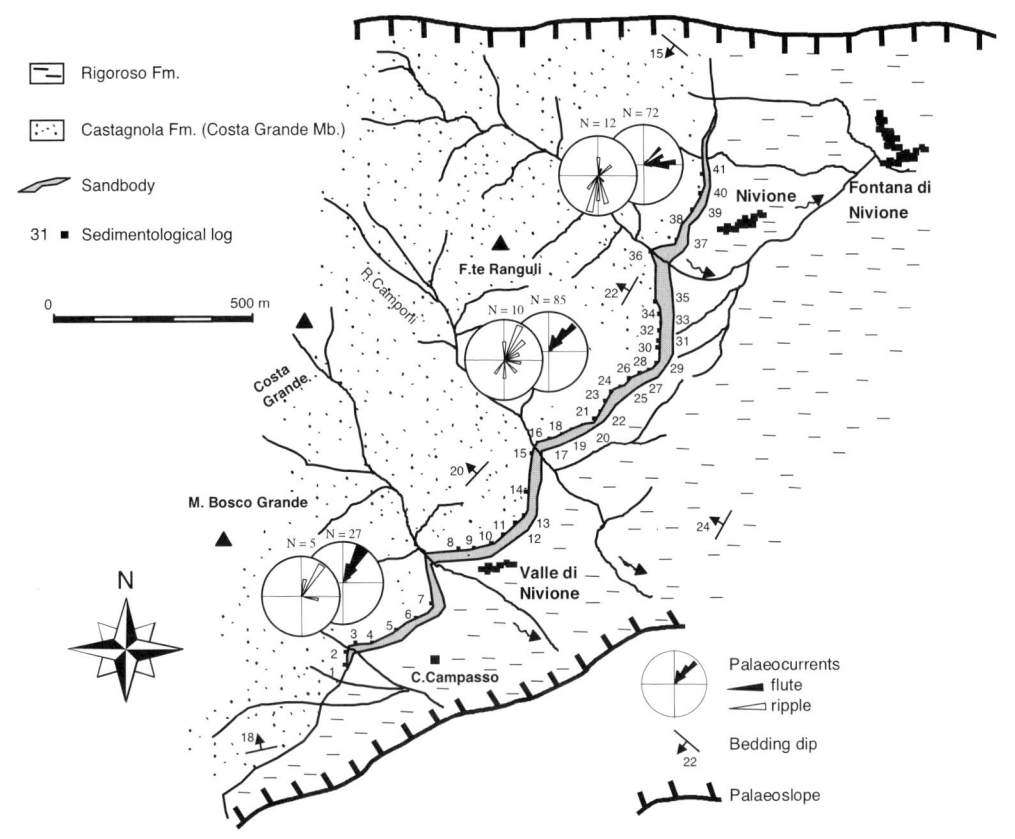

Fig. 5. Geological sketch map of the studied turbidite unit, showing location of the sand-body and logged sections. The figure shows palaeocurrent data taken from basal flutes (184 measurements) and ripples (27 measurements), and indicates that the currents flowed principally towards the NE. Close to the northern termination of the sand-body, a predominantly easterly directed trend is present at the base of the beds, whereas palaeocurrents from ripples higher in the bed (12 measurements) show a large spread and indicate a dispersal pattern, principally southerly directed, different from that deduced from the associated sole structures (after Felletti 2002).

By contrast, sedimentation within the uppermost part of the sand-body (sub-unit I) was controlled only by the northern basin margin, which induced trapping of the entire sedimentary load (normal grading and thick mudstone caps, lack of bioturbation in both mudstone caps and the underlying sandstones and the abundant climbing ripples suggest rapid deposition and presence of large-volume turbidity currents trapped by the confinement of the basin). As a consequence, sub-units B–H show a greater degree of heterogeneity than sub-unit I (Felletti 2002).

Database compilation

The sedimentological model that has been adopted is based on the description of 41 sedimentological logs (approximately 25–35 m apart) physically correlated and measured bed by bed. The number of sand and siltstone beds for each section ranges from 109 (in the northern depocentre) to 26 (towards the lateral termination of the sand-body). The 10,500 thickness and grain-size measurements collected correspond to the total number of depositional intervals that have been recognized, which represent the internal sedimentary structures of all the turbidite layers. Such an extensive dataset provided an ideal database to study the spatial distribution of the Hurst test.

The data were organized in a database file suitable for the statistical and geostatistical computation. For all the sections, three bed-by-bed measurements were considered for the Hurst statistic: coarse-division thickness (i.e. net thickness, in a single bed, of conglomerate + sandstone + siltstone), grain size score, and coarse-division thickness percentage (i.e. the

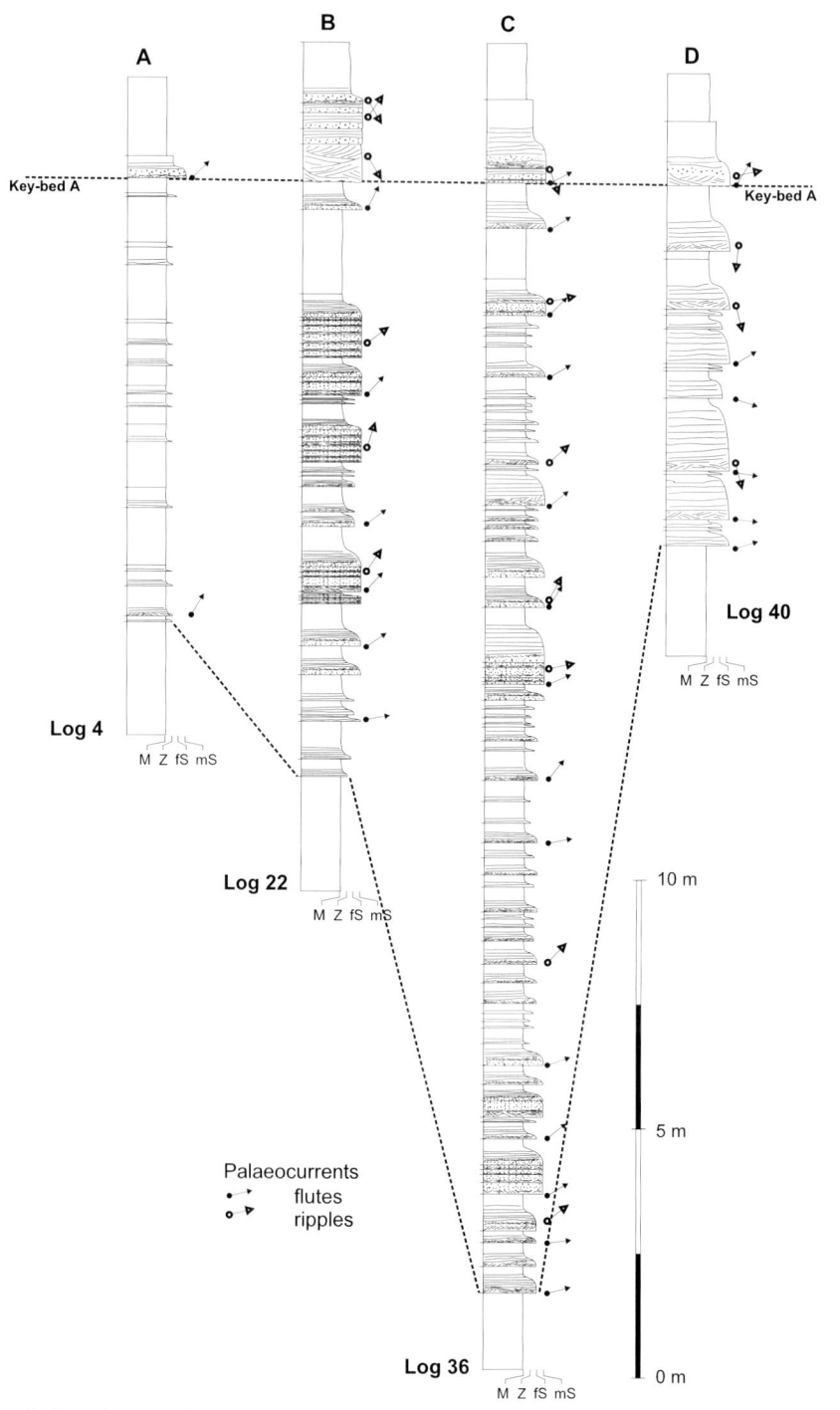

Fig. 6. Detailed stratigraphical logs with relative facies and palaeocurrent data of four different areas: (**A**) southern margin, (**B**) intrabasinal high, (**C**) northern depocentre and (**D**) northern margin. Note that towards the northern onlap termination of the studied unit, there is an overall gradual increase of the sand/mud ratio. Approaching the northern palaeoslope, the whole succession is characterized by thick sandstone beds with no obvious grain-size breaks (after Felletti 2002).

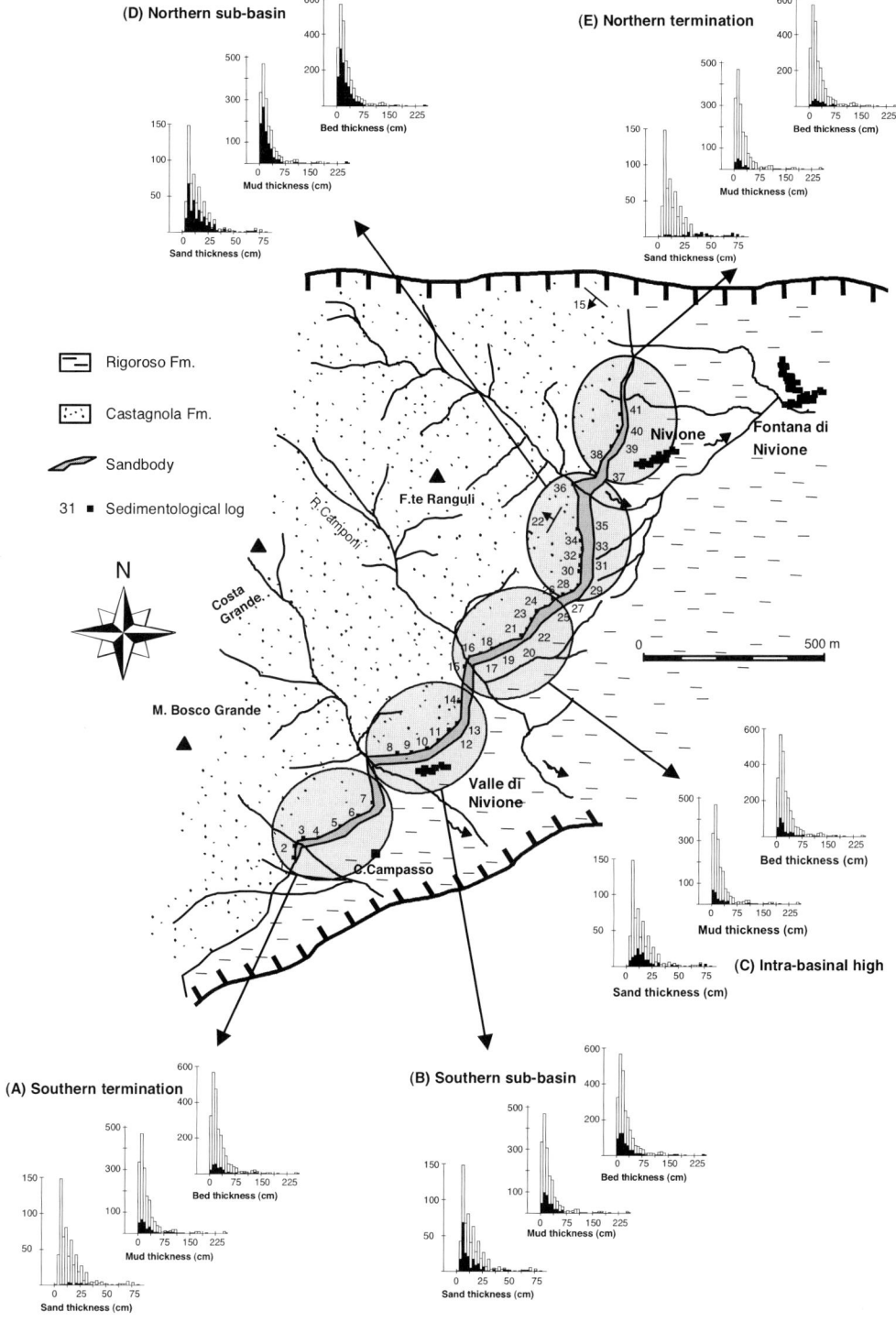

Fig. 7. Frequency distribution of sandy bed thickness, muddy bed thickness and total bed thickness of five different areas (black bars): (A) southern margin, (B) southern depocentre, (C) intrabasinal high, (D) northern depocentre and (E) northern margin. White bars represent the entire data set.

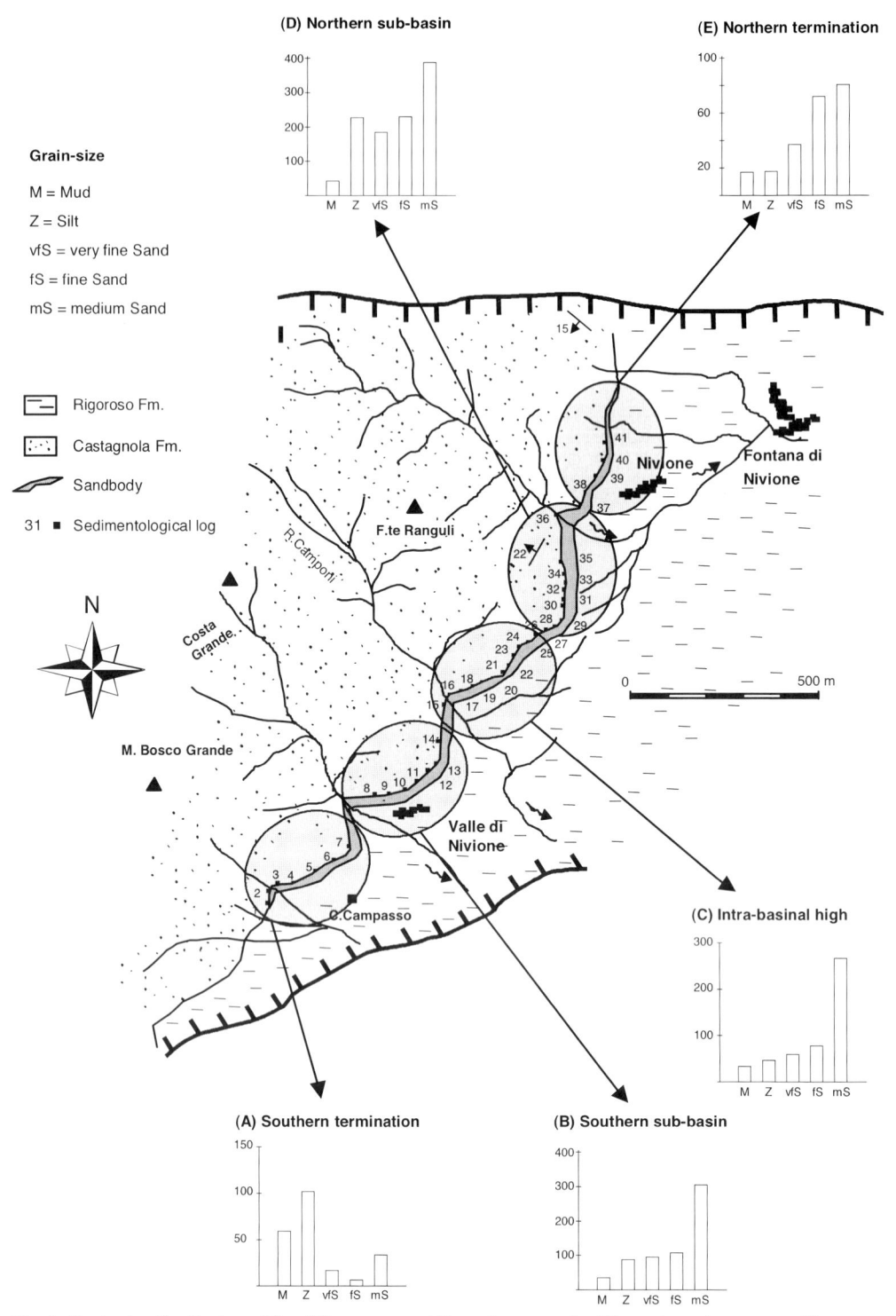

Fig. 8. Grain size distribution of five different areas: (**A**) southern margin, (**B**) southern depocentre, (**C**) intrabasinal high, (**D**) northern depocentre and (**E**) northern margin.

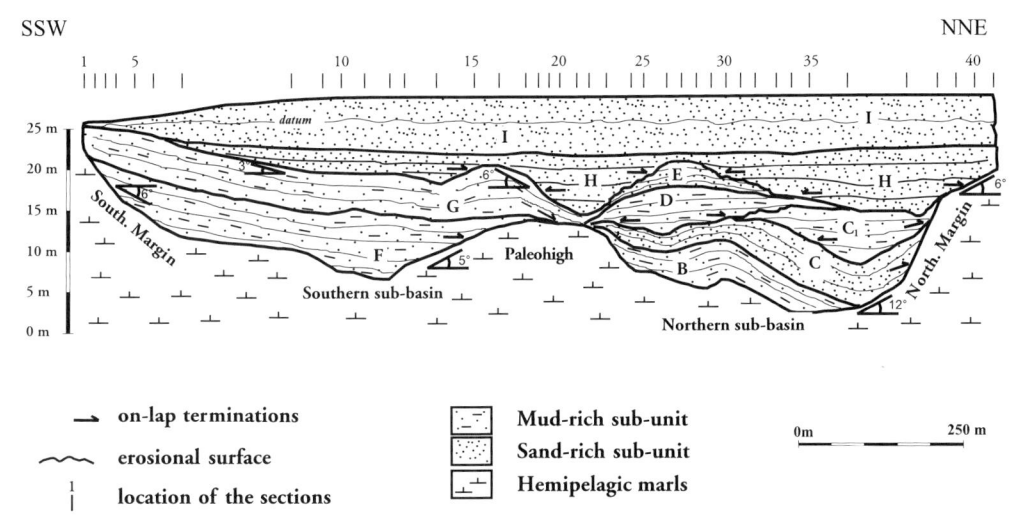

Fig. 9. Simplified cross-section of the studied turbiditic unit (sand-body A).

ratio of coarse division to overlying mudstones). The grain-size score was calculated for each bed using the same weights as proposed by Chen & Hiscott (1999). This score is based on the ϕ scale, where the score $= -4X$ (thickness proportion of pebble and granule division) $+ 0X$ (thickness proportion of coarse sand division) $+ 1.5X$ (thickness proportion of medium sand division) $+ 3X$ (thickness proportion of fine sand division) $+ 7X$ (thickness proportion of silt division). The numerical constant are midpoint ϕ size for each grain-size class. Hence the grain size score is an estimate of grain size for each bed (Chen & Hiscott 1999).

Results

Hurst test

All of the 31 logs measured in the depocentral areas of the basin passed the Hurst H test for the coarse-division thickness, coarse-division percentage and grain-size score at significance level $\alpha < 1\%$. This means that the number of sequences with $H \geq H_0$ in the 300 randomly shuffled sequences is smaller than 1% (H_0 is the Hurst H value of a measured turbidite section and H corresponds to the Hurst H value of a randomly shuffled sequence). These results indicates that there is strong clustering of all three parameters in these sections, and that each log is long enough to capture several clusters. Towards the northern and southern terminations of the sand-body, 7 of 10 logs passed the Hurst H test at a significance level slightly higher than 6% for the coarse-division percentage, and grain-size score, but <2% for the coarse-division thickness. Only 2 of 41 logs failed to pass the test at the 10% significance level for all the studied variables. These two logs, however, are located close to the southern termination of the studied lithosome, where the beds are thin, fine-grained, and not easily distinguishable from the underlying marlstones (Rigoroso Formation). The high rate of rejection of the null hypothesis suggests a long-term clustering of the studied variables and the presence of the Hurst phenomenon.

Figures 10–13 represent the results of the Hurst statistic computed using three variables and show the Hurst H for the 41 measured logs versus the standard deviation from the mean H for 300 random sequences generated by shuffling the field-based data.

From the analysis of the result of the Hurst statistics some observations are of note:

(1) The Hurst H values range from 0.6322 to 0.8276; the deviation from the mean H for 300 randomly shuffled sequences ranges from 0.0743 to 2.62 (Fig. 10). The high Hurst H values for the studied turbidite successions demonstrate that the high and low values of field measurements do cluster together, revealing a small-scale clustering of these successions.

(2) The clustering patterns in the cumulative departures from the mean are very similar for all three types of bed-by-bed measurement. This behaviour suggests a strong positive correlation between basal grain size and turbidite thickness (i.e. most thick-bedded bed clusters are also coarse-grained

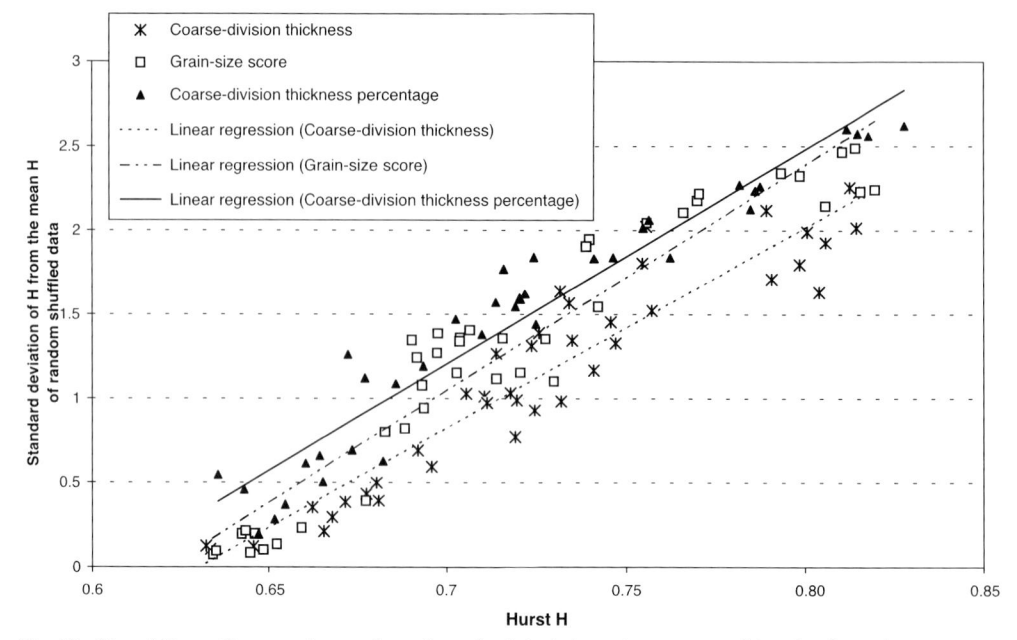

Fig. 10. Plot of Hurst *H* versus the number of standard deviations that separate this value from the mean *H* for 300 random sequences generated by shuffling the field-based data. The lines correspond to the linear regression lines for each studied variables.

and mudstone poor, and most thin-bedded bed clusters are fine-grained and muddy). The existence of this correlation has been recently analysed in detail by Talling (2001) within the Marnoso Arenacea Formation (North Italy). However, as shown by the regression lines, the plotted points corresponding to the values computed for the

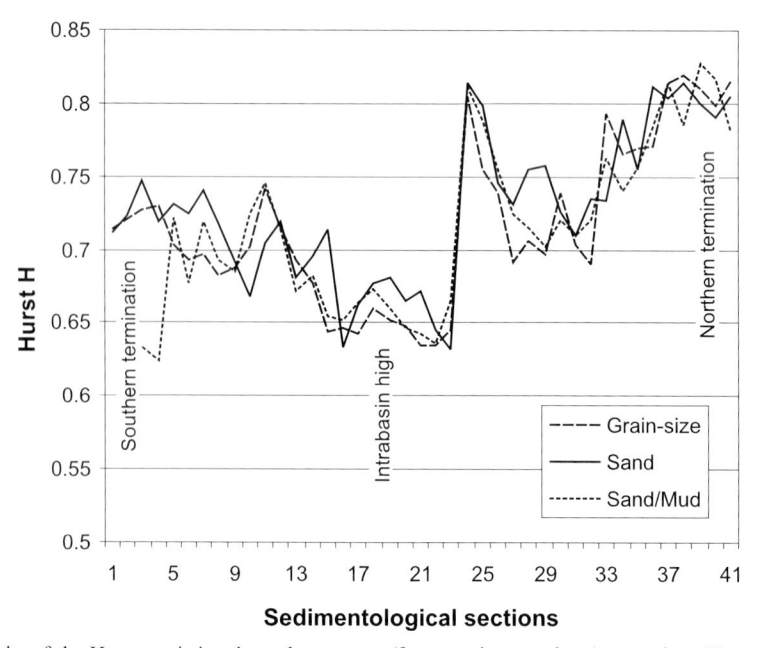

Fig. 11. Results of the Hurst statistics along the transect (from section number 1 to section 41).

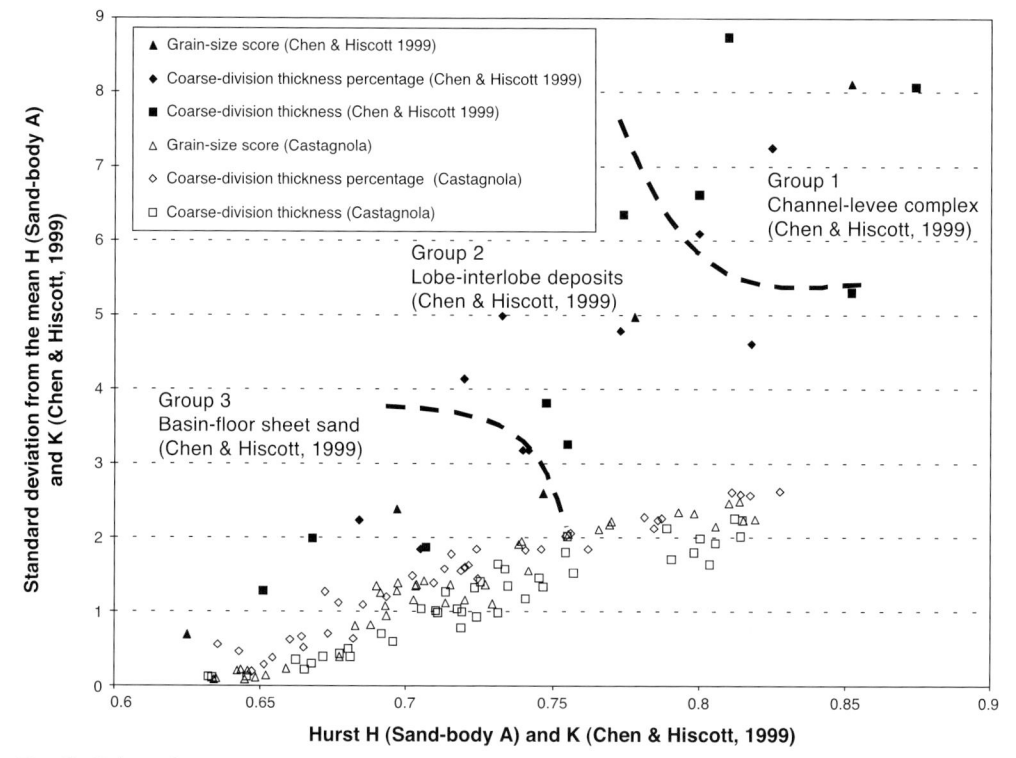

Fig. 12. Values of H, computed for the sand-body, are compared with the values shown by Chen & Hiscott (1999).

coarse-division thickness percentage are located in the upper portion of the scatter-plot whereas the values computed for the coarse-division thickness occupy the lower part (Fig. 10). This behaviour is probably related to the presence of the T_e division (mud-cap), the thickness of which is not fully correlated with the coarse-division thickness, resulting in a separation of the regression lines.

(3) If the H values computed for the sand-body are compared with the K values shown by Chen & Hiscott (1999), we can observe a different distribution of the plotted data points (Fig. 12). A smaller deviation from the mean H characterizes the studied dataset. As shown by Chen and Hiscott (1999), this behaviour can be explained by considering that values of H generally are somewhat smaller than values of K, and for random data are closer to 0.5 than are values of K. By contrast, different values of the Hurst test may indicate that it is difficult to fit the architecture of the studied sand-body into any rigid model (i.e. channel–levee complex; lobe–interlobe deposits; basin-floor sheet sand), because the clustering of the three studied variables is dependent on a range of factors (e.g. small basin, topographic confinement, tectonic activity).

(4) The Hurst test has revealed the existence of common clustering for the three studied variables. This clustering is believed to be associated with lateral and vertical facies variations in response to changes in position relative to basin floor topography and distance from source within the Castagnola basin. Figure 13 shows the results of the Hurst H statistics considering different areas: (a) the onlap termination of the sand-body toward the northern palaeoslope; (b) intrabasin high; (c) the southern termination of the sand-body; (d) southern depocentral area; (e) northern depocentral area.

Geostatistical analysis

The last point (4) above suggests that the strength of clustering, inferred from Hurst H values (thickness of sandstones and siltstones, grain-size score, and sand and silt thickness

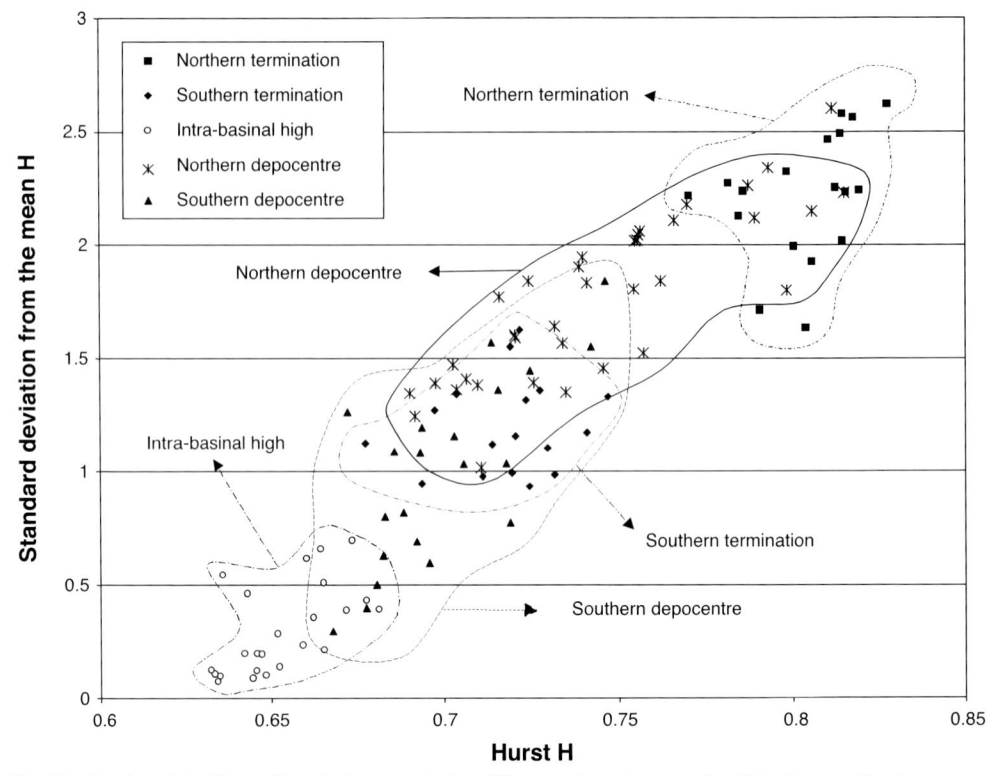

Fig. 13. Results of the Hurst H statistics considering different sub-environments within the sand-body.

percentage), provides an index to distinguish sub-environments within the basin. The next step is to estimate the spatial distribution of these values at unsampled locations. For this reason, a geostatistical analysis of Hurst H values has been performed and used to highlight gradual and abrupt transitions of the Hurst index within the basin. With reference to the spatial distribution of the Hurst statistics, it is necessary to be aware that similar H values can be derived for various stacking patterns (i.e. similar H values observed in an area of a basin are not definite proof of similar processes). In comparing the Hurst statistic spatially (by way of the contour maps in Figs 14–16), the interpretation of locally similar processes occurring in the five distinct areas of the basin has been based on field observation of similar stacking patterns.

In order to determine the correlation length of the three studied variables, a structural analysis of variograms has been performed. As the data lie principally along a line of section parallel to the main palaeocurrent trend, the estimation in two dimensions suffers from inadequate sampling perpendicular to this direction. Thus modelling the variogram and determining anisotropy ratios is associated with substantial uncertainty. Only two transects allow for computing the variogram, normal to the main palaeocurrent trend but, in this case, variograms are based on few data pairs and so the reliability of the results is lower if compared with the omnidirectional variograms.

Most points on the variograms shown are based on several tens of data pairs, documenting the reliability of the results that have been obtained. The behaviour of the experimental variogram curves indicates that the studied variables can be considered as *regionalized variables* (i.e. a sill and a range can be clearly distinguished) in the sense of Matheron (1971). This term emphasizes two apparently contradictory aspects of this type of variable: a random aspect, which accounts for local irregularities, and a structured aspect, which reflects large-scale tendencies. Comparison of the structure of omnidirectional experimental variograms shows that they have similar patterns, and indicates that these curves can be assembled into two groups (Figs 14–16):

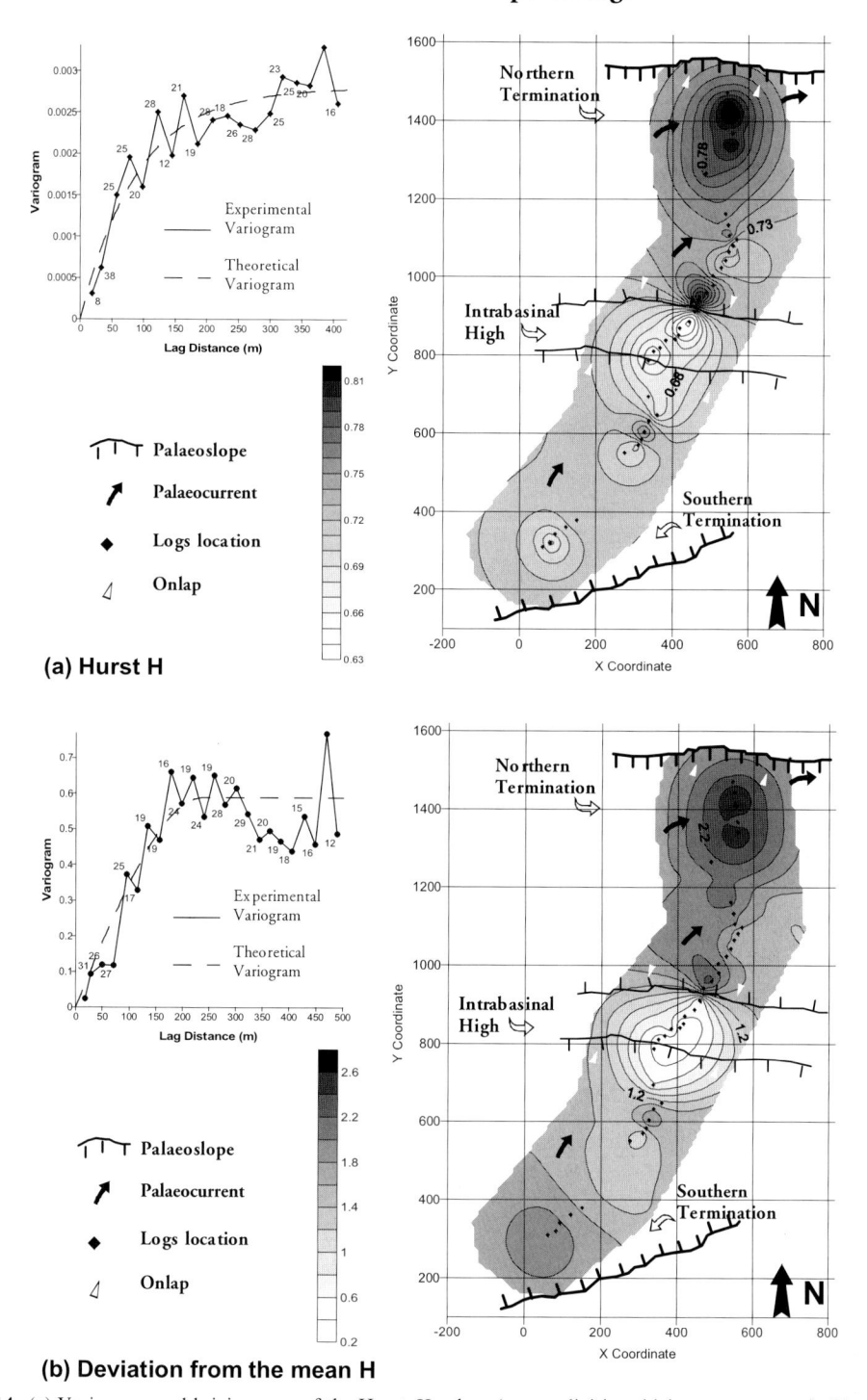

Fig. 14. (**a**) Variogram and kriging map of the Hurst H values (coarse-division thickness percentage); (**b**) variogram and kriging map of the standard deviation from the mean H (coarse-division thickness percentage).

Coarse-division thickness

(a) Hurst H

(b) Deviation from the mean H

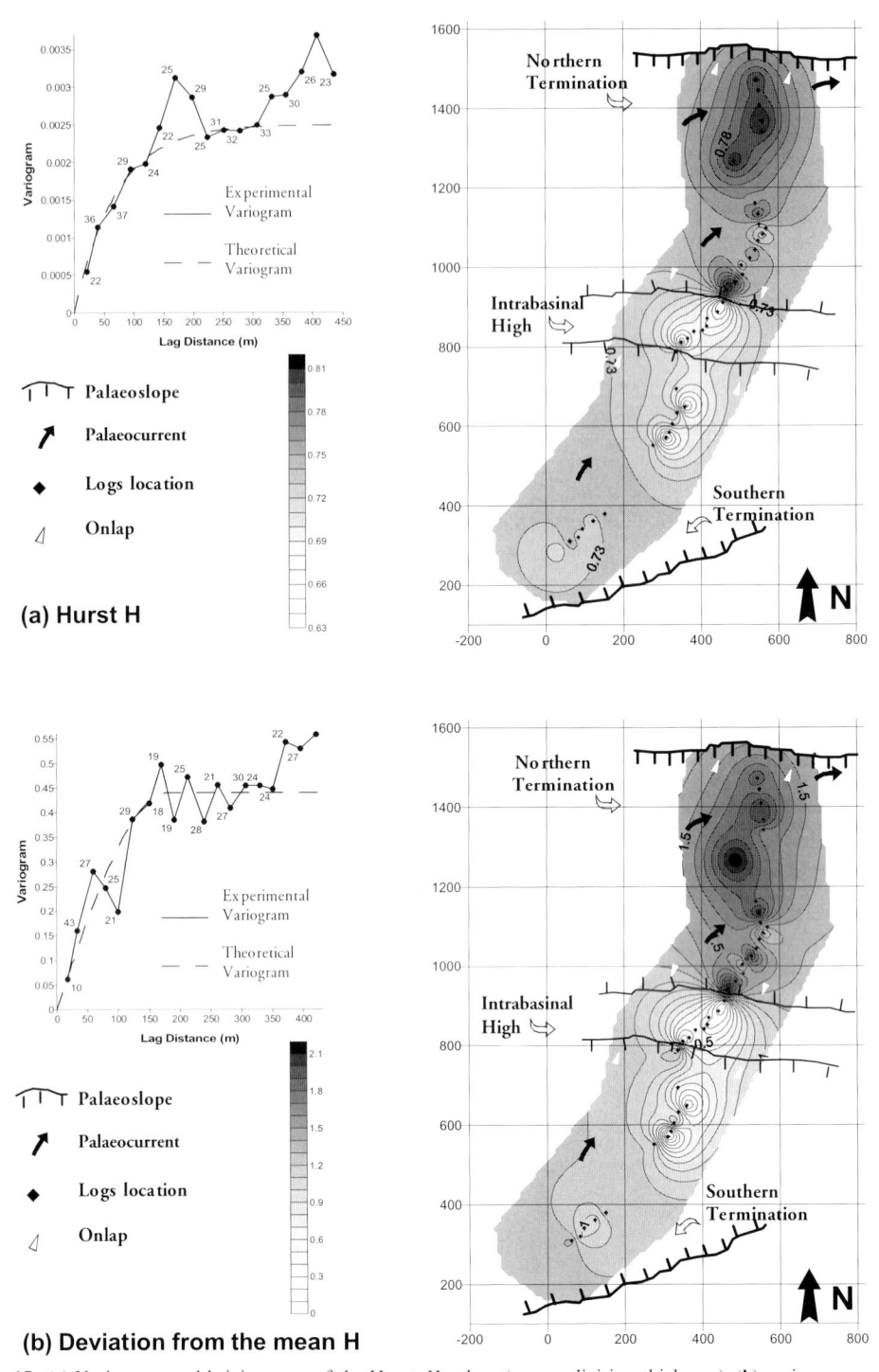

Fig. 15. (a) Variogram and kriging map of the Hurst H values (coarse-division thickness); (b) variogram and kriging map of the deviation from the mean H (coarse-division thickness).

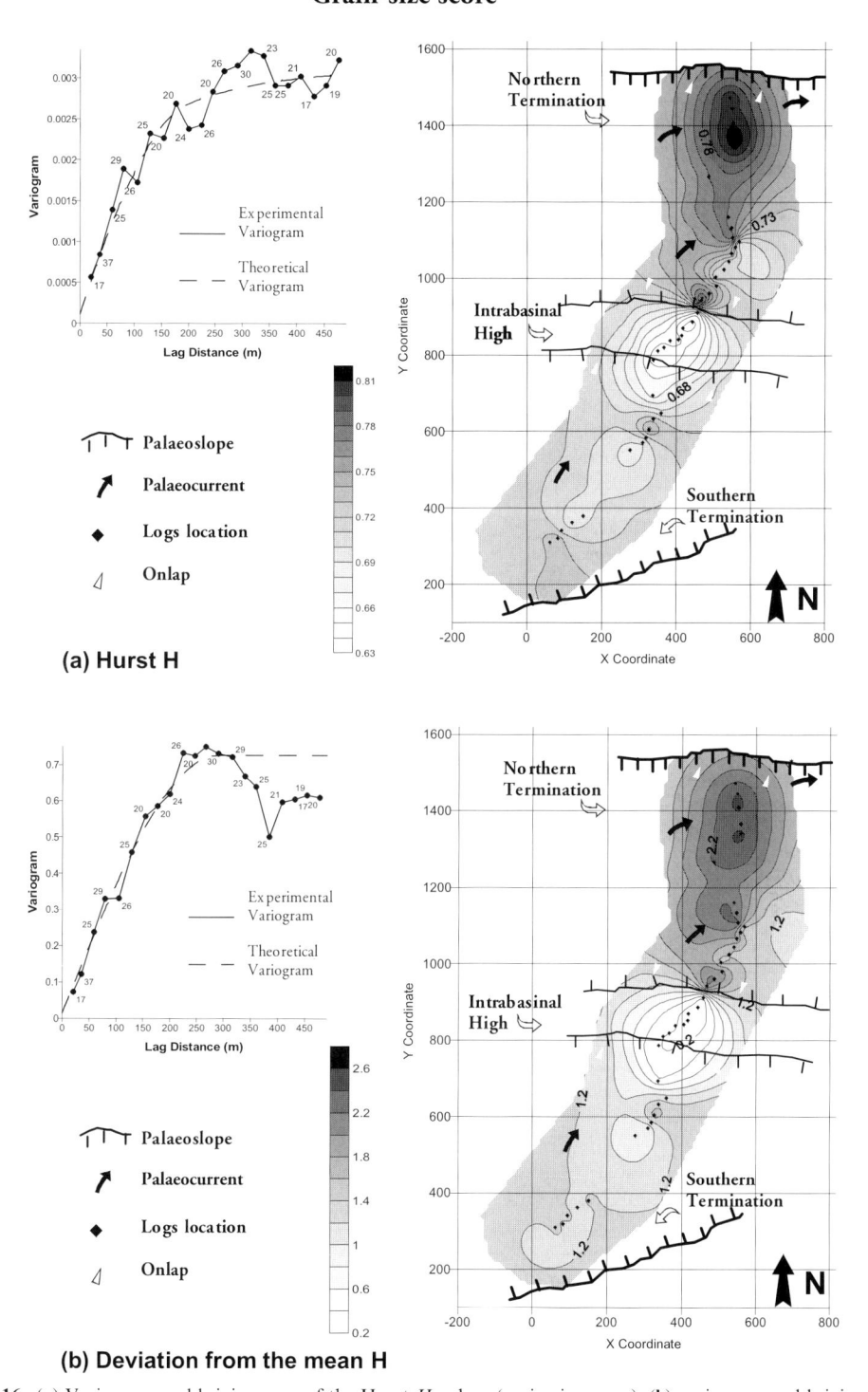

Fig. 16. (a) Variogram and kriging map of the Hurst H values (grain-size score); (b) variogram and kriging map of the standard deviation from the mean H (grain-size score).

(1) Variograms of the Hurst H values have been fitted with an exponential model: 150 m is the practical range for the three considered variables; the sill ranges from 0.0025 to 0.0029.
(2) Variograms of the deviation from the mean of H for random sequences have been fitted with a spherical model: 250 m is the range for the three considered variables; the sill changes from 0.45 (coarse division thickness) to 0.7 (grain-size score).

The absence of a prominent *nugget effect* (where the variogram does not tend to zero as the lag distance tends to zero) highlights the strong continuity of the studied variables at short distance. Cross-variograms of the Hurst H values confirm the existence of a strong positive correlation between the three studied variables.

The parameters obtained from the variogram structural analysis have been used to estimate, with a kriging method (*ordinary kriging*), the spatial distribution of the studied variables (Isaaks & Srivastava 1989; Armstrong 1998). The results are shown in Figures 14–16. As mentioned above, the data can only really assess spatial variation in Hurt statistics along the transect parallel to palaeoflow. For this reason, the estimation of the studied variables is extended only 150 m laterally to the section. This distance corresponds to the range computed for omnidirectional variograms of the Hurst H values, and it has been chosen assuming an isotropic behaviour of the variable. From the resulting contour maps, some observations are of note:

(1) High values of clustering and deviation from the mean H expected for random sequences are present towards the northern onlap termination of the studied unit, where the deposits appear predominantly sandy, and an overall and gradual increase of the amalgamated beds, sand/mud ratio and thickness occurs. This behaviour is related to the presence of a confining northern slope, 50° oblique to the main turbidite flow and at least 60 m high, that caused turbidity currents to lose energy and deposit a large proportion of their sediment load near the base of the slope where the nose of the current climbed the opposing counter slope (Felletti 2002). The above mechanism may explain the increase in the sand/mud ratio and the much thicker deposits found towards the northern onlap termination of the sand-body. Turbiditic beds have an abrupt lateral pinchout onto the palaeoslope from the base to the top; the upper and lower surfaces of these beds are both sharp, with the upper surface being locally bioturbated, with more intense bioturbation towards the north (i.e. up palaeoslope). On the other hand, the mudstone-rich intervals that separate the sandstone-rich packages show a marked decrease in thickness approaching the palaeoslope.

This irregular and non-periodic stacking of thin-bedded, fine-grained muddy bed clusters and thick-bedded, coarse-grained sandy bed clusters with no recognizable internal order produces the high degree of clustering of low and high values. Each cluster shows the development of different vertical stacking patterns and variable lateral facies arrangements, as a consequence of the position relative to the northern basin margins.

(2) Low values of clustering (H) and deviation from the mean H expected for random sequences are observed in correspondence of the submarine intrabasinal high. This relative high is located, in the present day, approximately 1 km south of the Villalvernia-Varzi Line. The outcrops are present from sections 18 to 23 and separate the basin floor into two minor sub-basins, with the maximum relief being around the location of log 23. In relation to the present-day geometry, a difference of 10 m exists between the top of the relative high (logs 18–23) and the northern depositional area of the sand-body (logs 35 and 36). The submarine high shows an asymmetric shape, with the northern slope (8° dip) steeper than that in the south (3° dip). Along the studied transect, the onlap termination of turbiditic beds onto the submarine high shows that the high grew before deposition of the sand-body. The unconformity between the underlying marlstones and the turbidites, with an angular discordance of approximately 8°, suggests that the relative high was deformed and tilted before the beginning of Castagnola turbiditic deposition. The maps (Figs 14, 15) show an abrupt transition from high to low values on the northern side of the palaeo-high, whereas, on the opposite side, the transition is more gradual. The low degree of clustering is explained by the absence, in this area, of an alternate stacking of thin-bedded, fine grained muddy bed clusters and thick-bedded, coarse-grained sandy bed clusters with no recognizable internal order. This behaviour is the result of the complex interactions between the relative low obstacle (intrabasinal high) and the gravity currents that caused abrupt change in bed thickness and grain size (Felletti

2002). In fact, turbidites entering this section of the basin may have been critically balanced between deposition and transport, and these features confer to the vertical succession a random character resulting in a low degree of clustering.

(3) The southern termination of the sand-body tends to have an intermediate/low H and a moderate/low departure from the mean of H for random sequences. This behaviour is related to similar thicknesses of the thin-bedded turbidites that dominate the southern part of the basin. In this area, the sandstone beds pinchout laterally, grading from the typical sand packets into laminated siltstones and mudstones, which themselves onlap the underlying hemipelagic marlstones. As the main palaeocurrent trend is towards the NE, the southern termination of the sand-body appears to be lateral to the centre of the basin where the turbidites were mainly deposited. The thin laminated turbidites are considered as a sediment drape on the southern margin of the basin due to suspension fallout from the finer cloud of the turbidity current tail (Felletti 2002). The typical facies association consists of very thin- to medium-bedded, very fine-grained turbidite beds (0.01–0.2 m thick) showing only Bouma T_c, T_d and T_e divisions. Internal sedimentary structures in the sandstone division include parallel laminae, cross-lamination (including ripple and climbing ripple lamination), and convolute laminations; the mudstone caps are apparently structureless, except for poorly defined pseudonodules in their lower thin part.

(4) The northern depocentre area is characterized by intermediate/high H and a moderate/high departure from the mean of H for random sequences gradually evolving downcurrent (northward) into higher values. This relatively high degree of clustering is principally related to the compensation effect (by depositional smoothing) and erosional scouring that produce abrupt thickness variation of the beds and the diachronous development of several sand packages (Felletti 2002). Each bundle consists of a group of individual thick sandstone beds separated by muddy intervals, and shows different vertical stacking patterns and different horizontal trends in facies arrangement. The beds are bipartite, with a lower, fine- to medium-grained sandstone grading up into a thick silty mudstone top, which is moderately bioturbated. Above the lowest division, a grain-size break may be defined by a millimetre-scale silt–mud layer that represents a minor interruption in the overall decrease in grain size throughout a bed. This silt–mud layer division typically consists of parallel laminae and current ripples. The lowest sandy division may be structureless or consist of parallel laminae and current ripples; the last sandy/silty division before the mudstone cap invariably consists of fine parallel laminae. These beds are analogous to Bouma-style turbidites, except that there is no T_b interval. The vertical repetition (from 2 to 7 times) of thinning-upward couplets of cross-laminated sandstone and massive sandstone intervals is typical of single turbiditic beds within sub-units H and C.

A number of changes occur towards the top of the sand-body: (a) sandstone beds progressively increase in thickness; (b) the grain size increases slightly from fine- to medium-grained sandstones; and (c) there is a decrease in the frequency of sandstone beds and an increase in the thickness of the intervening mudstones.

(5) The southern depocentre area tends to have low Hurst H and a weak departure from the mean value of H expected for random sequences, gradually evolving upcurrent (southward) into intermediate values. This behaviour results from a low degree of clustering of low and high values, and can be explained by the presence of regular and non-regular stacking of fine- to medium-grained sandy bed clusters and muddy bed intervals composed of thin-bedded siltstone, turbidite mudstone, and hemipelagic mudstone. The bedded turbidites (which dominate the southern part of the basin) have relatively similar thicknesses and grain-size. For this reason high Hurst exponents do not occur in this area, where sequences comprise only thin-bedded turbidites.

Conclusion

The present study confirms that the Hurst H test can be used as a valid statistical tool to quantify the degree of clustering of high and low values in vertical turbiditic sections within a small confined and tectonically mobile basin.

Only 2 of 41 logs failed to pass the test at the 10% significance level for all of the studied variables. The high rate of rejection of the null hypothesis suggests a long-term clustering of the studied variables and the presence of the Hurst phenomenon. The Hurst test has revealed the existence of common clustering for the three studied variables. This clustering is associated with lateral and vertical facies variations in response to changes in position relative to basin-floor topography and distance from source within the Castagnola basin.

Intermediate/high values of clustering are related principally to the compensation effect (by depositional smoothing) and erosional scouring that produce abrupt thickness variation of the beds and the diachronous development of several sand packages. *High values* of clustering are present towards the northern onlap termination of the studied unit, where the deposits appear predominantly sandy and an overall and gradual increase of the amalgamated beds, sand/mud ratio and thickness occurs. This behaviour is related to the presence of a confining northern slope that caused turbidity currents to lose energy and deposit a large proportion of their sediment load.

Low Hurst H can be explained by the presence of regular and non-regular stacking of fine- to medium-grained sandy bed cluster and muddy bed intervals composed of thin-bedded siltstone, turbidite mudstone, and hemipelagic mudstone. These bedded turbidites (which dominate the southern part of the basin) have relatively similar thicknesses and grain-size. *Low values* of clustering, observed in correspondence of the submarine intrabasinal high, are explained by the absence of an alternate stacking of thin-bedded, fine-grained muddy bed clusters and thick-bedded, coarse-grained sandy bed clusters with no recognizable internal order. This behaviour is the result of the complex interactions between the relative low obstacle (intrabasinal high) and the gravity currents that caused an abrupt change in bed thickness and grain size.

Combining facies characteristics with the statistical analysis, five sub-environments can recognized:

(1) the onlap termination of the sand-body toward the northern palaeoslope;
(2) the intra-basin high;
(3) the southern termination of the sand-body;
(4) southern depocentre area;
(5) northern depocentre area.

Geostatistical analysis has been performed in order to estimate the spatial distribution of the Hurst statistics within the studied area at unsampled locations, and has been used to highlight gradual and abrupt transitions of the Hurst index within the basin. This analysis confirms that the Hurst *H* value can be considered a *regionalized variable* in the sense of Matheron (1971). Comparison of the structures of the experimental variograms shows that they have similar patterns, and indicates that these curves can be assembled into two groups fitted with spherical and exponential curves.

These results confirm the conclusions illustrated by Chen & Hiscott (1999) and refine the position of group boundaries for some architectural elements of turbiditic confined basin (i.e onlap termination, sheet-like elements and depocentres). However, before utilizing this index in a predictive way, a better quantitative understanding of factors controlling the Hurst *H* test is necessary. The Castagnola basin is relatively small, confined and tectonically active, and the relationship between the Hurst *H* values and sub-environments may not be a simple one. A large number of sections and a wide database must be studied to get a better knowledge of the variability of this index. Bed-by-bed data from cores of subsurface reservoirs can be easily incorporated into this analysis. In addition, the Hurst statistic could be used to constrain stochastic facies simulation (i.e. sequential indicator simulation or truncated Gaussian simulation; Gallli & Beucher 1997), widely used for modelling reservoir heterogeneities.

The author thanks R. Bersezio and R. Gelati for their help during the fieldwork and for the numerous suggestions and constructive comments that improved that manuscript. The author is especially grateful to the University of Milan (Earth Science Department) and CNR for providing an opportunity to renew research into the sedimentology of siliciclastic deposits. D. Conybeare and P. Talling are thanked for their constructive reviews. Preparation of this paper was supported by a CNR (1998) fund to R. Bersezio.

References

ARMSTRONG, M. 1998. *Basic Linear Geostatistics*, Springer-Verlag, Berlin.

BARUFFINI, L., CAVALLI, C. & PAPANI, L. 1994. Detailed stratal correlation and stacking patterns of the Gremiasco and lower Castagnola turbidite systems, Tertiary Piedmont Basin, northwestern Italy. *In:* WEIMER, P., BOUMA, A. H. & PERKINS, B. F. (eds) *15th Annual Research Conference Submarine Fans and Turbidite Systems.* GCSSEPM Foundation 9–21.

BIELLA, G. C., GELATI, R., MEISTRELLO, M., MANCUSO, M., MASSIOTTA, P. & SCARASCIA, S. 1987. The structure of the upper crust in the Alps–Apennines boundary region deduced from refraction seismic data. *Tectonophysics*, **142**, 71–85.

BIELLA, G. C., CLARI, P. ET AL. 1992. Geometrie crostali al nodo Alpi/Appennino: conseguenze sull'evoluzione cinematica dei bacini neogenici. *Riassunti 76a Riunione estiva Società Geologia Italiana, Firenze*, 192–195.

CAVANNA, F., DI GIULIO, A., GALBIATI, B., MOSNA, S., PEROTTI, C. R. & PIERI, M. 1989. Carta geologica dell'estremità orientale del Bacino Terziario Ligure-Piemontese. *Atti Ticinensi di Scienze della Terra*, **32**.

CHEN, C. & HISCOTT, R. N. 1999. Statistical analysis of facies clustering in submarine-fan turbidite successions. *Journal of Sedimentary Research*, **69**(2), 505–517.

DI GIULIO, A. & GALBIATI, B. 1993. Escursione nell'estremità orientale del Bacino Terziario Piemontese. interazione tettonica-eustatismo nella sedimentazione di un bacino tardo-post orogenico. *In: 3° Convegno del gruppo informale di sedimentologia del C.N.R. Dip. di Scienze della Terra, Università di Pavia*, Salice Terme, 1–28.

DI GIULIO, A. & GALBIATI, B. 1998. Turbidite record of the motions along a crustal-scale tectonic line: the motions of the Villarvernia-Varzi line during the Oligocene–Miocene (North-Western Italy). *In: 15th International Sedimentological Congress*, IAS—Universidad de Alicante, Alicante, 293–294.

FEDER, J. 1988. *Fractals*. Plenum, New York.

FELLETTI, F. 1999. *Quantificazione e previsione delle variazioni di facies in una successione torbiditica. Integrazione tra analisi geologica e geostatistica (Fm. di Castagnola, Bacino Terziario Piemontes)*. PhD thesis, Università degli Studi di Milano, Milano.

FELLETTI, F. 2002. Complex bedding geometries and facies associations of the turbiditic fill of a confined basin in a transpressive setting (Castagnola Formation, Tertiary Piedmont Basin—NW Italy). *Sedimentology*, **49**, 645–667.

GALLI, A. & BEUCHER, H. 1997. Stochastic models for reservoir characterization: a user-friendly review. *In: Fifth Latin American and Caribbean Petroleum Engineering Conference and Exhibition*, SPE 38999, 1–11, Rio de Janeiro, Brazil.

GELATI, R. & FALLETTI, P. 1996. The Piedmont Tertiary Basin. *Giornale di Geologia*, **58**(1-2), 11–18.

GOOVAERTS, P. 1997. *Geostatistics for Natural Resources Evaluation*, Oxford University Press, New York.

HURST, A., VERSTRALEN, I., CRONIN, B. & HARTLEY, A. 1999. Sand-rich fairways in deep-water clastic reservoirs; genetic units, capturing uncertainty, and a new approach to reservoir modelling. *American Association of Petroleum Geologists Bulletin*, **83**(7), 1096–1118.

HURST, H. E. 1951. Long term storage capacity of reservoir. *Transactions of the American Society of Civil Engineers*, **116**, 770–808.

ISAAKS, E. H. & SRIVASTAVA, R. M. 1989. *Applied Geostatistics*, Oxford University Press, New York.

JOURNEL, A. G. & HUIJBREGTS, C. J. 1978. *Mining geostatistics*, Academic Press, San Diego, 600 pp.

LAUBSCHER, H. P., BIELLA, G. C., CASSINIS, R., GELATI, R., LOZEJ, A., SCARASCIA, S. & TABACCO, I. 1992. The collisional knot in Liguria. *Geologische Rundschau*, **81**(2), 275–289.

MATHERON, G. 1971. La théorie des variables régionalisées. *Les Cahiers du Centre de Morphologie Mathématique*, **5**, 212 pp.

MUTTI, E. 1992. *Turbidite Sandstones*. Eds Agip, Milan.

MUTTI, E. & SONNINO, M. 1981. Compensation cycles: a diagnostic feature of turbidite sandstone lobes. *In: IAS European Regional Meeting*, Bologna, Italy,

PLOTNICK, R. E. & PRESTEGAARD, K. L. 1995. Time series analysis I. *In:* MIDDLETON, G. V., PLOTNICK R. E. & RUBIN, D. M. (eds) *Nonlinear Dynamics and Fractals: New Numerical Techniques for Sedimentary Data*. Society for Sedimentary Geology (SEPM), Short Courses, **36**, 47–67.

SINCLAIR, H. D. 2000. Delta-fed turbidites infilling topographically complex basin: a new depositional model for the Annot Sandstones, SE France. *Journal of Sedimentary Research*, **70**(3), 504–519.

STOCCHI, S., CAVALLI, C. & BARUFFINI, L. 1992. I depositi torbiditici di Guaso (Pirenei centro meridionali), Gremiasco e Castagnola (settore orientale del BTP): geometria e correlazioni di dettaglio. *Atti Ticinensi di Scienze della Terra*, **35**, 154–177.

TALLING, P. 2001. On the frequency distribution of turbidite thickness. *Sedimentology*, **48**, 1297–1329.

WALLIS, J. R. & MATALAS, N. C. 1970. Small sample properties of H and K estimator of the Hurst coefficient h. *Water Resources Research*, **6**, 1583–1594.

WALLIS, J. R. & MATALAS, N. C. 1971. Correlogram analysis revisited. *Water Resources Research*, **7**(6), 1448–1459.

Reservoir modelling of the Hamitabat Field, Thrace Basin, Turkey: an example of a sand-rich turbidite system

D. M. CONYBEARE[1], S. CANNON[2], O. KARAOĞUZ[3] & E. UYGUR[3]

[1] *Roxar Limited, 14 Albert Street, Aberdeen, AB25 1XQ, UK*
(e-mail: david.conybeare@roxar.com)
[2] *Roxar Limited, Pinnacle House, 17–25 Hartfield Road, London SW19 3SE, UK*
[3] *Turkish Petroleum Corporation, Mustafa Kemal Mahallesi,*
2 Cadde No. 86 06520, Ankara, Turkey

Abstract: The inter-montane, fault bounded Thrace Basin of northwestern Turkey is one of the largest Tertiary sedimentary basins in Turkey. Tertiary sedimentation within the basin comprises interbedded fine- to coarse-grained clastics from a variety of depositional environments, muddy carbonates with local reef developments, and tuff horizons. A thick pile of Quaternary alluvial deposits overlies this marine-dominated succession. Accommodation space, and hence sedimentation, has been influenced by the structural history of the basin, reflecting the active compressional setting adjacent to the North Anatolian Fault System. The Hamitabat Field, located towards the northern basin margin, is the largest developed gas field in the Thrace Basin. The trapping mechanism is by an anticlinal structure bounded to the north by a northwest–southeast striking reverse fault and to the south by a set of antithetic normal faults with a general east–west strike orientation. The main reservoir is the middle to upper Eocene Hamitabat Formation, comprising sandstones deposited by episodic gravity flow events in a shallow marine shelf to slope environment, with sediment sourced from fluvial/deltaic systems to the north and northeast. Correlation, based mainly on upward-fining cyclical successions, indicates lobate depositional geometries and topographic control on sediment dispersal (compensation). Progradational shallow marine sandstones, which infill earlier topography, overlie the turbidite succession. A three-dimensional geocellular reservoir model of the field, based on well and seismic data, has been constructed. The model has been subdivided on the basis of the stratigraphical correlation and populated with facies objects with dimensions based on analogue data. These facies have then been populated with petrophysical parameters. The model has been upscaled and simulated fluid flow through the modelled volume has been compared with the production history. This history matching provides an opportunity to assess and refine the model and to determine a strategy for future field development.

Reservoir modelling has become a key activity in the development of hydrocarbon resources. The process of modelling pools a number of data types, and modelling software provides an array of tools for analysing the spatial distribution of data. This affords insights into the geometry of sedimentary systems, the distribution of their components, and the processes responsible for deposition. Dynamic production data and the simulation of fluid flow through the modelled reservoir offer additional information and means of assessing the veracity of the model by way of comparison with aspects of the production history for the field. In this paper we present an overview of the three-dimensional modelling of a turbidite reservoir from the Thrace Basin of northwestern Turkey, to demonstrate one approach to the modelling of a confined turbidite system. In doing so we shall illustrate a number of tools available to the reservoir modeller that permit analysis and visualization of data in three dimensions. This enhances the understanding of the geometry of such a system and the distribution of its components, allowing the creation of more accurate and realistic models.

Regional overview

The fault bounded Thrace Basin of northwestern Turkey trends broadly east–west and was formed by extension in late middle Eocene to latest Oligocene times (Perinçek 1991; Turgut *et al.* 1991; Fig. 1). Basement-related normal block faulting caused rapid subsidence, during which thousands of metres (up to 9000 m in the centre of the basin) of marine clastics were deposited. Many basement faults were reactivated and underwent strike-slip motion in late Miocene times. Miocene tectonism caused intense deformation and erosion at the south margin of the basin (Top Danismen Unconformity). The Tertiary sedimentary succession, overlying

From: LOMAS, S. A. & JOSEPH, P. (eds) 2004. *Confined Turbidite Systems.* Geological Society, London, Special Publications, **222**, 307–320. 0305-8719/04/$15.00 © The Geological Society of London 2004.

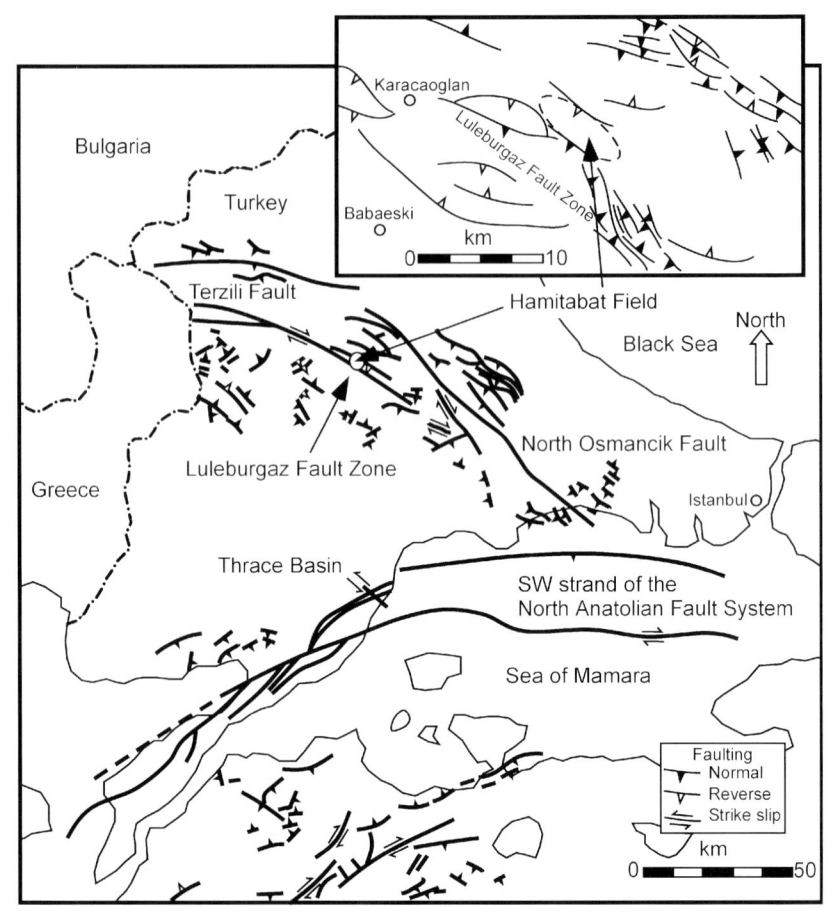

Fig. 1. Map of northwest Turkey illustrating the location of the Hamitabat Field in the context of major structural features (modified from Turgut *et al.* 1991).

Palaeozoic to Mesozoic metamorphic basement, comprises interbedded fine- to coarse-grained clastics from a variety of depositional environments, muddy carbonates with local reef developments and tuff horizons (Table 1). A thick pile of Quaternary alluvial deposits overlies the marine succession. Accommodation space and hence sedimentation has been influenced by the structural history of the basin, reflecting the active compressional setting adjacent to the North Anatolian Fault System.

Hamitabat Field

Oil and gas exploration and development has taken place in the Thrace Basin since the 1950s, and the Hamitabat Field is the largest developed gas field in the area, measuring approximately 7 km by 4 km. The field is operated by the Turkish Petroleum Corporation (TPAO) and is situated in the northern part of the basin, close to the basin margin (Fig. 1). The main reservoir is the middle to upper Eocene Hamitabat Formation, into which some 35 wells have been drilled on the Hamitabat Field structure, with well spacings of between approximately 400 m and 800 m. The total original gas in place is estimated at 5.2 billion m^3. The trapping mechanism for the Hamitabat Field is a folded structure parallel to the east–west Thrace strike-slip fault zone (Luleburgaz Fault zone). The fold is bounded to the south by normal faults striking WNW–ESE and resulting from extension in late middle Eocene to middle Oligocene times (major faults exhibit throws of about 150 m). The margin of the fold to the north comprises a reverse fault, the variation in

Table 1. *Stratigraphical subdivision of the Thrace Basin*

Age	Formation	Lithology
Oligocene–lower Miocene	Danismen	Interbedded shales, calcareous mudrocks, siltstones, unconsolidated silty sandstones, and coal intercalations
	Osmancik	Very fine to medium-grained sandstones interbedded with greenish-grey shales and siltstones
	Mezardere	Interbedded greenish-grey to green waxy shales, siltstones, calcareous mudrocks and fine-grained sandstones
Middle-upper Eocene	Ceylan	Interbedded dark grey to greenish shales, siltstones, calcareous mudrocks, argillaceous micritic limestones, very fine-grained sandstones, tuffs and other volcanics
	Sogucak	Shallow marine shelf limestones and deeper marine calcareous mudrocks
	Hamitabat	Interbedded grey sandstones, siltstones and dark grey shales
Palaeozoic–Mesozoic		Basement: various grades of metamorphism, metasediments, ophiolitic and serpentinite melange, igneous complexes and volcanics

the sense of fault displacement reflecting the structural complexity of this wrench fault system (Perinçek 1991).

Scope of work and dataset

The modelling of the Hamitabat Field was undertaken using seismic, well log and core data (392 m from three wells). The aim of the modelling process was to construct a structural model of the field, and to populate the modelled volume with facies and their characteristic petrophysical parameters prior to simulation of fluid flow. Flow simulation permitted projections to be made into the future, based on a number of different production scenarios.

Seismic interpretation

A seismic interpretation study was undertaken utilizing a number of two-dimensional seismic lines across the Hamitabat Field. The seismic data were used to generate a structure map of the top of the Hamitabat Formation and to identify a number of faults. Four horizons relating to extensive stratigraphical boundaries were identified (Top Danismen, Top Mezardere, Top Ceylan and Top Hamitabat). Whenever possible, events were autotracked within the seismic processing softwater package SeisWorks, but in certain cases, because of the poor data quality, a point-to-point interpretation approach had to be adopted. Synthetic seismograms were used to match the seismically resolved Top Hamitabat Formation event with the corresponding lithological boundary on the well logs, and to identify the time shift necessary to adjust the seismic to the synthetic data.

The final interpretation was imported into SeisVision to create a velocity model prior to depth conversion. The depth conversion was based on creating an interval velocity surface for each interpreted horizon. Checkshot data velocities and the seismic two-way time (TWT) recorded at 34 well locations were used to derive the control points necessary to produce a final velocity grid. Interval velocity maps and TWT maps for each surface were gridded and combined in SeisVision to produce a final depth map.

Faults were identified from the analysis of seismic data and were defined for the top Hamitabat Formation using fault polygons (defining the intersection of the horizon with the fault plane). These faults were incorporated into the three-dimensional modelling process to permit accurate volumetric analysis and also to assess transmissibility across fault planes during reservoir simulation.

The calibration of seismic attributes to geological parameters (e.g. facies and porosity) can prove useful in distributing these parameters within a model. The quality of the data in this case, however, was insufficient to produce meaningful seismic attributes for use in subsequent geological modelling.

Hamitabat Formation sedimentology

Initial deposition in the middle Eocene occurred during rapid subsidence of the Thrace Basin, during which thick sequences of turbidites, including the Hamitabat Formation, were deposited (Turgut *et al.* 1991). In the northern part of the basin, regional seismic data indicate deposition of an Eocene clastic slope wedge extending south from a northern shelf area. The shelf/slope break of this northern margin was controlled largely by the presence of the listric, normal North Osmancik fault to the north of the Hamitabat Field location. The geometry of sediments along this shelf margin suggests that sediments were line-sourced from the north and northeast rather than originating from a specific point on the shelf margin. The basin reached its largest extent in the late Eocene, which also represents a time of maximum transgression and a decrease in subsidence. Clastic sedimentation was reduced, resulting in the widespread deposition of carbonates forming the Sogucak Formation overlying the Hamitabat Formation sandstones.

Cores from three wells (H-25, H-27 and H-30) oriented northwest–southeast and totalling 392 m in length were described. The succession within the Hamitabat Formation is dominated by arkosic turbidite sandstones that are predominantly grey coloured and range from fine to coarse grained. The sandstones occasionally contain scattered lithoclasts (0.5–5 cm) of bioclastic limestone, lime mudstone (derived from shelf carbonates to the north) and black mudclasts, commonly at the base of sandstone units. The turbidite sandstones are commonly massive, with occasional dish and pillar and vertical pillar water escape structures, although normal grading is locally recognized (Facies B1; Mutti & Ricci-Lucchi 1975). Cross-bedding is rarely observed (Facies B2), and is generally associated with sharp-based units, occasionally with a coarse basal lag. Ripple and planar laminae are common in very fine to fine-grained micaceous sandstones in both sandy and muddy heterolithic intervals (Facies D1 to Facies D3). In the upper part of the Hamitabat Formation swaley/hummocky cross-strata are locally apparent. Mudrocks are rare, comprising dark grey laminated muds and silty muds.

The general lack of erosive bases with obvious relief, only occasional conglomeratic deposits and the degree of correlation observed between wells, coupled with depositional geometries, both from regional seismic data (Turgut *et al.* 1991) and isochore data (discussed later), suggest a lack of discrete channelization and more sheet-like deposition within the Hamitabat Field sandstones. However, as will be shown later, there is evidence for topographical confinement of sediments.

A number of early post-depositional features are observed within the cored intervals. Bioturbation, dominantly in the sandstone facies, is apparent throughout the succession, commonly occurring near the top of sandstones units. Trace fossils observed include *Planolites*, *Ophiomorpha* and *Thalassinoides*. Although indicative of onshore shallow marine conditions in the Mesozoic, there is evidence that the benthic organisms that created these traces had migrated to deeper-water offshore environments by the Late Cretaceous and Cenozoic in response to greater deepwater oxygenation (Jacobs & Lindberg 1998). Similar bioturbation has also been identified in other turbidite systems interpreted as being deposited in relatively deep water (e.g. Bruhn & Walker 1995). However, although it is difficult to be concise about the depth of sedimentation, there are indications that deposition occurred at shallower water depths in the upper part of the Hamitabat Formation, prior to the cessation of siliciclastic sedimentation and the deposition of shallow-water carbonates (Sogucak Formation). Hummocky cross-stratification indicates storm current reworking below fair-weather wavebase, and the results of this current activity is coupled with an increase in bioturbation. Soft sediment deformation, in the form of discrete slumped units displaying the local overturning of sedimentary units, is observed throughout the cored successions. These structures indicate slope instability and movement prior to significant lithification and may be in response either to depositional instabilities or to synsedimentary tectonism.

Reservoir properties

Diagenesis within the Hamitabat Formation has had a significant impact on the reservoir characteristics of the sandstones, particularly in greatly reducing permeability. Observations from petrographical analyses of a number of thin sections and scanning electron microscope secondary electron imaging of stub-mounted chip samples indicate the development of a number of authigenic minerals. Both detrital and authigenic clay minerals are common, the latter including chlorite and, less abundantly, illite and kaolinite. Poikilotopic carbonate cementation is common, and quartz cementation on detrital quartz grains is commonly observed. High resistivity values in both shallow and deep induction logs within cemented intervals suggest that carbonate cementation is pervasive rather than concretionary.

Reservoir properties were defined from wireline and core analysis data. Before petrophysical interpretation, all logs were matched to the density and neutron logs and standard environmental corrections were carried out. Wireline logs were also normalized to account for variation between logging runs and tool types. As a result of the arkosic, immature nature of the Hamitabat Formation sandstones, the gamma ray log was unsuitable for deriving the argillaceous matrix content. The 'shale volume' (V_{sh}) was therefore estimated using a density/neutron cross-plot. Porosities, both total porosity (independent of V_{sh}) and effective porosity (incorporating V_{sh}), were computed using the density trace. Both total and effective porosity were used in the water saturation calculations. Water saturation was determined using the Archie equation, and curves of saturation versus height were generated from an evaluation of the water saturation data from all wells. Two reservoir compartments were delineated from the saturation data in the crestal and eastern areas, separated by faults and displaying hydrocarbon/water contacts at differing heights (3000 m and 2900 m). For each of these two reservoir compartments, saturation curves were generated for porosities of greater and less than 6%.

Modelling facies

When defining a facies scheme for the purpose of reservoir modelling, a number of factors need to be taken into account. It is not practical to model facies at the scale of core observations, as, because of the constraints of model size, the layer thickness in the model is commonly greater than sedimentary bed thickness (in the case of the Hamitabat field model, a 1 m layer thickness was used). It is also problematic to model a large number of facies types, as this can considerably increase processing time. Additionally, to fully incorporate wireline log derived parameters, such as porosity, permeability and water saturation, into the modelling, a facies scheme needs to be applicable to the uncored, as well as the cored interval. The facies scheme should also be based on distinct petrophysical characteristics as, ultimately, the distribution of petrophysical parameters in three-dimensions is the requirement for subsequent flow simulation. As a result of these constraints, and considering the degree of uncertainty associated with the inter-well volume, a hybrid facies scheme, which combines depositional facies and incorporates the influence of diagenesis, has been developed.

On the basis of core and wireline observations, five facies were defined for the turbidite succession of the Hamitabat Formation (Table 2). As discussed above, a feature of the Hamitabat Formation is that the GR response of the sandstones is high (a consequence of immature, arkosic and clay/mica-rich sediments), which prohibits a differentiation between sandstones and mudrocks using GR alone. The generation of net sand and net pay curves from neutron/density logs, combined with a comparison of core-derived lithofacies, has shown that a cut-off of 4% porosity delineates net and non-net facies, and 40% V_{sh} distinguishes sandy from muddy lithofacies.

Initially, the sandstone facies were divided into two sub-facies (greater and less than 4% porosity) and modelled accordingly. However, comparison of flow simulation results with production data implied that, on the basis of this subdivision, the volume of connected (i.e. permeable) sandstone had been overestimated. Examination of resistivity, spontaneous potential (SP) and repeat formation test (RFT) data indicates that much of the sandstone with >4% porosity exhibits poor permeability characteristics. The wireline log characteristics that define these low-permeability sandstones are a lack of separation of the shallow and deep resistivity curves (indicating an insignificant penetration of drilling fluid) and the deflection of the SP log within a sandstone interval. This low permeability has been attributed to the detrimental impact of diagenetic clays (particularly chlorite) on permeability. The sandstone facies have therefore been divided into three subgroups based on the above criteria (Table 2).

Porosity data indicate that a cut-off of 25% V_{sh} differentiates clean sandstones and heterolithic sandstones (interbedded with siltstones and mudrocks). Hemi-pelagic, suspension fallout fines generally exhibit V_{sh} values of >40%.

The match between core-derived and log-predicted facies is not perfect, particularly as the facies observed form a continuum and are variably influenced by diagenesis. However, the facies scheme provides a means of differentiating facies and facies associations on the basis of petrophysical character, such that a facies log could be generated for the entire logged turbidite succession for 25 wells within the Hamitabat Formation.

Well correlation

In order to construct a stratigraphical framework within which to model facies and petrophysical characteristics, a correlation of the Hamitabat Formation was undertaken across the Hamitabat Field. On a broad scale, the boundary

Table 2. *Facies classification based on a combination of core and wireline data applicable to the entire wireline logged interval in each well*

Facies	V_{sh}	Porosity	Additional criteria	Lithofacies description
Facies 1 Clean sandstone	<25%	>4%	Positive neutron/density separation and evidence of a flushed zone around the borehole (separation in deep and shallow resistivity)	Fine- to coarse-grained turbidite sandstones
Facies 2 Low-permeability sandstone	<25%	>4%	Positive neutron/density separation, lack of flushed zone; low permeability indicated by RFT data; deflections in the SP curve	Fine- to coarse-grained turbidite sandstones exhibiting low permeability characteristics despite porosity values greater than 4% as a result of authigenic mineral growth
Facies 3 Cemented sandstone	<25%	<4%	Negative neutron/density separation	Fine- to coarse-grained turbidite sandstones displaying poor porosity and permeability characteristics resulting dominantly from carbonate cementation
Facies 4 Heterolithics	25–40%	>4%	Negative neutron/density separation	Fine-grained sandy heterolithic facies. Comprising dominantly Facies C and D sandstones (Mutti & Ricci Lucchi 1975)
Facies 5 Mudstone	>40%	<4%		Hemi-pelagic mudrock (and siltstone) facies

between the siliciclastics of the Hamitabat Formation and the overlying Sogucak Formation argillaceous limestones is easily recognized on wireline logs. The Hamitabat Formation can be divided into two large-scale units, termed Hamitabat A and Hamitabat B, with the reservoir interval contained in stratigraphically higher Hamitabat A. None of the wells penetrated down to the metamorphic basement underlying the Hamitabat B unit, so it is not possible to obtain a clear picture of patterns of sedimentation in this horizon, although the influence of basin topography during initial deposition of the Hamitabat formation has been identified from seismic data (Perinçek 1991). The pick for the top Hamitabat B has been identified in 13 wells, giving an average thickness of Hamitabat A of c. 200 m (ranging from 149 m to 231 m). Wireline data indicate an upward-fining succession at the top of the Hamitabat B unit, and the base of the Hamitabat A horizon marks the abrupt transition to coarser sandstone deposition, possibly representing a sequence boundary. The succession from the top of Hamitabat B to the top of the Hamitabat Formation reflects a transition to a shallower marine shelf (prodelta?) environment, with sediments continuing to be deposited below fair weather wavebase.

A number of distinctive successions, which commonly show an upward-fining trend, are observed in facies successions and V_{sh} data within the Hamitabat Formation. Although varying in character between wells both in a proximal to distal (northeast–southwest) orientation and also laterally along depositional strike, cyclical patterns can be traced across much of the field. On the basis of these stacking patterns, the Hamitabat A was subdivided using the recognition of potential parasequence boundaries and flooding surfaces. The recognition of cyclical successions was aided by the use of Cyclolog[R]. Cyclolog is a software package that provides a number of mathematical tools to identify and characterize cyclicity in curve data (Fig. 2). Prediction error filter analysis (PEFA) has been used as the primary tool in assisting correlations made from the wireline and facies dataset. PEFA calculates a curve of the difference between the actual value of a wireline curve and the value at the same point predicted by maximum entropy spectral analysis (MESA). MESA analyses the frequency content of the curve data within a specified window (in this case 10 m) stepped up through the data, and produces a model of the analysed data in the form of superimposed waveforms. This model is then used to predict the next

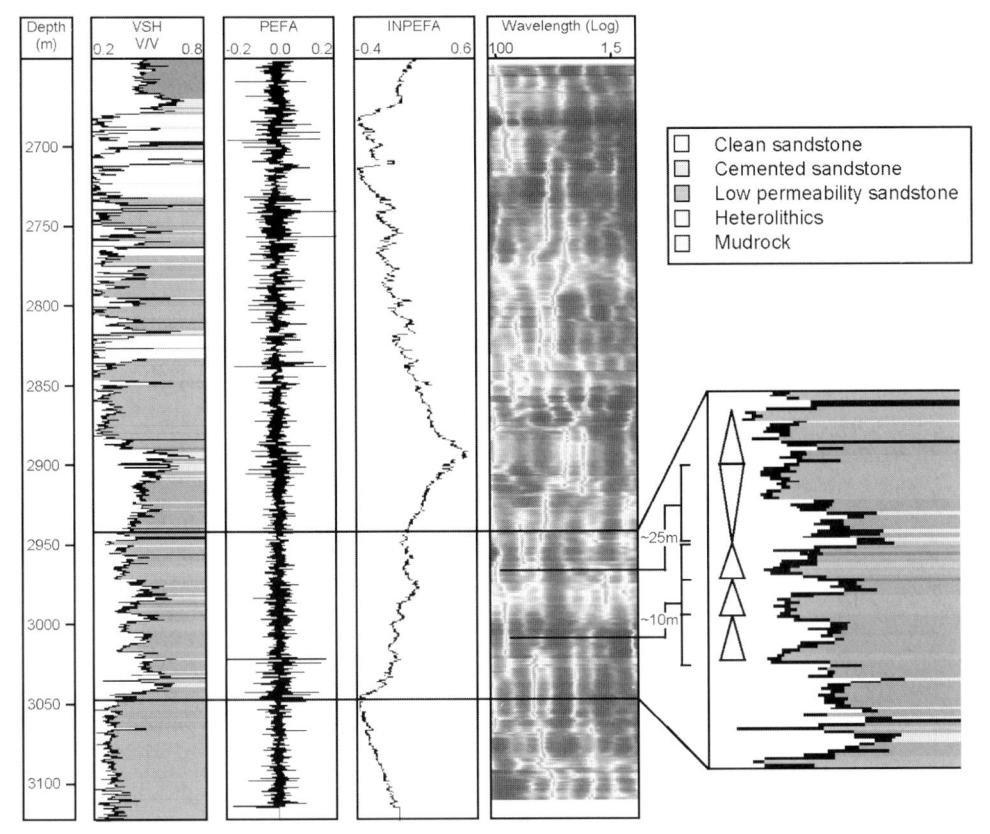

Fig. 2. Illustration of the identification of cyclicity from V_{sh} data using prediction error filter analysis and the analysis of cyclical wavelengths. The inset displays varying wavelengths of cyclicity within a section (lighter colours on the wavelength log indicate the presence of a particular wavelength). The large-scale cycles possibly relate to parasequences whereas the lower-wavelength cyclicity relates to bed-scale alternations.

value above the window, which is compared with the actual value. The prediction error is zero if the prediction is perfect; otherwise it is a non-zero value whose magnitude depends on the degree of error. Large positive or negative PEFA values, therefore, indicate discontinuity and possible stratigraphical breaks. By integrating the PEFA curve and plotting the area under the curve (INPEFA), consistent trends within the data can be evaluated. Additionally, a frequency spectrum within a data window can be plotted, defining the various cyclical wavelengths present in the curve data (Fig. 2).

The correlation exercise, incorporating the analysis of facies and wireline data, in conjunction with Cyclolog, defined seven laterally extensive correlatable surfaces within the Hamitabat A succession (A1 to A7 from the base of the interval). Of these, three (A1, A3 and A5) mark significant field-wide breaks in deposition and were used as stratigraphical boundaries within the three-dimensional geological model. However, all seven boundaries were utilized in the subsequent reservoir simulation. An example of the reservoir correlation rationale and the correlation surfaces in the Hamitabat A interval are displayed in Figures 3 and 4.

Stacking patterns and depositional geometry

The degree of correlation of stacking patterns across the field, coupled with core observation, indicates generally sheet-like depositional geometries for sand-bodies as opposed to highly channelized morphologies. To further elucidate the large-scale depositional geometry of the turbidite succession, isochore maps were constructed from the correlation (Fig. 5). Isochore maps for the four major intervals within Hamitabat A indicate lobe-like, confined morphologies

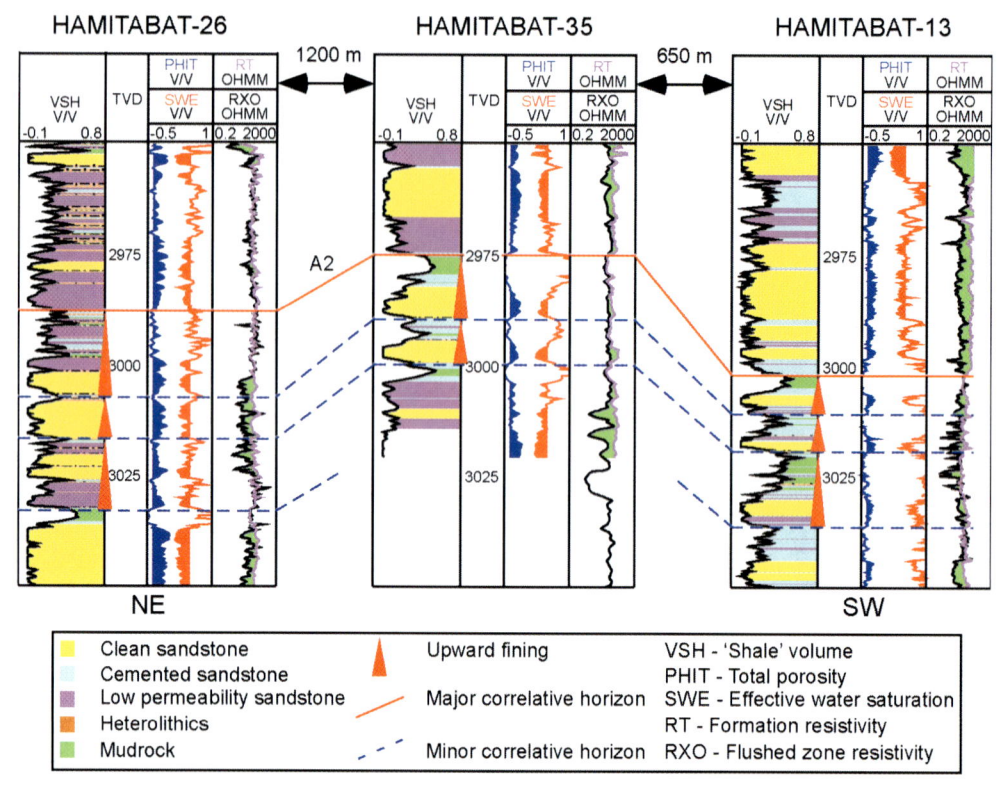

Fig. 3. Illustration of the correlation rationale based on stacking patterns. Dotted lines correlate a proximal to distal (northeast–southwest) series of subordinate upward-fining successions between major correlative horizons. These intervals display decreasing thickness and an increase in fine-grained facies content in the direction of sediment transport.

with a general northeast–southwest orientation—concurring with a fluvio/deltaic source area to the north and northeast. It is apparent that the location of lobes varies through the succession, with the axes of thickest deposition in one horizon generally coinciding with areas of relatively thinner, attenuated deposition in the underlying horizon (Fig. 5). This is interpreted as the response of large-scale sandstone deposition to earlier depositional topography (compensation).

The distribution of facies within these depositional lobes was evaluated by comparing the probability density curves for each facies (derived from the blocked well data) with the stratigraphic thickness of the lobe (Fig. 6). This plot shows that there is a general increase in the proportion of cleaner, coarser sandstones with increasing isochore thickness. Conversely, there is a general decrease in the proportion of mudrock and poorer-quality sandstone with greater lobe thickness. These observations are interpreted as resulting from higher-energy sedimentation in proximal and axial areas, with finer sediments being deposited in relatively distal and marginal locations. The proportion of cemented sandstone is relatively constant with respect to thickness, a result of the post-depositional formation of this facies.

Geological grid

Three-dimensional modelling was carried out using Reservoir Modelling System (RMS) software. The model encompasses the anticlinal feature forming the trapping mechanism for the field, ensuring that the hydrocarbon leg is fully incorporated on the basis of a hydrocarbon/water contact at 3000 m.

To construct the structural framework for the model, the top Hamitabat Formation surface and fault polygons derived from seismic data, and well points data from the well correlation,

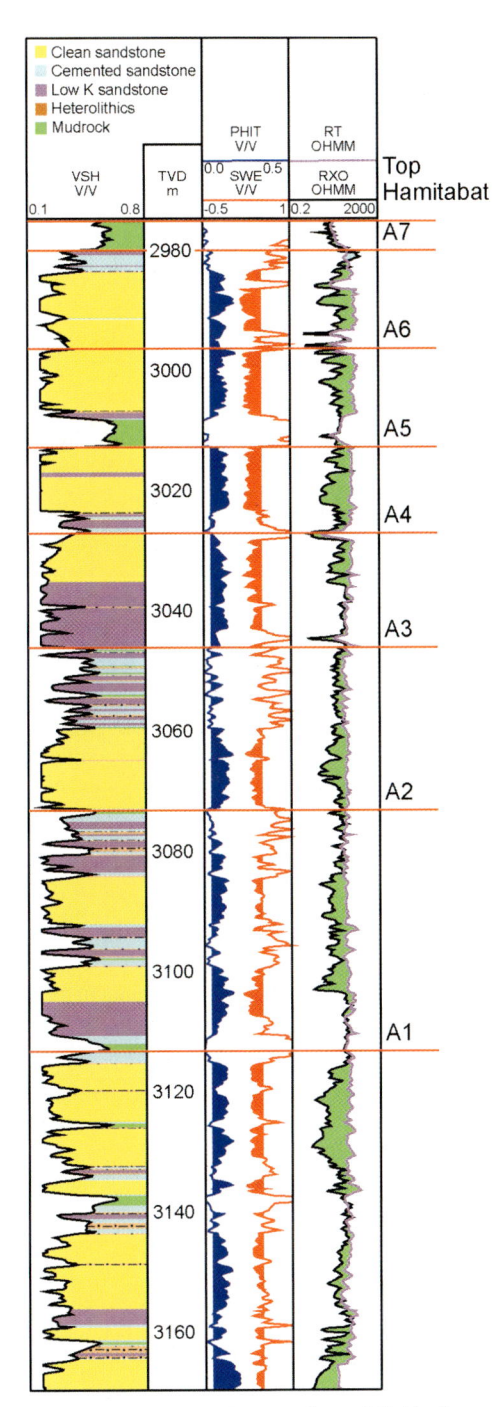

Fig. 4. Correlation surfaces used to subdivide the Hamitabat A interval.

were imported into the model database. Isochore maps, created from the well data, were used to construct the correlation surfaces. On the basis of these data, a multi-zone (subgrid) cellular geological grid, comprising approximately 6 million cells (50 m × 50 m × 1 m), was built using corner-point geometry (i.e. the coordinates of each corner of the grid cells are defined; Fig. 7). The grid was oriented northwest–southeast in alignment with the major fault strike orientation. An 'onlap' model was adopted to allow for the stacking patterns anticipated in the turbidite depositional realm, with the potential for the erosion and infilling of topography. A coarser vertical cell resolution was used for Hamitabat B, as this horizon is within the water leg and therefore of lesser importance for modelling and simulation. A summary of the main finescale grid parameters is given in Table 3.

The interpreted petrophysical data and facies coding were imported for each well and the data 'blocked' (i.e. up-scaled) from logging resolution to the resolution of the grid cells. The discrete facies log was blocked using the modal facies code for each cell penetrated by the well, and a check was made to ensure similar facies proportions for both the original (raw) and upscaled data. The continuous parameters (i.e. porosity and water saturation), were subsequently arithmetically averaged using the chosen cell facies code as a bias, i.e. only log values associated with the facies type assigned to a grid cell are used in the calculation.

Turbidite facies model

Based on an analysis of well and seismic attribute data, no well-defined relationship between facies and petrophysical parameters was established, and therefore the seismic amplitude data were not used in the simulation process.

The facies model was constructed using a geostatistical technique allowing the various facies to be modelled as objects with geometries and distributions that match the well data, the proportion of facies in each stratigraphical horizon and any apparent distribution trends. Sandbodies in the turbidite succession (incorporating clean sandstone, low-permeability sandstone and cemented sandstone) were modelled as elliptical objects. Elliptical objects were chosen as a proxy for the distribution of sand units deposited by non-channelized turbidity current flow, from a small area of initiation to an area of downflow expansion and final waning/deceleration. The range of dimensions of the modelled turbidite sand-bodies was derived from an analysis of sandstone unit thickness in the Hamitabat Formation and thickness/width/length data from a number of analogue studies in confined settings (Fig. 8). Whilst not completely matching the spread of sand-body thickness data observed in the Hamitabat Formation, these examples represent turbidite successions deposited in confined settings, both proximal and distal to source, displaying similar facies and cyclicity to those in the Hamitabat Formation. These examples therefore give an indication of the thickness/width and thickness/length ratios that could be anticipated.

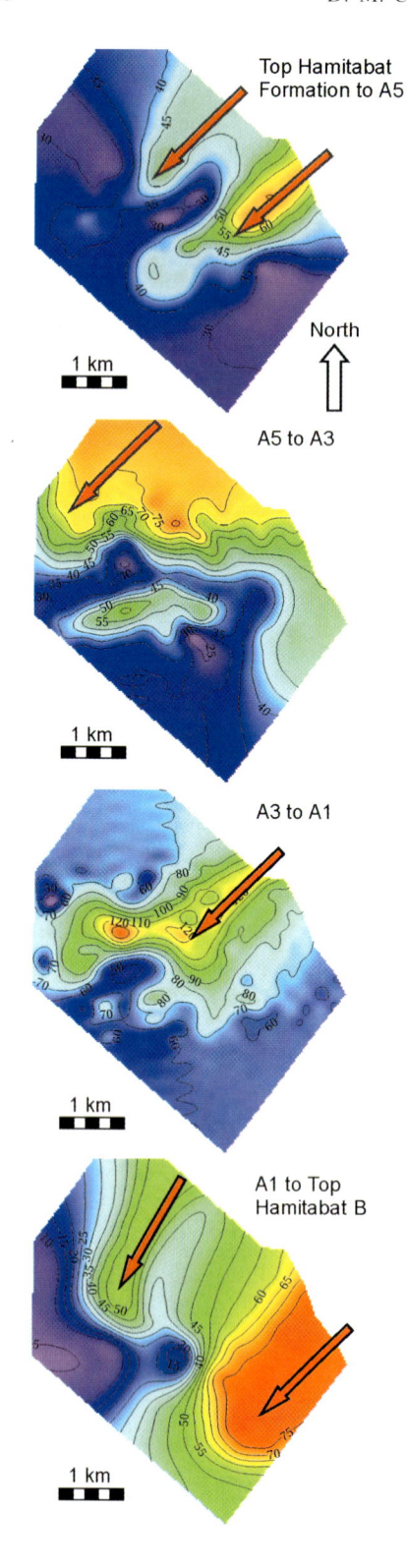

As described above, isochore maps for each of the correlatable zones indicate lobate depositional geometries trending northeast–southwest, in agreement with a provenance to the north and northeast. Sandstone objects within the turbidite succession have therefore been modelled using a dominant northeast–southwest orientation. The distribution of facies identified from probability density plots for each stratigraphical horizon was used within the facies modelling such that the statistical distribution of facies modelled compared with that observed. Also incorporated into the modelling were the proportions of each facies identified within individual stratigraphical horizons. An example of one layer from the facies model relative to the corresponding isochore map is shown in Figure 9.

Petrophysical model

Having established a modelled facies framework, the petrophysical parameters of porosity, permeability and water saturation were modelled for each facies. In order to populate the model with these petrophysical parameters, the data were analysed by evaluating their spatial distribution and facies relationship. Porosity was modelled using descriptive statistics generated for each facies, including semi-variograms that define the correlation distances of a parameter

Fig. 5. Evaluation of stacking patterns from isochore maps for each subgrid within the Hamitabat A interval. Red arrows illustrate the main axes of deposition within each horizon. The orientation of depositional axes is northeast–southwest, in agreement with the conceptual model of a sandstone source area to the north and northeast. The main areas of deposition switches between layers, with the loci of deposition in one layer generally coinciding with a zone of attenuated deposition in the underlying layer. The interpretation for this is the influence of depositional topography on subsequent sandstone deposition.

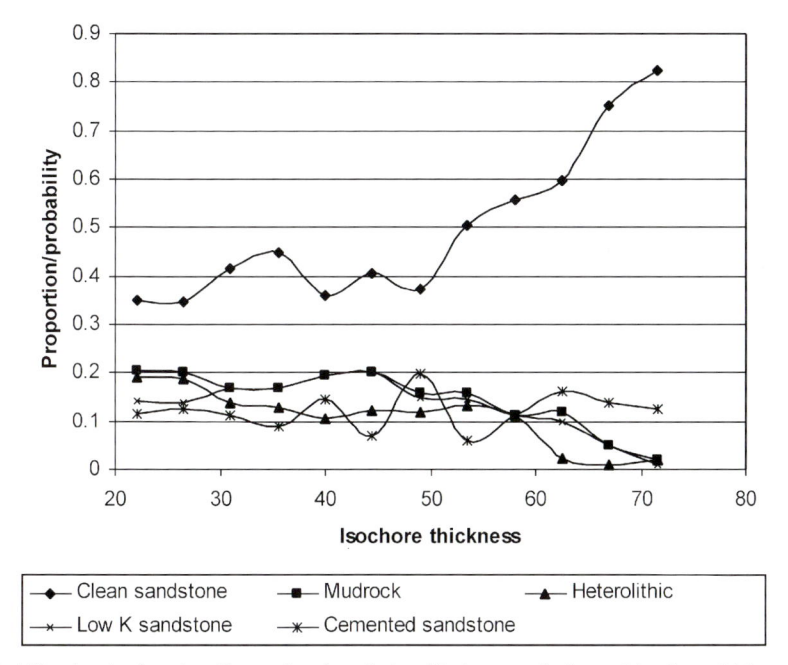

Fig. 6. Probability density function illustrating the relationship between facies and isochore thickness for the interval between the top Hamitabat B and Hamitabat A5 horizon.

away from well locations in three orthogonal orientations. Although the semi-variograms were quite well constrained in the vertical orientation, the definition of the range in the horizontal orientations was more ambiguous and therefore reference was also made to sand-body size from analogue data and the correlative distance observed across the wells. Permeability

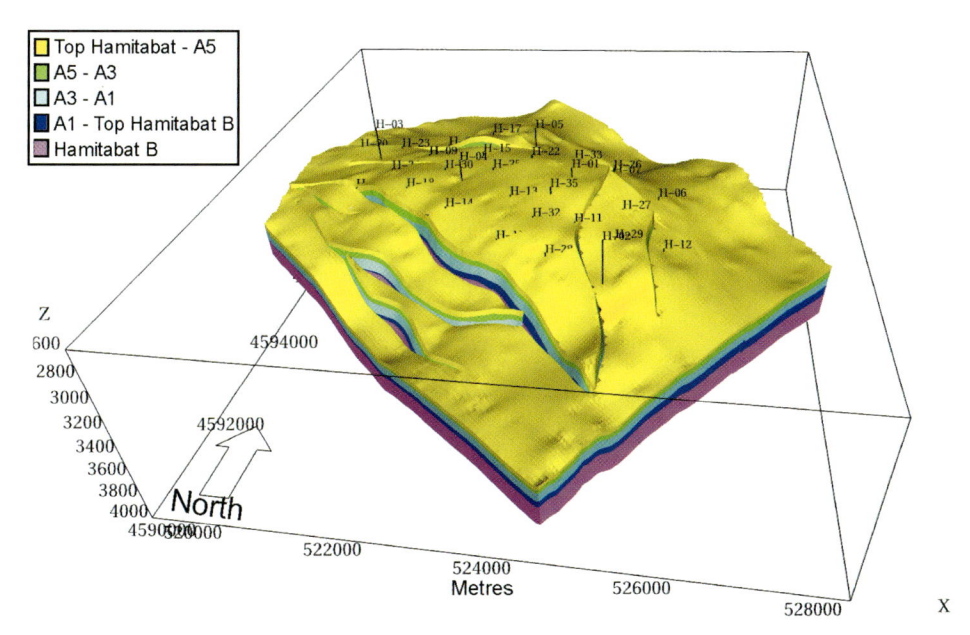

Fig. 7. Structural and stratigraphical model of the Hamitabat Field illustrating the five subgrids defined by major correlation surfaces and well locations on the crest of the structure.

Table 3. *Geological grid parameters*

Subgrid	Layer thickness (m)	Grid dimensions		No. of cells
1	1	Min X	524 800 m	X 130
2	1	Min Y	4 590 000 m	Y 174
3	1	X length	6 500 m	
4	1	Y length	8 700 m	
5	2	Rotation	50°	

was calculated from the modelled porosity using the expression

$$K_{air} = 0.03\, e^{0.35\Phi} \quad (1)$$

where K_{air} = horizontal air permeability and Φ = porosity.

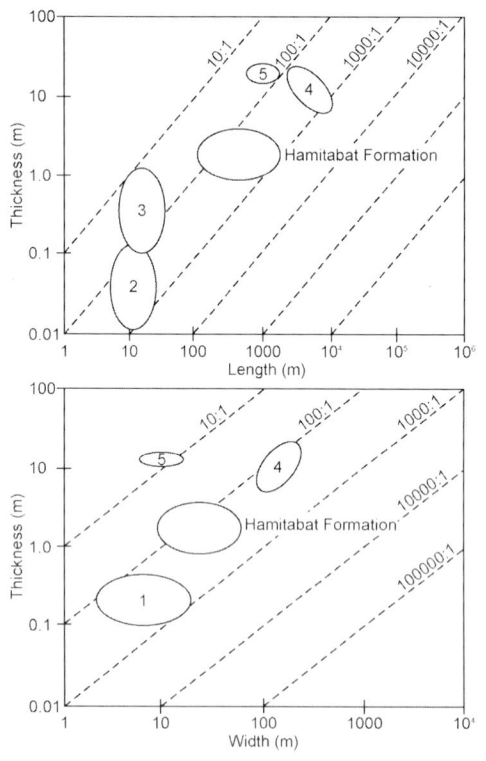

1. Castelnuovo Member – Smith 1995
2. Cloridorme Formation – Enos 1969a
3. Broto – Hecho Group – Schuppers 1995
4. Campos Basin – Bruhn & Walker 1995
5. Pliocene, Los Angeles Basin – Lowry et al. 1993

Fig. 8. Comparison of sandstone unit thickness from the Hamitabat Formation (derived with reference to wireline log and core data) with field analogue data from a number of confined turbidite systems. The extent of the ellipse for the Hamitabat Field defines the modelled sand-body size distribution.

This was derived from a cross-plot of core plug porosity and permeability. However, it was found during initial simulation that this relationship over-predicted permeability, and it was necessary in the history match to limit the maximum value from this correlation to 0.5 mD. This over-prediction is believed to result from the desiccation of core plugs prior to analysis reducing the negative impact of clay minerals on plug permeability. Water saturation data were assessed for two groupings of wells on the basis of well head flowing pressures. These data indicate two saturation–height relationships that were used to create a three-dimensional water saturation distribution, while retaining the degree of variability present in the original wireline dataset.

Reservoir simulation

Flow simulation involves the modelling of fluid flow through the modelled volume, initially to iteratively match the historical production data, and then to make predictions about future reservoir performance. The history-matching phase provides an opportunity to evaluate the veracity of the geological model, for example the distribution of facies and reservoir properties and the transmissibility of faults. The geological grid was upscaled to a coarser resolution, as the amount of computational time required to simulate flow using the geological model resolution would have proved impractical. Upscaling resulted in a reduction in grid size from approximately 6 million to c. 90 000 cells (approximately 100 m × 100 m × 16 m). Porosity and water saturation parameters from the geological grid were arithmetically averaged from to the simulation grid resolution. Permeability was upscaled using the diagonal tensor method, and the modal value was rescaled from the discrete facies dataset.

Sector and individual well models were studied to analyse reservoir behaviour in detail, prior to simulating the full field model. This modelling indicated that fault planes are apparently sealing. Also, it is of note that the correlation surfaces (defined either by relatively fine-grained sedimentation associated with potential flooding surfaces or by the influx of coarser sedimentation overlying an upward-fining succession) tend to act as extensive transmissibility baffles. The reservoir simulation also supported the general distribution of facies and associated petrophysical characteristics within the model.

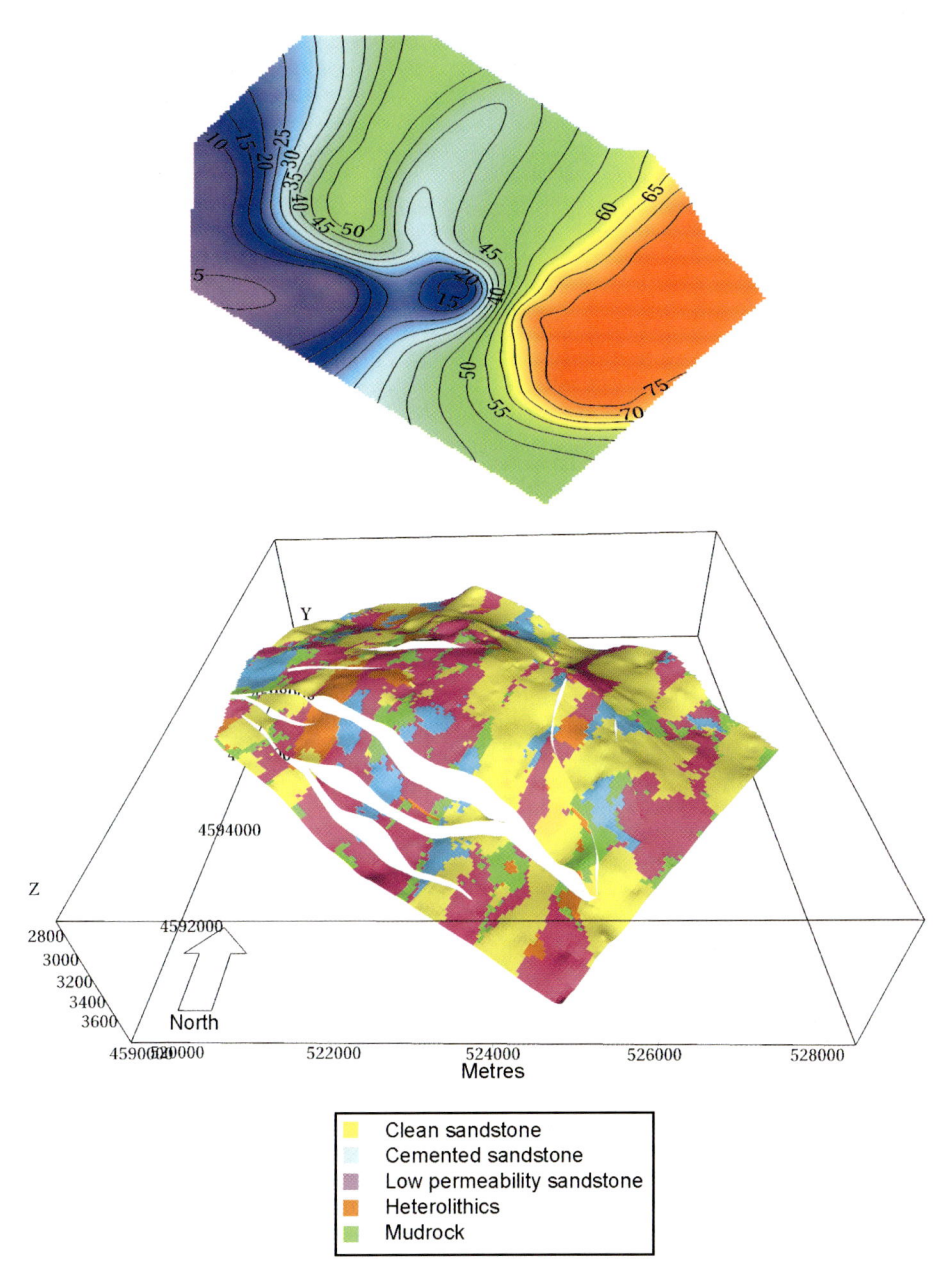

Fig. 9. Illustration of the turbidite facies distribution in relation to isopach thickness for the interval between the top Hamitabat B and Hamitabat A5 horizon, with sand-bodies displaying a dominant northeast–southwest orientation.

Conclusions

The tools available to the reservoir modeller can provide a greater understanding of the three-dimensional geometry and distribution of sediment deposition and corresponding petrophysical properties within a confined turbidite setting in the subsurface.

Following the designation of a facies scheme incorporating both core and wireline data, reservoir correlation was achieved by the identification of mainly upward-fining cyclical stacking patterns, which were elucidated by the use of Cyclolog to mathematically analyse cyclicity in the V_{sh} curve data.

Isochore maps created for major correlatable horizons indicate lobate geometries with axes oriented northeast–southwest. This is in agreement with the conceptual model of the Hamitabat Field area of the Thrace Basin, in which a fluvio-deltaic system to the north and northeast feeds sediment to the south and southwest into a shelf and ramp environment. Compensation stacking of lobes suggests confinement and the influence of depositional topography on subsequent sediment dispersion.

An analysis of the relationship between facies distribution and isochore thickness has been used to quantify facies distribution and to populate the model with facies bodies. Dimensions of facies have been derived from a comparison with outcrop analogue data.

Simulation of fluid flow through the modelled reservoir volume and comparison of simulated and actual production data allowed aspects of the model to be verified and iteratively adjusted. In particular, connectivity, permeability (associated with facies distribution) and fault transmissibility can be assessed in this manner. The association of fine-grained sediment above flooding surfaces and at the top of upward-fining successions is shown by flow simulation to form laterally extensive vertical permeability baffles.

The authors would like to thank S. Rae and D. Burkett for their invaluable input to the modelling and simulation aspects of the study respectively. Thanks are also due to A. Burrows in association with the seismic interpretation and D. Nio for the use of Cyclolog[R]. An initial script was enhanced by the constructive reviews of A. Gardiner and N. Satur.

References

BRUHN, C. H. L. & WALKER, R. G. 1995. High resolution sequence stratigraphy and sedimentary evolution of coarse-grained canyon-filling turbidites from the Upper Cretaceous transgressive megasequence, Campos Basin, offshore Brazil. *Journal of Sedimentary Research*, **B65**, 426–442.

ENOS, P. 1969. Anatomy of a flysch. *Journal of Sedimentary Petrology*, **39**, 680–723.

JACOBS, D. K. & LINDBERG, D. R. 1998. Oxygen and evolutionary patters in the sea: Onshore/offshore trends and recent recruitment of deep-sea faunas. *Proceedings of the National Academy of Science, USA*, **95**, 9396–9401.

LOWRY, P., JENKINS, C. D. & PHELPS, D. J. 1993. Reservoir scale sandbody architecture of Pliocene turbidite sequences, Long Beach Unit, Wilmington Field, California. *SPE 26440, Annual Technical Conference, Houston Texas*, 251–258.

MUTTI, E. & RICCI LUCCHI, F. R. 1975. Turbidite facies and facies associations. *In*: MUTTI, E. *et al.* (eds) *Examples of Turbidite Facies and Associations from Selected Formations in the Northern Appennines*. IX International Congress of Sedimentology, Nice-75, Field Trip A11, 21–36.

PERINÇEK, D. 1991. Possible strand of the North Anatolian Fault in the Thrace Basin, Turkey: an interpretation. *American Association of Petroleum Geologists Bulletin*, **75**(2), 241–257.

SCHUPPERS, J. D. 1995. *Characterisation of deep-marine clastic sediments from foreland basins*. PhD thesis, Delft University.

SMITH, R. D. A. 1995. Complex bedding geometries in proximal deposits of the Castelnuovo Member, Rocchetta Formation, Tertiary Piedmont Basin, Northwest Italy. *In*: PICKERING, K. T., HISCOTT, R. N., KENYON, N. H., RICCI LUCCHI, F. R. & SMITH, R. D. A. (eds) *Atlas of Deep Water Environments*. Chapman & Hall, London, 244–249.

TURGUT, S., TURKASLAN, M. & PERINÇEK, D. 1991. Evolution of the Thrace Sedimentary Basin and its hydrocarbon prospective. *In*: SPENCER, A. M. (ed.) *Generation, Accumulation, and Production of Europe's Hydrocarbons*. European Association of Petroleum Geoscientists, Special Publications, **1**, 415–437.

Index

Note: Page Numbers *in italics* refer to figures; **bold** numbers indicate tables.

abandoned meanders 93, 110–12, *111*
Aberystwyth Grits 217–18, *218*, *219*
acoustic data, Zaire valley 95–6, 100–1, *100*
Acquerino turbidite system 126
Adana Basin fans, Southern Turkey 4–5, 241–60
  Eastern and Western Fans *242*, 243–4
  facies and stratigraphy *242*
  feeder system channels 246–55
  lithofacies 244–6, **244**, *245*
  multisource feeder system *243*
  sedimentary logs *248–9*, *254*
  see also Cingöz Fan
aggradation 81, 168–9, *224*
amalgamated surfaces 10
Apennines *see* Northern Apennine foredeep basin
architecture 4, 23, 37–9
  basin-floor fans, Spitsbergen *200*, 201
  Caban-Ystrad Meurig System *216*
  down-channel changes 246–60
  Golo turbidite system 66–77
  'Macigno costiero' 265–78, *266–71*, *273*
  Pab Sandstone, Pakistan *165*, *167*, 169–71, 181–3
  sand-bodies 45–58
  slope-fans 178–9
avulsion, channel–levee complexes 75, 83

backfill 141
backstepping sequential evolution 171–2, 182
base-of-slope 15
basin subsidence models 153–6
basin-fill sediments, New Zealand 230–6
basin-floor fans
  'Macigno costiero' evolution 278, *279*, *281*
  Pab Sandstone 159–60, 169–71, *170*
    evolution 171–2, 181–3
  Spitsbergen 187–208, *192*
    clinoforms 12 and 14 194, *195*, *198*
    construction stages *205*, 206
    early formation 191–3, *192*
    facies 196–201, *197*
    Hyrnestabben *199*, *200*, *202–3*
    large-scale architecture *200*, 201
    relationship to host clinoforms 191–3
    tectonic control *205*, 206
    vertical and lateral facies trends 201–5
basins
  Adana Basin, Southern Turkey 241–60
  Central Tertiary Basin, Spitsbergen 187–208
  New Zealand Tertiary 229–40
  Northern Apennines foredeep 115–34, 261–83
  subsidence models 153–6
  Tabernas-Sorbas Basin, SE Spain 135–58
  Tertiary Piedmont Basin, NW Italy 285–305
  Thrace Basin, Turkey 307–20
  Welsh Basin 209–28

bathymetry
  deposition control 135
  Zaire deep-sea fan *94*, 95, *96*, 97–8, *99–100*
bedforms in channels 257
biofacies, Telychian *210*, 213–14, **213**
biostratigraphy
  calcareous nannofossils 118–19, *119*, *123*, 125
  Ordovician–Silurian graptolites 210, *211*, *212*
bioturbation 139, 310
bipartite beds 146, 147, 153, 155
Blackmount Formation 235–6, *235*, *236*
Bouma Sequences 10, 17, 236, *237*
bounding slope processes 37–9
braided deltas *163*, *179*
braided stream network 257–8
Brazilian slope 27
breccia 235
Burdigalian–Seravallian 241
bypass, Lower Pab turbidite system 168, 171, *172*

Caban-Ystrad Meurig System 215–17, *216*
calcareous nannofossils 118–19, *119*, *123*, 125
canyons
  Golo turbidite system *63–4*, 64, 70–1, *70*, *76*, *77*, 87
  infill 15
  Pab Sandstone, Pakistan 166, 181–2
  Tabernas-Sorbas Basin 141
  terraces 91–2
  Zaire deep-sea fan *94*, 97
cascade of silled sub-basins 30–4, *31*, 39
Castagnola Formation, NW Italy 5, 285–305, *288*, *289*
  bed thickness distribution *293*
  geological model 287–91
  geostatistical analysis 297–304
  grain size distribution *294*
  Hurst H values 295–7
  Hurst statistic methodology 285–7
  sand-body 288–91, *290*, *291*, 303
  stacking patterns 289, *295*, *298*, *302*
  stratigraphical logs *292*
Castiglione dei Pepoli turbidite system 127
Central Tertiary Basin, Spitsbergen 187–208, *188*
  clinoforms 12, 14 and 15 187–90, *189*
  stratigraphy 190–1, *190*
  tectonic setting 187–90, *188*
  turbidite facies 196–201
channel margin 263, **264**, *266*, 276–7, *278*
channel overbank flow 263, **264**
channel sands 15–16
channel-fill 263, **264**, 276–7
  feeder system 246–55
  Welsh Basin *216*, 217
channel–levee complexes *11*, *14*, 278

channel–levee complexes (*cont.*)
  avulsion 75
  controls on deposition 81
  Golo turbidites *63*, 64–6, 71–6, *72*, *73*, *77*
  lateral migration *72*, 75, 82–3, *82*
  longitudinal migration 75, *75*, 82–3
  Lower Pab turbidites 166, 168–75, *169–71*, *173*
  progressive migration 74–5
  supply shut-off 174–5
  Zaire valley 96, 105–7, *107*, 111–12
channel-lobe complex, Niger Delta slope *32*, 33–4
channel–lobe transition
  Golo turbidite system 71, *74*
  'Macigno costiero' **264**, 265, *266–7*, 276–7
  Pysgotwr Formation 220
channel-termination lobes 66
channels
  destabilization of walls 110
  feeder system 243–4, *243*, 246–55
  migration 110–11, *111*
chaotic deposits 274–5, 276, 277, 701
chronostratigraphy *123*
Cingöz Fan, Adana Basin *242*, 243–4
  channel sedimentation logs *249*
  feeder system channels 246–55, *248*, *250*, *251*
  physiography model 255
  source 246
clast-type ratios 246, *246*, 253, 255
clastic wedge 263, 309
clays 102–5
climate 12, 19
climatic–eustatic variation *84*
clinoforms, Spitsbergen 187–208
  12, 14 and 15 *189*, *194*, *195*, *198*
  five main components 191–3, *192*
  relationship to basin-floor fans 191–3
  sand-prone 191, 193
  scale and dimensions 193
  tectonic setting 187–90
clustering 302–3, 304
  see also Hurst statistics
coarse-grained turbidite systems *11*, 13–16
  see also sand-rich turbidite systems
collisional phases 263, 281
confinement 1–2, 53–6
  'Macigno costiero' basin 261, 280, 287
  Welsh Basin 217–18, *219*
conglomerates
  Cingöz Fan 247, *248*, *250–1*, 252–3
  cross-bedding 247, 252
  fanglomerates 137–9
  Turret Peaks Formation 233, 234–5
conglomeratic deepwater fan feeder system 241–60
connected tortuous corridor 25, *29*, 34–7, *35*, 39
containment 1, 150, 151, 155
  see also confinement
continental shelf, Golo system 63–4, *63*, 66, 68
continental slope 9

controls of deposition 2, 9–22, 115–34
  bathymetry 135
  climate 12, 19
  sea-level variations 12–13, 19, 80–1, 115, 131–2
  sediment character and processes 13
  tectonics 10–12, 19, 60–2, 82, 129–31, *130*, 141
  topography 81–2, 159
convergent baselap 25
convergent thinning 25
convolute bedding 196–9
Corsican east coast 59–89
  see also Golo turbidite system
couplets 146, 236
cross-bedding 178, *178*, 247, 252
currents *see* palaeocurrent directions; turbidity currents

debris flow deposits 10
deepwater fan feeder system 241–60
deepwater mass flow system 233
deformation 135, 143, *144*, 156
  see also faulting
Delaware Basin turbidite complex 18–19
deltas *see* fans
depocentres 129–31, 233, 303, 304
deposition
  adjacent to fault zone 146–53
  diachronous 289, 303, 304
  rates of 129–30, *129*
  syndepositional faulting 135–58
  third-order sequences 179–81
depositional architecture *see* architecture
depositional factors *see* controls of deposition
depositional geometry
  Golo turbidite system 45–58
  Hamitabat Formation 313–14, *316*, *317*
depositional models, lobe deposits 77–80, *78*
diachronous deposition, Castagnola Formation 289, 303, 304
diapiric interslope basins 18, *18*
dinocysts 191
distal lobes 71, *74*, *77*, **264**, 265, *266–7*, 277
distal slope fan *170*, 171, 175
distributary channel patterns 13, 25
down-channel changes, feeder channel fill 246–60
downlap surfaces 37–8, 70, 71

Eastern Fan *see* Cingöz Fan
El Cautivo fault zone 4, 135, *137*, *140*, 141–6
  adjacent turbidite deposition 146–53
  deformation 143, *144*
  impact on flow containment 155–6
  kinematics 143–6, *145*
  origin 146
  syndepositional movement 151–5, *154*
  wet argillaceous sediments 156
end member types 2, *11*, 14, 19
entry point processes 37

Eocene
  shelf-slope clinoforms 187–208
  successions 4
  turbidites 287
equilibrium profiles, flow character 28–30
eustatic curve 210, *212*
experimental work 2–3, 45–58
  flow efficiency 46–7
  partially blocked flows 53–4
  unobstructed flows 47–53
external controls *see* controls of deposition

facies 118, 129–32
  basin-floor fans 196–201, *197*
  Central Tertiary Basin, Spitsbergen 196–21, 196–201
  Golo turbidite system *78–9*
  Hamitabat Formation **311**
  'Macigno costiero' 263–5, **264**, *265*
  New Zealand small basins 230–1, *232*, 238
  Pab Sandstone, Pakistan *167*
  proximal to distal changes 255–60
  Tabernas-Sorbas Basin *138*
  trends in sandstone-bodies 201–5
  Upper Pab slope fan *180*
  vertical changes 257, 265
facies architecture 37–9
facies associations, 'Macigno costiero' 263–5, **264**, 266–8
facies model, Hamitabat Formation 315–16, *319*
Falterona turbidite system 120–6
fan valley *14*, 15
  see also upper-fan valley
fanglomerates, Tabernas-Sorbas Basin 137–9
fans 76, 77
  depositional architecture 159–85
  depositional elements **265**, 266–8
  fine-grained 15–16
  gravel-rich 13
  growth factors 80–2
  growth patterns 82–5, 87
  late-stage 191–3
  lobes 193–4
  multiple-sourced 243–4, *243*
  pre-existing morphology control 81–2
  sea-level variations 80–1
  tectonic control 82
  variations in characteristics 75–7
  Zaire deep-sea fan 93–8, *94*, *96*, *99–100*
  see also basin-floor fans; lobe deposits; slope fans; submarine fans
fault gouge fabric 143, *144*, 146
faulting
  control of basins 11, *12*
  normal 307
  seabed 155
  syndepositional 135–58
  see also El Cautivo fault zone

feeder system channels, Cingöz Fan 243–4, *243*
  Channel 1 246–53, *248*, *250*, *251*, 257
  channel fill architecture *250*
  Channels 2-4 253–5, *254*
  down-channel changes 255–60
fill-and-spill model 23, 30–2, *31*, 209
filled incised valleys 66, *68*
fine-grained (mud-rich) turbidite systems *11*, 13–16, 19
fine-grained sediment fraction 47, **48**, 52
fines-rich flows 52
flame structures *152*
flow character, equilibrium profiles 28–30
flow containment *see* containment
flow deflections 1, 38
flow efficiency 2, 45–6, 56, *56*
  experimental methods 46–7, **48**
  partial blocking effect 53–4
  unobstructed flows 47–53
flow types, 'Macigno costiero' **264**
flow volumes 27–8, 56
flute casts 236
flysch facies 233, *234*, 235, *235*, *236*, *237*
foredeep basins, Northern Apennines 115–34, 261–83

geometries
  depositional
    Golo turbidite system 45–58
    Hamitabat Formation 313–14, *316*, *317*
  internal, Golo turbidite system 66–71
  single flow deposit 49–51, *50*
geostatistical analysis 286, 297–303
Gerig Gwynion Grits System 214–15, *215*
Golo river, Corsica *60*, 62
Golo turbidite system 3, 59–89, *60*, *76*, *77*
  ancient systems comparison 86–7
  conceptual model 83–5, *84*
  continental shelf morphology 63–4, *63*, 66, *68*
  data and methods 63
  facies distribution *78–9*
  fan characteristics 75–7
  hydrologic/hydrodynamic context 62
  internal geometry 66–71
  lateral and longitudinal evolution 71–5
  modern studies 85–6
  morphology 60, *61*
  seismic-reflection facies 66, *67*
  stratigraphic and palaeogeographic evolution 77–80
  tectonic control 60–2
graded deposits 246, 247
grain-size distribution 45, 56, 109, *294*
graptolite biostratigraphy 210, *211*, *212*
gravel-rich fans 13
gravelly sandstones 252
gravels 247
gravity flows 182
growth patterns of fans 82–5
Gulf of Mexico slope *24*, 26, 32

gullies, Golo system *63*, 64, 66

half-graben 62, 229–30
Hamitabat Field, Turkey 307–20
  exploration 309
  reservoir modelling 307–20, *308*
  seismic interpretation 309
Hamitabat Formation 5
  facies classification 311–12, **311**
  facies model 315–16, *319*
  geological grid 314–15
  petrophysical model 316–18
  post-depositional features 310
  reservoir properties 310–11
  reservoir simulation 318
  sedimentology 309–10
  stacking patterns and depositional geometry
    313–14, *316*, *317*
  stratigraphy **308**
  structural model *317*
  turbidite facies model 315–16, *319*
  well correlation 312–13
healed slope 25, 29
hemipelagic deposits
  Castagnola Formation *290*
  Cingöz Fan 247, 250
  Pab Sandstone 171, 172, *173*
  Zaire valley 103, 109
Hurst statistics 285–305
  coarse-division thickness percentage *299–300*
  database compilation 291–5
  geostatistical analysis 297–303
  grain-size score *301*
  Hurst H test 295–7, *296–8*
hydraulic jumps 181, 257, 258, *258*, 259, 276
hydrocarbon prospectivity 238
Hyrnestabben, Spitsbergen
  basin-floor fan deposits *199*, *200*, *202–3*
  clinoforms *195*

ichnofossils 213–14, **213**
imbricated clasts 247, 252, 253
incised valleys 66, *68*, 181
initial flow density 47
initial sediment volume 47–9, *48*
inner levees 93, 106, 108, 111, *111*
interchannel deposit 263, **264**, 278
intermediate-distal lobes **264**, 265, *266–7*, 277
  see also distal lobes
Italy
  Castagnola Formation 285–305
  'Macigno costiero' turbidite system 261–83
  Northern Apennine foredeep basin 3, 5, 115–34,
    116

Kaplankaya Formation 246, 247
Karaisali Formation 246, 247
kinematics, El Cautivo fault zone 143–6, *145*

Kirthar fold belt, Pakistan 161–7, *161*
Kota Fan, southern Iceland 29–30, *30*
kriging 287, *299–301*, 302

Laki Range, Pakistan *160*, 161, *163*
laminated sandstones 196
lateral basin slope 221–3
lateral migration, channel–levee complexes *72*, 75,
  82–3, *82*
levees
  inner or confined 93, 106, 108, 111, *111*
  Pab Sandstone *173*, 174
  sandstones 16
  see also channel–levee complexes
Ligurian units 116–17, 125
liquefaction 199
lithofacies, Adana Basin 244–6, *244*, *245*
lithostratigraphy, turbidite systems *124*
lobe deposits
  depositional models 77–80, *78*
  Golo turbidite system 71, *74*, 76, 82
  sand-bodies 168, 172–4
lobe-fan fringe, 'Macigno costiero' **264**, 265, *266*,
  277
lobes 35, 193–4, 221, 249, 252
  channel-termination 66
  distal 71, *74*, 77, **264**, 265, *266–7*, 277
  Lower Pab *170*, 171, *172*
  Niger Delta slope *32*, 33–4
  progradation *267*, 277, *278*, 281
  proximal 71, *74*, 77, **264**, 265, *266–7*, 277
  sand-rich 168
  sandstone 215, 221, *224*
  tabular *216*, 217, 218
  Upper Pab *175*
  see also channel–lobe transition
Loma de los Baños Formation 139, *140*, 146–52
  Alfaro sub-basin 146–7, 151
  cliff section *148*
  lateral variations 147–9
  lithology 146, *147*
  origin and fault control 151–3
  palaeoflow 149–50, 153
  provenance 149
  transition to Verdelecho Formation 150–1
longitudinal migration, channel–levee complexes 75,
  82–3
Lower Pab turbidite systems
  depositional architecture *165*, 169–71, 181–3
  mud-rich 172–5, *174*, 183
  sand-rich 168–71, *169*, *170*, 183

Maastrichtian, Pab Sandstone, Pakistan 4, 159–85
'Macigno costiero' turbidite system 5, 261–83, *262*
  architecture stages 265–78, *266–8*
  correlation and architecture pattern *269–71*, *273*
  depositional system 278–81
  evolution *278*, *279*, 281

INDEX

facies and facies associations 263–5, **264**, 265, 266–8
flow types **264**
sedimentation logs *266–8*
syncollisional clastic wedge 263
MacIvor Formation 236, *237*
main channel complex *268*
marls 139, 141
Marnoso Arenacea turbidite systems 127–8
massive sandstones 233, 274–5
meander development 110–12, *111*, *112*
mid-slope fan *170*, 171, 175–8, *178*
migration of channels *72*, 74–5, 82–3, *82*, 110–11, *111*
minibasins *see* sub-basins
Miocene
    Adana Basin, Southern Turkey 241–60
    turbidite systems 3, 120–8, 287
Mississippi Fan 15
mobile salt structures 34–6, *34*
modelling facies, Hamitabat Formation 311–12
modern turbidite systems, Golo system 3, 59–89
Molinos Formation 137–9
morphology
    continental shelf 63–4, *63*, 66, *68*
    fan growth control 81–2
    slopes 159
    terraces in submarine valleys 97–105
mud-rich slope fans *174*
mud-rich turbidite systems *11*, 13–16, 19
    Lower Pab 172–5, *174*, 183
mudstones 10, 201, 233
multiple terraces *see* terraces
multiple-sourced fan 243–4, *243*

Neogene intramontane basins 3, 136, *136*
New Zealand small basins 4, 229–40
    basin size 237–8
    basin-fill sediments 230–6
    Blackmount Formation 235–6, *235*, *236*
    facies 230–1, *232*
    hydrocarbon prospectivity 238
    MacIvor Formation 236, *237*
    stratigraphy *232*
    tectonic setting *230*
    tectonics and facies diversity 238
    Turret Peaks Formation 231–5, *233*, *234*
normal block faulting 307
North Golo canyon 64, *65*, 66, *70*
Northern Apennine foredeep basin 3, 5, 115–34, *116*, 261–83, *262*
    geological setting 116–17
    structure *122*
    tectonic map *121*
    turbidite systems 120–8, 265–81
    see also 'Macigno costiero' turbidite system
northwest Borneo slope 26

oblique-dextral strike-slip fault 151, 153–5, *154*, 156

Oligocene
    foredeep turbidite systems 3, 120–8
    sand-rich turbidite systems 261–93
olistostromes 278
onlap *38*
    Castagnola Formation, NW Italy 289, 302, 304
    Golo turbidite system *68*, *69*, *70*, *71*
    lateral basin slope 223
    Loma de los Baños Formation *148*, *149*
    Peïra Cava Sandstone *17*
    Upper Pab sand-rich fan *177*
Ordovician–Silurian Welsh Basin fill 4, 209–28

Pab Range, Pakistan *160*, 161, *164*
Pab Sandstone, Pakistan 4, 159–85
    braided deltas *163*, *179*
    depositional system architecture *165*, 169–71, 181–3
    evolution 181–3
    Lower Pab mud-rich turbidite system 172–5, *174*, 183
    Lower Pab sand-rich turbidite system 168–71, *169*, *170*, 183
    palaeogeography 166–7
    reconstruction 167–8
    sequence architecture and facies *167*
    sequence stratigraphical analysis 179–81, *180*
    stratigraphy *176*, *178*
    turbidite systems *162–4*
    Upper Pab sand-rich turbidite system 175–9, *177*, *179*, 183
Pakistan *see* Pab Sandstone
palaeocurrent directions
    Cingöz Fan *242*
    'Macigno costiero' *279*, 280, 288
    Peïra Cava Sandstone *16*, 17
    Tabernas-Sorbas Basin *150*, 151, 153
    Telychian *222*
palaeoflow directions, Welsh Basin 215, *215*, 217–18
palaeogeography
    Golo turbidites evolution 77–80
    Northern Apennine foredeep basin *117*
    Pab Sandstone 166–7
    Telychian *219*, *222*
palaeotopography 209
passive margins 13, 159, 161, 166
pebbly sandstones 265, 270
Peïra Cava Sandstone 16–18
Permian 15
petrophysical model, Hamitabat Formation 316–18
Pianosa slope *61*, 66, *69*, *70*
Piedmont Basin, NW Italy 285–305, *288*, *289*
piggyback thrust sequence 120, 131
ponding 1, 25, 29
    see also containment
progradation
    lobes, 'Macigno costiero' *267*, 278, 281
    sand-rich slope fan 179, 181
    wedges 66–70, *68*

progressive migration, channel–levee complexes 74–5, 75
provenance, Loma de los Baños Formation 149
proximal lobes
  Golo turbidite system 71, *74*, *77*
  'Macigno costiero' **264**, *265*, *266–7*, 277
proximal slope fan 170–1, *170*, 178
proximal to distal facies changes 255–60
pseudo-hummocky cross-stratification 178, *178*
Pysgotwr Formation and lateral basin slope 220–3

ramp evolution 181
reservoir architecture 37–9
reservoir modelling, Hamitabat Field 307–20
Reynolds numbers 47
Rhuddnant Grits 218

salt highs, sediment distribution paths 34–7, *36*
sand delivery systems 191
sand-bodies
  architecture 45–58
  Castagnola Formation 288–91, *290*, *291*, 303
  Hyrnestabben, Spitsbergen *203*, 204
  lobe deposits 168, 172–3
  sheet-like 175–8, 204, 214, 313
  Tabernas-Sorbas Basin 139
sand-prone clinoforms 191
sand-rich feeder channels 236
sand-rich lobes 168
sand-rich ramps model 233–4
sand-rich turbidite systems *11*, 13–16, 19
  Hamitabat Formation, Turkey 307–20
  Lower Pab 168–71, *169*, *170*, 183
  'Macigno costiero', Italy 261–83
  Upper Pab 175–9, *177*, *179*, 183
sandstones
  convolute bedded 196–9
  facies trends 201–4
  levees 16
  lobes 215, 221, *224*
  massive 233, 274–5
  petrography 119
  petrology 125, *125*, 126
  thick-bedded 196
  thin-bedded 196
sandy channel complexes 168
sandy debrites 10
sandy overflow deposits 171
Sartanella Formation 139, *140*, 149
scaling, experimental work 47
scours *218*, 275–6, *275*, 277
sea-level curves 212, *212*
sea-level variations 12–13, 19, 115, 131–2
  fan growth 80–1, 83–5
  Ordovician–Silurian Welsh Basin 212–13
seabed faulting 135–58
sediment analogues 46, 47, **48**
sediment character 13, 19

sediment gravity flows 29
sediment transport 13, 182
sedimentary bodies
  distribution and fan characteristics 75–7
  internal geometry 66–71
  lateral and logitudinal evolution 71–5
sedimentary processes 13
sedimentary structures *265*, 270–8
sedimentation 80, 155–6
seismic interpretation, Hamitabat Field 309
seismic profiles
  Pianosa slope *69*
  sedimentary bodies *73*, *74*
  Zaire valley 98–100, *101*, *103*, 106–7
seismic-reflection facies, Golo system 66, *67*
sequence architecture, Pab Sandstone *167*
sequence stratigraphical analysis, Pab Sandstone 179–81, *180*
Shah Noorani, Pab Range *164*, *165*, *173*
shale 10
shallow canyons 62
sheet conglomerates 257
sheet turbidites 146
sheet-like sand-bodies 175–8, 204, 214, 313
shelf, sand-delivery system 191
shelf break 66, *68*, 181
shelf width 159
shelf-slope clinoforms *see* clinoforms
shelfal limestones 246
silled sub-basins 24–5, *26–7*
  areal extent *29*
  cascade model 30–4, *31*, 39
  end-member scenarios 32–3
  sand-rich and mud-rich stages *31*
siltstones 201, 250, 252, 253
Silurian *see* Ordovician–Silurian Welsh Basin fill
single flow deposit geometries 49–51, *50*
slides *see* slumps/slides
slope fans 160, 166
  architecture 178–9
  evolution 182–3
  mud-rich *174*
  prograding sand-rich 179
  reconstruction *179*
  Upper Pab sand-rich 175–9, *177*, *179*, 183
slopes 9
  base-of-slope 15
  bounding slope processes 37–9
  Brazilian slope *27*
  continental slope 9
  Gulf of Mexico slope *24*, 26, 32
  healed slope 25, *29*
  lateral basin slope 220–3
  morphology 159
  Niger Delta slope 26, *26*, *32*, 33–4
  northwest Borneo slope *26*
  Pianosa slope *61*, 66, *69*, 70
  see also topographically complex slopes

slumped thin-bedded turbidites 276
slumps/slides 92–3, 107, 151–3, *152*, **264**, 277
small-scale basins, New Zealand 229–40
sole structures 168, 215, *215*, *219*
source tectonics 131
South Golo canyon *63*, *64*, *65*, *70*, 75–6, *77*
South Golo channel *63*, 64–6, *65*
Spitsbergen, basin-floor fans 4, 187–208
square flume tank 46
stacking patterns 4, 32, 129, 178
   Castagnola Formation 289, *295*, 298, 302
   Hamitabat Formation 313–14, *316*
   'Macigno costiero' 270, *270*
Stagno turbidite system 126–7
statistical methods 285–305
Storvola, Spitsbergen, clinoforms *189*, *195*, *198*
stratal architecture *see* architecture
stratigraphical logs *292*
structural confinement, Rhuddnant Grits 218
structure, Northern Apennines 2, *122*, *262*
Stuart Mountains, New Zealand *233*, *234*
sub-basins 9–10, 11–12, *12*
   see also silled sub-basins
submarine fans
   Adana Basin, Turkey 241–4
   clustering 285
   New Zealand 230–1
   Turret Peaks Formation 233–5
submarine intrabasinal high *299–301*, 302, 304
submarine valleys 64
   terraces 91–114
subsidence 153–5
supply shut-off, channel-levee complexes 174–5
suspension density 47, *48*, 50
suspension volume *48*, 51–2, *55*
syn-tectonic deepwater sedimentation 155–6
syncollisional clastic wedge 263
syndepositional structures
   faulting, Tabernas-Sorbas Basin 135–58, *154*
   tilting 289
   Welsh Basin 210

Tabernas-Sorbas Basin, SE Spain 3–4, 135–58
   basin fill 136–7
   depositional evolution 141, *142*
   Gordo megbed *149*, 151, *152*
   stratigraphy 137–41, *138*
   subsidence and syn-tectonic deposition 153–6
tabular depositional lobes *216*, 217, 218
Tanqua Karoo 15–16
Te Anau Basin, New Zealand 229–30, *231*
tectonic control 10–12, 60–2, 141
   basin-floor fans *205*, 206
   facies diversity 238
   fans 82
   foredeep basins *121*, 129–31
   Golo turbidite system 60–2
   New Zealand small basins 237–8

Telychian
   lateral basin slope 221–3
   palaeogeography *219*, *222*
   sea-level change 212–13
   trace fossils 213–14, **213**
terminology 9–10, 24–5
terraces in submarine valleys 91–114
   abandoned meanders 93, 110–12, *111*
   creation of available space 110–11
   detailed morphology 97–105
   dying 110
   filling and incision interpretation *92*, 107–8
   inner levees 93, 106, 108, 111, *111*
   interpretation models 91–3, *92*
   slump/slide interpretation 92–3, *92*, 107
   turbidity currents overflow 108, 109, *109*
   valley incision 92, 107–8
Tertiary
   New Zealand fans 229–40
   Piedmont Basin, NW Italy 285–305, *288*, *289*
   Thrace Basin, Turkey 307–20
   see also Eocene; Miocene; Oligocene
thalweg, submarine valleys *100*, 105–6, *105*, 110, *112*
thickening–coarsening upward sequence 271, 274, 276, 277, 281
thin-bedded turbidites
   Castagnola Formation 302–3, 304
   'Macigno costiero' 271–3, *272*, *274*, 276, *278*
   Welsh Basin 221, *224*
third-order depositional sequences 179–81
Thrace Basin, Turkey 307–20
   see also Hamitabat Field; Hamitabat Formation
thrust system, piggyback progradation 120, 131
thrust-top basin 280
toolmarks 120
toplap surfaces *68*, *70*, *71*
topographically complex slopes 24–43
   cascade of silled basins model 30–4, *31*, 39
   connected tortuous corridor model 25, *29*, 34–7, 39
   recognition criteria 39
   structure growth rates 30
   types 25–7
topography
   confining effects 53–6
   control 81–2, 159
   syndepositional fault 151–3
Torrente Carigiola turbidite system 126, *131*
Tortonian to Lower Messinian fill *138*, 141, 153
trace fossils 213–14, **213**
traction carpets 247, 270, 275
transgression 181
transport 13, 182
tributary channel patterns 25
tripartite beds 149
turbidite classification *118*
turbidite complexes 10, 118

turbidite flows (recent) 109–10
turbidite sequences *see* Bouma Sequences
turbidite stages 118, 120, 126–7
turbidite substages 118
turbidite systems 10, 118
  Acquerino 126
  biostratigraphy 118–19, *119*, *123*
  Castagnola Formation 285–305
  Castiglione dei Pepoli 127
  chronostratigraphical distribution *123*
  coarse- and fine-grained *11*, 13–16
  confined 1–7
  deposition adjacent to active fault 146–53
  depositional rates 129–30, *129*
  Falterona 120–6
  Hamitabat Formation 307–20
  lithostratigraphy *124*
  Lower Pab mud-rich 172–5, *174*, 183
  Lower Pab sand-rich 168–71, *169*, *170*, 183
  'Macigno costiero', Italy 261–83
  Marnoso Arenacea 127–8
  petrography and petrology 119, 125, *125*, 126
  stacking patterns 129
  Stagno 126–7
  study methods 117–20
  syndepositional faulting 135–58
  Tabernas-Sorbas Basin 137–41
  tectonic arrangement *122*
  Torrente Carigiola 126, *131*
  Upper Pab sand-rich 175–9, *177*, *179*, 183
  Welsh Basin *210*, 214–30, *214–16*, *219*
turbidity currents 233, **264**
  containment 150, 151
  deposits 10
  entry point and bounding slope processes 37–9
  flow efficiency 47–56
  overflow terraces 108, 109, *109*
  overloading 28
Turkey
  Adana Basin 241–60
  Hamitabat Formation turbidite system 307–20
Turret Peaks Formation 231–5, *233*, *234*
Tuscan and Umbria-Marche units 116–17

unconfined flows 55–6
unobstructed flows 47–53
Upper Pab sand-rich turbidite system 175–9, *177*, *179*, 183

upper-fan valley 97–105, *98*, *99*, *103*, 107–10

valleys
  fan valley *14*, 15
  incision and filling 66, *68*, 92, 107–8, 181
  submarine 64
  upper-fan valley 97–105, *98*, *99*, *103*, 107–10
  see also terraces in submarine valleys; Zaire valley
variogram analysis 286–7, 298–303, *299–301*
Verdelecho Formation 139–41, *140*, *149*, 150–1, *152*
vertical facies changes 257, 265
vertical stacking 32, 74, 76
Vicchio Marl 125

Waiau Basin, New Zealand 230, *231*
wedges 66–70, *68*, 115–16, 263, 309
well correlation 312–13
Welsh Basin fill 4, 209–28
  analogues application 223–5
  biofacies *210*, 213–14
  lateral basin slope 221–3
  stratigraphy 210–13
  turbidite systems 214–23, *214*
    Aberystwyth-Rhuddnant systems 217–19, *219*
    Caban-Ystrad Meurig System 215–17, *216*
    Gerig Gwynion Grits System 214–15, *215*
    Pysgotwr Formation and lateral basin slope 220–3
Western Fan, Adana Basin 243
western Niger Delta slope 26, *26*, *32*, 33–4
wet argillaceous sediments, deformation 156

Zaire Canyon *94*, 95, 97
Zaire deep-sea fan
  cores 96
  location and bathymetry 93–5, *94*, *96*, 97–8, *99–100*
  survey data 95–6, *101*
Zaire valley 3, 91–114
  acoustic profiles 100–1, *104*
  channel–levee systems 96, 105–7, *107*, 111–12
  core results 101–4, *103–4*
  morphology 96–7
  recent evolution 112–13, *112*
  recent turbidite flows 109–10
  seismic profiles 98–9, *99*, *101*, *103*, 106–7
  upper-fan valley 97–105, *98*, *99*, *103*, 107–10
  see also terraces in submarine valleys